AF326995

AAPG Treatise of Petroleum Geology

The American Association of Petroleum Geologists
gratefully acknowledges and appreciates the leadership and support
of the AAPG Foundation in the development of the
Treatise of Petroleum Geology

TABLE OF CONTENTS

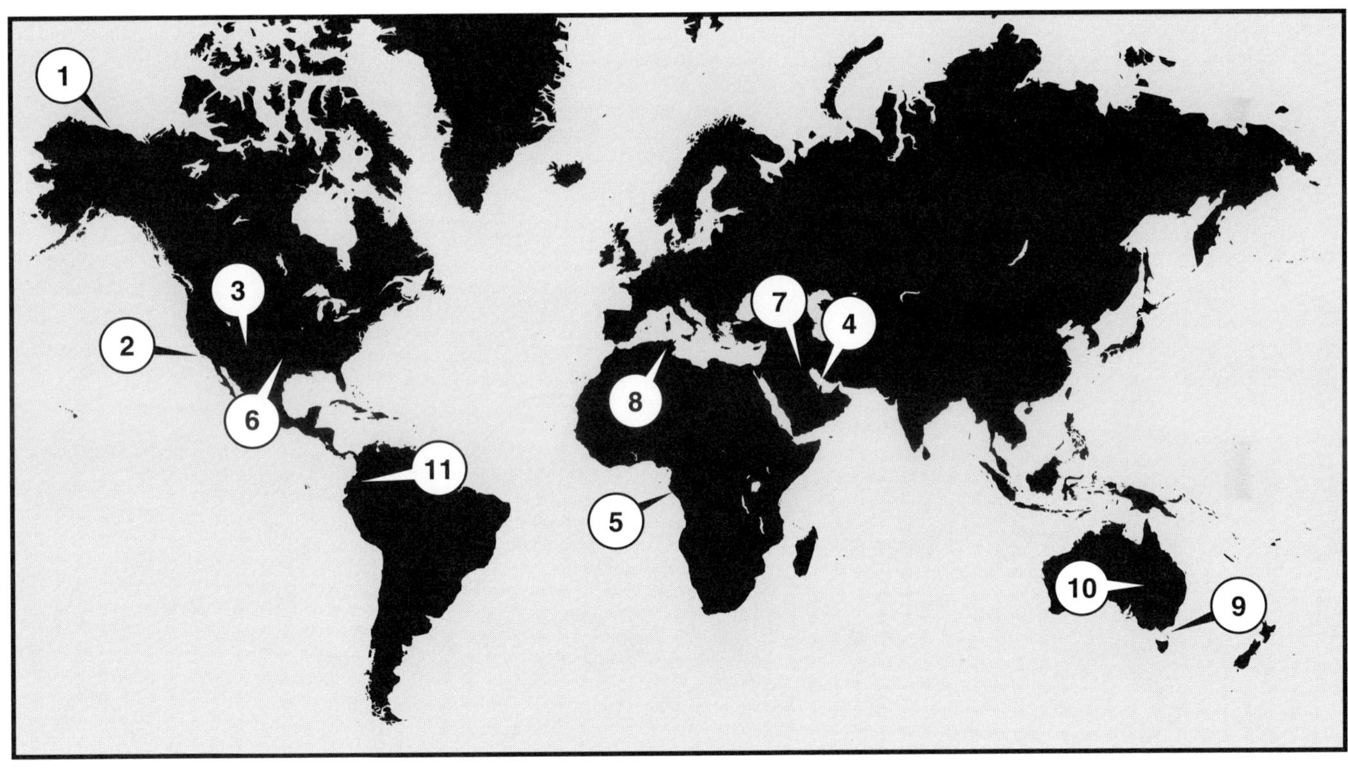

Eric A. Rudd
Floyd F. Sabins, Jr.
Nahum Schneidermann
Peter A. Scholle
George L. Scott, Jr.
Robert T. Sellars, Jr.
Faroog A. Sharief
John W. Shelton
Phillip W. Shoemaker
Synthia E. Smith
Robert M. Sneider
Stephen A. Sonnenberg
William E. Speer
Ernest J. Spradlin
Bill St. John
Philip H. Stark
Richard Steinmetz
Per R. Stokke
Denise M. Stone
Donald S. Stone

Doug Strickland
James V. Taranik
Harry Ter Best, Jr.
Bruce K. Thatcher, Jr.
M. Ray Thomasson
Jack C. Threet
Bernard Tissot
Donald F. Todd
M. O. Turner
Peter R. Vail
B. van Hoorn
Arthur M. Van Tyne
Ian R. Vann
Harry K. Veal*
Steven L. Veal
Richard R. Vincelette
Cecil von Hagen
Fred J. Wagner, Jr.
William A. Walker, Jr.
Anthony Walton

Douglas W. Waples
Harry W. Wassall, III
W. Lynn Watney
N. L. Watts
Koenradd J. Weber
Robert J. Weimer
Dietrich H. Welte
Alun H. Whittaker
James E. Wilson, Jr.
John R. Wingert
Martha O. Withjack
P. W. J. Wood
Homer O. Woodbury
Walter W. Wornardt
Marcelo R. Yrigoyen
Mehmet A. Yukler
Zhai Guangming
Robert Zinke

* Deceased

American Association of Petroleum Geologists Foundation
Treatise of Petroleum Geology Fund*

Major Corporate Contributors
($25,000 or more)

Amoco Production Company
BP Exploration Company Limited
Chevron Corporation
Exxon Company, U.S.A.
Mobil Oil Corporation
Oryx Energy Company
Pennzoil Exploration and Production Company
Shell Oil Company
Texaco Foundation
Union Pacific Foundation
Unocal Corporation

Other Corporate Contributors
($5,000 to $25,000)

Ashland Oil, Inc.
Cabot Oil & Gas Corporation
Canadian Hunter Exploration Ltd.
Conoco Inc.
Marathon Oil Company
The McGee Foundation, Inc.
Phillips Petroleum Company
Transco Energy Company
Union Texas Petroleum Corporation

Major Individual Contributors
($1,000 or more)

John J. Amoruso	Roy M. Huffington
Thornton E. Anderson	J. R. Jackson, Jr.
C. Hayden Atchison	Harrison C. Jamison
Richard A. Baile	Thomas N. Jordan, Jr.
Richard R. Bloomer	Hugh M. Looney
A. S. Bonner, Jr.	Jack P. Martin
David G. Campbell	John W. Mason
Herbert G. Davis	George B. McBride
George A. Donnelly, Jr.	Dean A. McGee
Paul H. Dudley, Jr.	John R. McMillan
Lewis G. Fearing	Lee Wayne Moore
Lawrence W. Funkhouser	Grover E. Murray
James A. Gibbs	Rudolf B. Siegert
George R. Gibson	Robert M. Sneider
William E. Gipson	Estate of Mrs. John (Elizabeth) Teagle
Mrs. Vito A. (Mary Jane) Gotautas	Jack C. Threet
Robert D. Gunn	Charles Weiner
Merrill W. Haas	Harry Westmoreland
Cecil V. Hagen	James E. Wilson, Jr.
Frank W. Harrison	P. W. J. Wood
William A. Heck	

The Foundation also gratefully acknowledges the many who have supported this endeavor with additional contributions.

*Based on contributions received as of April 23, 1991.

PREFACE

The Atlas of Oil and Gas Fields and the Treatise of Petroleum Geology

The *Treatise of Petroleum Geology* was conceived during a discussion held at the annual AAPG meeting in 1984 in San Antonio, Texas. This discussion led to the conviction that AAPG should publish a state-of-the-art textbook in petroleum geology, aimed not at the student, but at the practicing petroleum geologist. The textbook gradually evolved into a series of three different publications: the Reprint Series, the Atlas of Oil and Gas Fields, and the Handbook of Petroleum Geology. Collectively these publications are known as the *Treatise of Petroleum Geology*, AAPG's Diamond Jubilee project commemorating the Association's 75th anniversary in 1991.

With input from the Advisory Board of the Treatise of Petroleum Geology, we designed this set of publications to represent, to the degree possible, the cutting edge in petroleum exploration knowledge and application: the Reprint Series to provide useful and important published literature; the Atlas to comprise a collection of detailed field studies that illustrate the many ways oil and gas are trapped and to serve as a guide to the petroleum geology of basins where these fields are found; and the Handbook as a professional explorationist's guide to the latest knowledge in the various areas of petroleum geology and related disciplines.

The Treatise Atlas is part of AAPG's long tradition of publishing field studies. Notable AAPG field study compilations include *Structure of Typical American Fields*, published in 1929 and edited by Sidney Powers; and Memoir 30, *Giant Fields of 1968–1978*, published in 1981 and edited by Michel T. Halbouty. The Treatise Atlas continues that tradition but introduces a format designed for easier access to data.

Hundreds of geologists participated in this first compilation of the Atlas. Authors are from all parts of the industry and numerous countries. We gratefully acknowledge the generous contribution of their knowledge, resources, and time.

Purpose of the Atlas

The purpose of the Atlas is twofold: (1) to help exploration and development geologists become more efficient by increasing their awareness of the ways oil and gas are trapped, and (2) to serve as a reference for both the petroleum geology of the fields described and the basins in which they occur.

Imagination is the primary tool of the explorationist. Wallace E. Pratt once said that the unfound field must first be sought in the mind. In part, what is imagined is based on what is remembered; memory is the direct link to what is created in the mind. To create ideas that lead to the discovery of new fields, the mind of the geologist builds from its knowledge of petroleum geology. To that end, the Atlas of field studies will be a primary source for locating much of the information necessary for creating prospects and will provide a connection to the phenomenon of oil and gas traps.

Next to the firsthand experience of having prospects tested with the drill bit, studying the many facets and concepts of developed fields is perhaps the best way for the geologist to develop the ability to create plays and prospects. Also, familiarity with the many ways oil and gas are trapped allows the geologist to see through the noise inherent to exploration data and to close gaps in that data.

Format of the Atlas

To facilitate data access, all field studies in the Atlas follow the same format. Once users become familiar with this format, they will know where to look for the information they seek. Different fields from different parts of the world can be easily compared and contrasted.

The following is a generalized format outline for field studies in the Atlas:

Location
History
    Pre-Discovery
    Discovery
    Post-Discovery
Discovery Method
Structure
    Tectonic History
    Regional Structure
    Local Structure
Stratigraphy
Trap
    General Description
    Reservoir(s)
    Source(s)
Exploration Concepts

Criteria for Inclusion of a Field

Fields described in the Atlas are selected using two main criteria: (1) trap type, and (2) geographic distribution. Our ultimate goal for the Atlas is to include a field study from each major petroleum-producing province and to include an example of each known trap type. Size or economic importance are not, of themselves, criteria. Many fields that are not giants are included because they are geologically unique, because they are significant examples of geological investigation and original thinking, or because they are historically important, having led to the discovery of many other fields.

Grouping of Fields into Separate Volumes

We considered several ways to group fields in these volumes. We chose trap type because the purpose of the Atlas is to make exploration geologists more effective oil and gas trap finders, regardless of where they search for traps.

Grouping oil and gas field studies into separate volumes by trap type is a difficult exercise. We decided to group the fields into volumes by designating them as structural or stratigraphic traps. Most traps are a combination of both structure and stratigraphy. Some traps are obviously more a consequence of one than the other, but many are not. The continuum that exists between purely stratigraphic and purely structural traps is what makes grouping difficult. A further complication is that many fields contain more than one trap type.

Papers Selected for *Structural Traps V*

This volume in the Atlas of Oil and Gas Fields series contains studies of fields with traps that are mainly structural in nature. Stratigraphy controls the extent of the reservoir in the traps of several fields, but overall, the main trapping features within the group of fields in this volume are structural. Distribution of the fields of this volume is worldwide. All of the fields except West Puerto Chiquito and Ruston are giants.

Authors classified nine of the fields as anticlines or domes. Closure for the trap of Endicott field, Alaska, is a combination of an anticlinal nose, unconformity truncation, and faulting. West Puerto Chiquito, in the San Juan basin, New Mexico, could have been placed either in a volume on stratigraphic traps or, as we have done, in a volume on structural traps. The field has a fractured shale reservoir, and the *same* shale reservoir also served as the source. Fractures are concentrated in a regional monocline at points along the axis of the synclinal flexure. The top and bottom seal rocks are a ductile facies of the shale that serves as reservoir. The updip lateral seal, not far from the outcrop, is made by mineralization and fill of the fractures. Depending on how one views the trap, one could argue that it is stratigraphic because neither an anticline nor a fault is part of the main trapping mechanism. Alternatively, one could argue that the fractures were formed by structural forces and that therefore it is a structurally controlled trap.

Point Arguello field is similar to West Puerto Chiquito in that it, too, has a fractured reservoir that served in addition as the source rock. But unlike the former, the oil is trapped in an anticlinal structure in Point Arguello field.

Dukhan, Sendji, and Ruston are domes generated by salt movement. Sendji and Ruston are found in passive margin basins, and Dukhan is located in a foredeep basin. Sendji and Ruston are moderate in size, 120 million bbl of oil and 614 bcf of gas, respectively, whereas Dukhan is a world-class giant at 4.5 billion bbl of oil.

Until the middle 1960s, explorationists did not expect much oil and gas to be found in Australia. People thought that the rocks were too old, too mature, and too nonmarine, and that rock sections in the basins were too thin. This volume has two important Australian fields that proved otherwise: Tirrawarra, onshore in the Cooper basin, and Snapper, offshore in the Gippsland basin. There is little doubt now that the oil in both fields was not only sourced from nonmarine sediments but even from coals—something most geologists thought was impossible until recently.

A unique feature of the Atlas is the exploration history of each field in the study. What seems obvious today usually was not obvious when these fields were discovered. Many times what was expected was not what was found. Geologists discovered these fields by creating concepts based on information limited by the technology available at the time. Drilling and discovery show how closely concept matches reality. Knowing the history of discovery may help explorationists realize that problems, seemingly insoluble at one time, were eventually solved. It is also instructive to learn of the sequence of thinking that solved these problems.

Careful study of these fields will enhance the prospect generator's knowledge base, and consequently, his or her ability to apply that knowledge toward future prospecting.

Norman H. Foster
Edward A. Beaumont, Editors

Endicott Field—U.S.A.
North Slope Basin, Alaska

J. L. WICKS
M. L. BUCKINGHAM
J. H. DUPREE
BP Exploration (Alaska) Inc.
Anchorage, Alaska

FIELD CLASSIFICATION

BASIN: North Slope
BASIN TYPE: Passive Margin
RESERVOIR ROCK TYPE: Sandstone
RESERVOIR ENVIRONMENT OF
   DEPOSITION: Fluvial

RESERVOIR AGE: Mississippian
PETROLEUM TYPE: Oil
TRAP TYPE: Truncated and Faulted
   Anticlinal Nose

LOCATION

The Endicott field is located in the Beaufort Sea of North Alaska (Figure 1) 8 mi (13 km) east of the Prudhoe Bay field and approximately 40 mi (64 km) east of the Kuparuk field. Endicott is being produced from two artificial gravel islands, 4 mi (6 km) offshore, in 2 to 14 ft (0.5–4 m) of water. It is the first offshore Arctic oil development. Seven delineation wells and 66 development wells have been drilled to date. Upon completion, the total investment in the project will approximate $1 billion.

Endicott is the third largest oil field in Alaska and the ninth largest field in North America in terms of undepleted reserves. The Endicott field produces from the Mississippian Kekiktuk formation (Figure 2). Total hydrocarbon volumes are estimated at 1 billion barrels of oil in place, with approximately 400 million stock tank barrels (STB) of recoverable reserves; gas in place includes 730 billion standard cubic feet (scf) in the gas cap, and 750 billion cubic feet (bcf) of solution gas. Eighty-eight million STB of oil have been recovered to date. Endicott production began in October 1987, and the current rate is over 100,000 barrels of oil per day. Water injection and gas reinjection are essential for the proper exploitation of reserves.

HISTORY

Endicott field leases were acquired in three Alaska State Lease sales held between 1967 and 1979. The field covers 16,944 ac (6857 ha) and includes all or parts of nine leases. British Petroleum (BP) leased the northwestern portion of the field area at Sale 18 held in 1967. Acreage over the south-central area was obtained by Arco/Exxon and Union/Amoco in Sale 23 in 1969. Sale 23 was held shortly after the discovery of the Prudhoe Bay field and consequently the cost of acreage increased significantly. After the Endicott discovery well was completed in 1978, only two open leases remained over the area of the accumulation. Both tracts were leased in 1979 at State Sale 30. One lease, located in the northwestern section of the field, was acquired by Sohio along with several Alaska native corporations. Arco/Exxon/ Union leased the other, which is at the far eastern periphery of the field. Arco sold the majority of their leases to Exxon in 1984. Table 1 summarizes the working interest ownership of the Endicott field.

Endicott field leases were consolidated into the Duck Island Unit in 1978 (Figure 3). The Unit has been expanded three times and contracted twice. It will contract to the initial Endicott Participating Area in November 1992 if no action is taken by the owners to preserve the affected leases within the Unit.

In 1976, BP began exploratory drilling in the Endicott area with the initiation of the Sag Delta well program. Sag Delta 1 discovered oil in the Carboniferous Lisburne formation (Figure 2), and Sag Delta 2 was planned to delineate the accumulation. Sag Delta 2 and Sag Delta 3 were drilled concurrently in the 1976–1977 drilling season and were taken to basement in order to test the Kekiktuk sands. Although these sands had been encountered in previous North Slope wells, they lacked significant porosity and were not considered as reservoir quality sandstones. In the Sag Delta 3 well proposal, Area Geologist Wilf Bischoff recommended drilling the Kekiktuk interval because, as he stated “. . . it is possible that under a different depositional and

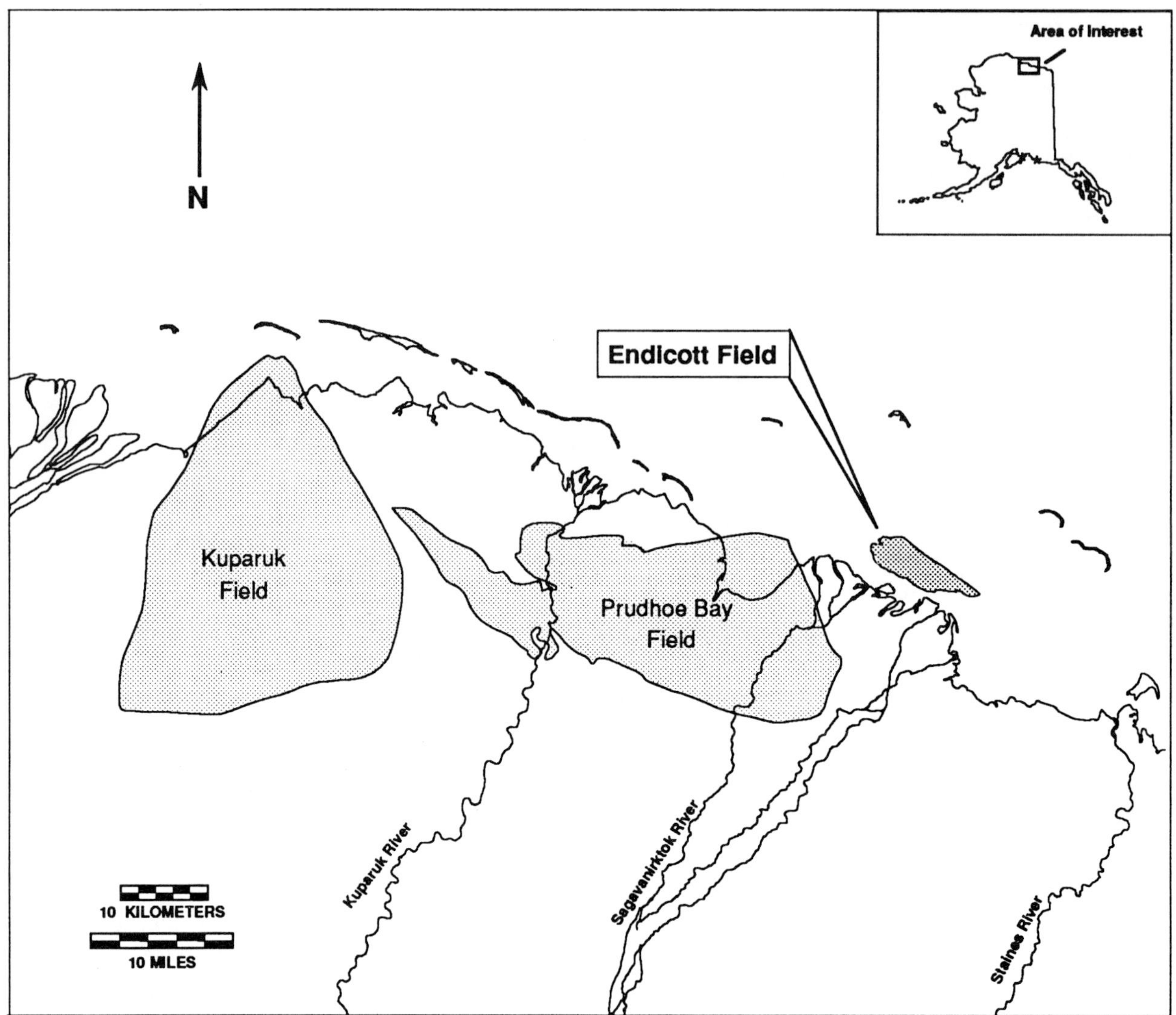

Figure 1. Location of Endicott field relative to other major North Slope producing fields.

diagenetic scenario, silicification may not have taken place."

Although the Lisburne formation was wet in Sag Delta 2, both Sag Delta 2 and Sag Delta 3 contained oil-stained Kekiktuk sands of reservoir quality. A time structure map of the top Kekiktuk was subsequently generated by John Fisher and a location was chosen for the Sag Delta 4 well.

The discovery well, Sag Delta 4, spudded in March 1977. The well could only be drilled to a total depth of 2514 ft (766 m) in that season. Drilling resumed the following year under Sohio operation and a discovery was made that totaled 363 ft (111 m) of gross oil pay and 38 ft (12 m) of gross gas pay in the Kekiktuk formation. Sag Delta 4 tested at 2473 BOPD. The Endicott field was further delineated with the drilling of Duck Island 1 by Arco and Exxon the following year. Duck Island 1 encountered 358 ft (109 m) of gross oil pay and 67 ft (20 m) of gross gas pay in the Kekiktuk, confirming a substantial discovery.

DISCOVERY METHOD

Seismic and well data were the primary exploration tools used in the discovery of the Endicott field. A time structure map on the top Lisburne formation was generated early in the leasing phase as lease acquisition was driven primarily by the Lisburne target. Figure 4 shows the Lisburne structure map as it appeared in 1975, prior to the drilling of the Sag Delta 1 well.

Kekiktuk well data were scarce because the formation had only been encountered in a few wells in the vicinity. Prior to the drilling of the Sag Delta wells, the Kekiktuk section seen in the North Slope

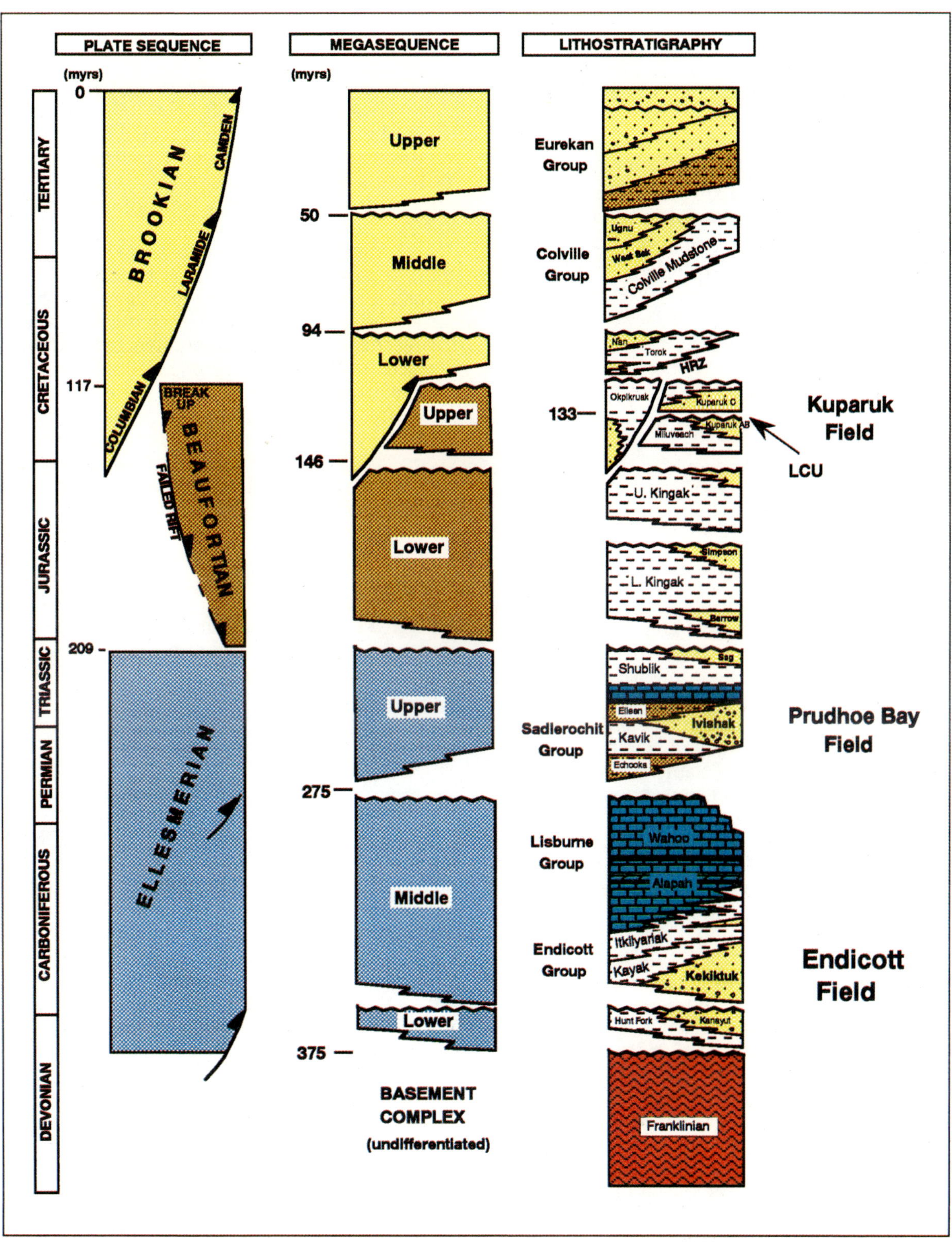

Figure 2. Generalized stratigraphy and plate sequences of North Alaska. (Modified from Hubbard et al., 1987.)

Table 1. Endicott participating area voting interest (%).

Amoco Production Co.	10.4940
Arco Alaska, Inc.	0.0234
BP Exploration (Alaska) Inc.	56.7825
Cook Inlet Region, Inc.	0.6456
Doyon, Limited	0.1291
Exxon Corporation	21.0206
Nana Regional Corporation, Inc.	0.3874
Union Oil Company of California	10.5174

subsurface had been well cemented. In addition, the Kekiktuk sandstone section exposed at the Eastern Brooks Range outcrops was also tight. The prospect was based on seismic analysis that revealed a slight divergence of Kekiktuk section beneath the Lisburne. This was recognized as a possible thicker sand section, with potential for better reservoir quality than previously encountered.

TECTONIC HISTORY AND REGIONAL STRUCTURE

Mississippian

Hubbard et al. (1987) have summarized the major tectonic events in the geologic history of North Alaska. They separate the post-Devonian section into three seismically defined plate sequences: the Ellesmerian, the Beaufortian, and the Brookian, which refer to prerift, synrift, and postrift, respectively (Figure 2). Other accounts of North Slope tectonic history have been presented by Bird (1981), Tailleur (1969, 1973), and Tailleur et al. (1970).

Ellesmerian uplift over much of North Alaska and north of the present-day coastline is evidenced by thick deposits of the Ellesmerian sequence observed today. Figure 5 shows an isopach map of the Endicott Group with some of the major Mississippian tectonic features of Northern Alaska. Kekiktuk formation coarse clastic deposits appear to have been deposited within deep local basins associated with these structural features. An example of this is the half-graben basin at the Endicott field that was controlled by northwest-southeast faulting. Fault geometry and associated basins suggest an extensional component existed, possibly associated with Mississippian wrench faulting (Hubbard et al., 1987).

Facies relationships, paleocurrent data from the northeastern Brooks Range (Nilsen et al., 1980), and subsurface logs indicate that Kekiktuk deposits on the central North Slope were derived from a major tectonic feature to the north of the present coastline. This is represented by the Mikkelsen high on Figure 5, which together with the Colville high bounded the Prudhoe basin during Kekiktuk times. Figure 6 is an east-west cross section illustrating the thickening of the Endicott deposits from the Colville high to the vicinity of Endicott field.

Post-Mississippian

The first major post-Mississippian uplift to affect the Endicott Group occurred at the end of the middle Ellesmerian. This uplift appears to have been controlled by a major tectonic event north of the present coastline, possibly related to the collision of the Siberian plate with Pangea that occurred at approximately 275 Ma (Smith, 1987). It resulted in a major unconformity that eroded portions of the Lisburne and older rocks over most of the North Slope. This is clearly seen at Seal Island A-1 and A-4 and the No Name Island 1 wells where both the Itkilyariak and Kekiktuk of the Endicott Group are truncated by the 275 m.y. unconformity (Figures 5 and 7).

Opening of the Canada basin during the Beaufortian was the most significant post-Mississippian tectonic event to affect North Slope geology. Rifting and accompanying regional unconformities developed during the Jurassic and Early Cretaceous (i.e., Lower Cretaceous unconformity or LCU) and were responsible for further truncation of the Ellesmerian over the central portion of the North Slope. Rift-related faulting, which occurred north of the present-day Arctic coastline, left behind the rift shoulder as a high (Barrow arch). Later, Brookian subsidence by tectonic loading associated with the formation of the Brooks Range resulted in the development of the Colville trough. The compensating peripheral bulge accentuated the high on a portion of the Barrow arch. It is this high, known as the Kuparuk/Prudhoe high (Hubbard et al., 1987), that eventually controlled the major hydrocarbon accumulations on the North Slope. These relationships are summarized in the regional cross section in Figure 7.

STRATIGRAPHY

The stratigraphy of North Alaska has been summarized by Hubbard et al. (1987) and is presented here in Figure 2. Pre-Mississippian metasediments consisting of argillite, quartzite, and carbonate define the economic "basement" on the central North Slope. This is overlain by the northern sourced clastics and carbonates of the Ellesmerian sequence (Mississippian–Triassic). Primary reservoirs in the sequence are the fluvial Kekiktuk sandstones, the Lisburne shelf carbonates, and the fluvial deltaic Ivishak sandstones. Shales and carbonates of the Shublik Formation are the major hydrocarbon source rocks in the sequence.

The Ellesmerian is overlain by silts and shales of the Jurassic Kingak formation which, together with the Miluveach and Kuparuk formations, make up the Beaufortian sequence. The Kingak is a northerly sourced shelf and slope deposit while the Miluveach shales and Kuparuk sandstones are thought to have been locally sourced from the Barrow arch. Organic shales in the lower portion of the Kingak are a

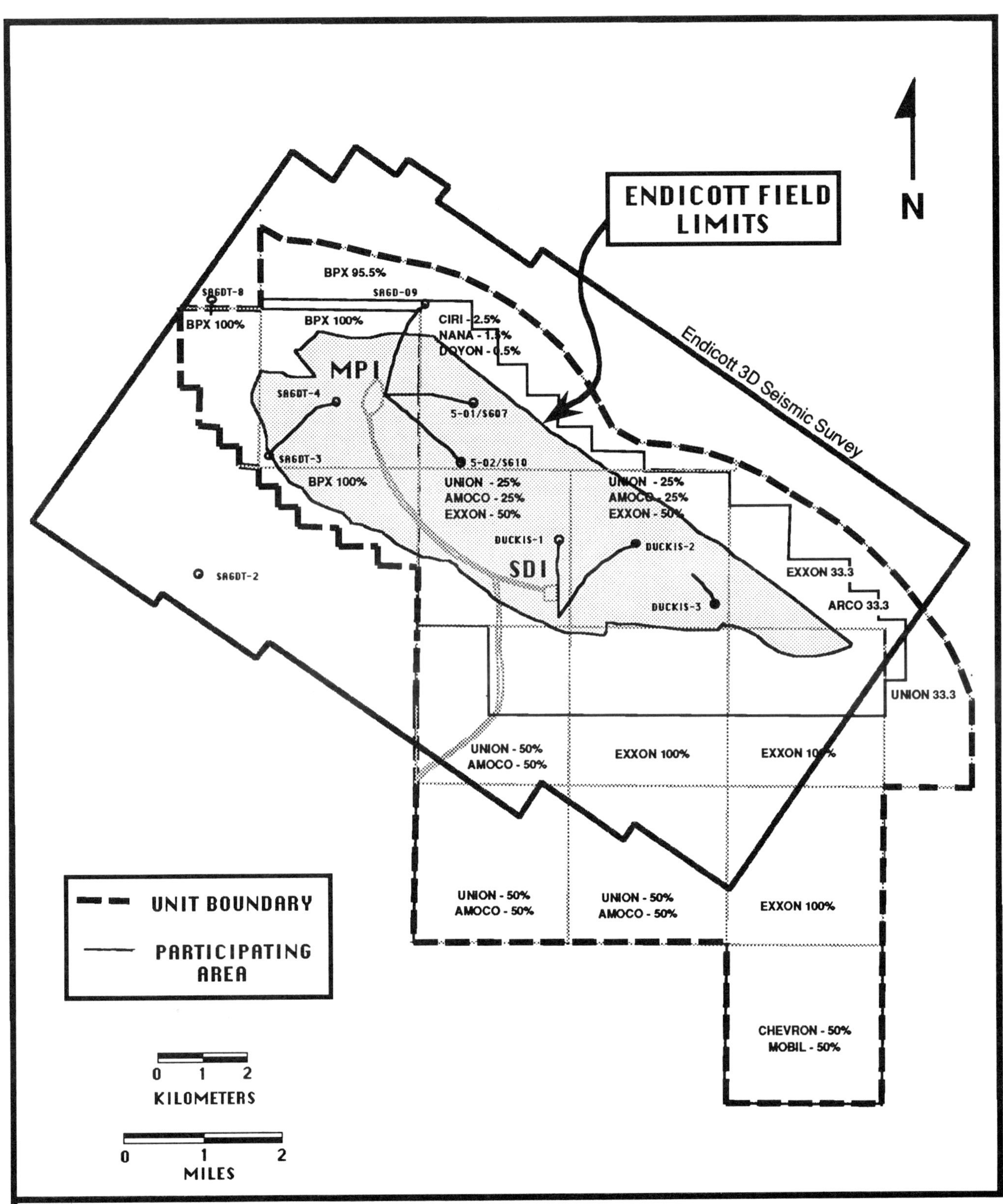

Figure 3. Duck Island Unit established in 1978. MPI is the main production island. SDI is the satellite drilling island. Endicott exploration wells are shown.

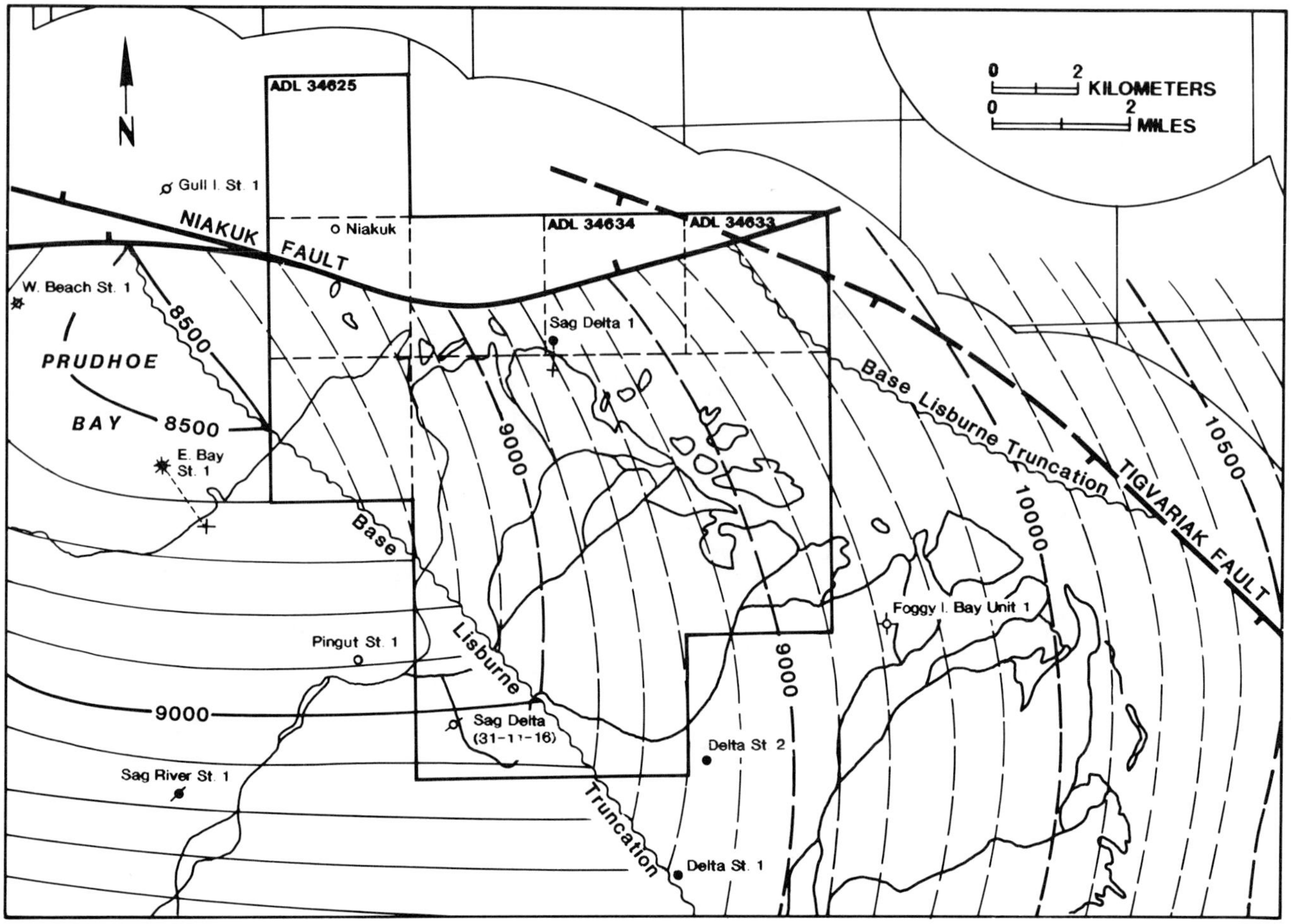

Figure 4. Top Lisburne structure map prior to drilling Sag Delta 1 (1975). The Lisburne carbonate was the primary exploration target on which the acreage was leased.

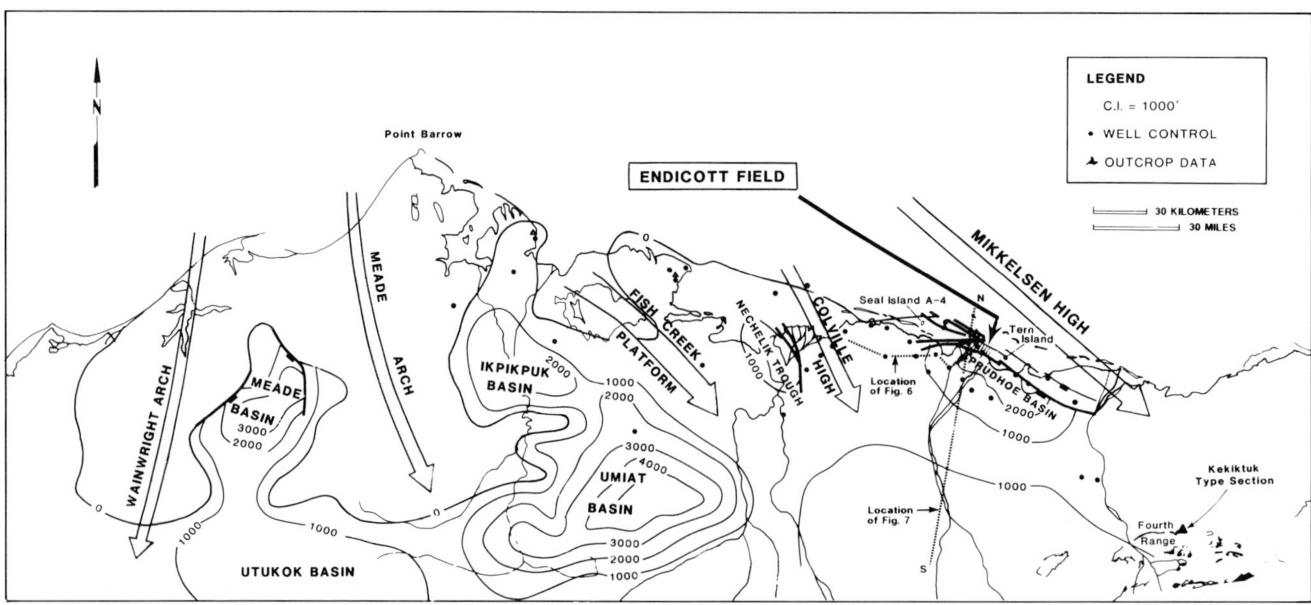

Figure 5. Generalized Endicott isopach map of North Alaska with major Mississippian tectonic features. Arrows indicate assumed plunge direction.

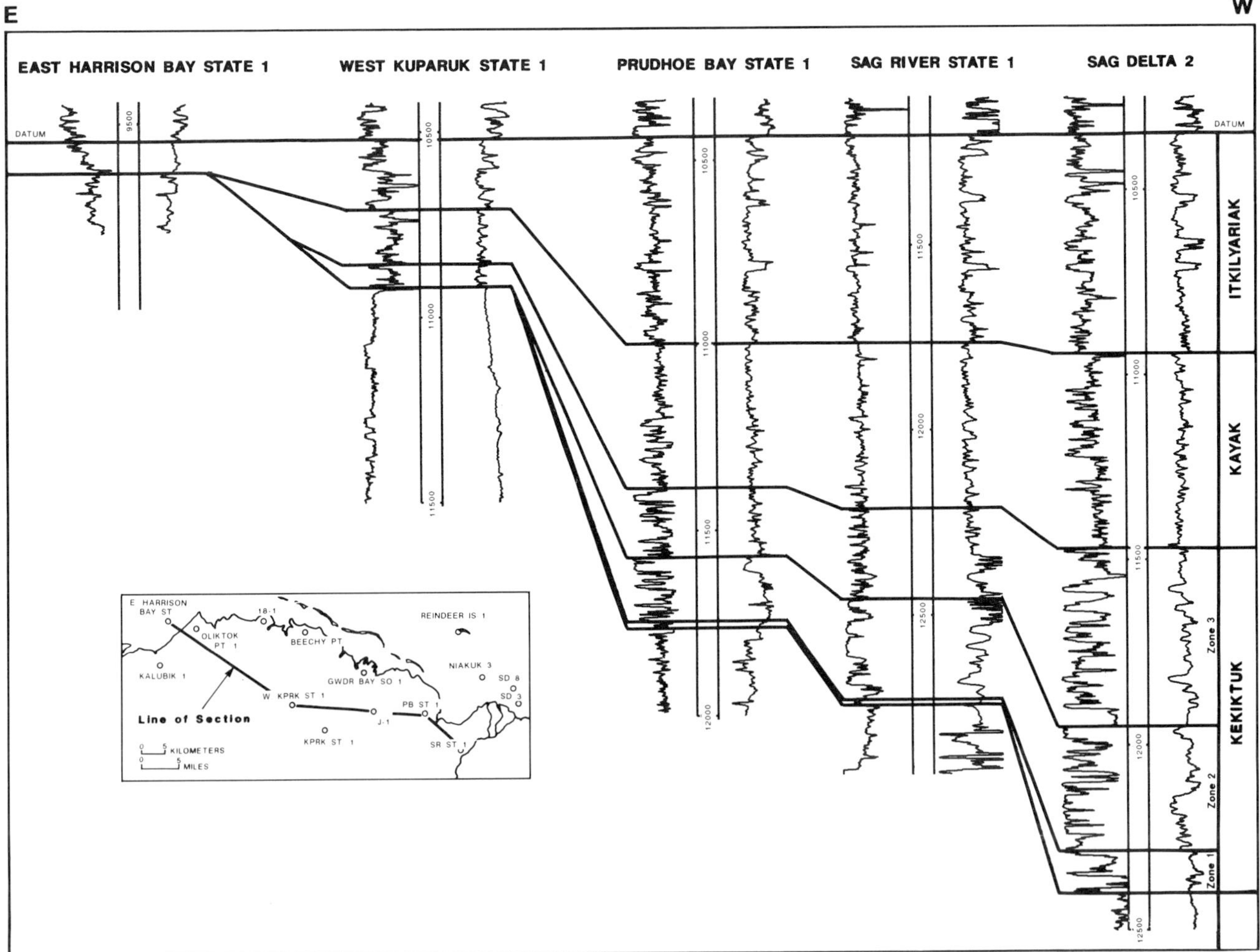

Figure 6. An east-west cross section illustrating thickening of Endicott deposits from the Colville high to the vicinity of the Endicott field. Log curves are gamma ray (left) and deep induction (right). They are shown here to illustrate correlations, and calibration scales are not included.

significant hydrocarbon source and the Kuparuk is a producing reservoir on the North Slope.

Clastics of the Brookian (Lower Cretaceous–Recent) overlie the Beaufortian. They were shed from the Brooks Range to the south and prograded northward filling the Colville trough and the Canada basin. These deposits consist of shelf and slope shales, sand (minor), and siltstones as well as sands and coals of fluvial-deltaic origin. The lower Brookian HRZ (high radioactivity zone of Carman and Hardwick, 1983) is recognized as an important source on the North Slope. There are a number of potential reservoir rocks in this sequence, although production has only been established in the West Sak and Ugnu formations.

TRAP

The structural geometry of the Endicott field is a faulted antiform with a southeasterly plunge. The Endicott hydrocarbon accumulation is trapped to the northeast and north by the Tigvariak and Niakuk faults (Figure 8). Structural dip closure provides the trapping mechanism for the southwestern and southeastern parts of the field. Topseal to the Kekiktuk, where truncated by the LCU, is provided by the HRZ shale (Figure 9). Locally, isolated pods of the Cretaceous Put River sandstone are sandwiched between the HRZ and the Kekiktuk along the unconformity. In the south, where the Kekiktuk is not truncated by the LCU, the top seals are shales within the Itkilyariak Formation that overlie the Kekiktuk.

Updip along the northeastern boundary, the Tigvariak fault juxtaposes the Kekiktuk against impermeable pre-Mississippian metasediments. In the north, the Niakuk fault juxtaposes Kekiktuk against Ellesmerian and Cretaceous sediments, including Kuparuk and Ivishak sands. If communication across this fault occurs, a greater closure would be realized stratigraphically or by northward dip.

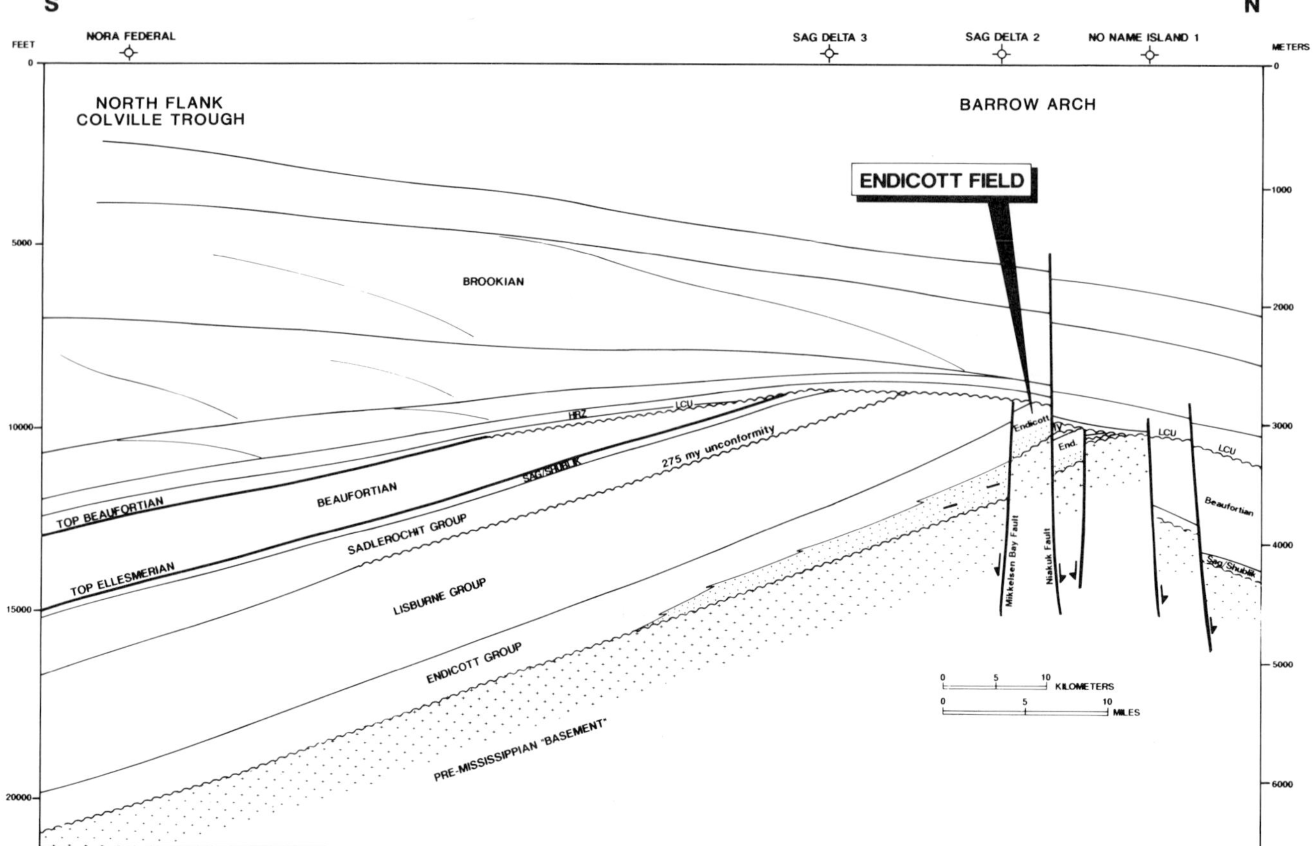

Figure 7. North-south regional cross section of North Alaska, showing the Endicott Group from the Colville trough to the Barrow arch. Location of the cross section (N-S) is shown on Figure 5.

Reservoir

Regional Reservoir Stratigraphy

The Endicott Group is distributed throughout most of northern Alaska. Surface exposures in the northeastern Brooks Range and subsurface deposits beneath the Arctic Coastal Plain are considered to be autochthonous (Moore and Nilsen, 1984; Nilsen, 1981; Nilsen and Moore, 1982). However, Endicott outcrops of the central and western Brooks Range are allochthonous, having been transported an unknown distance by northward-verging thrust sheets. Relationships between surface exposures of the central and western Brooks Range and the subsurface to the north have yet to be resolved.

Within the subsurface of North Alaska, the Endicott Group consists of three formations: the Kekiktuk, Kayak, and Itkilyariak formations (Bird and Jordan, 1977) (Figure 10). The fluvial Kekiktuk formation is overlain by a transgressive sequence including Kayak shales and the Itkilyariak siltstones, shales, and limestones.

The producing reservoir at the Endicott field is the Kekiktuk formation, which is defined by a type section located near Lake Peters (Figure 5) in the northeastern Brooks Range (Brosgé et al., 1962). At the type locality, the Kekiktuk contains late

Tournaisian sandstone, siltstone, coal, and pebble/cobble conglomerates of probable fluvial valley fill origin. Within the Endicott field, the Kekiktuk formation consists of multiple fining-upward successions of sandstone, siltstone, shale, and coal deposited by a Visean age fluvial system. The Kekiktuk at both locations unconformably overlies pre-Mississippian metamorphic basement.

During Mississippian times, the northwest-southeast-trending Colville and Mikkelsen highs bordered a basin stretching from the Prudhoe Bay area, in the west, to at least the Fourth Range in the east. Drainage was primarily toward the southwest (Woidneck et al., 1987). The Kekiktuk valleys were dominated by coarse-grained, braided river deposits during periods of high tectonic activity. During quiescent times, meandering streams and fine-grained overbank deposits prevailed (Melvin, 1987a).

Kekiktuk deposition in the vicinity of the field apparently began in the Tern Island area (Figure 5). Palynology within the Tern Island wells suggests a very early Visean time of deposition. Most of the upper Kekiktuk is eroded in the Tern Island wells. To the southeast in the Mikkelsen Bay State 1, the full Kekiktuk section reaches a thickness of 1646 ft

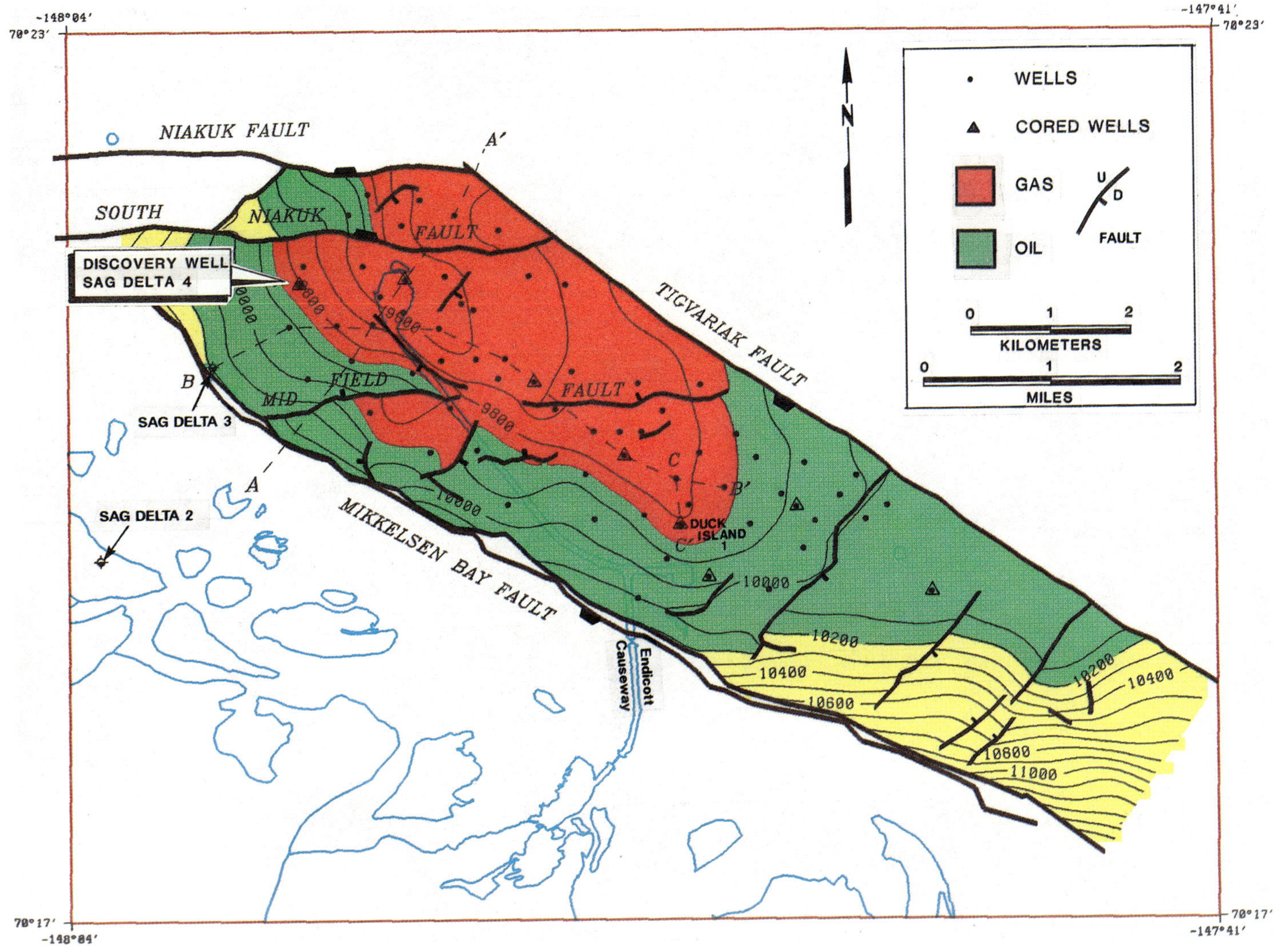

Figure 8. Structure map of the top Kekiktuk at Endicott field. (A–A′, B–B′, C–C′ are cross section lines for Figures 9, 11, and 14, respectively.)

(502 m) as opposed to 789 ft (240 m) within the Endicott field. Seismic evidence suggests thinning of the basal Kekiktuk section westward toward the field.

Field Stratigraphy

Within the Endicott field, the Kekiktuk formation is divided into three zones and five subzones based on lithology and genetic origin (Figures 9 and 10). Detailed core and log analyses have resulted in identification of four categories of fluvial-related depositional facies: (1) channel sandstones, (2) natural levee siltstone deposits, (3) flood-plain shales, and (4) backswamp deposits (coals and shales) (J. Melvin, personal communication, 1987).

Zone 1 is the lowermost Kekiktuk interval lying unconformably on Devonian metasediments. Lithologically, it consists largely of shale, siltstone, numerous coals, and minor sandstones. Facies relationships suggest that deposition occurred within a low-lying swamp setting with numerous lakes and traversing streams (Melvin, 1987b). Overall gamma log response within zone 1 is typically a coarsening-upward trend. The blocky gamma ray signature at the base of zone 2 marks the top of zone 1 (Figure 11). A strong seismic reflector (labeled P4 on Figure 12A) exists within zone 1 as a result of thick, laterally extensive coals.

Zone 2 represents the best producing interval within the Endicott field. In contrast with zone 1, it contains multiple, stacked successions of medium- to coarse-grained sandstones that are interpreted as having formed within a braided stream depositional setting (Melvin, 1987a). Backswamp and/or lacustrine shales with variable degrees of continuity exist within the mostly sandy subzones of 2A and 2B. Zone 2 is recognized on electric logs by its blocky gamma ray signature (Figure 11).

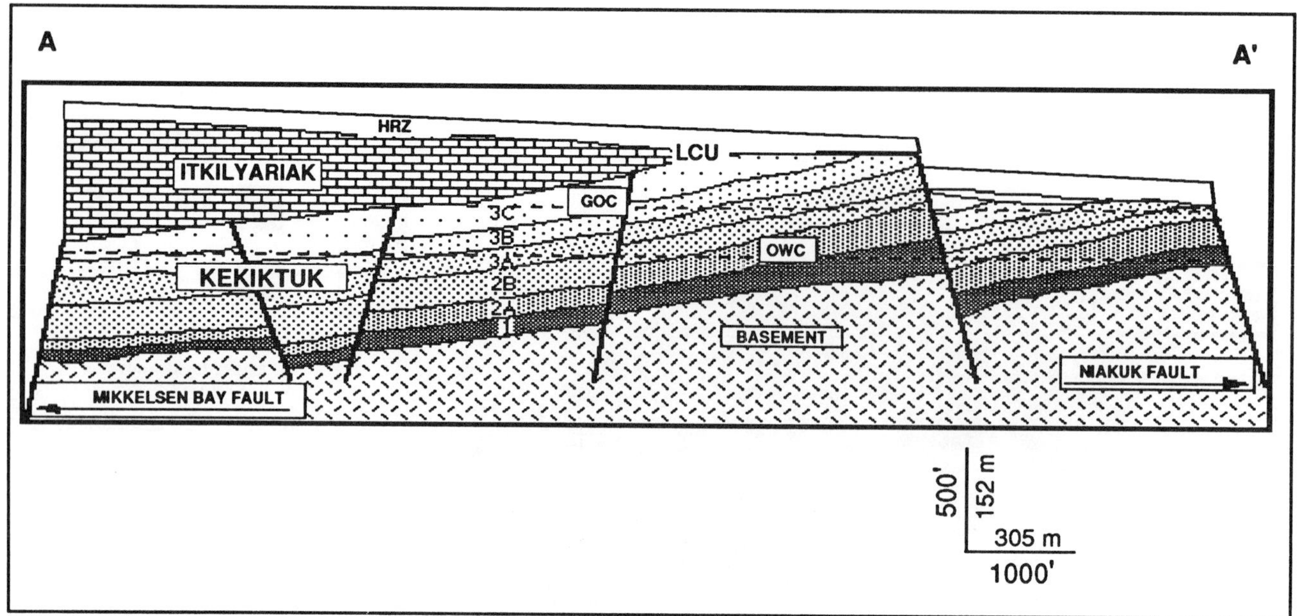

Figure 9. Dip section through Endicott field illustrating the trap. Subzones 1 through 3C, within the Kekiktuk, are shown with relative thicknesses schematically represented. The Kekiktuk and Itkilyariak are shown truncated by the Lower Cretaceous unconformity (LCU). The gas-oil contact (GOC) and oil-water contact (OWC) are shown as dashed lines. See Figure 8 for location of section A–A'.

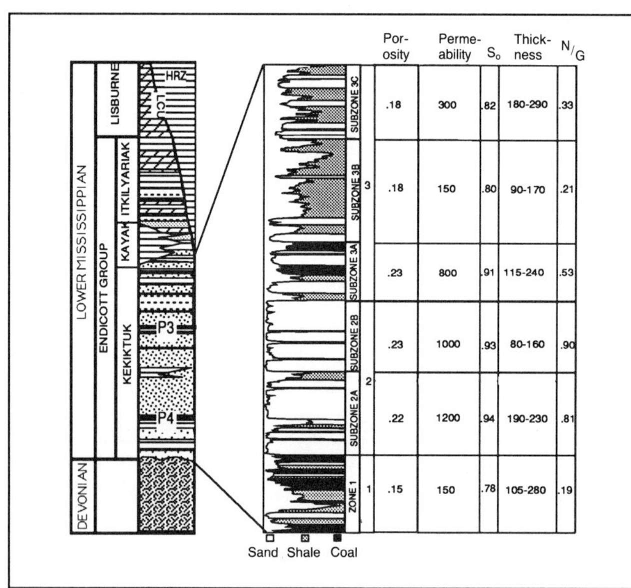

		Porosity	Permeability	S_o	Thickness	N/G
SUBZONE 3C		.18	300	.82	180–290	.33
SUBZONE 3B	3	.18	150	.80	90–170	.21
SUBZONE 3A		.23	800	.91	115–240	.53
SUBZONE 2B		.23	1000	.93	80–160	.90
SUBZONE 2A	2	.22	1200	.94	190–230	.81
ZONE 1	1	.15	150	.78	105–280	.19

Figure 10. Endicott type-log (gamma ray) and section (left), with average zonal oil sand properties (right).

and some sandstone separates the blocky sandstones within subzone 2A from those in subzone 2B. The top of this shale defines the top of subzone 2A. Subzone 2B is lithologically very similar to 2A and is also interpreted as a channelized braided stream system. It is distinguished only by the increased fining-upward trends toward the top that Woidneck et al. (1987) interpret as amalgamated point bar deposits within a waning braided system.

Zone 3 is considerably more diverse than zone 2, consisting of a series of well-defined fining-upward packages of conglomerate, sandstone, siltstone, shale, and coal (Figure 14). Zone 3 is divided into 3 subzones—3A, 3B, and 3C—that are largely representative of a meandering stream depositional system.

Subzone 3A consists of several fining-upward intervals of sandstone, siltstone, shale, and coal. Coals within this subzone give rise to one of the main mappable seismic reflectors within the Kekiktuk formation (Figures 12A and 12B), referred to collectively as the "P3" seismic horizon (Figures 11 and 14). Subzone 3A sands have a higher degree of lateral continuity relative to sands within subzones 3B or 3C. However, subzone 3A sandstones are isolated vertically by continuous shales and coals. These vertical permeability barriers are a constant consideration in the depletion planning. Subzone 3B consists mostly of siltstone, shale, and minor amounts of sandstone (Figure 14). Predicting stratigraphic distributions for the few sandstones present within this subzone is difficult owing to the discontinuities involved. Minimal field reserves are

Subzone 2A consists of over 90% moderately to well-sorted multistory units of massive to cross-stratified sandstone with a high degree of lateral continuity (Figure 13). Mudstone intervals (typically less than 3 ft [1 m] in thickness) punctuate the stacked channel successions and represent vertical accretion within inactive parts of the fluvial system. A thick, laterally extensive lacustrine shale interbedded with siltstone

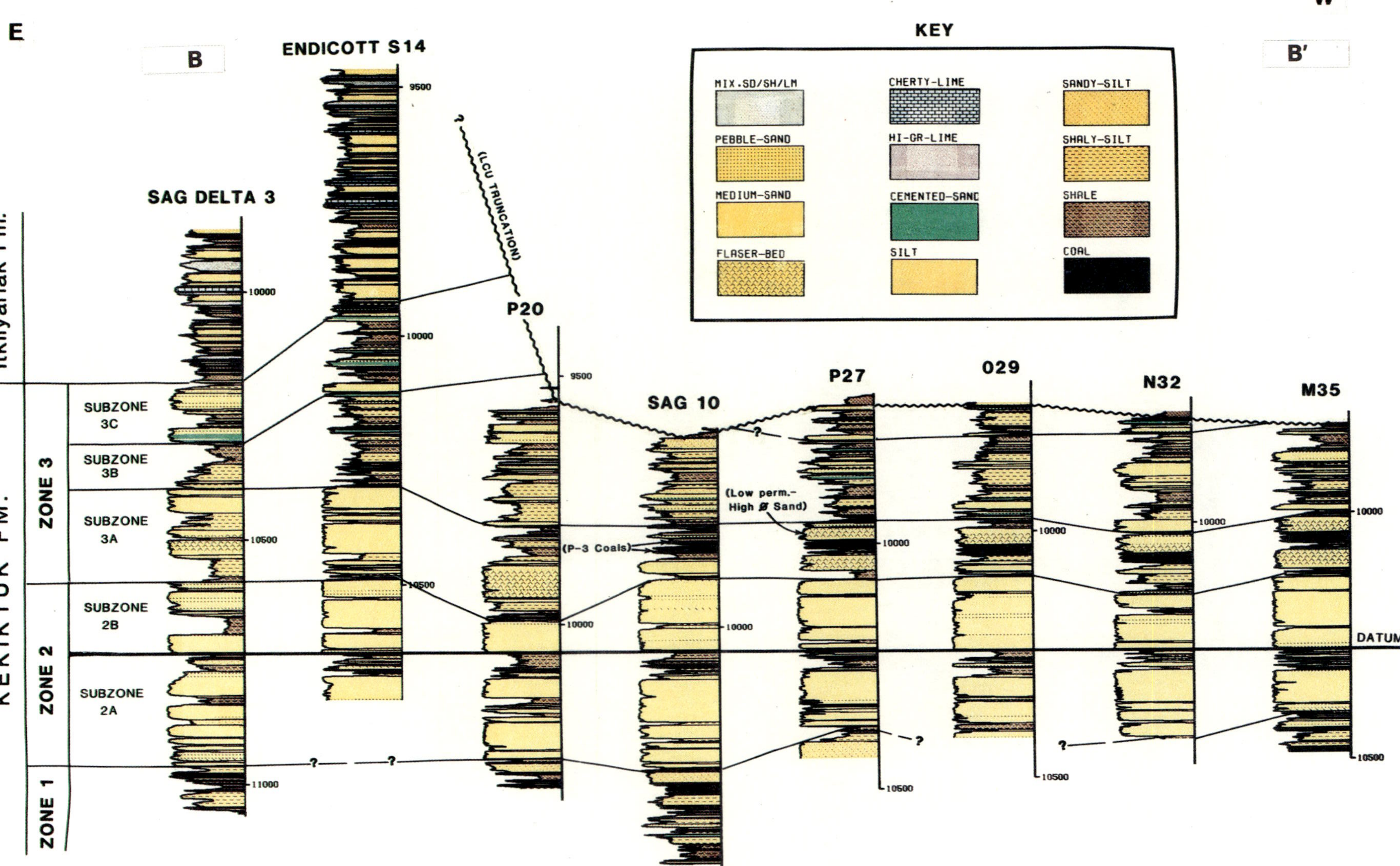

Figure 11. Stratigraphic cross section (strike line) of the Kekiktuk formation illustrating subzone thickness variations, Lower Cretaceous unconformity (LCU) truncation, and log facies correlation (see Figure 8 for section location B-B').

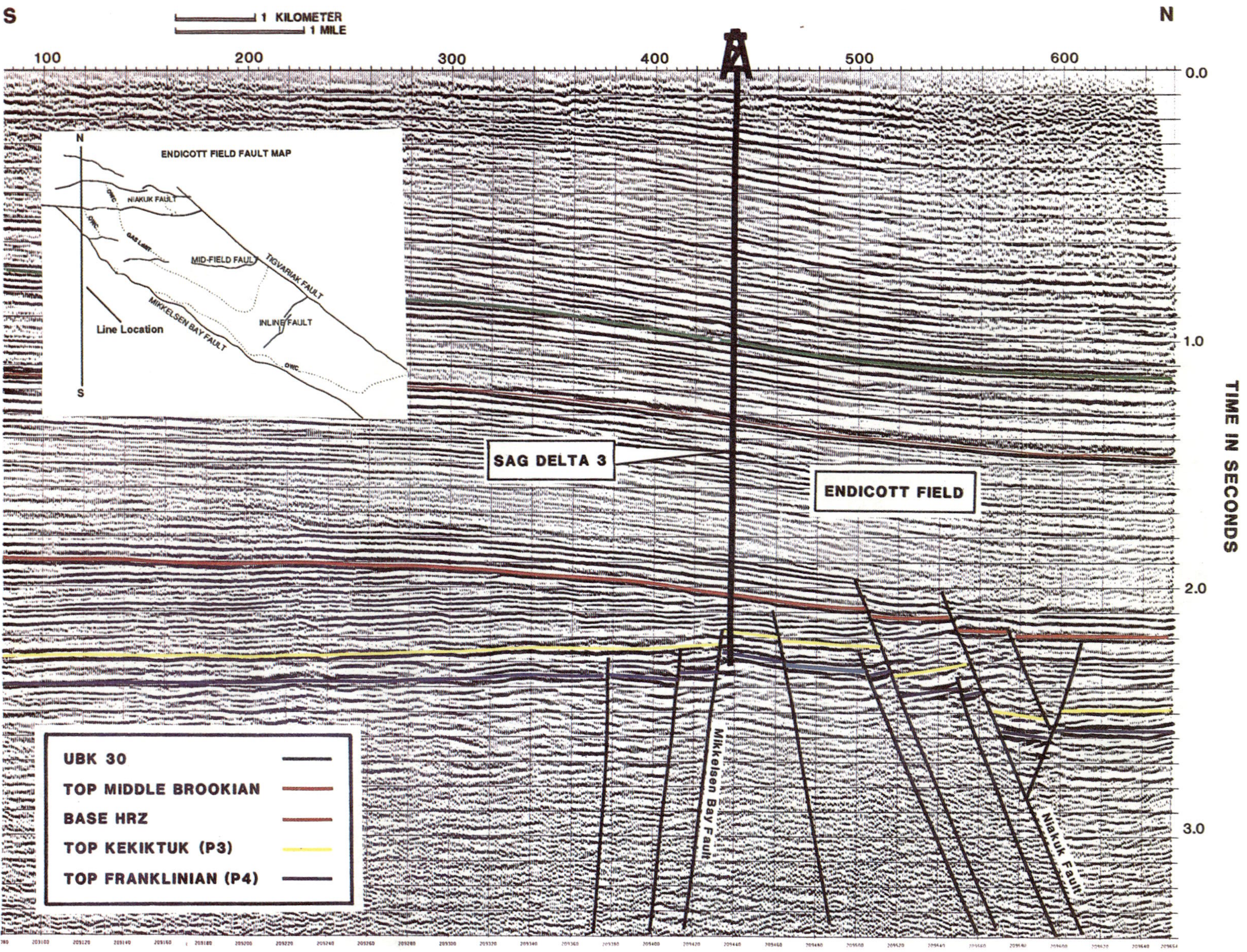

Figure 12A. Seismic line LB85-210 through the western portion of the Endicott field, showing mappable seismic reflectors P3 and P4.

12

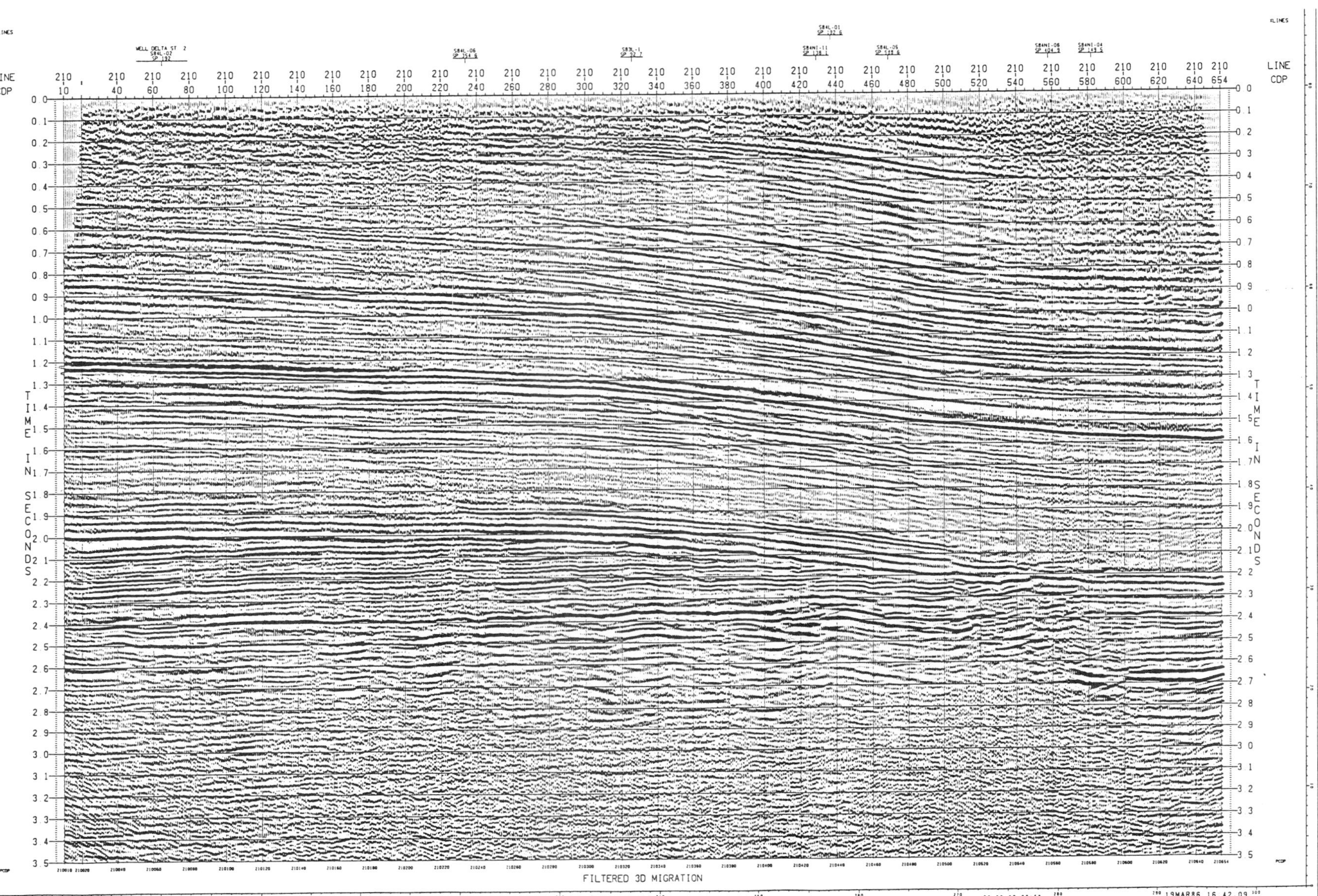

Figure 12B. Seismic line LB85-210 uninterpreted.

Figure 13. Core photo taken from subzone 2A of the Endicott 0-29 well. This photo illustrates the stacked nature of sandstones with variable grain sizes typical of zone 2 of the Kekiktuk formation. Enlarged photos to the right show an erosional scour overlain by medium- to coarse-grained cross stratified sandstone. Thin intervals of silt and shale are rare but do exist (12,487.3 ft). Large-scale cross lamination is also illustrated below 12,494 ft.

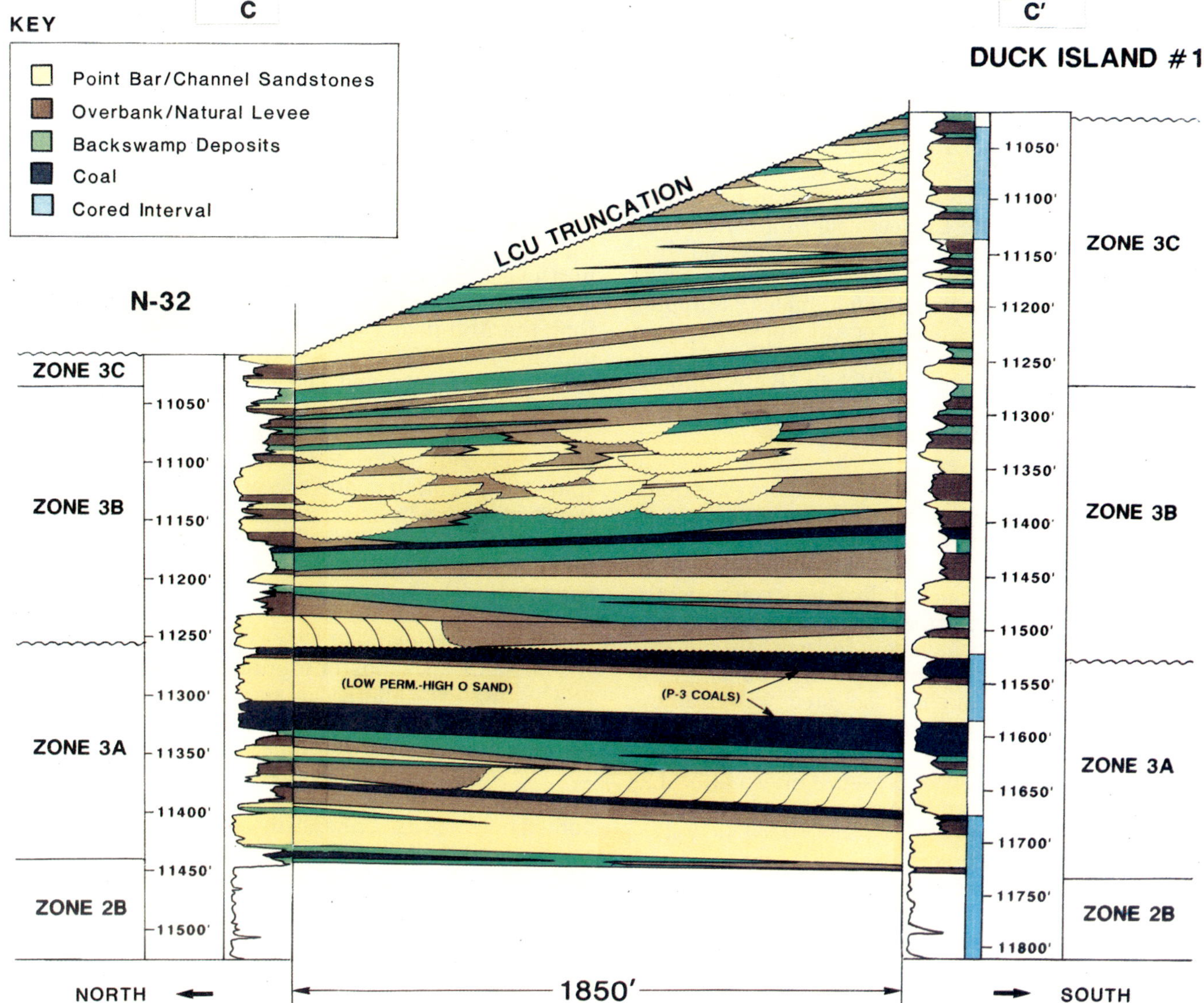

Figure 14. Stratigraphic cross section of the Kekiktuk formation, zone 3, illustrating the Endicott color facies scheme as determined from core and log response. Gamma ray log curve shown for correlation purposes, and no scale is shown (see Figure 8 for location C–C').

found within subzone 3B. Subzone 3C consists of a series of distinct fining-upward, individual, and stacked channel sandstones with interbedded conglomerate, siltstone, and shale deposits. The existence of multiple fining-upward cycles coupled with diverse lithologic assemblages suggests deposition within a meandering stream depositional environment on a low-lying alluvial plain.

Methods of Analysis

Most of the subsurface information at the Endicott field is extracted from well logs. A customary suite of open-hole electric logs run in each well consists of a dual lateralolog, microspherically focused log, and gamma ray run in conjunction with the density/neutron tool. Sonic and dipmeter logs are run as needed. Well log results are the primary means of estimating net pay, determining fluid contacts, and making well to well correlations. In addition, pulse neutron logs are run to establish a baseline for monitoring production.

Core has been recovered through all or part of the Kekiktuk formation in eight Endicott wells (Figure 8). Routine core analysis has been performed on core plugs cut at 1 ft intervals. Detailed, in-house, sedimentological studies have provided insights into the Kekiktuk depositional environments. Facies descriptions from the core compared to logs has led to a computer synthesis of facies in uncored wells (Sakuri and Melvin, 1988). Results of the latter technique can be seen in Figures 11 and 14. This technique has significantly enhanced well correlation and reconstruction of depositional environments within the Endicott field.

A 3D seismic survey was shot over the Endicott field in 1983 (Figure 3). Such a dense dataset lends itself to a wide range of manipulations that enable detailed subsurface interpretation in this area. Methods of analysis include seismic inversion, horizon flattening, horizontal time slices, multidirectional traverses, and seismic movies.

Well log data and seismic information have been integrated to map the Endicott subzones. Additionally, each individual subzone is subdivided into correlatable layers, where appropriate, in order to map individual sandstones and shales. Mapping porosity, net to gross, and net thickness for each subzone layer has been valuable in establishing sand continuity trends in subzones 3C and 3B. Continuous and semicontinuous shales in zone 2 have also been mapped. Estimates of shale widths, lengths, and types have been derived from the sedimentological studies using core data. This information coupled with well log data has allowed for estimates of effective horizontal and vertical permeabilities used for reservoir modeling studies.

Cased-hole pulsed nuclear logs and compensated neutron logs are run in the field to identify gas and water movement mechanisms. Pulsed neutron logs have identified gas underrunning, water cusping, and waterflood breakthrough in selected wells. These data are sometimes helpful in confirming the continuity of shale layers.

A significant amount of pressure information has been collected with open-hole wireline measurements and cased-hole measurements as discussed by Walker et al. (1989). The open-hole measurements have been particularly informative because they allow measurements in subzones that will not be accessible after the well is cased. These data have been used to study the connectivity of the subzones and sequences. Interpretation of pressure information concluded that subzones 3C and 3B, where net to gross can vary from 0 to 0.60 from well to well, have approximately 95% connectivity.

Faults as Barriers to Production

The Endicott reservoir is complicated by a series of minor extensional faults that parallel the major trends which delineate the field. The northwest-southeast-trending faults are Mississippian and older and become more numerous with depth, while east-west-oriented faults are thought to be Cretaceous and younger. Many of the minor faults have little or no effect on production. Several wells have been completed on the north side of the South Niakuk fault, for instance, where reservoir pressures indicate communication exists with the rest of the field.

A series of east-west-trending, down to the south extensional faults in the central part of the field are collectively known as the "Mid-field Fault system" (Figure 8). The faults appear to be discontinuous and do not significantly offset the sands. However, pressure tests in wells on either side of this fault system have indicated it is a sealing barrier that separates the field into two distinct areas. Due to

the lack of pressure communication across the system, waterflood plans were modified to allow for consistent pressure support in both areas. Lack of present-day pressure communication is attributed to stratigraphic variation, as well as mylonite, mineralization, or tar development along the fault planes.

Stratigraphic trends tend to vary considerably across the Mid-field Fault system. For example, the P3 coals that are prevalent on the southeast side of the faults are reduced across the faults to the northwest. Sands within subzone 3A pinch out near the same fault system. One possible explanation is that the Mid-field Fault system was active during deposition of subzone 3A, resulting in a topographic barrier that restricted the distribution of various environments of deposition.

As a result of interpreting production trends and individual well performance, it has become apparent that some smaller northwest–southeast-trending faults may also be transmissibility barriers to flow. The impact of these faults on field performance is being continually evaluated and will be more apparent in the future.

Rock Properties

The Kekiktuk sandstone found in the Endicott field is the highest quality reservoir rock currently producing on the North Slope. Fieldwide petrophysical characteristics are summarized by zone in Figure 10. The main features of the reservoir are the coarse-grained sands described above and secondary porosity. Porosity enhancement is believed to be related to the LCU, which provided Ellesmerian rocks with access to either meteoric waters during surface exposure at Early Cretaceous times or to organic acids from overlying shales some time after burial. The fluids apparently dissolved carbonate cements that filled most of the pore space in the rock, creating the excellent porosity-permeability characteristics observed today (Figure 15).

The clean sandstone intervals contained in subzones 3A, 2B, and 2A are mineralogically greater than 95% silica in the form of cement and grains, with the remaining fraction being made up of kaolinite and traces of carbonaceous material (Figure 16). The measured matrix densities of core plugs are between 2.62 and 2.65 gm/cc. Matrix densities below 2.65 are the result of the carbonaceous material and kaolinite. Clean sand intervals in subzones 3C and 3B have measured matrix densities between 2.65 and 2.66 gm/cc. The increase in matrix density is due to more abundant dolomite and pyrite in these sands.

Reservoir Conditions and Fluid Properties

Fluids within the Endicott reservoir include gas cap gas, an oil leg, and a tar mat that is underlain by an aquifer. There is often an intermediate aquifer between the oil leg and tar mat (Figures 17 and 18). The gas-oil contact (GOC) is a planar, horizontal surface at 9854 ft (3003 m) subsea. The oil-water contact (OWC) is also mapped as a planar surface

Figure 15. Photomicrographs illustrating the porosity/permeability characteristics of the Kekiktuk formation. Both samples display very good secondary porosity. Sample (A) is from Sag Delta 10, 12,351 ft, and shows quartz grains (white and gray) with remnant siderite (black and brown). The porosity (blue) is 28% and permeability is 1310 md. Sample (B) is from Sag Delta 10, 12,562 ft, and shows quartz grains with almost no remnant siderite; note the subhedral voids left by the dissolution of siderite. The porosity is 25% and permeability is 1830 md.

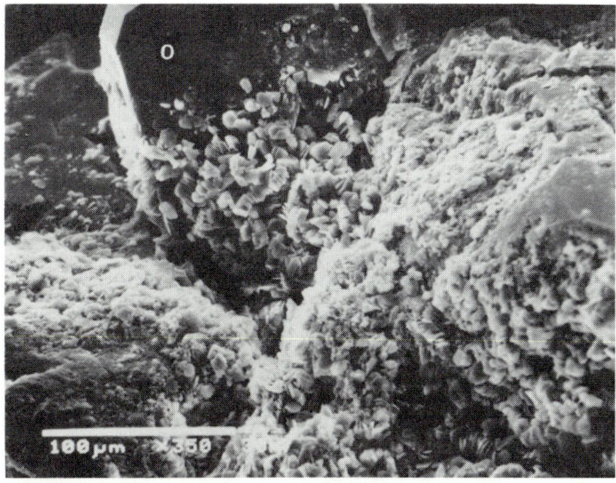

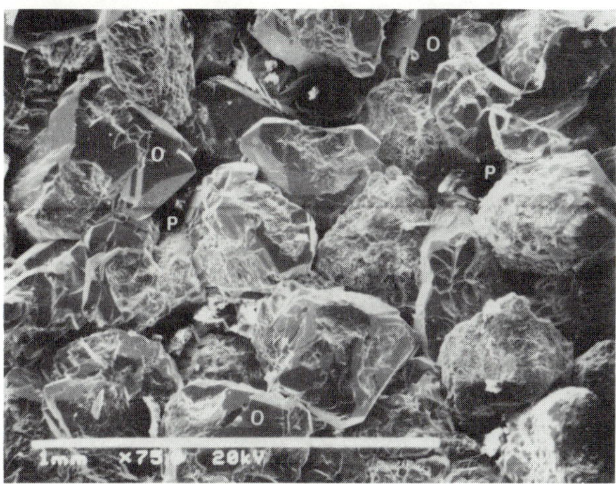

Figure 16. (A). SEM photograph from within the Kekiktuk subzone 2a in the Q35 well at 12,178 ft. The photograph identifies kaolinite (shown as euhedral platelets in the central and lower right) as the primary clay occluding the pore space. Note the diagenetic quartz overgrowths (O) as the primary cementing agent. (B) SEM photograph of typical reservoir rock in Kekiktuk subzone 3a in Q35 at 10,680 ft. The rock consists of fine- to medium-grained, sorted quartz grains. Note the euhedral quartz overgrowths (O) and the open pore throats (P).

at 10,195 ft (3107 m) subsea. Variance of fluid contacts within each well is generally plus or minus 10 ft (3 m), which is within the error tolerance of the directional survey.

The Endicott gas cap is up to 260 ft (79 m) thick and contains an estimated 730 bcf of original gas in place. The majority of the gas cap is confined to the northwestern area of the field.

Within the reservoir, the oil zone is 339 ft (103 m) thick. Fluid properties of the oil column are based on compositional analysis and PVT measurements from bottom hole oil samples taken under pressure in four wells. The reservoir is considered to be saturated.

The oil column is initially saturated at a bubble point of 4855 psia. Initial field gas oil ratio is 750 scf/stb. Oil gravity averages 23° API but is known to vary with depth from 21.1° to 23.6° API. Oil viscosity measurements indicate increases from 0.86 cp to 3.16 cp with depth in the oil column. Most of the oil column has an average oil viscosity of 1 cp with the higher viscosities coming from the interval 60 ft (18 m) above the base of the oil column. It indicates a trend toward slightly heavier crudes at the base of the oil column. A summary of field fluid properties is contained in Table 2.

The oil column is underlain by either a tar mat or an intermediate aquifer and tar mat. Where it exists, the intermediate aquifer is up to 120 ft (37 m) thick. The bulk of the tar mat that lies below the intermediate aquifer exists between 10,300 ft

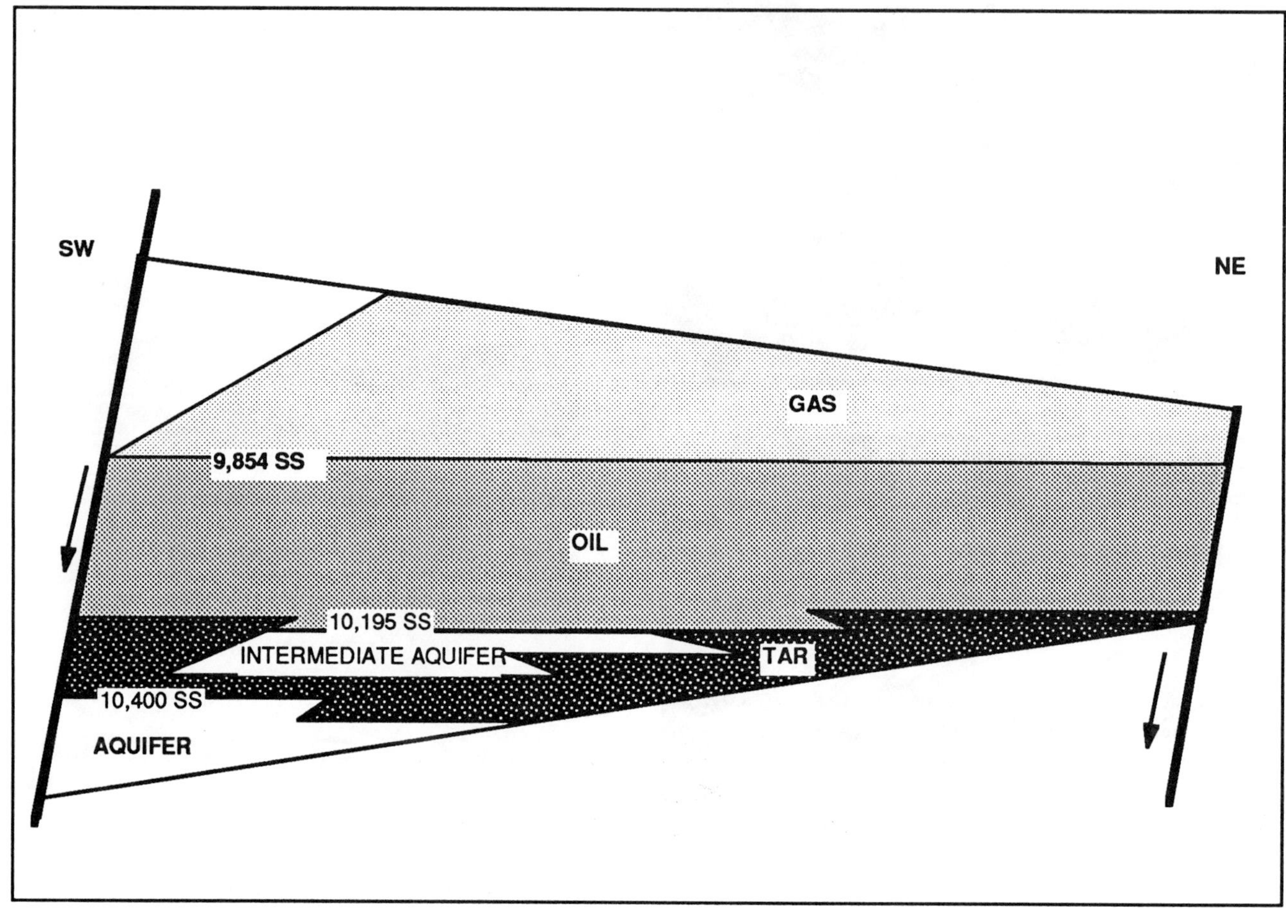

Figure 17. Schematic representation of the fluid column at the Endicott field.

(3139 m) subsea and 10,420 ft (3176 m) subsea. The tar saturation is measured to be in excess of 93%. The tar interval therefore is a strong barrier that reduces the amount of aquifer pressure support available to the reservoir. Reservoir simulation models and field pressure measurements confirm that the tar mat does have a low transmissibility to water. Residual oil saturations in the intermediate aquifer between the base oil column and tar mat are approximately 10–15%.

Depletion Planning

Reservoir description has played a key role in the depletion planning of the Endicott field. Field surveillance data and geologic models have indicated that each subzone can be treated as an individual reservoir. Further subdivision can occur between subzones in the northwest versus southeast areas of the field in consideration of the sealing Mid-field Fault system. The primary depletion plan involves optimization of oil recovery through an efficient waterflood program in each subzone and on both sides of the Mid-field Fault.

In most cases, the oil accumulation in each subzone is overlain by gas and underlain by water. Wells generally are placed and perforated to achieve the maximum standoff from gas and water contacts. Since future field oil offtake may be limited by surface gas handling capacity, it is essential that free gas production be curtailed. Water injection was recognized from modeling to be critical for maximizing recovery; approximately two-thirds of the reserves are associated with secondary recovery. An effective waterflood program will also maintain reservoir pressure through voidage replacement, therefore reducing the free gas production. Because of these benefits, the Endicott full field waterflood became operational within two years of production startup.

The projected field life of Endicott is 21 years, with about the first 6 years of production at a plateau rate of 100,000 BOPD.

Source

Endicott field oil was sourced by the Cretaceous HRZ and probably by the Triassic Shublik and Jurassic Kingak formations. The most significant contribution is from the HRZ; this oil-source

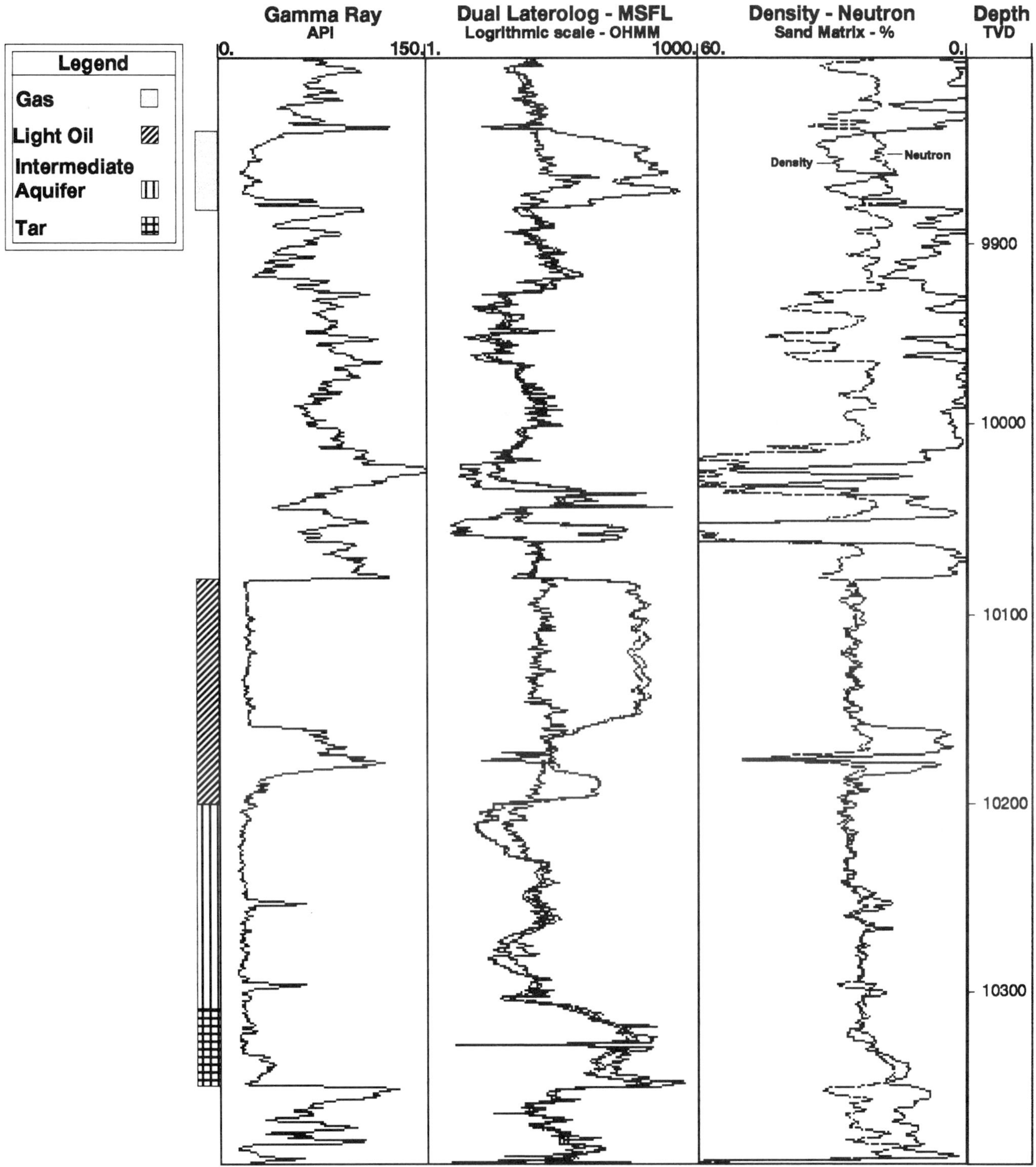

Figure 18. Characteristic Kekiktuk log response related to different fluid types. Endicott well P-21.

correlation is based on isotopic composition and biomarker fingerprints that are shown in Figures 19 and 20. Marginally mature HRZ shales are in direct communication with the Kekiktuk over a large portion of the field. However, most of the HRZ oil probably migrated into the trap from downdip along the subcrop to the southeast where the HRZ is more mature. No direct reservoir/source rock communication exists between the Kekiktuk reservoir and the Shublik/Kingak source rocks. However, the Shublik

Table 2. Endicott fluid properties.

Bubble Point Pressure	4855 PSIA
Reservoir Temperature	219 DEG F
Oil API Gravity	23°
Viscosity—Oil	1 cp
Viscosity—Water	0.3 cp
Viscosity—Gas	0.03 cp
Solution GOR	750 scf/STB
Oil Volume Factor	1.35 RB/STB
Gas Volume Factor	0.68 RB/MSCF
Gas Compressibility	0.97
Gas Specific Gravity	0.845

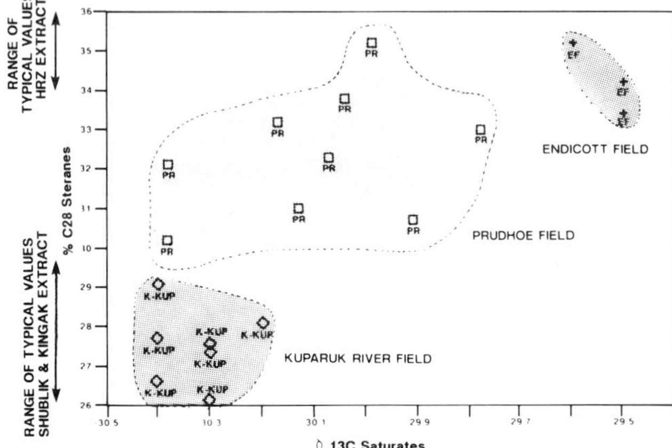

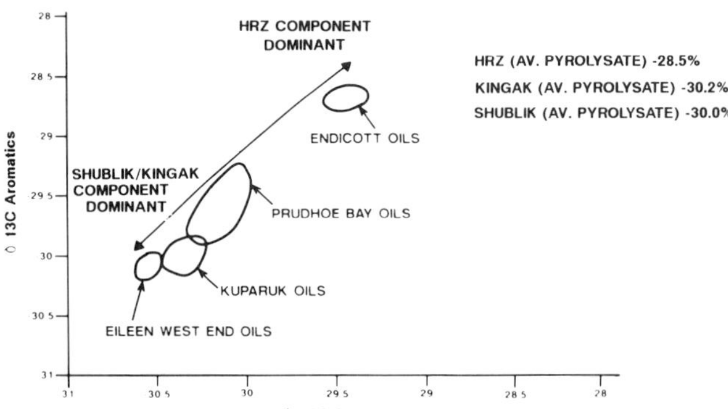

Figure 19. Sterane isotope cross plot (top) showing the spread of oil compositions for major accumulations on the North Slope. Isotopic cross plot for major oils (bottom) showing the Endicott to be heaviest. Pyrolysate isotopic values (listed) suggest a mixture of Shublik/Kingak and HRZ (dominant) oil at Endicott.

and Kingak formations are the primary sources for the Prudhoe Bay field and paleocommunication could be invoked to explain access of those hydrocarbons to the trap at Endicott.

Isotopic composition of the gas cap suggests that it contains a significant component generated from coal in addition to that yielded from the sources discussed above. The Kekiktuk coals are the obvious source of the coal gas. Late entry of this gas is postulated as being responsible for formation of the tar mat at Endicott.

PRODUCTION AND EXPLORATION CONCEPTS

Fluvial geometries and depositional trends are extremely complex and difficult to predict in the subsurface. Efficient recovery of reserves in fluvial reservoirs requires a rigorous integrated application of all petrotechnical disciplines. State of the art tools such as 3D seismic and production testing methods are an important part of this effort. Optimal recovery of oil at Endicott field will occur only through exploiting a three-dimensional understanding of the depositional and structural framework. Under these circumstances alone, pressure maintenance and secondary recovery methods will be most effective.

Endicott field would never have been discovered had the geoscientists involved accepted, as all outcrop and well data suggested, that the Kekiktuk was not a reservoir quality rock. This oil field is associated with a unique combination of circumstances responsible for the accumulation:

- Coarse-grained clastics deposited into the Prudhoe basin during Mississippian.
- Cretaceous rifting that was responsible for the trap and led to the development of secondary porosity at the LCU.
- Deposition of the HRZ seal and source.
- Subsidence of the Colville trough source kitchen.

Reservoir, source, and trap for the Endicott accumulation were provided by these elements. Although it is difficult to expect to find duplication of the exact sequence of events responsible for the Endicott oil accumulation elsewhere, the concept of oil exploration in ancient rift systems is to some degree universally applicable. The rift-related regional unconformity provided a portion of the trap, a migration route for hydrocarbons, and the facilitation of porosity enhancement. In addition, rift shoulder highs and rift-related fault blocks are prime foci for oil migration.

ACKNOWLEDGMENTS

Permission to publish this paper has been granted by BP Exploration and the Endicott field partners. The authors wish to acknowledge all the people who have done technical work on the Endicott field. This paper is largely a summary of their work. The authors wish to thank A. Franz, J. Melvin, B. Metzger, N. Piggott, and S. Johnson for specific technical contributions used here. D. I. Rainey, D. G. Roberts, and M. H. Brownhill critically reviewed the manu-

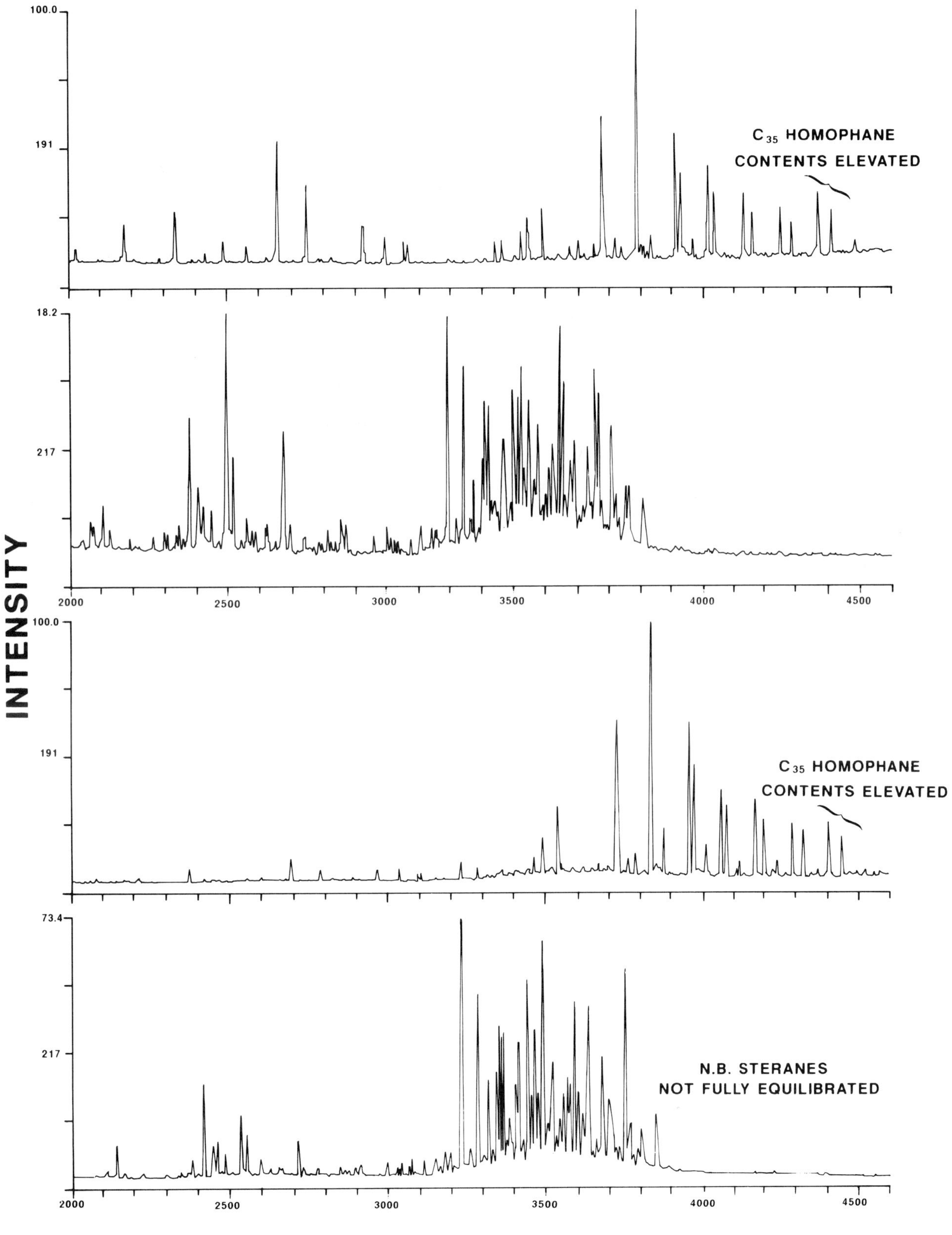

Figure 20. Sterane and triterpane distribution for Endicott oil from Sag Delta 10 at 12,730 ft (top) and an HRZ extract from Duck Island #3 at 10,400 ft (bottom), showing oil/source correlation.

21

script and provided numerous suggestions. Special thanks to L. Rooney and L. Larson for drafting.

REFERENCES CITED

Bird, K. J., 1981, Petroleum exploration of the North Slope in Alaska, U.S.A.: U.S. Geological Survey Open-File Report 81-227, 43 p.

Bird, K. J., and C. F. Jordon, 1977, Lisburne Group (Mississippian and Pennsylvanian), potential major hydrocarbon objective of Arctic Alaska: American Association of Petroleum Geologists Bulletin, v. 61, p. 1493–1512.

Brosgé, W. P., J. T. Dutro, Jr., M. D. Mangus, and H. N. Reiser, 1962, Paleozoic sequence in the eastern Brooks Range, Alaska: American Association of Petroleum Geologists Bulletin, v. 46, p. 2174–2198.

Carman, G. J., and P. Hardwick, 1983, Geology and regional setting of the Kuparuk oil field, Alaska: American Association of Petroleum Geologists Bulletin, v. 67, n. 6, p. 1014–1031.

Hubbard, R. J., S. P. Edrich, and R. P. Rattey, 1987, Geologic evolution and hydrocarbon habitat of the Arctic Alaska Microplate, in I. Tailleur and P. Weimer, eds., Alaska North Slope geology: The Pacific Section, Society of Economic Paleontologists and Mineralogists, p. 797–830.

Melvin, J., 1987a, Sedimentological evolution of Mississippian Kekiktuk Formation, Sagavanirktok Delta area, North Slope, Alaska (abs.), in I. Tailleur and P. Weimer, eds., Alaska North Slope geology: The Pacific Section, Society of Economic Paleontologists and Mineralogists, p. 60.

Melvin, J., 1987b, Fluvio-paludal deposits in the lower Kekiktuk Formation (Mississippian), Endicott Field, northeast Alaska, in F. G. Ethridge, R. M. Flores, and M. D. Harvey, eds., Recent developments in fluvial sedimentology: Society of Economic Paleontologists and Mineralogists Special Publication 39, p. 343–352.

Moore, T. E., and T. H. Nilsen, 1984, Regional variation in the fluvial Upper Devonian and Lower Mississippian (?) Kanayut Conglomerate, Brooks Range, Alaska: Sedimentary Geology, v. 38, p. 465–497.

Nilsen, T. H., 1981, Upper Devonian and Lower Mississippian redbeds, Brooks Range, Alaska, in A. D. Miall, ed., Sedimentation and tectonics in alluvial basins: Geological Association of Canada Special Paper 23, p. 187–219.

Nilsen, T. H., and T. E. Moore, 1982, Fluvial-facies model for the Upper Devonian and Lower Mississippian (?) Kanayut Conglomerate, Brooks Range, Alaska, in A. F. Embry and H. R. Balkwill, eds., Arctic geology and geophysics, Proceedings of the Third International Symposium on Arctic Geology: Canadian Society of Petroleum Geologists Memoir 8, p. 1–2.

Nilsen, T. H., T. E. Moore, and W. P. Brosgé, 1980, Paleocurrent maps for the Upper Devonian and Lower Mississippian Endicott Group, Brooks Range, Alaska: U.S. Geological Survey Open-File Report 80-1066, scale 1: 1,000,000.

Sakuri, S., and J. Melvin, 1988, Facies discrimination and permeability estimation from well logs for the Endicott field: Paper FF Presented at the 29th annual SPWLA Logging Symposium, San Antonio, Texas, June 1988.

Smith, D. G., 1987, Late Paleozoic to Cenozoic reconstructions of the Arctic, in I. Tailleur and P. Weimer, eds., Alaska North Slope geology: The Pacific Section, Society of Economic Paleontologists and Mineralogists, p. 785–795.

Tailleur, I. L., 1969, Rifting speculation on the geology of Alaska's North Slope: Oil and Gas Journal, v. 67, n. 39, p. 128–130.

Tailleur, I. L., 1973, Probable rift origin of Canada Basin, in Max. G. Pitcher, ed., Arctic geology: American Association of Petroleum Geologists Memoir 19, p. 526–535.

Tailleur, I. L., and W. P. Brosgé, 1970, Tectonic history of northern Alaska, in U. L. Adkinson and M. M. Brosgé, eds., Proceedings of the Geological Seminar on the North Slope of Alaska: American Association of Petroleum Geologists Pacific Section, p. E1–E19.

Walker, M., J. H. Dupree, and H. L. Hellman, 1989, Formation tester applications in the Endicott field of Alaska: SPWLA 30th Annual Logging Symposium, Denver, Colorado, June 1989

Woidneck, K., P. Behrman, C. Soule, and J. Wu, 1987, Reservoir description of the Endicott field, North Slope, Alaska, in I. Tailleur and P. Weimer, eds., Alaska North Slope geology: The Pacific Section, Society of Economic Paleontologists and Mineralogists, p. 43–59.

Appendix 1. Field Description

Field name .... *Endicott field*

Ultimate recoverable reserves .... *350 MMSTB; 1480 bcf*

Field location:

    Country .... *U.S.A.*

    State .... *Alaska*

    Basin/Province .... *North Slope (Colville trough)*

Field discovery:

    Year first pay discovered .... *M. Carboniferous (U. Mississippian) Visean Kekiktuk sandstone   1978*

Discovery well name and general location:

    First pay .... *Sag Delta #4 well Sec. 35, T12N, R16E, 702′ EWL, 801′ NSL, UPM*

Discovery well operator .... *Sohio Petroleum Company*

IP in barrels per day and/or cubic feet or cubic meters per day:

    First pay .... *2473 BOPD; 1819 MCFGPD*

All other zones with shows of oil and gas in the field:

Age	Formation	Type of Show
Cretaceous	*Colville Group*	*Mudlog, samples*
Mississippian	*Itkilyariak Formation*	*Mudlog, samples*

Geologic concept leading to discovery and method or methods used to delineate prospect
The potential for secondary porosity development in conjunction with a seismically delineated structural/ stratigraphic trap.

Structure:

    Province/basin type

    Tectonic history
Pre-Mississippian uplift and unconformity led to northern (present-day) sourced Ellesmerian basin. Regional Neocomian rift related uplift truncated Ellesmerian over Central North Slope. Provenance shifted to the south as Brooks Range sourced post-Albian sediments. Deposition of Brookian sediments led to regional northeast tilting.

    Regional structure
The field is located on the crest of the Barrow arch with the Colville trough to the southeast.

    Local structure
Truncated, faulted anticlinal nose, plunging south and southeast.

Trap:

    Trap type(s)
Truncated, faulted anticlinal nose with dip down to the south and west and fault trapped to the north and northwest. Truncation is sealed by organic-rich HRZ shale and Itkilyariak shale.

Basin stratigraphy (major stratigraphic intervals from surface to deepest penetration in field):

Chronostratigraphy	Formation	Depth to Top in ft
Tertiary	*Sagavanirktok Formation*	*−2514*
Upper Cretaceous	*Colville Group*	*−5492*
Lower Cretaceous	*HRZ (informal)*	*−9153*
Mississippian	*Itkilyariak Formation*	*−9425*
	Kekiktuk Conglomerate	*−10,043*
Pre-Mississippian	*"Basement"*	*−10,832*

Reservoir characteristics:

Number of reservoirs *1*

Formations .... *Kekiktuk*

Ages .... *Visean (lowermost Upper Mississippian)*

Depths to tops of reservoirs .... *–10,043 (varies by area)*

Gross thickness (top to bottom of producing interval) .... *789 ft*

Net thickness—total thickness of producing zones

    Average .... *160 ft*

    Maximum .... *300 ft*

Lithology

Fine- to coarse-grained quartz sandstone, moderate to well sorted; clast supported conglomerate

Porosity type .... *Intergranular porosity greatly enhanced by dissolution of carbonate cement*

Average porosity *22%*

Average permeability .... *1000 md*

Seals:

Upper

    Formation, fault, or other feature .... *HRZ shale, Itkilyariak Formation*

    Lithology .... *Shales*

Lateral

    Formation, fault, or other feature .... *Niakuk and Tigvariak faults, in addition to above*

    Lithology

    At the Niakuk fault, Kekiktuk reservoir is juxtaposed against Ellesmerian sands, silts, and shales; at the Tigvariak fault, the Kekiktuk is juxtaposed against meta sediments (primarily argillite)

Source:

Formation and age .... *HRZ, Shublik, Kingak*

Lithology .... *HRZ, organic shale; Shublik, phosphatic shale, siltstone, limestone; Kingak, shale and siltstone*

Average total organic carbon (TOC) .... *HRZ, 5.0%; Shublik, 2.1%; Kingak, 1.7%*

Maximum TOC .... *HRZ, 8.0%; Shublik, 5–6%; Kingak, 5.0%*

Kerogen type (I, II, or III) .... *HRZ, type II/III; Shublik, primarily type II; Kingak (LVU), type II/III*

Vitrinite reflectance (maturation) .... $R_o = 0.48–1.1$

Time of hydrocarbon expulsion .... *Late Cretaceous–Recent(?)*

Present depth to top of source .... *–9153 ft (discovery well)*

Thickness .... *HRZ, 100–300 ft; Shublik, 100–200 ft; Kingak, 0–2000 ft*

Potential yield .... *2–4 trillion bbl*

Appendix 2. Production Data

Field name .... *Endicott field*

Field size:

    Proved acres *8700*

    Number of wells all years *73*

    Current number of wells *66*

    Well spacing .... *NA*

    Ultimate recoverable .... *350 million STB*

    Cumulative production .... *88 million bbl*

    Annual production .... *36.5 million bbl*

    Present decline rate *0%*

Initial decline rate *0%*
Overall decline rate .... *NA*
Annual water production .... *10.95 million bbl*
In place, total reserves .... *1000 million bbl; 730 bcf gas cap; 750 bcf solution gas*
In place, per acre-foot .... *Variable*
Primary recovery .... *115.5 million bbl*
Secondary recovery .... *234.5 million bbl*
Enhanced recovery .... *NA*
Cumulative water production .... *10.6 million bbl*

Drilling and casing practices:

Amount of surface casing set .... *2500 ft*
Casing program
30-in. drive pipe 161 ft TVD; 13⅜-in. 72#/ft 2500 ft TVD; 9⅝-in. 47#/ft 9800 ft TVD; 7-in. 29#/ft 10,500 ft TVD
Drilling mud .... *Lignosulfate based mud*
Bit program .... *NA*
High pressure zones .... *None*

Completion practices:

Interval(s) perforated .... *Kekiktuk sandstone*
Well treatment .... *None*

Formation evaluation:

Logging suites .... *Dual laterolog, micro*
Testing practices .... *RFT*
Mud logging techniques .... *Cuttings description*

Oil characteristics:

Type .... *Asphaltic*
API gravity .... *23°*
Base .... *Sulfur-rich asphaltic intermediate crude*
(40% saturates, 28% aromatics, 25% polars, 7% asphaltines)
Initial GOR .... *750 SCF/STB*
Sulfur, wt% *1.0*
Viscosity, SUS .... *1 cp*
Pour point .... *NA*
Gas-oil distillate *8%*

Field characteristics:

Average elevation .... *10,000 ft*
Initial pressure .... *4890 psi*
Present pressure .... *NA*
Pressure gradient .... *0.33 psi/ft*
Temperature .... *210°F*
Geothermal gradient .... *1.7°F/ft*
Drive .... *Solution gas*
Oil column thickness .... *Max logged 300 ft*
Oil-water contact .... *10,195 ft*
Connate water *5–10%*
Water salinity, TDS .... *30,000 mg/L*
Resistivity of water .... *0.08 at res. temp.*
Bulk volume water (%) *1.32%*

Transportation method and market for oil and gas:
Trans-Alaska pipeline.

Point Arguello Field—U.S.A.
Santa Maria Basin, Offshore California

W. E. MERO
Chevron Overseas Exploration, Inc.
San Ramon, California

FIELD CLASSIFICATION

BASIN: Santa Maria
BASIN TYPE: Wrench
RESERVOIR ROCK TYPE: Chert (Fractured)
RESERVOIR ENVIRONMENT OF DEPOSITION: Pelagic, Deep Water Marine

RESERVOIR AGE: Miocene
PETROLEUM TYPE: Oil
TRAP TYPE: Faulted Anticline

TRAP DESCRIPTION: Single structure, related to wrenching, that can be divided into two doubly plunging anticlines separated by a saddle

LOCATION

The Point Arguello oil field, located in federal waters, is at the southern margin of the offshore Santa Maria basin (Figure 1). The field is about 8 mi (13 km) off the California coast in water depths ranging between 400 and 1100 ft (122–335 m). This major offshore discovery is a large, complex anticlinal structure (Figure 2) whose primary reservoir consists of fractured cherts and porcellanites of the middle and upper Miocene Monterey Formation. Crain et al. (1985) first described the discovery and early delineation phase of the field's development.

Figure 3 is a geologic cross section along the northwestern-trending axis of the Point Arguello structure. This simplified section shows the two main pools in the Monterey reservoir and the stratigraphic onlap and thinning southeastward onto the Amberjack high.

The Point Arguello field is part of a northwest-southeast anticlinal trend of Monterey oil pools within the offshore Santa Maria basin. To the east the Santa Barbara Channel has several fields with significant Monterey oil reserves, the largest being the Hondo and the South Ellwood fields (Figure 1).

The Point Arguello field is held by four federal offshore leases. Chevron, Phillips, and Union Pacific Resources hold leases P-0316, P-0450, and P-0451. Texaco, Pennzoil, Sun, and Koch hold adjacent lease P-0315 (Figure 2). The total productive area is estimated to be around 6000 ac (2428 ha) of which approximately 3600 ac (1457 ha) are Chevron-operated. Texaco is the operator for the remaining acreage. The stage of the development drilling was completed in 1989. Three production platforms and the onshore facilities are now in place.

Owing to the confidential status of much of the offshore activity, this paper will focus on the Chevron-operated portion of the Point Arguello field incorporating published data from the other field participants.

Although many unknowns remain, the estimated ultimate recovery may be over 300 million bbl (180+ million bbl from the Chevron-operated portion of the field). The field is ranked as the 509th largest commercial oil accumulation by Carmalt and St. John (1986).

HISTORY

Pre-Discovery

Although offshore California oil production dates back to 1898 when the first U.S. offshore oil well was drilled at Summerland from a pier east of Santa Barbara (Figure 1), the first significant offshore lease sale did not occur until 1956. Early exploration geologists recognized that the productive trends of the oil-rich Ventura basin would extend into the Santa Barbara Channel. After the 1956 California tidelands lease sale, drilling quickly led to a discovery and construction by Chevron of the first West Coast offshore platform at the Summerland field.

Additional drilling led to the discovery of 12 new oil and gas fields from two groups of Tertiary sandstone reservoirs in state waters. The deepest production comes from the Vaqueros, Sespe, Alegria, and Gaviota formations of Paleogene age as illustrated by the stratigraphic column in Figure 4. Vaqueros sandstones are an especially attractive exploration objective with high porosities and recovery rates.

Shallow reservoirs produce from highly porous, very prolific Pliocene sands whose oil source is

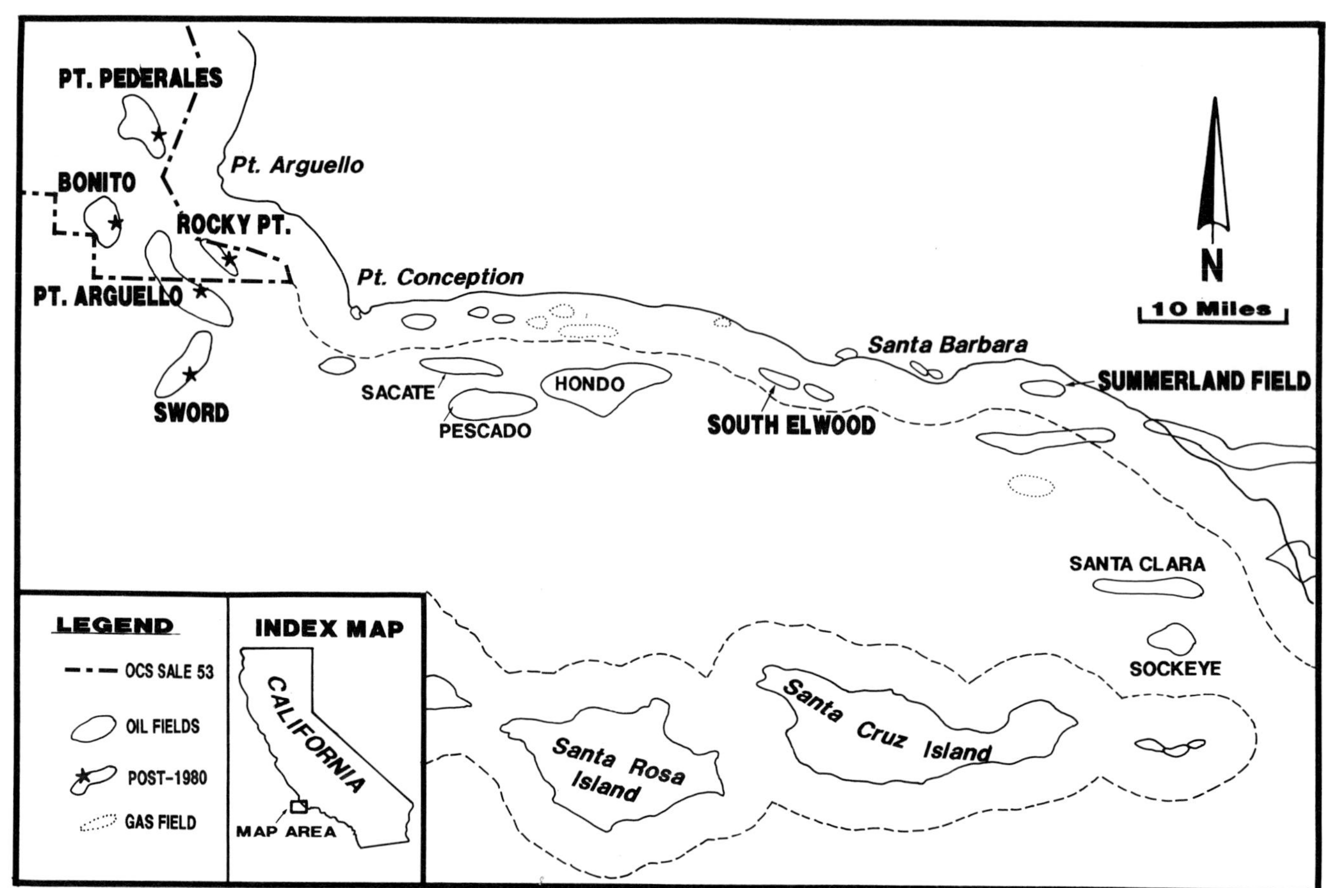

Figure 1. Santa Barbara–Santa Maria offshore index map illustrating the location of the Point Arguello Field. The Monterey Formation oil fields discovered since 1980 are starred. Oil fields in the Santa Barbara Channel discussed in the paper are labeled.

believed to be from the underlying Miocene Monterey shales, cherts, and carbonates. During the initial exploration of these state leases, the Monterey formation was probably not tested owing to its onshore reputation of steep decline rates and low gravity, high-sulfur oils.

The first federal sale of exploratory acreage within the Santa Barbara Channel was held in 1968. Over $602 million was spent by the industry to acquire approximately 350,000 ac (137,800 ha). Within the following four years, eight new fields were discovered in federal waters. Of these, five were Monterey discoveries: Hondo (1969), Pescado (1970), Sacate (1970), Sockeye (1971), and Santa Clara (1971), as shown in Figure 1. Chevron participated in each and used this additional knowledge of the offshore Monterey potential in the exploration program that led to the Point Arguello discovery.

In the onshore Santa Maria basin (Figure 5), fractured Monterey reservoirs have produced for over 87 years. In 1901 oil was discovered at the Orcutt field, which soon led to a number of Monterey discoveries (Regan et al., 1949). Exploration and development activities continue today with a total Monterey and Monterey-sourced production of around 948 million bbl of oil and 933 bcf of gas.

However, until the five outer continental shelf (OCS) discoveries, the Monterey Formation was not considered a major exploration target in the offshore. Onshore it commonly produced low gravity (average 12–14° API), high-sulfur oils at rates that were often uneconomic in the higher cost offshore environment. From its experience in the Santa Barbara Channel, Chevron found that the Monterey Formation could produce higher gravity oil at commercial flow rates if the fractured reservoir was highly siliceous and had been subjected to above-normal subsurface temperatures during oil generation and entrapment.

Discovery

After many years of political indecision following the 1969 Santa Barbara oil spill, OCS Federal Lease Sale 48 was scheduled for 29 June 1979. The sale area covered almost all the remaining unleased federal acreage within the Santa Barbara Channel. Chevron interpreters constructed a series of regional

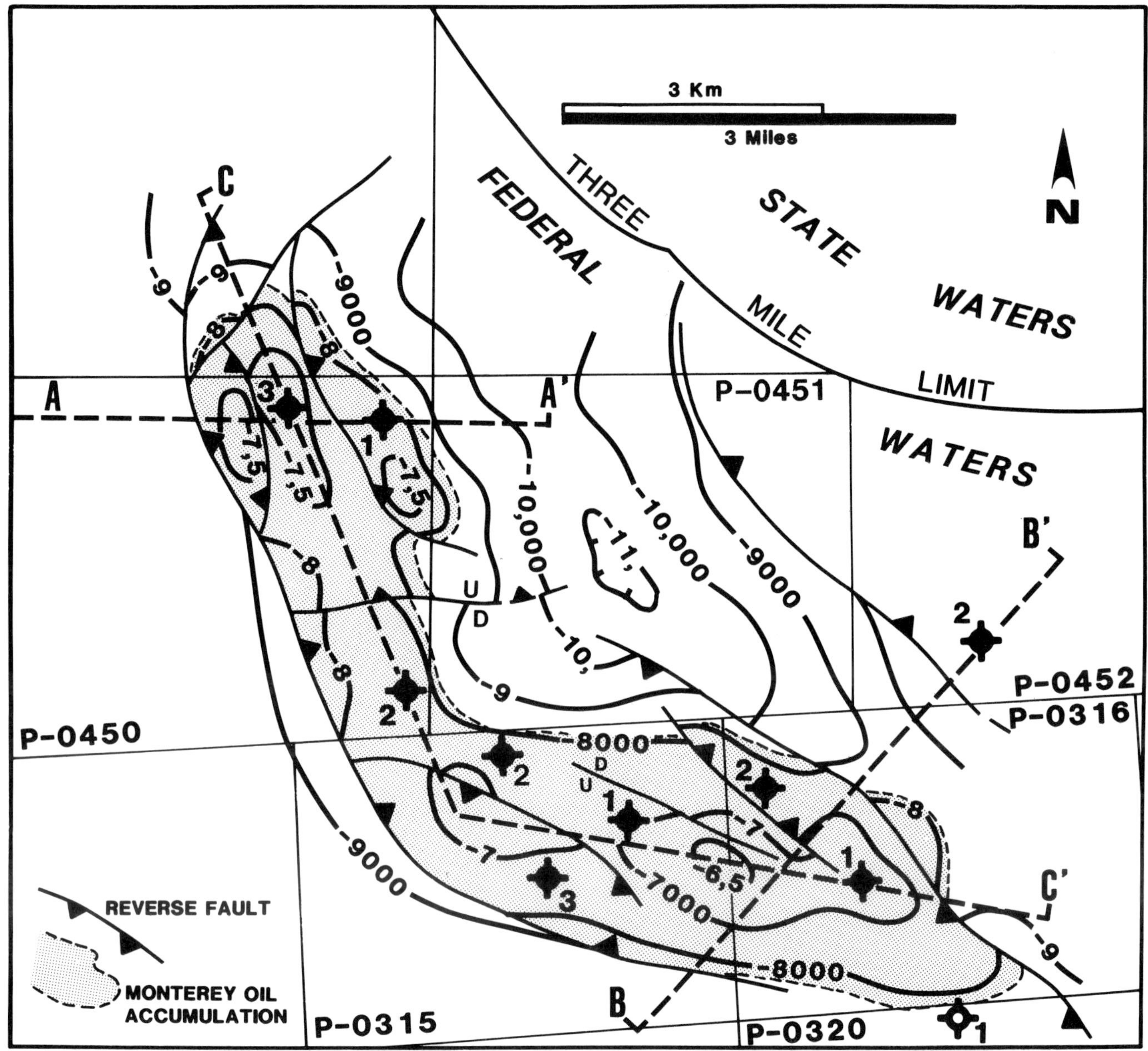

Figure 2. Structure map on the top of the Monterey reservoir illustrating the basic structural configuration and major faults of the Point Arguello structure. Contour interval is 1000 ft. See Figure 14 for seismic line A–A', Figure 15 for seismic line B-B', and Figure 3 for structural cross section C-C'.

seismic time, depth, and isopach maps over the entire Santa Barbara Channel.

The first Chevron map of the Point Arguello prospect was a seismic time map drawn by Dan O'Halloran and named the Lamprey Prospect. While this map of a large, low amplitude fold looked interesting at the time, it was only one prospect among many.

In order to map the Lamprey Prospect, a seismic interpreter had to traverse from wells at least 15 mi (24 km) away across faults and poor seismic data zones. The drilling of the COST well (Continental Offshore Stratigraphic Test), the OCS-CAL-78-164-1, by the oil industry in 1978 placed well control within 4 mi (6.4 km) of the Point Arguello structure (Figure 5). The productive Oligocene and Eocene sands of the Santa Barbara Channel (Figure 4) were missing as a result of nondeposition or stripping. However, oil-stained Monterey shales and cherts were present, and strong oil and gas shows were detected in overlying Miocene and Pliocene sands. While Federal law prohibits testing, the COST well established that the area was rich in hydrocarbons and had a higher than normal temperature gradient (McCulloh and Beyer, 1979).

Early regional stratigraphic studies by Chevron indicated that in the offshore Point Arguello area the Monterey diatomites had been converted into a

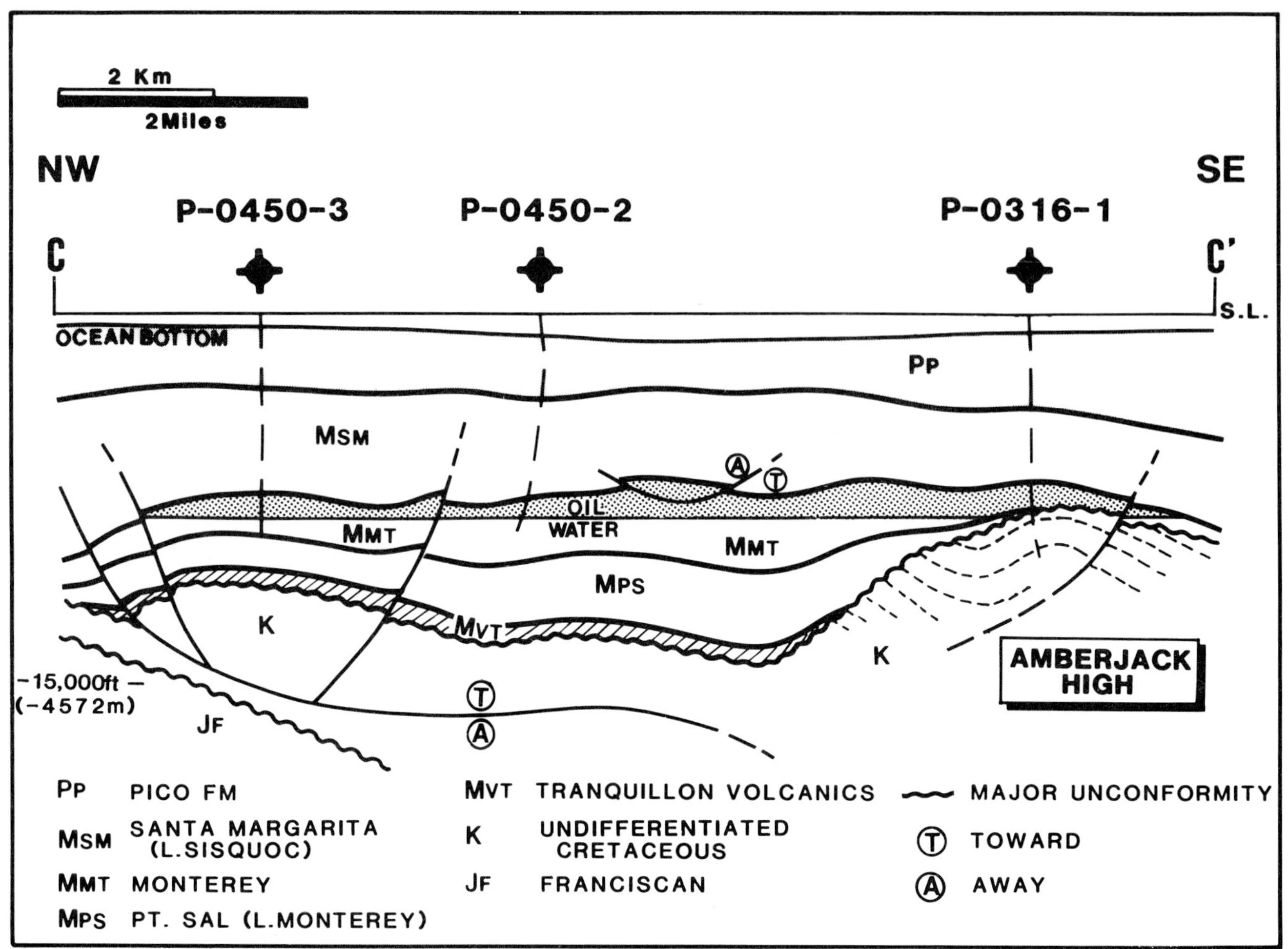

Figure 3. Northwest-southeast structural geologic cross section approximately along the axis of Point Arguello structure. See Figure 2 for location.

dense, brittle quartz phase and were, therefore, highly prone to fracturing. Calculated formation temperatures indicated the Monterey should have generated moderate-gravity oils, and biodegradation of any potential oil accumulations should have been prevented (Figure 6).

Figure 7 is a top of the Monterey seismic structure map made by Chevron before OCS Sale 48. While seismic control was relatively weak (Fischer, 1987) and the detailed interpretation was to be changed later, this first depth map indicated that a large, poorly defined anticlinal trend extended into the open acreage north of the OCS Sale 48 area. Chevron and its bidding partners, Phillips, Impkemix, and Union Pacific Resources, recognized that it was critical to secure a strong land position on the Lamprey–Point Arguello play, one of the largest untested anticlinal trends that remained in California waters.

Based on onshore geologic mapping by Dibblee (1950) and weak seismic control, we projected the Vaqueros sandstones into the Point Arguello Prospect area. A seismic structure map believed to be on the top of the Vaqueros Formation (Figure 8) indicated that lease P-0316 is structurally higher than lease P-0315. Two possible objectives, the Monterey and Vaqueros, were potential targets within lease P-0316 while only the Monterey appeared to be a major objective in lease P-0315. Therefore, the Chevron bidding group decided to focus on acquiring OCS lease P-0316.

Chevron and its partners acquired lease P-0316 for $36.6 million in June 1979. The Chevron et al. P-0316-1 discovery well was spudded 25 November 1980, approximately five months before the OCS Sale 53. The well was quickly drilled to a total depth of 9621 ft (2932 m) in Upper Cretaceous shales. Unfortunately, the Vaqueros sandstone objective was missing by nondeposition. The Monterey was cherty and had only traces of oil with gold fluorescence and yellow-white cut. Based on past experience in the Santa Barbara Channel, we suspected the presence of movable oil and a well-fractured reservoir. Casing was set and five successful tests were made in the Monterey–Point Sal formations, proving the entire

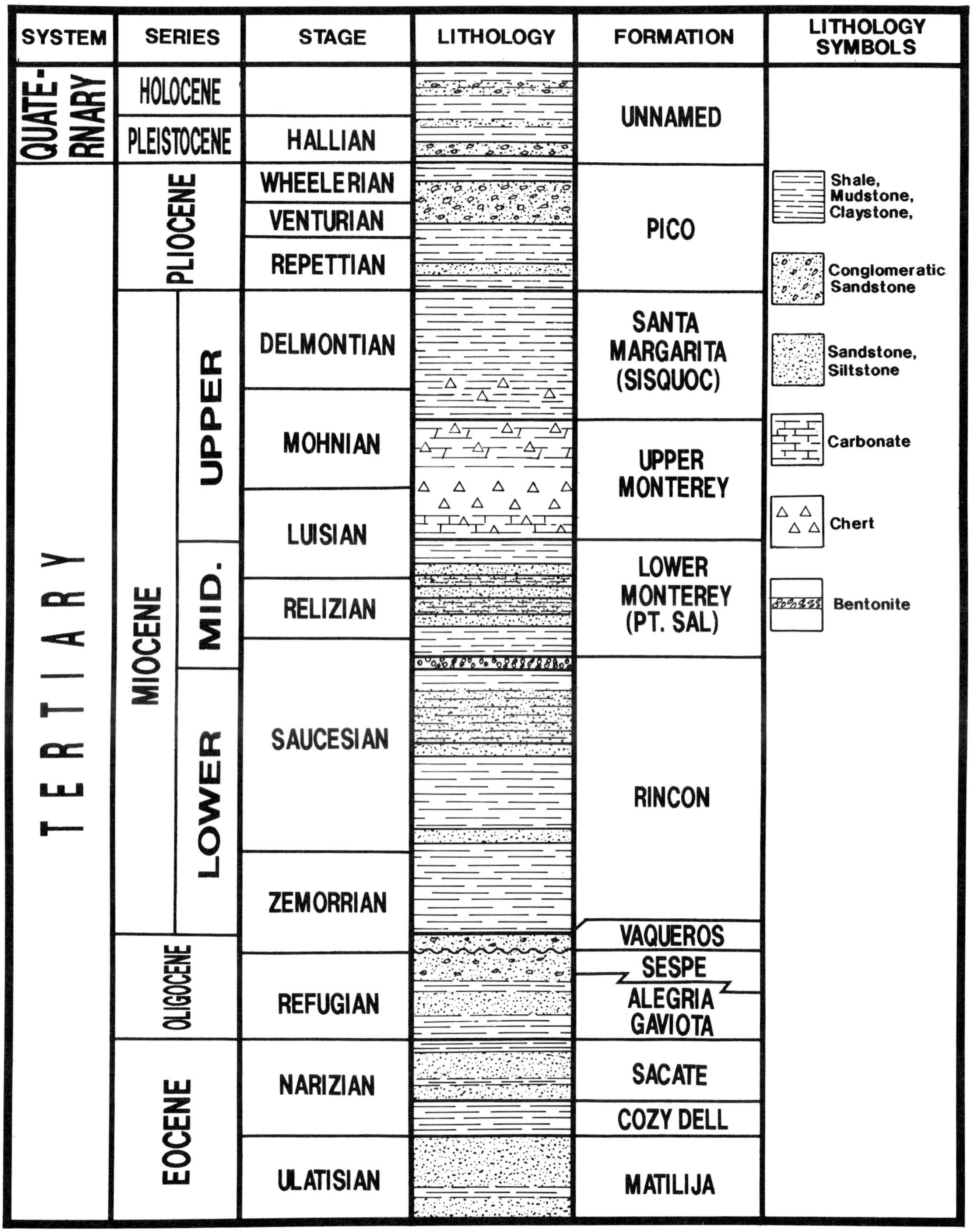

Figure 4. Columnar section showing the general stratigraphic relationships within the northern Santa Barbara Channel. Pliocene sands and the Vaqueros Formation were the primary offshore exploration objectives before the discovery of commercial Monterey oil reserves in the Hondo field in 1969.

SYSTEM	SERIES	STAGE	LITHOLOGY	FORMATION
QUATERNARY	HOLOCENE			UNNAMED
QUATERNARY	PLEISTOCENE	HALLIAN		UNNAMED
TERTIARY	PLIOCENE	WHEELERIAN		PICO
TERTIARY	PLIOCENE	VENTURIAN		PICO
TERTIARY	PLIOCENE	REPETTIAN		PICO
TERTIARY	MIOCENE (UPPER)	DELMONTIAN		SANTA MARGARITA (SISQUOC)
TERTIARY	MIOCENE (UPPER)	MOHNIAN		UPPER MONTEREY
TERTIARY	MIOCENE (UPPER)	LUISIAN		UPPER MONTEREY
TERTIARY	MIOCENE (MID.)	RELIZIAN		LOWER MONTEREY (PT. SAL)
TERTIARY	MIOCENE (LOWER)	SAUCESIAN		RINCON
TERTIARY	MIOCENE (LOWER)	ZEMORRIAN		RINCON
TERTIARY	OLIGOCENE	ZEMORRIAN		VAQUEROS
TERTIARY	OLIGOCENE	REFUGIAN		SESPE
TERTIARY	OLIGOCENE	REFUGIAN		ALEGRIA / GAVIOTA
TERTIARY	EOCENE	NARIZIAN		SACATE
TERTIARY	EOCENE	NARIZIAN		COZY DELL
TERTIARY	EOCENE	ULATISIAN		MATILIJA

LITHOLOGY SYMBOLS

- Shale, Mudstone, Claystone,
- Conglomeratic Sandstone
- Sandstone, Siltstone
- Carbonate
- Chert
- Bentonite

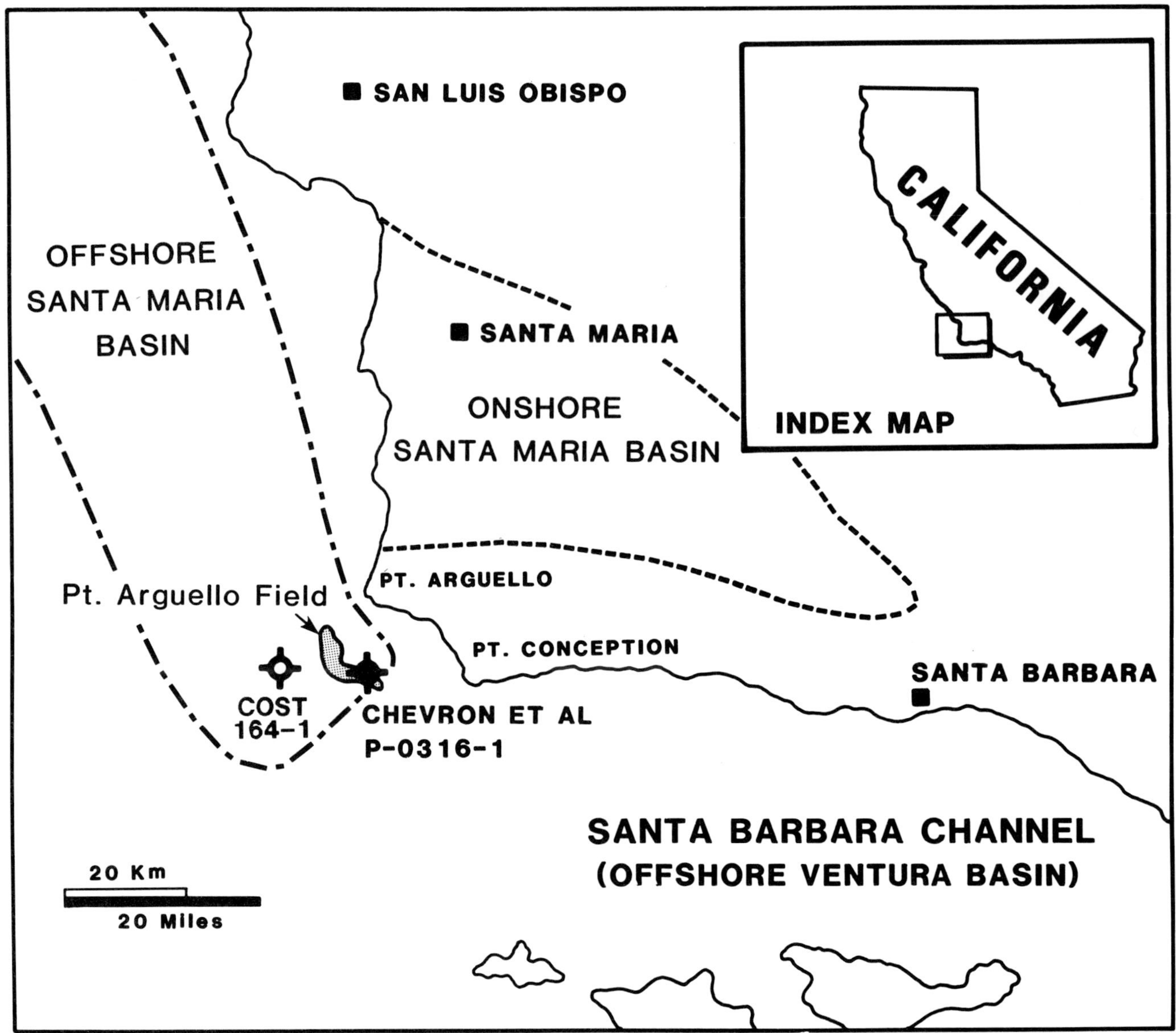

Figure 5. Index map illustrating the location of the discovery well, Chevron et al. P-0316-1, and the COST-1. The shape and orientation of the onshore and offshore Santa Maria basins are also shown.

Monterey section was productive. Figure 9 is the P-0316-1 electric log through the Monterey with a summary of the test results. Early calculations by the reservoir engineers suggested that the section could have flowed at a wide-open rate of nearly 40,000 BOPD.

These tests confirmed 1075 ft (328 m) of net pay with a combined controlled test rate of 6580 BOPD and 1680 mcf gas/day. Oil gravities ranged from 18.2° to 21.4° API. However, the total length of the oil column was still unknown because the base of the productive interval was cut off by the tight underlying Cretaceous sandstones and shales. The postulation that the true oil-water contact (OWC) was at least 7800 ft (2377 m) subsea placed considerable reserves in open P-0450 block, scheduled to be part of the upcoming OCS Sale 53.

Figure 10 is the revised top of the Monterey seismic structure map based on the new control from the Chevron et al. P-0316-1. The new map indicated that the Monterey accumulation would extend across Texaco's adjacent lease P-0315 into the open P-0450 block. When rumors circulated that the well logs were for sale, undercover security agents were sent to investigate. They returned empty-handed, much to the relief of those planning the bid strategy for the upcoming OCS Sale 53. Encouraged by the flowing tests, oil gravities, Monterey stratigraphy, and the revised structural interpretation, Chevron and Phillips, bidding as 50–50 partners, won lease P-0450 for $333.6 million.

On 24 April 1982, the Chevron-Phillips P-0450-1 was spudded 4.8 mi (7.7 km) northwest of the 316-1 discovery well. Two successful tests were made

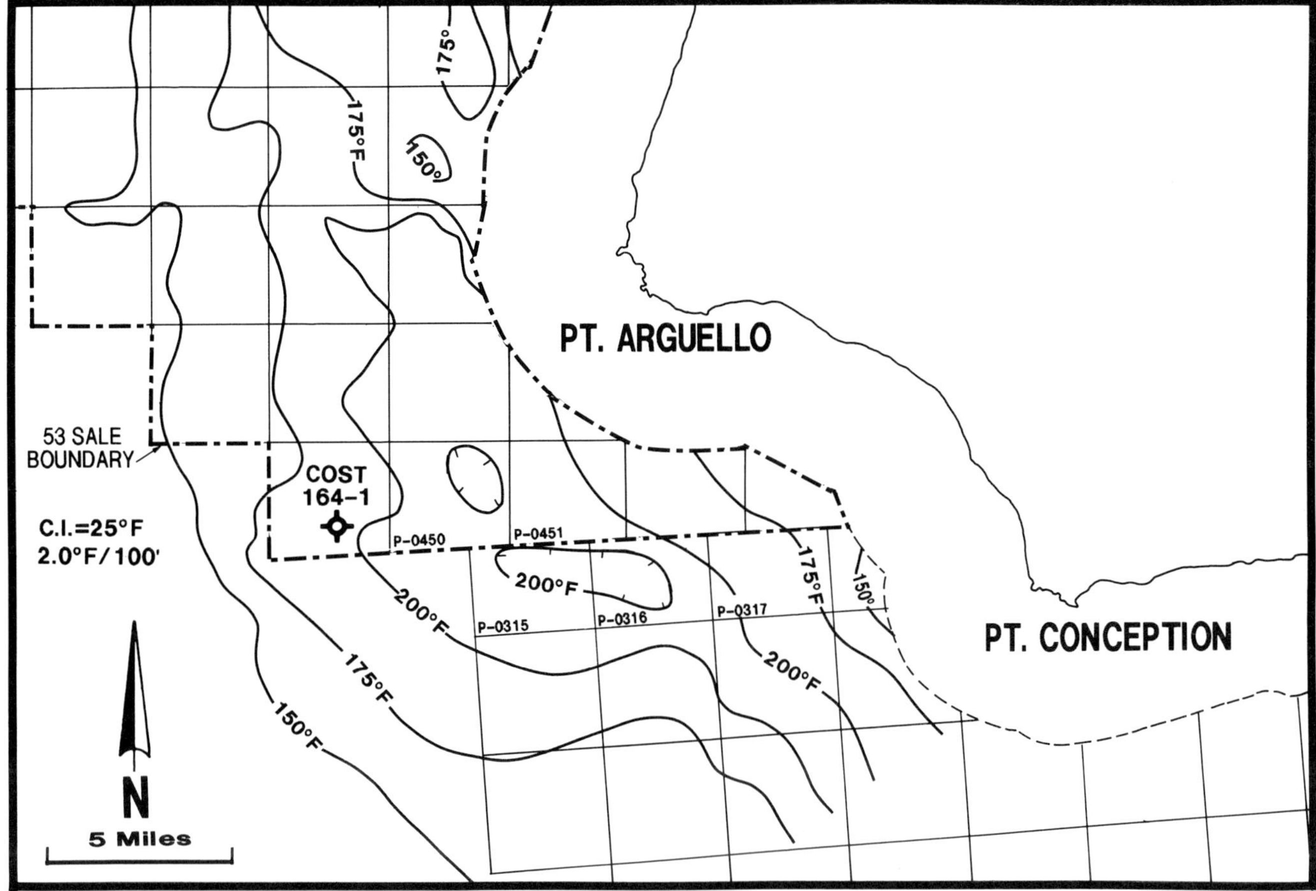

Figure 6. Point Arguello preliminary (original) subsurface temperature map at the top of the Monterey section. Drawn soon after OCS Sale 48, this extrapolation successfully predicted the location of maximum API° Monterey oil generation within the upcoming OCS Sale 53 area. Contour interval is 25°F.

in the upper Monterey that flowed at a combined rate of about 2400 bbl/day of 31.4° and 33.2° API oil. The higher oil gravities were a pleasant surprise and indicated that the Chevron-Phillips P-0450-1 was a new-pool discovery separate from the southern pool discovered by the Chevron et al. P-0316-1 (Figure 3). The oil-water contact in this newly discovered northern pool was found at 8150 ft (2485 m) subsea. Delineation drilling and mapping of both the southern and northern pools now began.

While the bulk of the reserves appear to be within the Monterey Formation, minor gas and oil may be produced from the upper Miocene Santa Margarita (lower Sisquoc) Formation and from Pliocene sandstones. Exploratory concepts and planning dating back over a decade had finally resulted in a major oil discovery.

Post-Discovery

Figure 11 shows the delineation wells that established the productive limits of the Point Arguello field. The first delineation well in the P-0316 lease, Chevron et al. P-0316-2, was spudded in May 1982

and bottomed in Lower Cretaceous shales at a total depth of 11,416 ft (3480 m). Three flowing tests were made in the Monterey Formation. The combined flow rates totaled 3620 bbl/day of 18.5° to 21° API oil. The oil-water contact of the southern pool was finally established at approximately 8235 ft (2511 m) subsea.

The Chevron-Phillips P-0450-2 tested the anticlinal plunge in the southwest corner of lease P-0450. The well spudded in October 1982 and bottomed in the Monterey after drilling to a total depth of 9277 ft (2828 m). Mechanical problems prevented testing all but a short interval near the base of the Monterey oil column, where 16° API oil flowed at 220 bbl/day.

Also during October 1982, the Chevron-Phillips P-0450-3 drilled into Point Arguello's higher-gravity northern pool. The well reached a total depth of 8881 ft (2707 m), bottoming in the Monterey Formation. The Monterey was tested at a combined rate of around 4900 bbl/day of 32° to 34° API oil. In most delineation wells the flow rate tests were restricted because of the limited barge capacity to hold the produced fluids.

Three successful wells were drilled by the Texaco group (Texaco Inc., Pennzoil Oil & Gas, Inc., Sun Exploration and Producing Co., and Koch Industries,

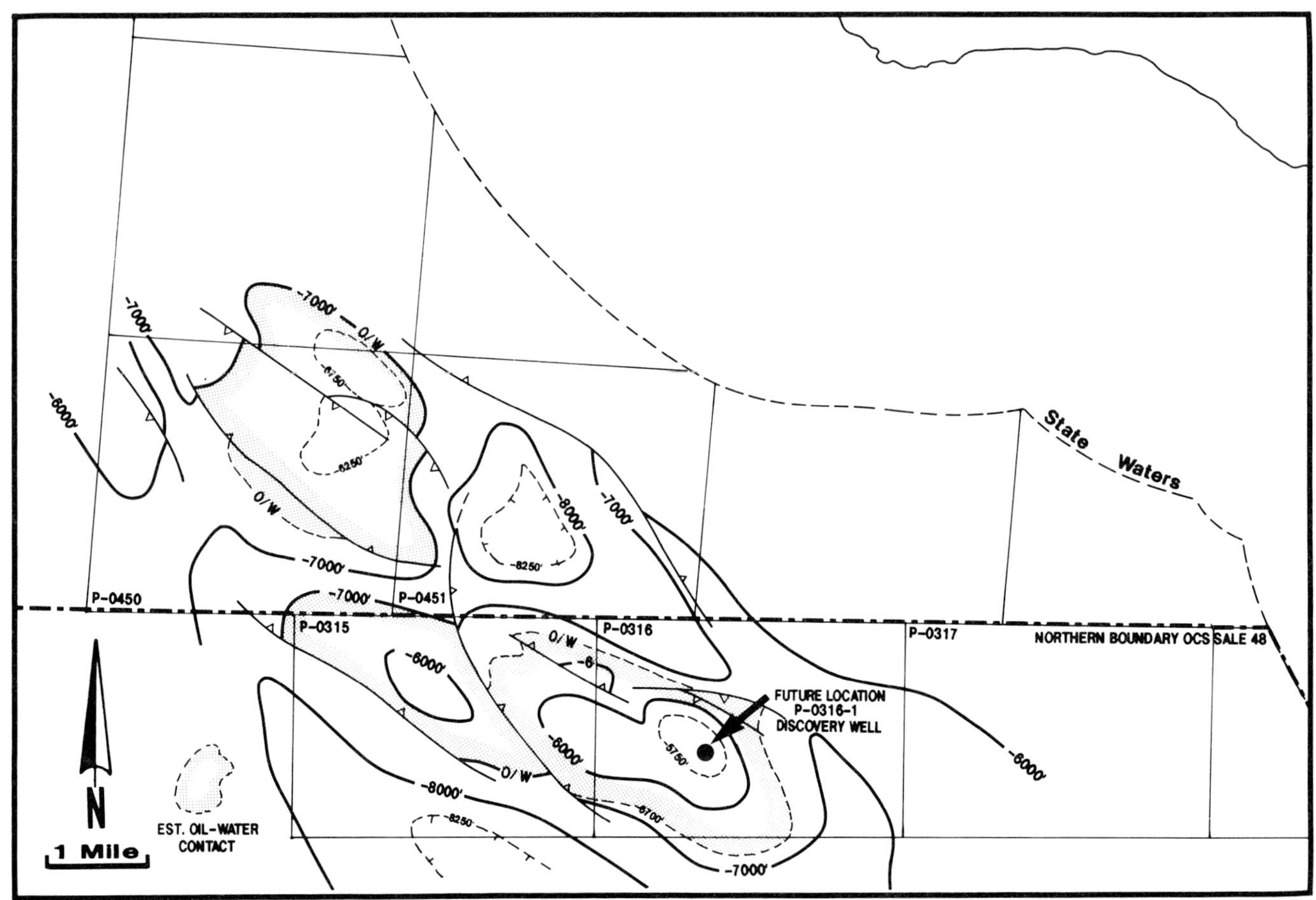

Figure 7. Chevron top of the Monterey structure map drawn (1979) before OCS Sale 48. The map indicated that the Point Arguello anticlinal trend extended into open acreage north of the proposed OCS Sale 48. The early interpretation also suggested that the anticlinal trap was structurally highest in Block P-0316. The contour interval is 1000 ft.

Inc.) in lease P-0315-1 and tested as much as 4200 bbl/day of 21° to 23° API oil from a single test within the Monterey. The P-0315-2 spudded in January 1982 and reportedly tested an individual zone within the Monterey for more than 3000 bbl/day of 17° to 18° API oil. On 19 July 1982, Texaco spudded the P-0315-3 and drilled to a total depth of 8454 ft (2577 m). Up to 1900 bbl/day of 11° to 23° API oil was reportedly tested from the Monterey.

The eight successful wells drilled on the Point Arguello structure delineated a giant Monterey oil discovery composed of two separate oil pools with differing oil gravities and oil-water contacts. The southern pool also contains a small primary gas cap with a gas-oil contact at 6620 ft (2018 m) subsea. After drilling the initial delineation wells and conducting three-dimensional (3-D) seismic surveys by both the Chevron and Texaco groups, reservoir studies indicated total recoverable oil reserves of over 300 million bbl.

Chevron operates two platforms (Hidalgo and Hermosa) while Texaco is the operator for one platform (Harvest) in the Point Arguello field. Figure 12 illustrates the location of the three installed production platforms and the oil- and gas-gathering pipeline to the facilities east of Point Conception at Gaviota. Chevron, Phillips, and Union Pacific completed Platform Hermosa, a 48-slot platform in 603 ft (184 m) of water and Platform Hidalgo, a 56-slot platform in 430 ft (131 m) of water. Texaco built Platform Harvest, a 50-slot production platform in 675 ft (206 m) of water.

The development of nearby Monterey oil fields is also proceeding. In April 1987 the Point Pedernales field went on stream when Unocal began commercial oil production from Platform Irene for about 16,000 bbl/day of 15° to 17° API oil. The 100 to 200 million bbl field will peak at about 20,000 bbl/day.

North of the Point Arguello field is the Rocky Point oil field in leases P-0451 and P-0452 (Figure 1). Delineation wells tested 24.9° to 34.8° API oil from the fractured Monterey. The total net pay ranges up to 1450 ft (442 m) with combined test rates as high as 3719 BOPD.

Production was originally planned to begin from proposed Platform Hacienda in 1992. The current political barriers to oil and gas production at Point Arguello will delay development of the Rocky Point

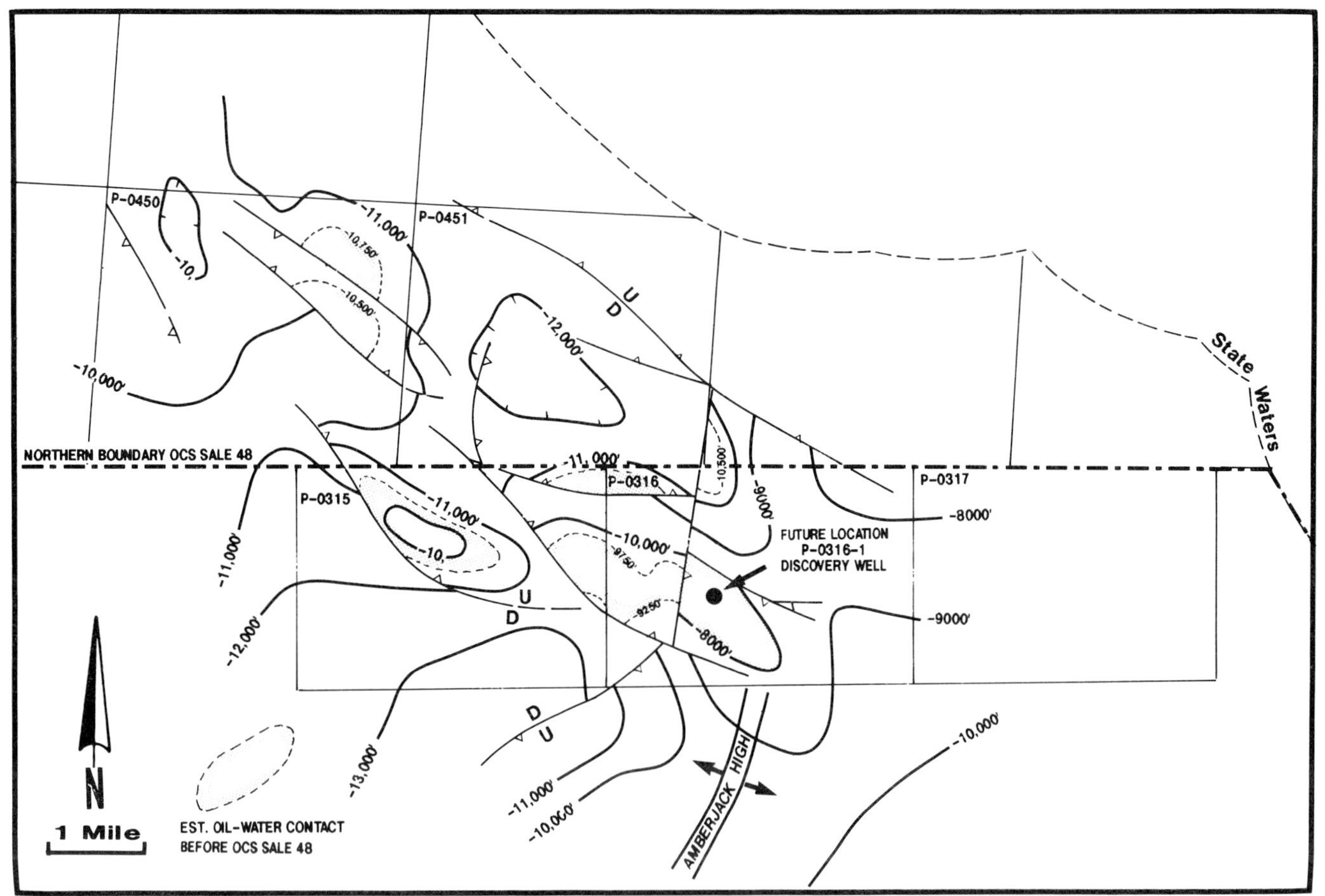

Figure 8. "Vaqueros" structure map drawn (April 1979) before OCS Sale 48. While no Vaqueros sands were ever found, this early structural interpretation does show the superposition of the Arguello structure across the older northeast-trending Amberjack high. The contour interval is 1000 ft.

field. Water depth only averages 300 ft (92 m) but the *Oil and Gas Journal* (1987) reports that Chevron is also waiting to see how the fractured Monterey reservoirs produce at the Point Arguello field before proceeding at Rocky Point.

For the explorationist, economics play as important a role as technical factors. Depressed oil prices in the 1980s put production of the Bonito field, located 5 mi (8.1 km) northwest of the Point Arguello field, further into the future (*Oil and Gas Journal*, 1987). The accumulation is in more than 1100 ft (335 m) of water. Initial test rates were 1400 bbl of 20.3° API oil/day and 1443 bbl of 22.7° API oil/day from two intervals. A 3-D seismic survey was shot as an aid in planning for the field's future development.

The Sword field, discovered by Conoco in 1982, is a giant offshore field with in-place Monterey reserves of over 1 billion bbl of low gravity oils (8.5° to 10.5° API) located in water depths ranging from 800 to 2000 ft (244–610 m). Individual oil production test rates from the Monterey were reported as 788 and 1865 bbl/day. The Santa Margarita–Sisquoc Formation tested at 1877 bbl/day. The operator may have injected special surfactants to aid in the flow of the heavy oils.

The trap is a northwest-trending, faulted anticlinal drape fold formed on a culmination of the Amberjack high immediately south of the Point Arguello field (Figure 1). While a 3-D seismic survey has been shot, the problem in developing economically a fractured reservoir of low gravity crude in deep water remains under study (Ballard, 1988).

TECTONIC HISTORY

The offshore Santa Maria basin is composed of several narrow north–northwest-trending subbasins bounded by basement uplifts (Figure 13). The Santa Lucia high forms the western margin and the right-lateral oblique-slip Hosgri fault system marks the eastern margin of the offshore Santa Maria basin (Hoskins and Griffith, 1971). Between the Santa Lucia high and the Hosgri fault is a large basement ridge noted on Figure 13 as the Central basin high.

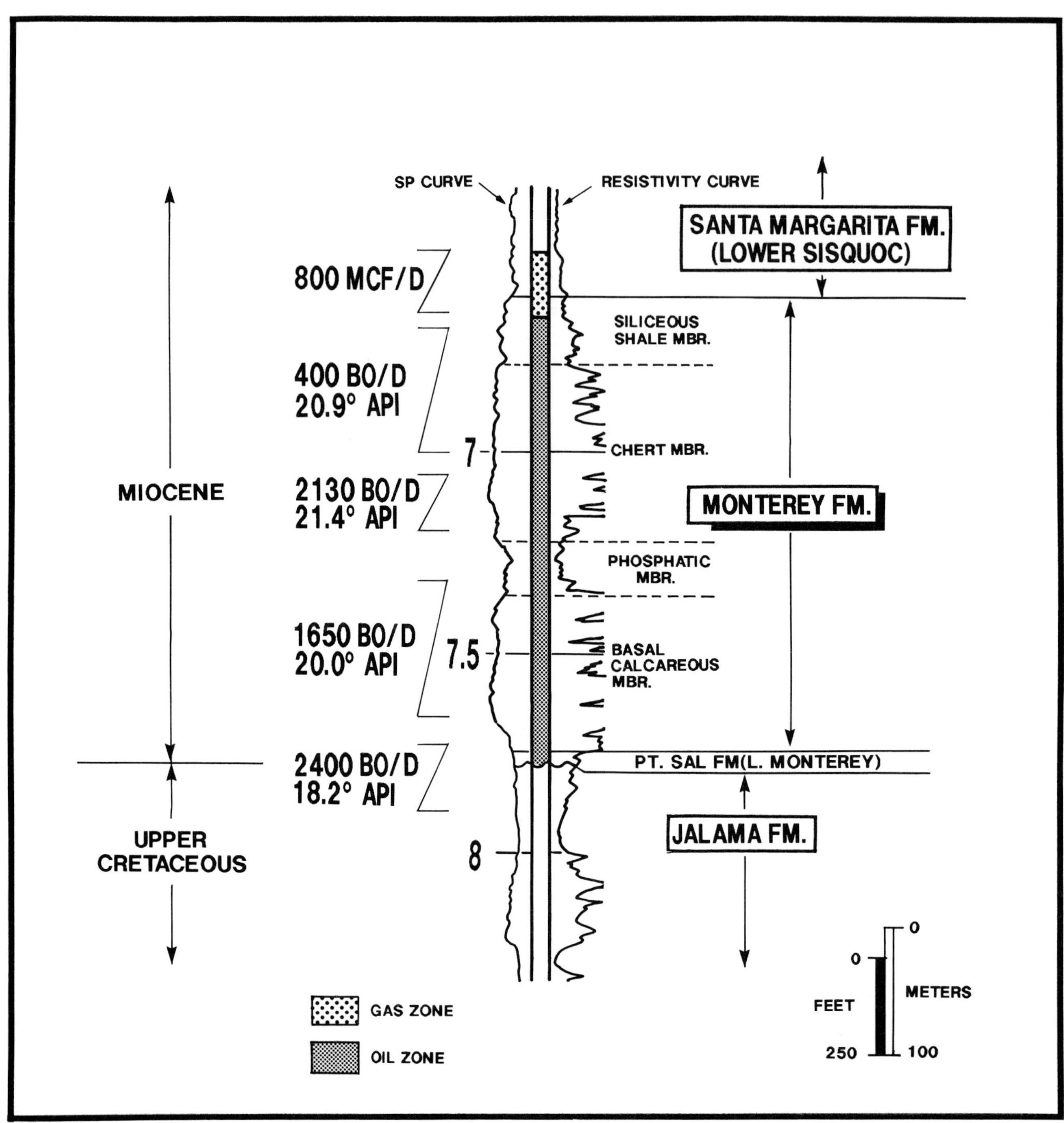

Figure 9. Electric log and drill-stem tests of Monterey Formation in P-0316-1 well (May 1981). The log response to the various Monterey members is shown. Log is marked in 500 ft intervals.

The author considers the southeastern boundary of the offshore Santa Maria basin to be a northeast-trending basement ridge, informally named the Amberjack high. In the Santa Barbara Channel, structural trends are generally east-west. From the Amberjack high west, the structural trends are mostly north-northwest, subparallel to the Hosgri fault system. The northwest-trending Point Arguello structure is superimposed across the older northeast-trending Amberjack high and extends into a small depocenter containing up to 15,000 ft (4600 m) of Neogene rocks immediately west of the Amberjack high.

The offshore Santa Maria basin is classified as 332 by Bally and Snelson (1980) (i.e., California-type basin related to episutural megashear system developed on an accretionary wedge). An arc-trench system developed in Late Jurassic time (Maxwell, 1974) as the Farallon plate was subducted beneath the North American plate.

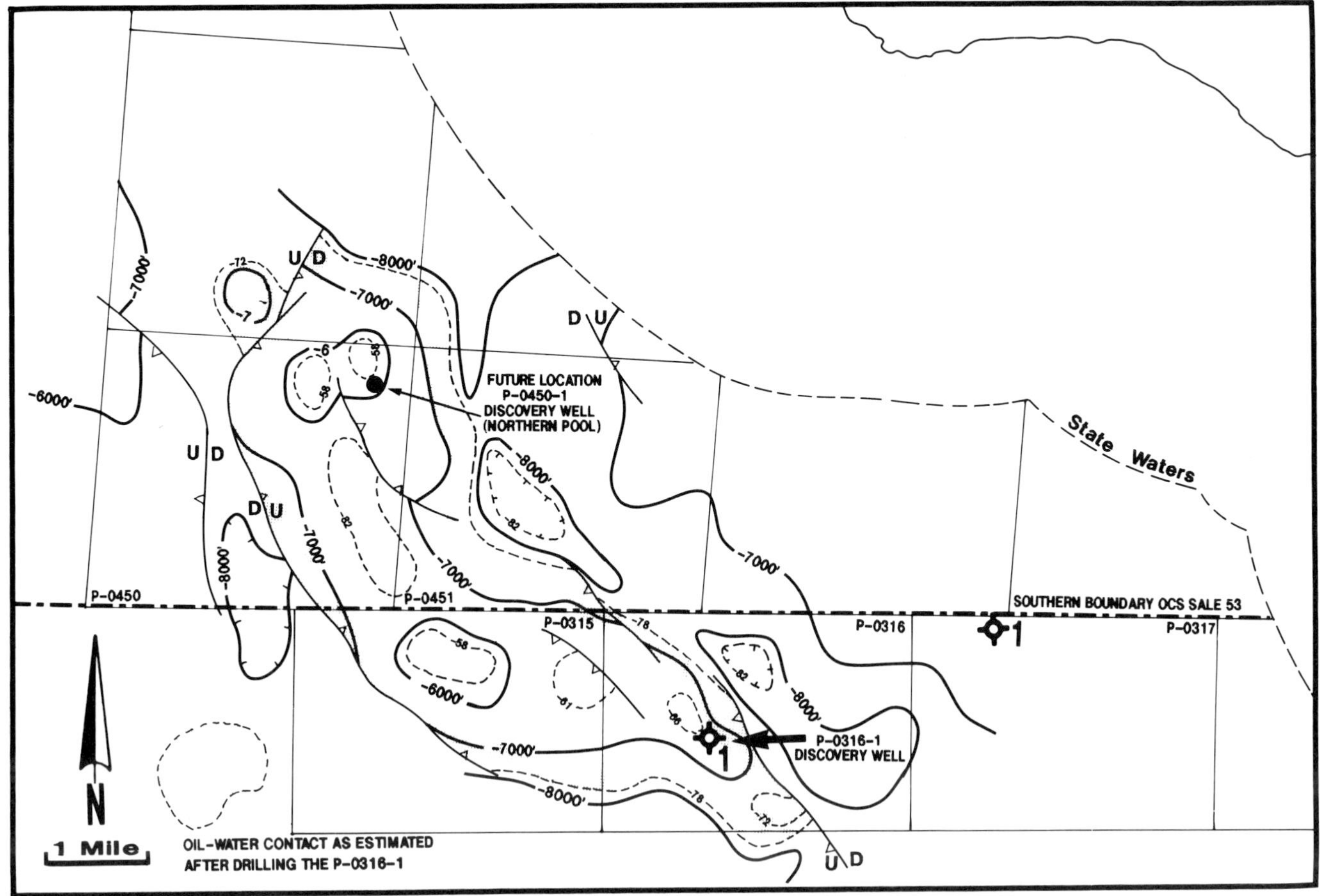

Figure 10. Monterey structure map constructed after drilling the Chevron et al. P-0316-1. The oil-water contact is at the assumed structural closure. Note how a successful discovery well and increased seismic control optimized the interpretation as compared to the first Monterey map in Figure 7. Contour interval is 1000 ft.

Widespread marine regression and erosion followed the spreading center-trench collision during the late Oligocene. The collision is believed to have initiated strike-slip faulting and deformation of the continental margin (Blake et al., 1978). Luyendyk and Hornafius (1987), basing their concept on paleomagnetic measurements, suggest that in southern California the deformation may have taken the form of a general clockwise rotation of individual crustal blocks. Block rotation may have helped form many pull-apart basins including both the offshore and onshore Santa Maria basins.

During the early Miocene, volcanics and associated marine clastics were deposited within the subbasins forming between the growing basement uplifts shown in Figure 13. Throughout most of the middle and late Miocene, the subbasins were areas of thick biogenetic deposition in an anaerobic environment. The small Neogene depocenter west of the Amberjack high acted as a major hydrocarbon-generating center for the Point Arguello accumulation. One can refer to Smith et al. (1979), Moore and Karig (1976), and Atwater (1970) for detailed discussion of the trench-slope basin model and plate-tectonic evolution of the western continental borderlands.

LOCAL STRUCTURE

Figure 2 is a simplified Monterey structure map of the Point Arguello anticlinal structure. The fold is approximately 9 mi (14.5 km) long and 2 mi (3.2 km) wide. In general, the structure can be divided into two faulted, doubly plunging anticlines separated by a saddle in lease P-0450.

The location of seismic lines A–A', B–B', and structure cross section C–C' are also shown on Figure 2. A–A' (Figure 14) through the northern portion of the Point Arguello field is an east–west-oriented time-migrated seismic line. Complex folding and reverse faulting is shown as well as the Chevron-Phillips P-0450-3 and P-0450-1, the northern pool discovery well. The seismic and well control indicate that faulted "rabbit ears" structures have formed on the flanks on the Point Arguello structure.

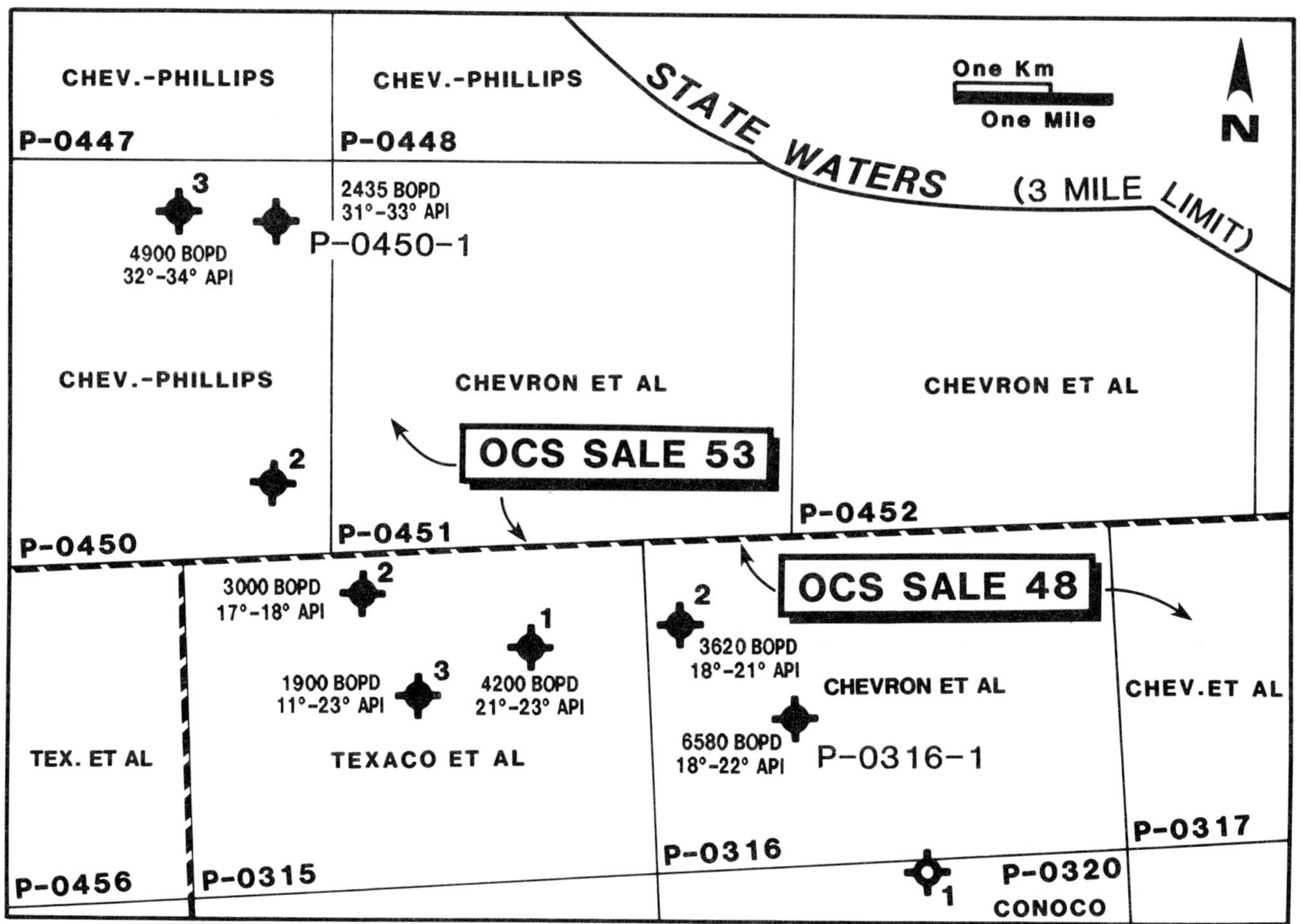

Figure 11. Location map illustrating the position of the initial delineation wells in Point Arguello field, with federal sale boundaries and lease blocks noted. The combined test totals are shown.

B–B′ (Figure 15) is a northwest–southwest-oriented time-migrated seismic line passing approximately 1800 ft (549 m) northwest of the Chevron et al. P-0316-1 discovery well for the southern pool. Both flanks of the anticline are cut by reverse faults. At the north end of B–B′, the Rocky Point field, a faulted anticlinal trap, has been tested at rates from 150 to 1629 bbl/day of 24° to 35° API oil from the Monterey and Santa Margarita formations.

Structure cross section C–C′ (Figure 3) shows both the structural and stratigraphic relationships between the northeast-trending Amberjack high and the north–northwest-trending Point Arguello field. Illustrated are the stratigraphic onlap and thinning of the Monterey and older Miocene units onto the Amberjack high as well as a simplified picture of the cross-faulting between P-0450-3 and P-0450-2, which may form the barrier between the northern and southern oil pools.

The Amberjack high may have developed as early as the Oligocene and continued to grow through much of the Miocene. Luyendyk and Hornafius (1987) suggested that the Amberjack high and the associated Neogene depocenter may have evolved during the Miocene rotation of the western Transverse Ranges.

Wrench-faulting on the Hosgri fault system (Figure 13) may have initiated the northwest-trending Point Arguello fold trend during the late Miocene. Periods of structural growth continued throughout the Pliocene.

STRATIGRAPHY

Figure 16 is a generalized stratigraphic column of the Tertiary section in the Point Arguello field as described by Crain et al. (1985). The oldest rocks are a Cretaceous age interval of tight sandstones and shales. Lower Miocene volcanics, volcaniclastics, and fine-grained marine clastics unconformably overlie the Mesozoic section. These lower Miocene units grade into upper Miocene dolostones; finely laminated, highly organic mudstones; and siliceous, biogenetic rocks deposited in a sediment-starved basin. During the late Miocene, an increase in silt and mud deposition began. By the Pliocene–Pleistocene, sands appear in the section in response to uplift and erosion of the nearby coast ranges. East of the Amberjack high the stratigraphy is typical of the

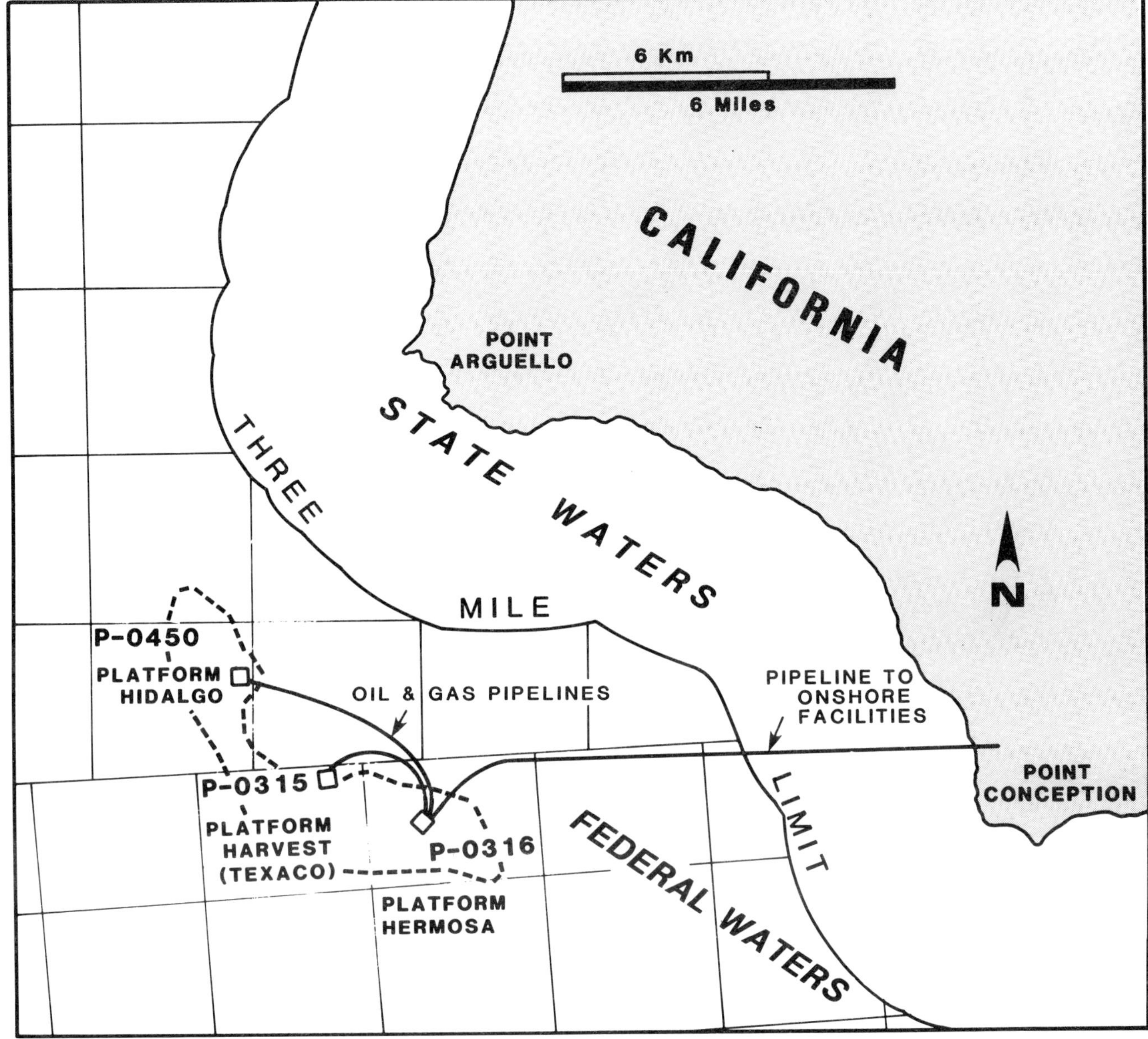

Figure 12. Map illustrating locations of production platforms and pipelines in Point Arguello field.

Santa Barbara Channel (Figure 4). West of the Amberjack high, in the Point Arguello field, the stratigraphy resembles the onshore Santa Maria basin.

Cretaceous

Espada Formation

Within lease P-0316, wells have penetrated Lower Cretaceous (Neocomian) dark-brown marine shales and occasional thin, fine-grained arkosic sandstones. These rocks are tentatively assigned to the Espada Formation, after similar rocks mapped onshore by Dibblee (1950). The Espada Formation is assumed to unconformably overlay the metasedimentary subduction melange of the Jurassic–Cretaceous Franciscan assemblage, a minor fractured reservoir producing Monterey sourced oil at both offshore Point Pedernales and onshore Santa Maria Valley fields.

Jalama Formation

The Chevron et al. P-0316-1 bottomed in Upper Cretaceous (Turonian to Campanian) dark gray-brown, marine, silty shales interbedded with light buff, fine- to medium-grained, light arkosic sandstones onlapped by lower Miocene sandy siltstones. Dibblee (1950) mapped similar Cretaceous rocks onshore as the Jalama Formation.

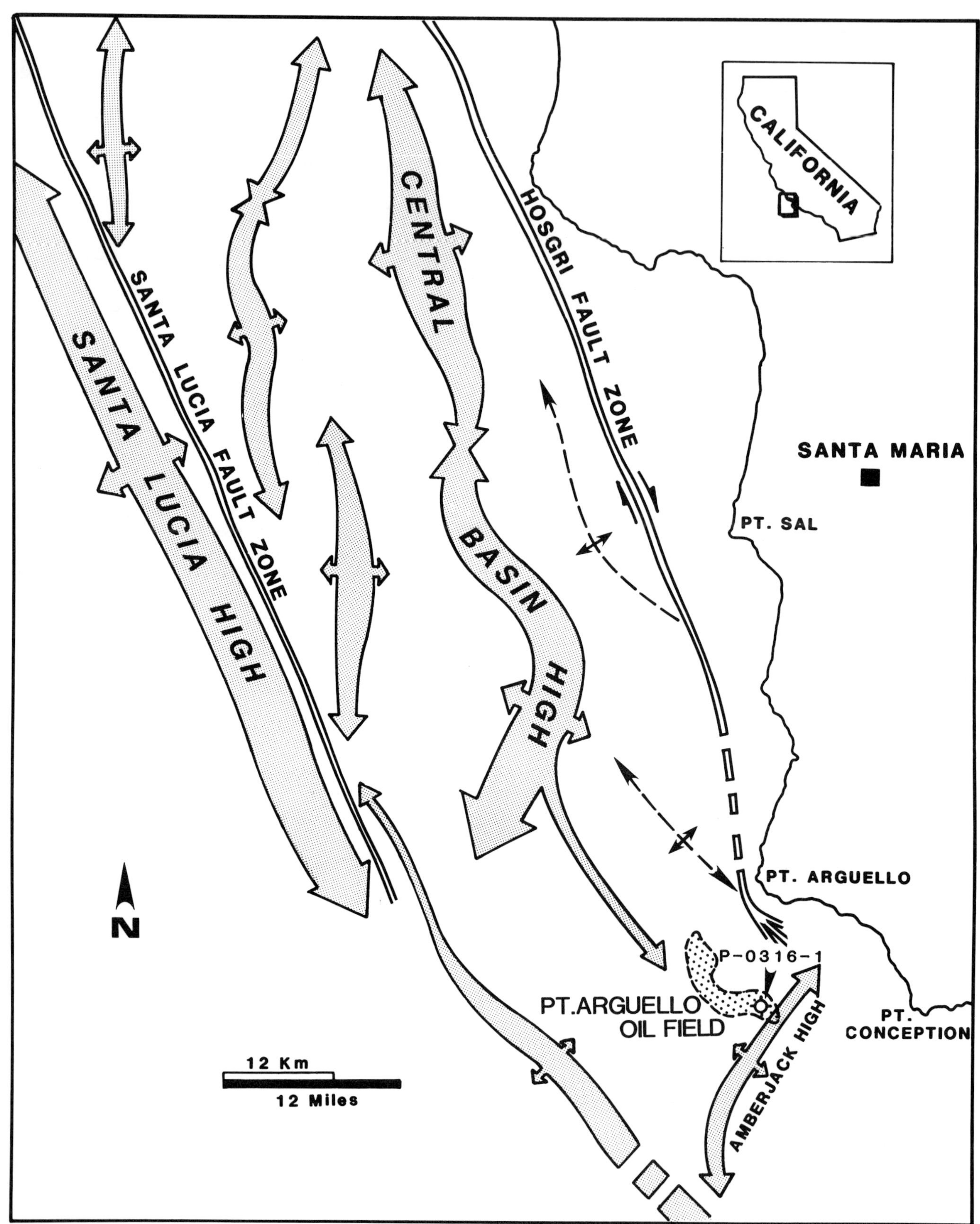

Figure 13. Major tectonic elements, offshore Santa Maria basin. Generalized basement highs and plunge directions are shown.

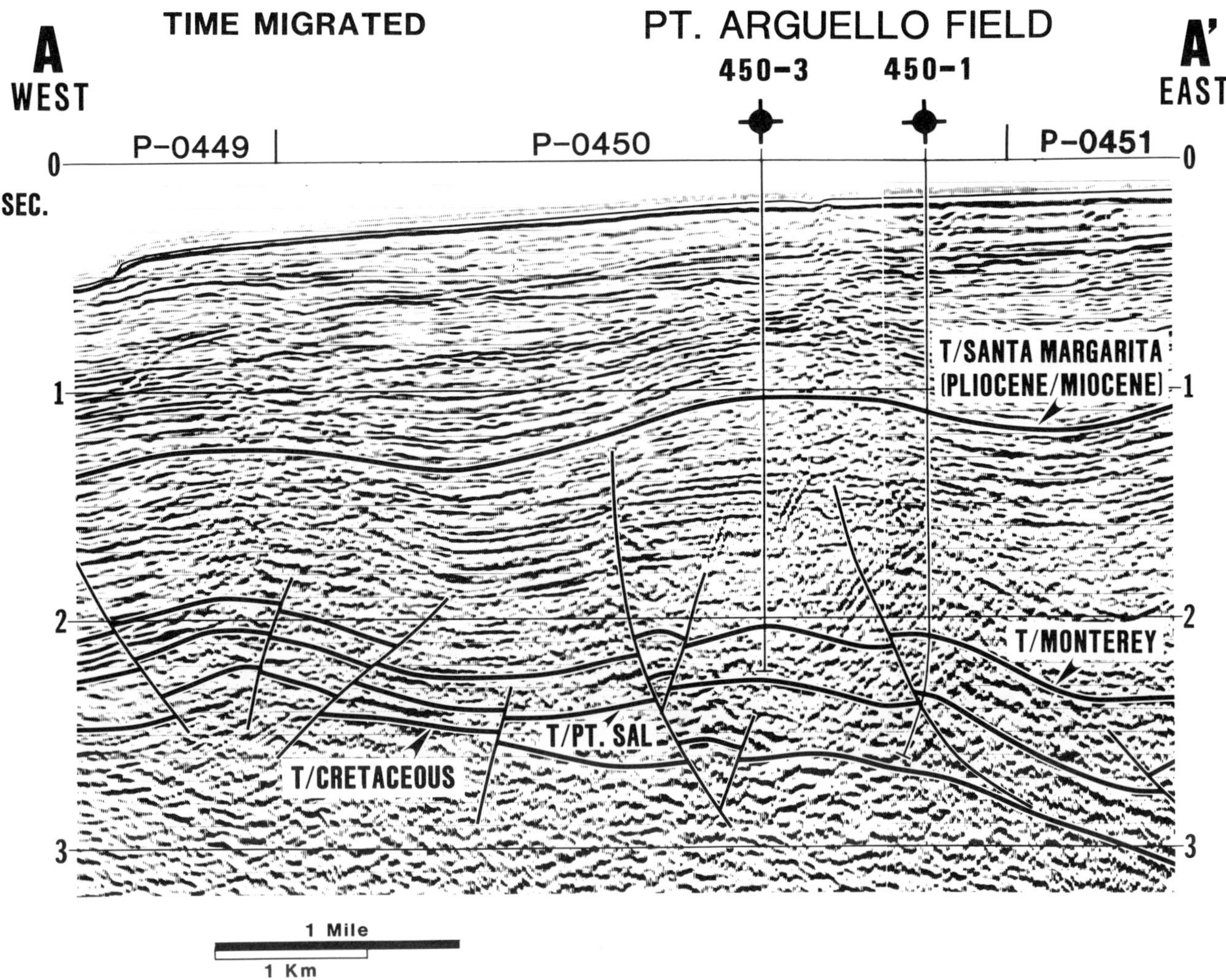

Figure 14. Time-migrated seismic section passing near Chevron-Phillips P-0450-3 and through Chevron-Phillips P-0450-1. See Figure 2 for location.

Tertiary

Lower Miocene Tranquillon Volcanics

The Tranquillon Volcanics, a marine lower Miocene (Saucesian Stage) series of rhyolitic tuffs, unwelded pumice, tuffaceous volcaniclastics, sandstones, and siltstones, is the oldest Tertiary unit found in the Point Arguello field. Over 700 ft (213 m) of Tranquillon Volcanics were penetrated in lease P-0450 and are unconformably overlain by the lower Miocene Point Sal Formation. In the Santa Barbara Channel the Tranquillon is represented by a thin tuffaceous zone at the top of the Rincon formation, usually less than 100 ft (30.5 m) thick (Figure 4). The Tranquillon Volcanics are missing by onlap and/ or truncation over the Amberjack high (Figure 3).

Middle Miocene

Point Sal Formation (lower Monterey)

The Point Sal Formation is predominantly a lower Miocene (Relizian Stage) limy marine mudstone with interbedded dolostones. Onshore, Dibblee (1950) mapped this unit as the lower Monterey. It is roughly coeval to part of the lower calcareous-siliceous member of the Monterey as defined by Isaacs (1980). The top of the Point Sal Formation may be as young as middle Miocene (Luisian Stage) and is transitional to the overlying, more siliceous Monterey Formation, diagrammatically shown by the correlation section in Figure 17. The Point Sal Formation can be as much as 2500 ft (762 m) thick in the Point Arguello field but pinches out westward onto the Amberjack high.

41

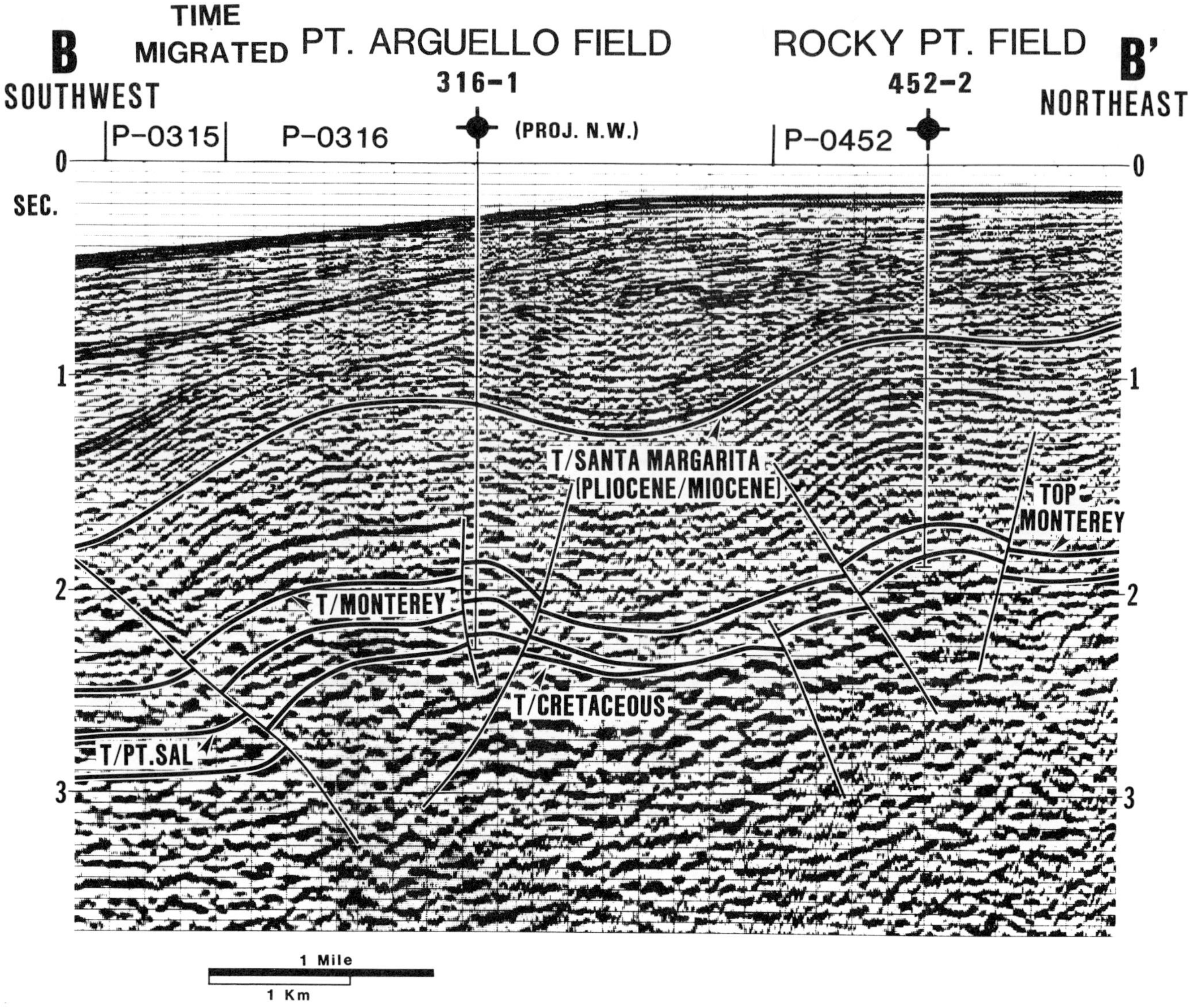

Figure 15. Time-migrated seismic section passing northwest of Chevron et al. P-0316-1 discovery well. See Figure 2 for location.

Fractured dolostones near the base of the Point Sal Formation tested a small amount of 31.7° API oil in Chevron-Phillips P-0450-1. The Chevron et al. P-0316-1 may have tested a significant amount of oil from a thin Point Sal section as indicated in Figure 9. However, a thin section of lower Monterey cherts and dolostones may have contributed most of the oil in DST #5. While not as rich as the Monterey, the Point Sal Formation is an important oil producer from sandstone turbidites in a number of onshore Santa Maria fields (Santa Maria Valley, Casmalia, Orcutt, and Cat Canyon). The Point Sal–lower Monterey Formation within the offshore Hondo field in the Santa Barbara Channel also contains several oil-charged sandstone reservoirs.

Middle and Upper Miocene

Monterey Formation

The Monterey Formation is both the primary reservoir and source rock at the Point Arguello field and in the offshore and onshore Santa Maria basins. By 1986 the onshore Santa Maria area had produced nearly 800 million bbl of Monterey oil. New Monterey discoveries in the offshore Santa Maria basin are

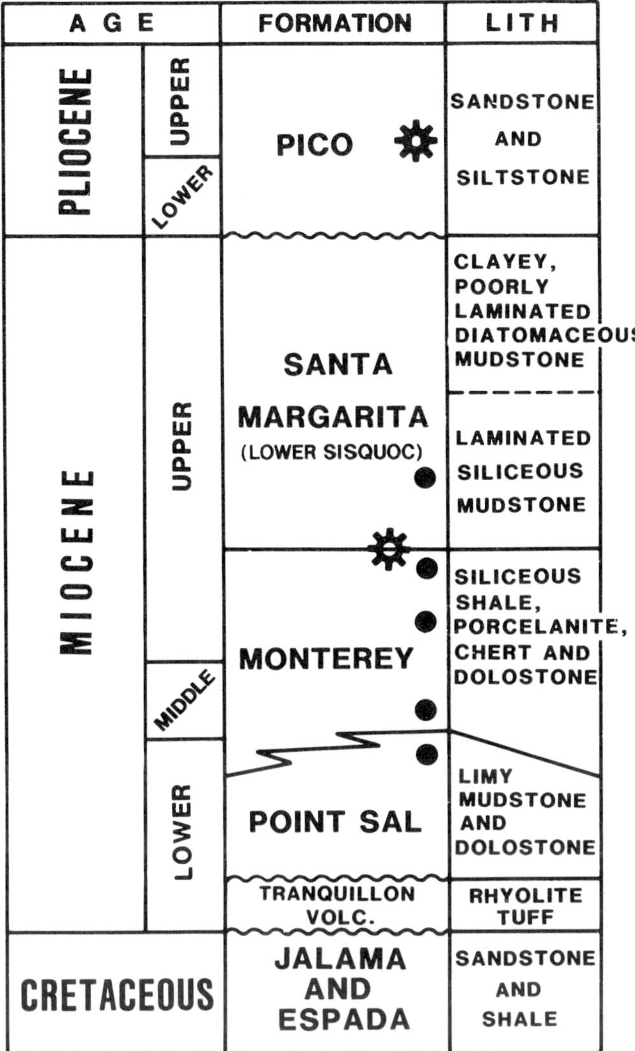

Figure 16. Generalized stratigraphic column of Tertiary section penetrated to date by Chevron et al. wells in Point Arguello field. Column illustrates Chevron's stratigraphic interpretation of Miocene and Pliocene section. Successfully tested zones indicated by the black dots.

estimated to have total reserves of as much as 2 billion bbl (Skillin, 1989), although much of the oil is low gravity and in deep water, making the fields uneconomic under present conditions. At least 300 million bbl in producible Monterey reserves are estimated to be present in the Santa Barbara Channel (Ogle et al., 1987).

Figure 18 is a Monterey isopach map that shows the stratigraphic effect of the Amberjack high and the Neogene depocenter that persisted throughout the Miocene. The Monterey Formation ranges in age from early through late Miocene (Relizian to late Mohnian age) and in thickness from 2800 to less than 1000 ft (853–305 m) in the Point Arguello field.

The onshore Monterey Formation has been subdivided into four members by Chevron (MacKinnon, 1983) after similar schemes by Isaacs (1980) and Pisciotto and Garrison (1981), shown in Figure 17 as (1) lower calcareous-siliceous, (2) phosphatic, (3) upper calcareous-siliceous, and (4) siliceous members. Within the Point Arguello field, the Monterey reservoir is also divided by Chevron into four members. Figure 17 demonstrates the correlation between the offshore and onshore Monterey facies. Along the coast, the top of the Monterey is difficult to pick, but it is usually mapped at the contact between the finely laminated siliceous rocks and the overlying bioturbated, siliceous mudstones. Chevron dates this contact as slightly below the Pliocene–Miocene boundary by careful diatom sampling. A thin, phosphatic nodule conglomerate at this boundary is sometimes present, representing a short period of nondeposition.

The offshore subdivisions are the (1) basal calcareous, (2) phosphatic, and (3) chert and siliceous shale members. The basal member is primarily composed of dolostones, porcellanites, and mudstones. The phosphatic member is primarily a dolomitic phosphatic mudstone, often with abundant dolostone nodules, and is highly organic (10–18% TOC). The chert member (the onshore upper calcareous-siliceous member) is dominantly chert with interbedded dolostones and porcellanites. This is the most fractured and productive reservoir in the Point Arguello field. The siliceous shale member (the onshore siliceous member) is commonly composed of finely laminated siliceous mudstones and porcellanites. The top of the siliceous shale member is picked at the increase in the spontaneous potential and resistivity on the dual induction log (Figure 9). Although a convenient mapping horizon, Ogle et al. (1987) believe this marker is a diagenetic facies without regional time significance.

Miocene–Pliocene

Santa Margarita (lower Sisquoc) Formation

In the Santa Barbara Channel and offshore Santa Maria, the Santa Margarita Formation is predominantly light-gray, diatomaceous, coarsely laminated mudstones and bedded dolostones. The marine section is of late Miocene age (late Mohnian–early Delmontian age), conformably overlying the Monterey Formation. This section is equivalent to the upper Miocene portion of the Sisquoc Formation as described by Dibblee (1950) and Woodring and Bramlette (1950). The Santa Margarita Formation thins over the Point Arguello structure to less than 4500 ft (1372 m) and thickens off the structure to more than 7000 ft (2134 m). A minor unconformity may be present at the top of the Santa Margarita, representing a period of nondeposition and mild structural growth in a deep marine environment. A

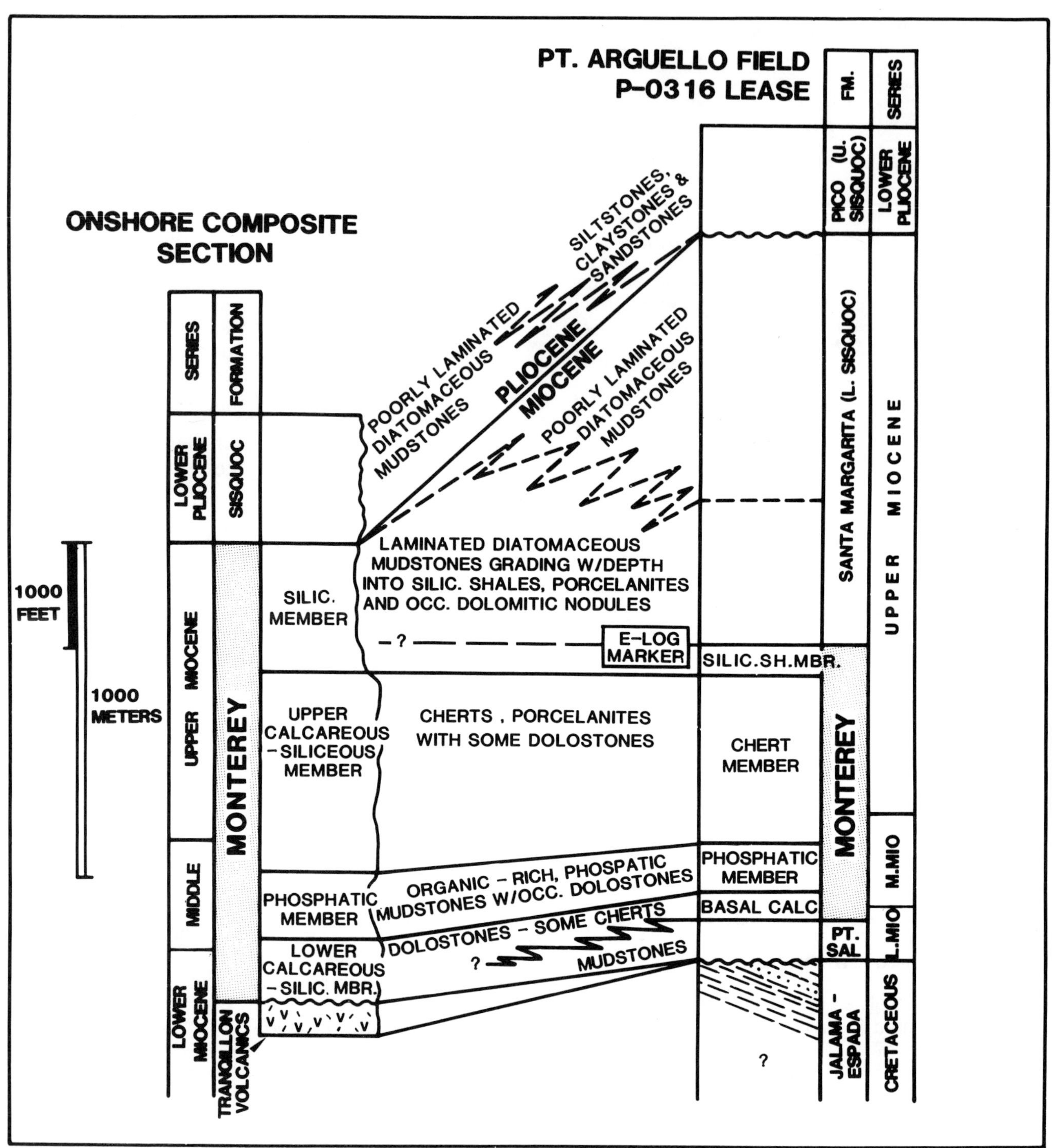

Figure 17. Correlation between composite onshore Monterey section and generalized section in Point Arguello field (P-0316 lease). Simplified facies relationships are also indicated.

few successful drill-stem tests have been made in fractured, dolomitic zones but the producibility appears to be low. Fractured Santa Margarita Formation may be an additional reservoir at the undeveloped Rocky Point field. The lower Sisquoc Formation produces in the onshore Santa Maria Valley and Cat Canyon fields.

Pliocene

Pico (Upper Sisquoc, Foxen, and Careaga) Formation

Chevron puts all of the Pliocene section within the Pico Formation, a correlation from the Santa Barbara Channel (Figure 4). The Pico Formation in the Point

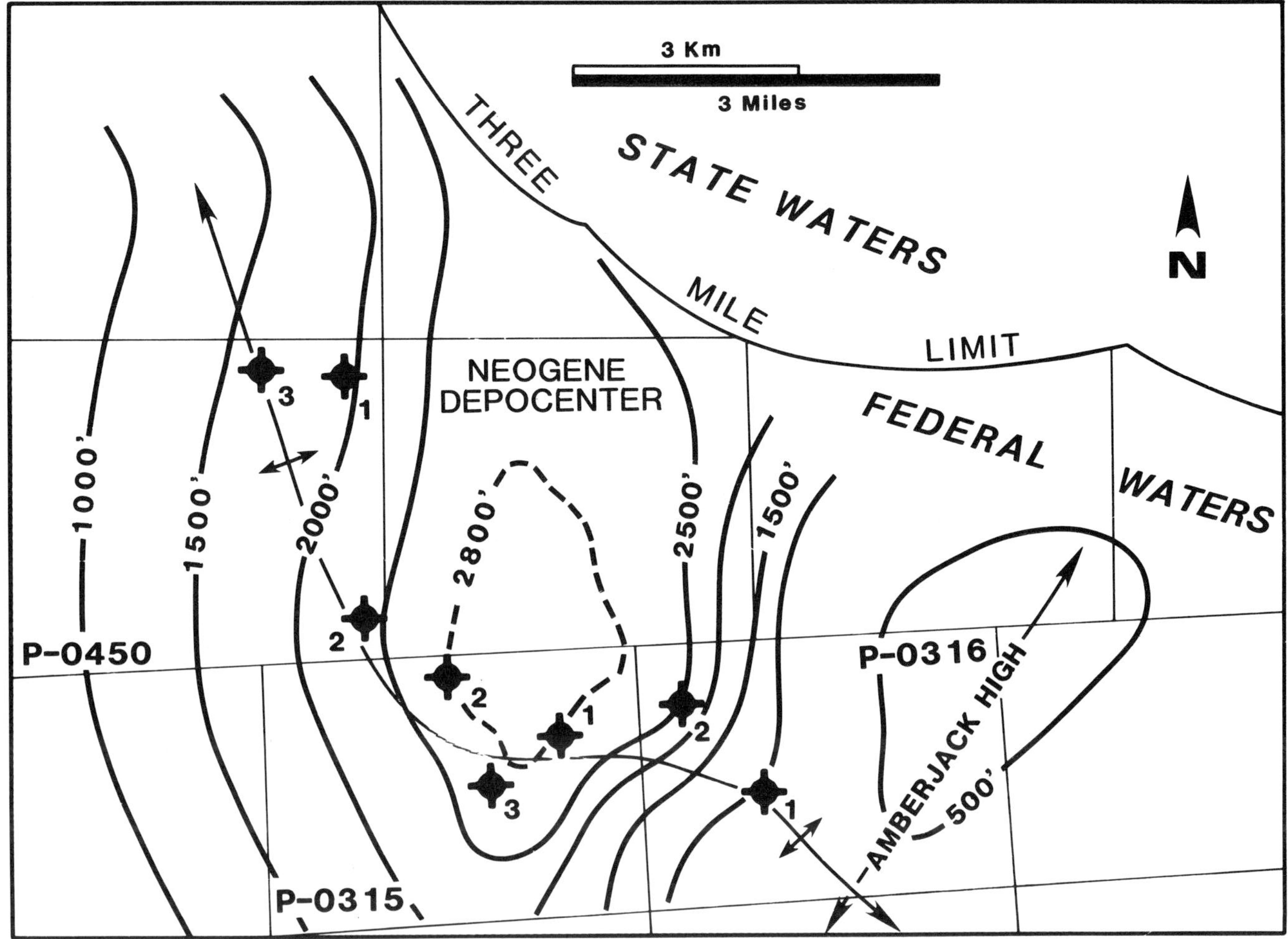

Figure 18. Monterey isopach map over Point Arguello field. The map demonstrates the influence of Amberjack high on Miocene-age Monterey depositional trends. Narrow north–northeast-trending depocenter was the probable generating center for higher gravity Monterey sourced oils from the late Miocene to present.

Arguello field area is about 3000 ft (914 m) thick. During the late Pliocene, turbidities deposited channel-like facies of poorly indurated lithic sandstones and pebbly conglomerates across the Point Arguello structure. Most of the Pico Formation is composed of dark gray-brown siltstones and claystones and is equivalent to the lower Pliocene upper Sisquoc and upper Pliocene Foxen and Careaga formations, as commonly mapped in the onshore Santa Maria basin.

In lease P-0315 Texaco et al. discovered a small gas accumulation from shallow Pliocene sandstones. Seismic amplitude anomalies or "bright spots" can be seen within the Pliocene section. They usually represent noncommercial gassy sands and siltstones.

TRAP

The trap is an anticlinal closure enhanced by faulting. Reflection seismic surveys show reverse faults cutting both flanks and the northwest plunge of the Point Arguello structure (Figure 2). Faulting within the structural saddle (Figure 3) may provide the southern closure for the northern oil pool.

Seismic line A–A' (Figure 14) demonstrates the northeast-dipping reverse fault that forms the southern boundary of the Point Arguello field. Seismic line B–B' (Figure 15) shows southwest-dipping reverse fault that breaks the northeast flank and enhances the closure of Point Arguello's southern pool. The maximum closure is about 2400 ft (732 m), and the maximum Monterey hydrocarbon column is around 1800 ft (549 m) thick.

Without detailed well control, accurately mapping the Monterey trap is difficult because of the poor reflection coefficient of the Santa Margarita (lower Sisquoc)–Monterey contact. The early seismic mapping (Figures 7 and 10) was derived by mistakenly traversing the more coherent basal events within the Santa Margarita Formation.

The Santa Margarita (lower Sisquoc) is the overlying seal. The Santa Margarita siltstones and mudstones often have irregular oil-stained fracture zones extending several hundred feet above the Monterey reservoir. In the Point Arguello field strong oil shows below the present oil-water contacts suggest that the oil column was originally deeper. The Santa Margarita–lower Sisquoc caprock within the opal-CT and quartz phases has local fracture zones charged with oil hundreds of feet above the Monterey reservoir.

Although no seeps are believed present over the Point Arguello field, they are commonly associated with known Monterey accumulations (Vernon et al., 1963; Priestaf, 1979). Unless the Santa Margarita–lower Sisquoc seal is thick and unfaulted, a Monterey oil accumulation is quickly lost. In the Santa Barbara Channel the oil-charged Pliocene sandstones overlying the Santa Margarita–lower Sisquoc Formation are almost all believed to be sourced from the more deeply buried Monterey Formation, a belief based on biomarker studies by Chevron and other companies.

Thinning of the Santa Margarita over the Point Arguello structure indicates that the anticlinal trap began to develop during the late Miocene while middle Pliocene–lower Pleistocene age compression gave the Point Arguello structure its present closure.

RESERVOIR

The depth of the southern pool in the Point Arguello field is around 6500 ft (1982 m) with an oil column of slightly over 1600 ft (488 m). A small primary gas cap in the southern pool has a maximum thickness of around 200 ft (61 m). The northern pool, located in a separate faulted anticlinal trap, is approximately 1000 ft (305 m) deeper at 7500 ft (2287 m) and has a maximum oil column of about 650 ft (198 m). No primary gas cap is present in the northern pool.

The principal fractured reservoir facies at the Point Arguello field are the clay-free glassy cherts and more clayey porcellanites, a microcrystalline siliceous rock with a matte luster similar to unglazed porcelain. The major producing interval appears to be the chert member (Figure 9), composed mainly of cherts and porcellanites.

As the clay content of the rock increases, the reservoir quality decreases. Siliceous mudstones of the siliceous shale member and mudstones of the phosphatic member are poor reservoirs but are often oil-prone source rocks. The 1 to 3 ft (0.3–1 m) dolomitic beds are usually poorly fractured. The carbonate-rich basal calcareous member is not considered a major producer in the Point Arguello field.

DEPOSITION AND DIAGENESIS

The Monterey Formation in the Point Arguello field was deposited in a marine deep-water, anaerobic environment. The Monterey is largely composed of diagenetically altered foraminiferal and diatomaceous oozes. Organic-rich calcareous phosphatic mudstones are common. During Monterey deposition, the offshore Santa Maria basin was divided into a series of clastic starved subbasins between the basement highs shown on Figure 13. Finely laminated biogenetic layers formed where anoxic bottom environments prevented bioturbation (MacKinnon, 1989).

Eventually as basins to the east filled, fine clastics spilled westward causing clays to mix with the diatomaceous sediments of the upper Monterey. By the late Miocene, oxygen-rich bottom conditions returned to the Point Arguello depocenter resulting in the deposition of the thick-bedded, bioturbated, clayey Santa Margarita (lower Sisquoc) Formation.

Diagenesis converted the finely laminated biogenetic-clayey Monterey sediments into cherts, porcellanites, siliceous siltstones, mudstones, and dolostones. With increasing time and temperature, the original diatomites were first converted from amorphous silica (opal A) to the more ordered crystalline phase, cristobalite-tridymite (opal-CT), as analyzed by X-ray diffraction (Murata et al., 1975; Pisciotto, 1981). Figure 19 is a scanning electron microscope (SEM) photomicrograph of unaltered diatomites illustrating the porous (60–70%), low permeability, low density pre-diagenetic deposition. The change from opal A to opal-CT occurs between the present temperatures of 105°F and 138°F (40.6–58.9°C) in the Point Arguello field. Figure 20 is a SEM photomicrograph of porcellanite in the opal-CT phase. The microporosity commonly decreases to around 20–30% in porcellanites.

The last diagenetic step at Point Arguello was a phase change from opal-CT to quartz, occurring within the present subsurface temperature range of 150°F to 190°F (65.5–87.8°C). Figure 21 is a SEM photomicrograph of a glassy chert in the quartz phase. Impermeable but most prone to fracturing, chert zones are the primary reservoirs within the Monterey. The Monterey in the Point Arguello field is entirely within the quartz phase. In other onshore and offshore fields, opal-CT porcellanites, cherts, and siliceous mudstones also form productive, fractured Monterey reservoirs.

Dolostones occur as beds, lenses, and nodules. Thin sections from cores and outcrop samples indicate dolostones formed by replacement and pore filling of the siliceous sediments. Close examination of relict laminations within the dolostones and compaction features in the surrounding rocks suggests dolostones formed in both the early precompaction and postcompaction stages of diagenesis.

Mudstones containing blebs or fine laminations of phosphatic material occur as erratic, thin interbeds throughout the Monterey but are thickest in the phosphatic member. Generally, the mudstones contain the highest organic content (as much as 18% TOC) in the Monterey section. The mudstones are

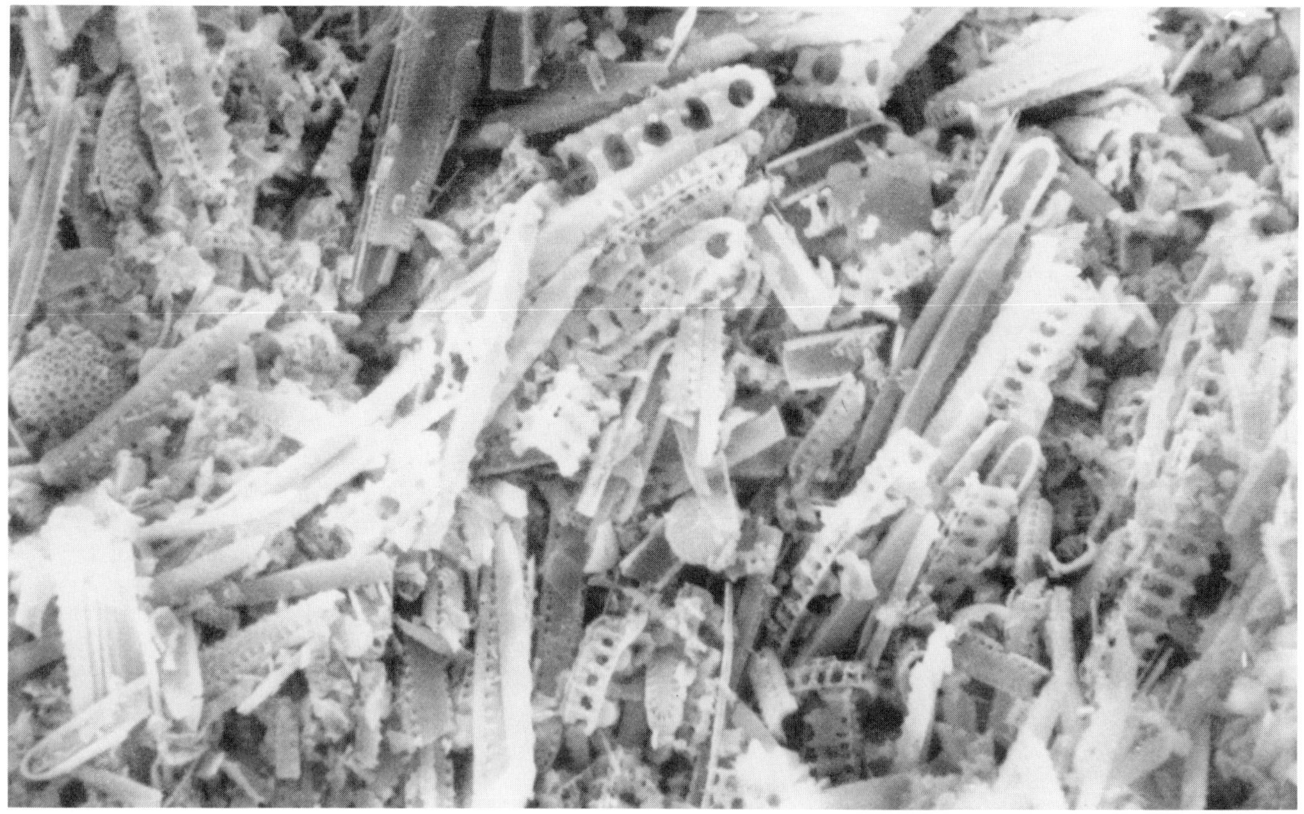

Figure 19. Scanning electron microscope (SEM) photomicrograph of unaltered diatomites (opal A). Scale bar = 10 microns. Bulk density = 1.70–2.04. Porosity = 60–70%. Velocity = 5800–7800 ft/sec (1800–2400 m/sec). During burial higher temperatures diagenetically alter this unstable biogenetic material into cherts, porcellanites, and siliceous shales.

calcareous with nodules and thin interbeds of dolostones, porcellanites, and cherts.

Finely laminated siliceous mudstones increase near the top of the Point Arguello Monterey reservoir, forming the bulk of the siliceous shale member. Thin-bedded cherts, porcellanites, and occasional dolostone nodules are also sometimes present.

RESERVOIR CHARACTER

Crain et al. (1985) reviewed the general porosity and permeability of the Point Arguello field as determined from the initial delineation wells. Fish (1989) reported that limited interference tests suggest good lateral permeability. The Monterey reservoir also has good vertical continuity as demonstrated by shut-in pressure measurements in a number of field wells.

Bulk permeabilities have been calculated from drill-stem test (DST) pressure data. Permeability to oil ranges from less than 1 m/ft (3.3 md/m) to more than 3 darcys/ft (9.8 darcys/m). Measured permeabilities on core plugs suggested a matrix permeability of 0.1 md or less for most of the Monterey reservoir facies (Crain et al., 1985).

Variations in fracture and not intercrystalline matrix permeability will control the production rates in the Point Arguello field. It is expected that the decline curves will be similar to the onshore Monterey decline curves, about 45% the first year, 30% the next, and 15% thereafter. The typical Monterey decline rate is assumed by most reservoir engineers to be caused by a dual porosity system. The large and intermediate fracture systems may source the initial flush production. As oils in the major fracture networks are quickly produced, production rapidly declines and then stabilizes as microfractures slowly recharge the large fracture network.

Fracture density varies according to the argillaceous content of the rock. Fracturing is least in the mudstones, siliceous shales, and clayey dolostones and is best developed in the glassy cherts, pure dolomites, and porcellanites. Core recoveries from the most productive intervals are poor, forcing formation evaluation specialists to base their interpretation of the Monterey fracture zones on outcrops, test data, and wireline logs. In the cores fractures are often parallel and at high angle to the bedding. The natural

Figure 20. Scanning electron microscope (SEM) photomicrograph of porcellanite in the opal-CT (christobalite-tridymite) phase. Porcellanites can be in both opal-CT and quartz phases. Scale bars = 10 microns. Bulk density = 2.24–2.33. Porosity = 20–30%. Velocity (p) = 10,300–13,900 ft/sec (3140–4250 m/sec).

fractures can be open, partially open, filled with some porosity, and filled with no porosity.

The bit will occasionally drop several feet within chert or dolostone lost circulation zones (Fish, 1989). Dilation breccias (Redwine, 1981; Roehl, 1981) are mapped onshore but have not been cored in the Point Arguello field. However, Fish (1989) reported that wells have occasionally encountered up to 50 ft (15 m) of intensely fractured cherts. These intervals, marked by increased penetration rates and good production rates, have calculated isotropic interclast porosity of as much as 25% and may represent fault-associated fracture zones or dilation breccias.

Normal chert-rich zones can have an average permeability of up to 300 md/ft (984 md/m) but often contain thin intervals of highly brecciated rocks. The unbrecciated layers often average around 45 md/ft (148 md/m) and are believed connected to the erratic fracture zones by occasional high-angle, crosscutting fractures. Porcellaneous zones are commonly around 30 md/ft (98 md/m) and average about two fractures/ft (6.6 fractures/m). Unfractured porcellanites have less than 1 md/ft (<3.3 md/m) of permeability.

It appears that most fracturing is tectonically induced. Published data from the Santa Barbara Channel's South Ellwood field (Belfield et al., 1983) suggest that northeast-southwest tension fracture may be present. Outcrop measurements (Carpenter et al., 1984) indicate that fractures occur as conjugate sets of both shear and extension systems, depending on their fold position.

Total reservoir porosity ranges between 10% and 20%, with fracture porosity averaging between 1% and 2% of the reservoir volume (Crain et al., 1985). Some secondary vugular porosity has been described in both dolostones and cherts. Primary matrix microporosity in the Monterey reservoir is in the 0% to 15% range.

HYDROCARBON CHARACTER

The Monterey oil produced from the Point Arguello field is classified as aromatic-intermediate. The southern pool is gravity segregated, with around 11°

Figure 21. Scanning electron microscope (SEM) photomicrograph of glassy chert in the quartz phase. Scale bar = 10 microns. Bulk density = 2.40–2.56. Porosity 0–10%. Velocity (p) = 14,000–19,000 ft/sec (4270–5800 m/sec). The major Point Arguello oil reserves are probably contained in the fractured and brecciated Monterey cherts. Cherts contain little or no clay and can be either in the opal-CT or quartz phases.

API at the base to about 23° API at the top of the Point Arguello field—a segregation common in many Monterey reservoirs. The API gravity of Monterey oils is roughly inverse to the sulfur content. Sulfur content varies from about 0.8% in the 30° API oils (northern pool) to as much as 5% in the 11° to 23° API oils (southern pool). A plot based on Monterey oil analyses by Orr (1986) and Magoon et al. (1983) illustrates the general relationship between oil gravity and percentage sulfur content (Figure 22).

A representative gas chromatogram is shown in Figure 23. No biodegradation is present. Normal paraffins are well developed with a slight even-carbon preference among the normal paraffins in the C_{18}–C_{30} range. The heavier oils in the southern pool contain thermally unstable porphyrins, suggesting an accumulation of thermally immature oil. The carbon isotope ratio ($^{13}C/^{12}C$) of the whole crude averages around -22.7 ‰ PDB.

The oil is undersaturated with a solution-gas content of 400 ft³/bbl. Gas from the P-0316 lease has a gross heating value of 1183 Btu/ft³. H_2S is about 0.58 mole %, and methane is around 82 mole %. CO_2 and N_2 average 4% and 0.43%, respectively. Gas from the northern pool has a gross heating value of 1077 Btu/cubic ft. H_2S is about 0.05 mole %, and methane averages about 89 mole %. CO_2 and N_2 average 2.6% and 1.24%, respectively (Crain et al., 1985).

HYDROCARBON GENERATION

All the hydrocarbons found in the Point Arguello field are believed to have been generated within the Monterey and Point Sal formations on the basis of carbon isotopes and biomarker studies (Crain et al., 1985). Understanding the critical factors that permit the early generation and trapping of large hydrocarbon accumulations of oils with widely diverse API gravities is important to any Monterey exploration program.

The total organic carbon (TOC) of the Monterey Formation in the Point Arguello field averages about 3%. Phosphatic mudstones vary but commonly

49

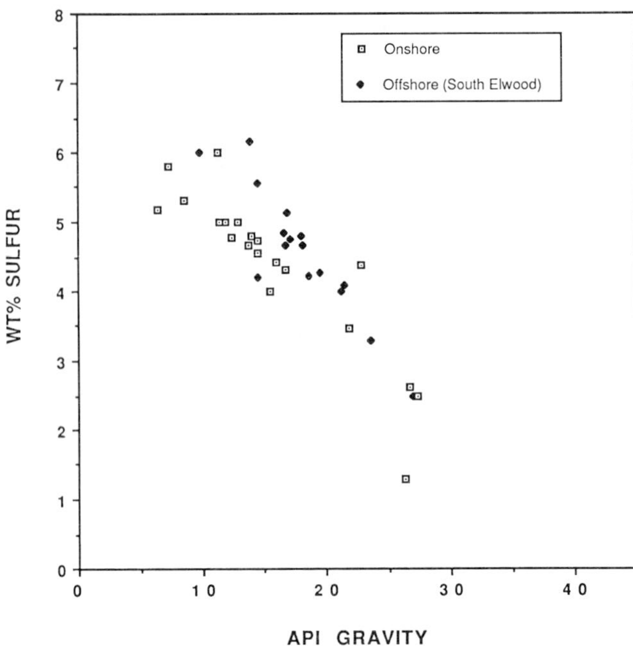

Figure 22. Graph illustrating the relationship between Monterey oil gravities and sulfur content, onshore and offshore. While there is a small variation between areas, the correlation is good enough to be useful in predicting Monterey oil gravities before drill-stem testing.

contain between 10% and 18% TOC. The Monterey kerogens are classified as oil-prone high sulfur type II with a sulfur content of up to 9% in the Point Arguello field.

Most of the oil-prone Monterey kerogens in the Point Arguello field have an average atomic hydrogen/carbon ratio of about 1.25. Generally, the Monterey section at Point Arguello field has a hydrogen index of 600–700 (mg H/g TOC) and a hydrocarbon potential of between 160 and 1200 bbl/ac-ft. From 95% to 98% of the kerogens examined are sapropelic-amorphous derived from algal material. Little or no humic organic material is present.

The small organic-rich oil-generating "kitchen" sourcing the Point Arguello field is located in the Monterey depocenter shown in Figure 18. The location of the Point Arguello structure within the generating center is important because it provided the short migration path to the Point Arguello trap necessary for forming a large oil accumulation from upper Tertiary age source rocks. The higher than normal heat flow (Blackwell, 1979) and rapid burial within the Neogene depocenter put early Miocene Monterey source rocks into the oil generation window by late Miocene to early Pliocene.

In addition, the sulfur-rich Monterey kerogens are cracked into thermally immature, heavy hydrocarbons at a lower maturation level than normal type II kerogens (Orr, 1986). Oil generation from the sulfur-rich Monterey kerogens begins at vitrinite reflectance maturation levels of between 0.3% and 0.4% R_o (Petersen and Hickey, 1983) or at a time-temperature index (TTI) value of 3.0 or less (Waples, 1987), a maturation level normally considered thermally immature.

Figure 24 is a generalized geohistory plot (SIN-WELL 2.0, Chevron Proprietary software) from the Point Arguello Neogene depocenter. Typical Monterey kerogen activation energies were used to model the hydrocarbon generation window after the calculation techniques described by Tissot and Welte (1984). The decrease in heat flow in coastal California since the Oligocene has been modeled by Heasler and Surdam (1985). Using similar thermal decay assumptions, the present heat flow is calculated at about 88 mw/m² , a value compatible with published heat flow data (Blackwell, 1979). The temperature gradient is about 2.6°F/100 ft (48°C/km), similar to the temperature gradient in the nearby COST well (McCulloh and Beyer, 1979).

The geohistory plot shows continuous sedimentation since the deposition of the Point Sal (lower Monterey) Formation. The Conversion Index (CI) is a percent of the kerogen's generative potential that has been converted to hydrocarbons. A CI of 10% and 90% defines the approximate beginning and end of the oil-generating window. Thermal modeling indicates that the Monterey entered the oil window between 7 and 6 m.y.a. Today the oil generating window is within the Monterey and Point Sal formations. From the Pliocene–Pleistocene to the present, high static Monterey Formation temperatures could have thermally cracked the earlier formed, immature heavy oils. This may be the source of the 30+° API gravity oil found in the Point Arguello and nearby Rocky Point fields.

EXPLORATION CONCEPTS

The Point Arguello field is part of a regional play or trend that can be partly defined as a number of anticlinal-fault enhanced traps in close proximity to a major oil-generating center. Monterey rocks are chemically unstable and within this trend, rapid burial, high heat flow, and high temperature gradients quickly altered the Monterey into fracturable reservoirs of cherts, porcellanites, and siliceous shales. The low clay content of this siliceous Monterey "fairway" produced a high percentage of cherts, the most productive of the fractured reservoirs. The regional Monterey trend is also rich in oil-prone organic matter that converts to hydrocarbons under lower maturation levels than normal type II kerogens.

The factors that created the oil accumulation in the Point Arguello field also are common to many fields in the onshore Santa Maria and San Joaquin basins that produce from the Monterey or Monterey equivalents. Early oil generation, short migration paths, siliceous fractured reservoirs, and structural

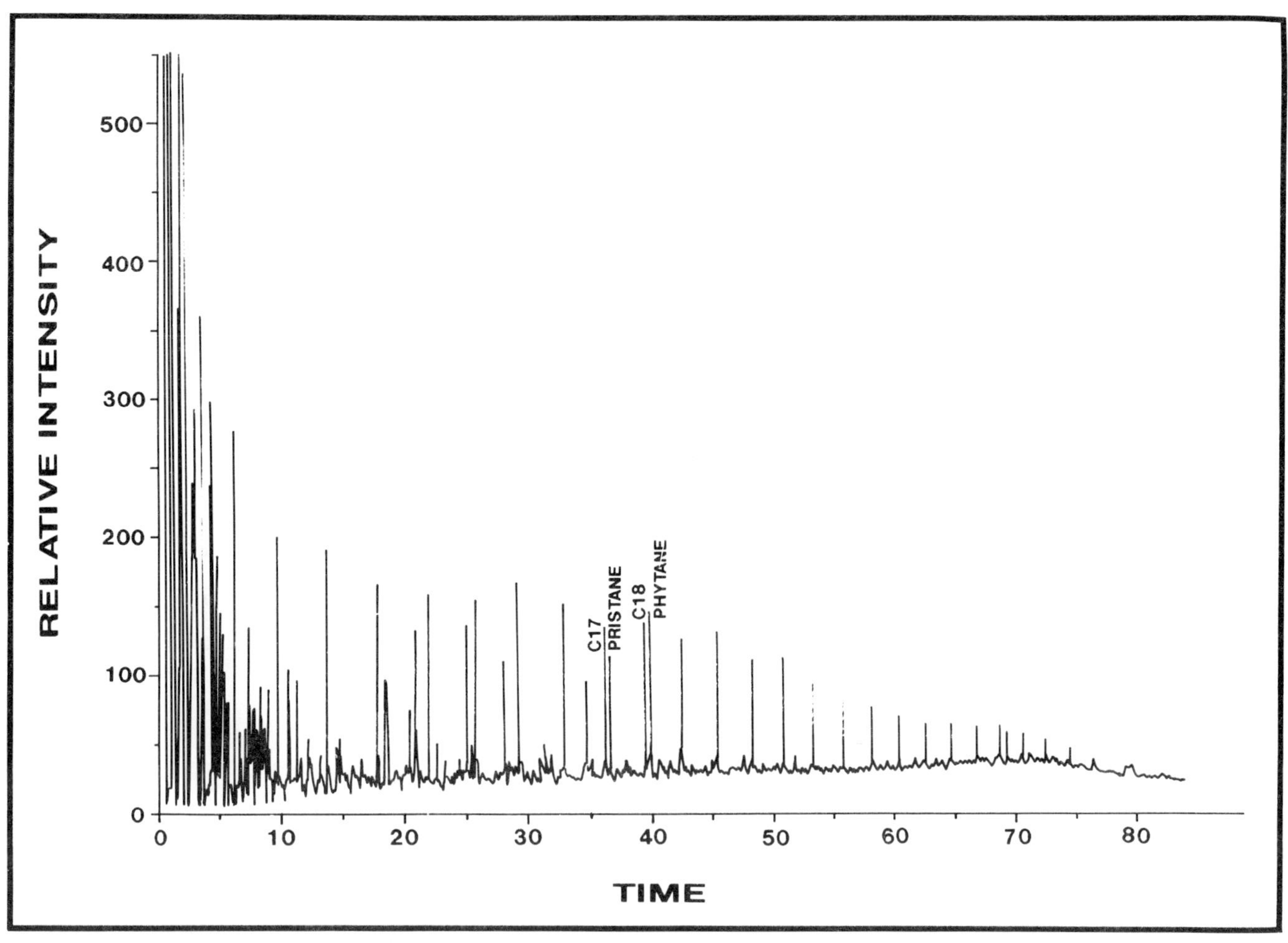

Figure 23. Typical gas chromatogram of a Monterey oil sample from the southern pool of the Point Arguello field.

traps are common to a number of Monterey producing areas throughout southern and central California.

Most of the large anticlinal Monterey traps have now been tested except in those California offshore basins closed to exploration. Therefore, large remaining Monterey oil accumulations will probably be found in stratigraphic-structural traps. The first step in an exploration program will be to identify cherty, low clay Monterey trends close to or within high API oil gravity, low sulfur oil-generating source areas. The seismic response to sharp diagenetic siliceous phase boundaries around the flanks of regional uplifts may identify potential stratigraphic traps formed at the boundary between opal A and the fracturable CT and quartz phases (Ogle et al., 1987). A thick overlying Santa Margarita–Sisquoc seal is desirable in order to help prevent biodegradation and vertical loss of Monterey hydrocarbons. Traps have also been formed by near-surface asphaltic seals caused by biodegradation, water washing, and loss of volatiles.

The successful exploration for commercial Monterey accumulations requires the integration of geology, stratigraphy, geochemistry, and geophysical disciplines. Billions of in-place barrels of oil have been discovered in the Monterey during the past decade. Three-dimensional seismic surveys have become a common tool for planning development of offshore Santa Maria fields. However, velocity determinations for stacking and migration purposes can be extremely difficult because of steep dips (over 45°) and rapid lateral velocity changes common over the offshore structures. At the Point Arguello field, Phillips Petroleum developed reprocessing techniques that resulted in a smoother stacking velocity model. A layered velocity model from borehole and stacking velocities was also constructed allowing a more successful one-pass 3-D migration of seismic events (Fischer, 1987).

The reservoir engineer as well as the explorationist will play an increasingly important role in expanding our producible reserves from these erratically fractured reservoirs. Horizontal drilling techniques and long-term flowing tests may be necessary to properly evaluate and economically develop many of these Monterey reserves.

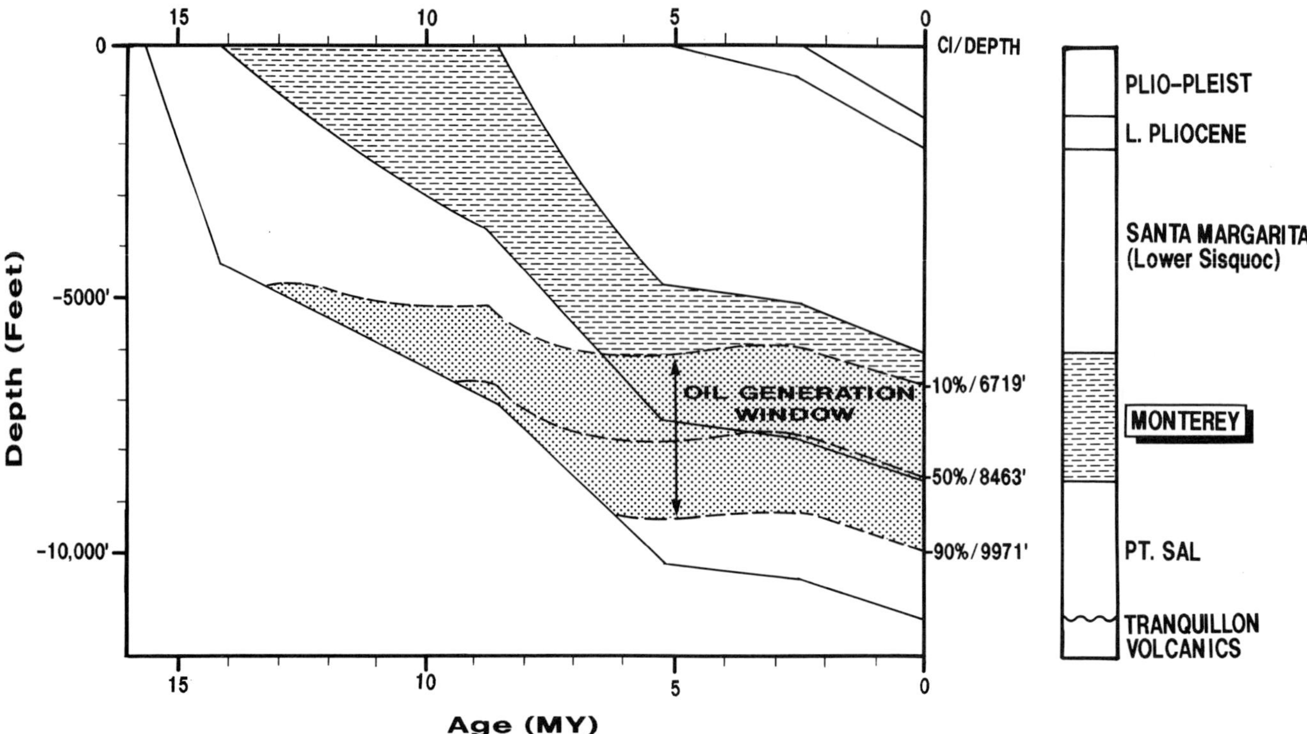

Figure 24. A generalized, computer-derived Lopatin-style burial diagram that models the thermal maturation of the Monterey source rock within the Neogene depocenter. The activation energies of sulfur-rich, type II Monterey kerogen were used to estimate the oil-generating window. The computer program considers the effect of compaction on the burial history.

ACKNOWLEDGMENTS

The author thanks the Chevron Corporation for permission to publish this paper. Numerous Chevron personnel assisted in this update of the Point Arguello discovery. Special thanks are due to T. C. MacKinnon for his petrographic and stratigraphic advice, M. H. Pytte for help in the geochemical interpretation, and R. Stoddard who, together with all the Chevron development geologists, contributed information on current development practices. In addition, thanks are due to Ms. Shamsi Moussighi who was responsible for most of the illustrations and my wife, Kathleen, who typed and proofed the final draft and has followed me around the world in the search for hydrocarbons.

REFERENCES

Atwater, T., 1970, Implications of plate tectonics for the Cenozoic tectonic evolution of western North America: Geological Society of America Bulletin, v. 81, p. 3513-3536.

Ballard, J. H., 1988, Sword field, offshore California: challenges in making this giant oil field commercial (Abs.): American Association of Petroleum Geologists Bulletin, v. 72, n. 3, p. 374.

Bally, A. W., and S. Snelson, 1980, Realms of subsidence, *in* Facts and principles of world petroleum occurrence: Canadian Society of Petroleum Geologists Memoir 6, p. 9-94.

Belfield, W. C., J. Helwig, P. LaPointe, and W. K. Dahleen, 1983, South Ellwood oil field, Santa Barbara Channel, California, a Monterey Formation fractured reservoir, *in* C. M. Isaacs and R. E. Garrison, eds., Petroleum generation and occurrence in the Miocene Monterey Formation, California: SEPM Pacific Section, p. 213-221.

Blackwell, D. D., 1979, Heat flow and energy loss in the western United States: Geological Society of America Memoir 152, p. 175-208.

Blake, M. C., Jr., R. H. Campbell, T. W. Dibblee, Jr., D. G. Howell, T. H. Nilsen, W. R. Normark, J. C. Vedder, and E. A. Silver, 1978, Neogene basin formation in relation to plate-tectonic evolution of San Andreas fault system, California: American Association of Petroleum Geologists Bulletin, v. 62, p. 344-372.

Carmalt, S. W., and B. St. John, 1986, Giant oil and gas fields, *in* M. T. Halbouty, ed., Future petroleum provinces of the world: American Association of Petroleum Geologists Memoir 40, p. 11054.

Carpenter, A. B., W. E. Seixas, and S. L. Hicks, 1984, Results of preliminary study of fracture spacing in outcrops, Point Arguello area: Chevron Oil Field Research Co. unpublished technical memorandum, 49 p.

Crain, W. E., W. E. Mero, and D. Patterson, 1985, Geology of the Point Arguello discovery: American Association of Petroleum Geologists Bulletin, v. 69, n. 4, p. 537-545.

Dibblee, T. W., 1950, Geology of southwestern Santa Barbara County, California-Point Arguello, Lompoc, Point Conception, Los Olivos, and Gaviota quadrangles: California Division of Mines Bulletin 150, 95 p.

Fischer, T., 1987, Discovery of the Point Arguello oil field from a geophysical perspective: Geophysics: The Leading Edge of Exploration, v. 6, n. 10, p. 16-21.

Fish, J. L., 1989, Formation evaluation of the fractured Monterey formation, Pt. Arguello field, California, *in* T. C. MacKinnon, ed., Oil in the California Monterey Formation: 28th

International Geological Congress Field Trip Guide Book T311, American Geophysical Union, p. 45–49.

Heasler, H. P., and R. C. Surdam, 1985, Thermal evolution of coastal California with application to hydrocarbon maturation: American Association of Petroleum Geologists Bulletin, v. 69, n. 9, p. 1386–1400.

Hoskins, E. G., and J. R. Griffith, 1971, Hydrocarbon potential of northern and central California offshore, *in* Future petroleum provinces of the United States—their geology and potential: American Association of Petroleum Geologists Memoir 15, p. 212–228.

Isaacs, C. M., 1980, Diagenesis in the Monterey Formation examined laterally along the coast near Santa Barbara, California: U.S. Geological Survey Open File Report 80-606, 329 p.

Luyendyk, B. P., and J. S. Hornafius, 1987, Neogene crustal rotations, fault slip, and basin development in southern California, *in* R. V. Ingersoll and W. G. Ernst, eds., Cenozoic basin development of coastal California, Rubey Volume VI, p. 259–283.

MacKinnon, T. C., 1983, Photographic album of the Monterey Formation in the Santa Barbara–Lompoc–Santa Maria coastal areas, California: Chevron Oil Field Research Co. unpublished technical memorandum, 45 p.

MacKinnon, T. C., 1989, Origin of the Miocene Monterey formation in California: Oil in the California Monterey Formation, 28th International Geological Congress Field Trip Guidebook T311, American Geophysical Union, 50 p.

Magoon, L. B., and C. M. Isaacs, 1983, Chemical characteristics of some crude oils from the Santa Maria Basin, California, *in* C. M. Isaacs and R. E. Garrison, eds., Petroleum generation and occurrence in the Miocene Monterey Formation: Pacific Section Society of Economic Paleontologists and Mineralogists, p. 201–211.

Maxwell, J. C., 1974, Anatomy of an orogen: Geological Society of America Bulletin, v. 85, p. 1195–1204.

McCulloh, T. H., and L. A. Beyer, 1979, Geothermal gradients, *in* Geologic studies of the Point Conception deep stratigraphic test well 1 OCS-CAL 78-164, outer continental shelf, southern California, U.S.: United States Geological Survey Open File Report 79-1218, p. 43–48.

Moore, G. F., and D. E. Karig, 1976, Development of sedimentary basins on the lower trench slope: Geology, v. 4, p. 693–697.

Murata, K. J., and R. R. Larson, 1975, Diagenesis of Miocene siliceous shales, Temblor Range, California: Journal of Research, U.S. Geological Survey, v. 3, p. 553–566.

Ogle, B. A., W. S. Wallis, R. G. Heck, and E. B. Edwards, 1987, Petroleum geology of the Monterey Formation in the offshore Santa Maria/Santa Barbara areas, *in* R. V. Ingersoll and W. G. Ernst, eds., Cenozoic basin development of coastal California, Rubey Volume VI, p. 382–406.

Oil and Gas Journal, 1987, Development off California hobbled, 13 July, v. 85, n. 28, p. 26–27.

Orr, W. L., 1986, Kerogen/asphaltene/sulfur relationships in sulfur-rich Monterey oils: Organic Chemistry, Advances in Organic Chemistry 1985, Part 1, Petroleum Geochemistry, p. 499–516.

Petersen, N. F., and P. J. Hickey, 1983, Evidence of early generation of oil from Miocene source rocks, California coastal basins, *in* C. M. Isaacs and R. E. Garrison, eds., Petroleum generation and occurrence in the Miocene Monterey Formation, California: SEPM Pacific Section, p. 226.

Pisciotto, K. A., 1981, Diagenetic trends in the siliceous facies of the Monterey shale in the Santa Maria region, California: Sedimentology, v. 28, p. 547–571.

Pisciotto, K. A., and R. E. Garrison, 1981, Lithofacies and depositional environments of the Monterey Formation, California, *in* R. E. Garrison and R. G. Douglas, eds., The Monterey Formation and related siliceous rocks of California: SEPM Pacific Section, p. 97–122.

Priestaf, I., 1979, Natural tar seeps and asphalt deposits of Santa Barbara County: California Geology, v. 32, n. 8, p. 163–169.

Redwine, L. E., 1981, Hypothesis combining dilation, natural hydraulic fracturing, and dolomitization to explain petroleum reservoirs in Monterey shale, Santa Maria area, California, *in* R. E. Garrison and R. G. Douglas, eds., The Monterey Formation and related siliceous rocks of California: SEPM Pacific Section, p. 221–248.

Regan, L. J., Jr., and A. W. Hughes, 1949, Fractured reservoirs of Santa Maria district, California: American Association of Petroleum Geologists Bulletin, v. 33, n. 1, p. 32–51.

Roehl, P. O., 1981, Dilation brecciation—a proposed mechanism of fracturing, petroleum expulsion and dolomitization in the Monterey Formation, California, *in* R. E. Garrison and R. E. Douglas, eds., The Monterey Formation and related siliceous rocks of California: SEPM Pacific Section, p. 285–315.

Skillin, R. H., 1989, Monterey development-onshore and offshore Santa Maria basins, *in* T. C. MacKinnon, ed., Oil in the California Monterey Formation: 28th International Geological Congress Field Trip Guidebook T311, American Geophysical Union, 55 p.

Smith, G. W., D. G. Howell, and R. V. Ingersoll, 1979, Late Cretaceous trench-slope basins of central California: Geology, v. 7, p. 303–306.

Tissot, B. P., and D. H. Welte, 1984, Petroleum formation and occurrence, Second edition: New York, Springer-Verlag, 699 p.

Vernon, J. W., and R. A. Slater, 1963, Submarine tarmounds, Santa Barbara County, California: American Association of Petroleum Geologists Bulletin, v. 47, n. 8, p. 1624–1627.

Waples, D. W., 1987, Predicting thermal maturity, *in* N. H. Foster and E. A. Beaumont, eds., Geologic basins I, classification, modeling, and predictive stratigraphy: American Association of Petroleum Geologists Treatise of Petroleum Geology, Reprint Series, No. 1, p. 249–282.

Woodring, W. P., and M. N. Bramlette, 1950, Geology and paleontology of the Santa Maria district, California: U.S. Geological Survey Professional Paper 222, 185 p.

Appendix 1. Field Description

Field name .... *Point Arguello field*

Ultimate recoverable reserves .... *300 million bbl*

Field location:

    Country .... *U.S.A.*

    State .... *California*

    Basin/Province .... *Offshore Santa Maria basin*

Field discovery:

    Year first pay discovered .... *Miocene Monterey Formation   1981*

    Year second pay discovered .... *Upper Miocene upper Monterey Formation   1982*

    Third pay .... *Monterey Formation 1982*

Discovery well name and general location:

    First pay .... *OCS-P0316 No. 1, 8.5 mi south of Point Arguello, California*

    Second pay .... *OCS-P0450 No. 1, 4.5 mi northwest of P0 No. 1*

    Third pay .... *OCS-P0315 No. 2, 3 mi west of P0316 No. 1*

Discovery well operator .... *Chevron USA*

    Second pay .... *Chevron USA*

    Third pay .... *Texaco Inc.*

IP in barrels per day and/or cubic feet or cubic meters per day:

    First pay .... *6580 BOPD and 1680 mcf/day*

    Second pay .... *20 BOPD*

    Third pay .... *2200 mcf/day*

All other zones with shows of oil and gas in the field:

Age	Formation	Type of Show
Pliocene	*"Pico"*	*Oil and gas*
Late Miocene	*Santa Margarita (lower Sisquoc)*	*Oil and gas*
Cretaceous	*Jalama*	*Oil*

Geologic concept leading to discovery and method or methods used to delineate prospect, e.g., surface geology, subsurface geology, seeps, magnetic data, gravity data, seismic data, seismic refraction, nontechnical:

Seismic mapping revealed a deep untested anticlinal trend. There were two major exploration objectives, the fractured Monterey shale and the Vaqueros sandstone. Onshore geologic mapping indicated the Oligocene age Vaqueros sandstone was a potential objective at Point Arguello. Regional stratigraphic mapping suggested the Monterey would be in the quartz phase and highly fracturable. Subsurface temperature studies also indicated that producible Monterey oils might be present creating an economically attractive prospect.

Structure:

    Province/basin type .... *Bally 332; Klemme III B b*

    Tectonic history

An arc-trench system developed in Late Jurassic time. During the late Oligocene, a spreading center-trench collision occurred. Wrench faulting began, producing pull-apart basins along the continental margin. From the early Miocene through the early Pliocene, the Point Arguello subbasin developed. The Point Arguello anticlinal structure began forming during the late Miocene and grew until the late Pliocene in response to wrench faulting along the Hosgri fault system.

    Regional structure

Three to four miles southwest of the Hosgri fault system.

    Local structure

The Point Arguello field is formed from two faulted, doubly plunging, northwest-trending anticlines.

Trap:

Trap type(s)

Two known anticlinal traps composed of a single pay with closure increased by sealing faults.

Basin stratigraphy (major stratigraphic intervals from surface to deepest penetration in field):

Chronostratigraphy	Formation	Depth to Top in ft*
Pliocene	*Pico Formation (upper Sisquoc)*	*0–1000*
Miocene-Pliocene	*Santa Margarita (lower Sisquoc)*	*3000–4000*
Middle-upper Miocene	*Monterey*	*6000–8000*
Middle Miocene	*Point Sal (lower Monterey)*	*7800–10,000*
Lower Miocene	*Tranquillon Volcanics*	*12,000–13,000*
Lower Eocene	*Juncal(?)*	*7500–9000*
Upper Cretaceous	*Jalama*	*7900–14,000*

**From ocean bottom.*

Reservoir characteristics:

Number of reservoirs

Formations

Ages .... *Middle to upper Miocene*

Depths to tops of reservoirs .... *6000–8000 ft subsea*

Gross thickness (top to bottom of producing interval) .... *500–2800 ft*

Net thickness—total thickness of producing zones

    Average .... *700 ft*

    Maximum .... *1800 ft*

Lithology .... *Finely laminated siliceous cherts, porcellanites, siltstones, mudstones, and dolostones*

Porosity type .... *Microporosity and fracture porosity*

Average porosity .... *15% (fracture porosity = 1–2%*

Average permeability .... *Highly variable (0.1–3000+ md)*

Seals:

Upper

    Formation, fault, or other feature .... *Santa Margarita (lower Sisquoc) Formation*

    Lithology .... *Predominantly light-gray, diatomaceous mudstones and occasional dolostones*

Lateral

    Formation, fault, or other feature .... *Anticlinal reversal, high-angle reverse faults and fracture variabilities*

    Lithology .... *NA*

Source:

Formation and age .... *Monterey and Point Sal (lower Monterey) formations (early to late Miocene)*

Lithology .... *Laminated, organic-rich, siliceous shales and phosphatic mudstones*

Average total organic carbon (TOC) *3.0%*

Maximum TOC *18%*

Kerogen type (I, II, or III) .... *IIS (sulfur rich)*

Vitrinite reflectance (maturation) .... *NA; primary vitrinite is very rare in the Monterey section*

Time of hydrocarbon expulsion .... *Late Miocene to present*

Present depth to top of source .... *6000–11,000 ft subsea (1829–3354 m)*

Thickness .... *500–5300 ft (152–1616 m)*

Potential yield .... *400–600 bbl/ac-ft (avg.)*

Appendix 2. Production Data

Field name .... *Point Arguello*

Field size:

    Proved acres .... *6000 (2428 ha)*

    Number of wells all years .... *9 delineation wells (P&A);*
production planned from 3 drilling platforms (up to 96 wells)

    Current number of wells .... *39 wells drilled to date; suspended until permitting problems resolved*

    Well spacing .... *160 and 80 ac*

    Ultimate recoverable .... *Approx. 300 million bbl*

    Cumulative production .... *None*

    Annual production .... *Shut in*

    Present decline rate .... *NA*

        Initial decline rate .... *Unknown; estimated from onshore Monterey production*
to be about 45% in first year

        Overall decline rate .... *Expected to be around 15%/year*

    Annual water production .... *NA*

    In place, total reserves .... *2.2–2.5 billion bbl*

    In place, per acre-foot .... *400–500+ bbl/ac-ft*

    Primary recovery .... *300 million bbl*

    Secondary recovery .... *Under study*

    Enhanced recovery .... *Under study*

    Cumulative water production .... *NA*

Drilling and casing practices:

    Amount of surface casing set .... *500 ft below mudline*

    Casing program .... *24-in. conductor to 1150 ft; 18⅝-in. to 1800 ft; 13⅜-in. to 3000 ft;*
9⅝-in. to T/Monterey Formation and 7-in. to T.D.

    Drilling mud .... *Low solid, nondispersant gel*

    Bit program .... *Conventional rock bit*

    High pressure zones .... *None*

Completion practices:

    Interval(s) perforated .... *Upper Monterey Formation*

    Well treatment .... *Selected perforations through casing or liner*

Formation evaluation:

    Logging suites .... *Dual induction, neutron density, sonic, NGT, EPT dipmeter, FMS, mud log, and*
measurement-while-drilling (MWD) log

    Testing practices .... *Tests are normally conducted behind casing over a 50 to 200 ft interval;*
after perforation, packer is set and drill-stem test conducted

    Mud logging techniques .... *FID chromatograph; standard procedures*

Oil characteristics:

    Type .... *Aromatic-intermediate class*

    API gravity .... *11–34°*

    Base .... *11° (Southern Pool), 31° (Northern Pool)*

    Initial GOR .... *400 SCF/bbl*

    Sulfur, wt% .... *0.8% (30° API oils) to 5% (11–23° API oils)*

    Viscosity, SUS .... *1–4 cp at 250°F (avg.)*

    Pour point .... *15–30°F (avg.)*

    Gas-oil distillate .... *Not present*

Field characteristics:

Average elevation .... *Water depths from 400 to 1100 ft*
Initial pressure .... *3400 psi*
Present pressure .... *Same as above; no production to date*
Pressure gradient .... *0.465 psi/ft*
Temperature .... *230-240°F at T/Monterey*
Geothermal gradient .... *2.5-2.8°F/100 ft*
Drive .... *Gas solution, gravity drainage*
Oil column thickness .... *1800 ft*
Oil-water contact .... *8235 ft subsea (Southern Pool), 8150 ft subsea (Northern Pool)*
Connate water *50%*
Water salinity, TDS .... *13,000-26,000 ppm Cl (highly variable)*
Resistivity of water .... *0.3 ohms at 75°F*
Bulk volume water (%) .... *7.5% (approx.)*

Transportation method and market for oil and gas:
Planning for transport by pipeline to serve California market.

West Puerto Chiquito—U.S.A.
San Juan Basin, New Mexico

ALBERT R. GREER
Benson-Montin-Greer Drilling Corp.
Farmington, New Mexico

RICHARD K. ELLIS
Leede Exploration
Englewood, Colorado

FIELD CLASSIFICATION

BASIN: San Juan
BASIN TYPE: Cratonic Sag
RESERVOIR ROCK TYPE: Shale (Fractured)
RESERVOIR AGE: Cretaceous
PETROLEUM TYPE: Oil
TRAP TYPE: Fractured Reservoir
RESERVOIR ENVIRONMENT OF DEPOSITION: Marine
TRAP DESCRIPTION: Fracture system isolated from outcrop by calcite vein-filling and without connection to water

LOCATION

The West Puerto Chiquito field is located in Rio Arriba County, New Mexico, approximately 80 mi (130 km) southeast of the town of Farmington. The field is situated on the southeast flank of the San Juan basin (Figure 1). Surface use is administered by the Santa Fe National Forest, Bureau of Land Management, Jicarilla Indian Reservation, and various fee interests. Elevations range from 7000 to 8600 ft (2134–2621 m) in rugged canyon and mesa topography.

Immediately to the west lie Gavilan Mancos (oil), Gavilan Pictured Cliffs (gas), and West Lindrith Gallup–Dakota (oil) fields, while Boulder Mancos (oil) and East Puerto Chiquito Mancos (oil) fields are located north and northeast, respectively (Figure 2).

Depending on the future method of depletion, particularly as to the degree that pressure maintenance can continue to be utilized, ultimate recovery of West Puerto Chiquito is expected to be 15 to 20 million bbl of oil and 15 to 20 bcf of gas.

HISTORY

Early Exploration

Early exploration for fractured Niobrara oil reservoirs in the San Juan basin focused on the Hogback monocline, which forms the basin margin on the west, north, and east sides. Verde field, discovered by C. M. Carroll in October 1955, is located on the monocline on the west side of the basin. The pool was initially developed on 80 ac spacing and later partially infilled to 40 ac spacing with no apparent increase in recovery. Verde produced approximately 7.5 million bbl of oil from fractures formed in the Niobrara shale at depths ranging from 2000 to 4500 ft (609–1372 m). East Puerto Chiquito field (Figure 2), discovered by Intex Oil in February 1960, was developed by Benson-Montin-Greer Drilling Corp. on 160 ac spacing and has produced approximately 4 million bbl of oil at depths ranging from 1500 to 4000 ft (457–1219 m). East Puerto Chiquito produces from fractures developed in the Niobrara on a northwest-plunging nose crossing the monocline on the east side of the basin. Boulder field (Figure 2) was discovered by P-M Drilling Co. in May 1961 and has produced approximately 1.7 million bbl of oil on 80 ac spacing at depths ranging from 3400 to 4500 ft (1036–1372 m). Boulder produces from the fractured Niobrara interval along the monocline bounding the east side of the basin.

All three fields benefited from gravity drainage, which significantly increased recovery. Gravity drainage occurred because of shallow depths and the resulting low volume of gas in solution, steep dips, and low withdrawal rates (prorated allowables at Verde and Boulder and voluntary restrictions at East Puerto Chiquito).

Despite efficient recovery under gravity drainage, Boulder was overdrilled on 80 ac spacing, resulting in poor economics. As a consequence, major oil companies discontinued Niobrara exploration on the east side of the San Juan basin.

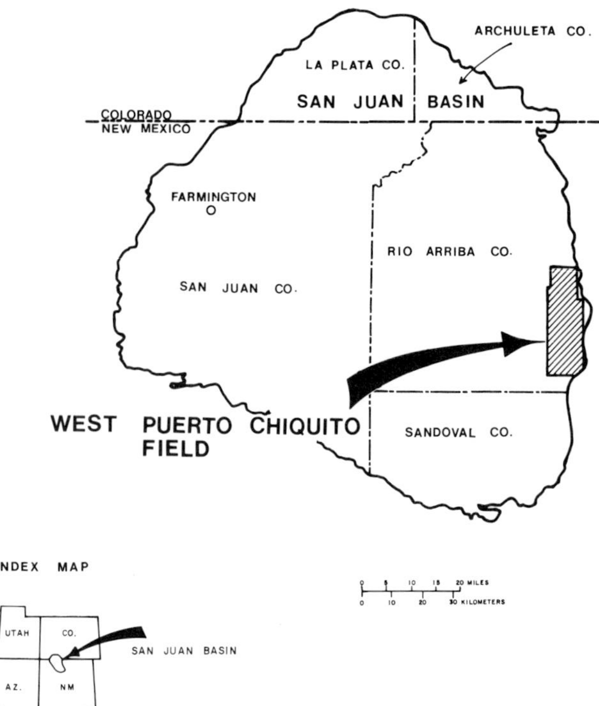

Figure 1. Location map of the San Juan basin showing the West Puerto Chiquito field.

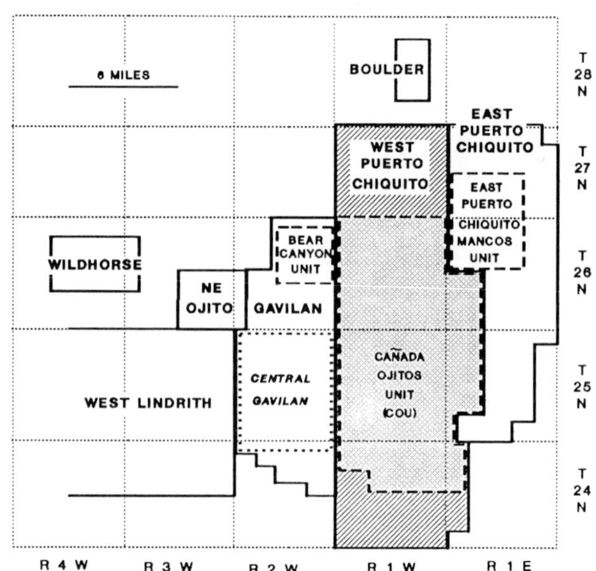

Figure 2. Map showing the locations of fractured Niobrara pools on the east side of the San Juan basin.

Field Discovery

West Puerto Chiquito field was discovered by Bolack and Greer in July 1963 at the Canada Ojitos Unit (COU) No. 2 (K-13; NE SW Sec. 13, T25N, R1W). Drilled to a total depth of 6022 ft (1835 m), the "C" zone of the Niobrara (Figure 3) was hydraulically fractured with 111,000 lb of sand and 85,620 gal. of oil in an open-hole interval from 5976 to 6022 ft (1821–1835 m). The well was completed on pump for 95 BOPD. The initial reservoir pressure was 1620 psig at a datum of +1195 ft (364 m). The discovery well was drilled to evaluate a large acreage block and test fractured reservoir development along a synclinal flexure parallel to the Hogback monocline.

Post-Discovery Activity

Cores from wells drilled early in the life of the West Puerto Chiquito Pool indicated the producing interval to contain no effective matrix porosity (Appendix 1). Production and interference testing (Appendices 2 and 3) confirmed that all porosity was fracture porosity and that per-acre volumes of oil in place were low: 1500 to 2000 bbl/ac. Although production capacities of some of the wells were high (1000 to 3000 BOPD per well), the solution gas drive recovery of this fractured reservoir was expected to be low. This coupled with the small volume of oil in place would result in very low primary recoveries, in the range of 100 to 150 bbl of oil/acre. In contrast, Boulder field recovery, supplemented by gravity

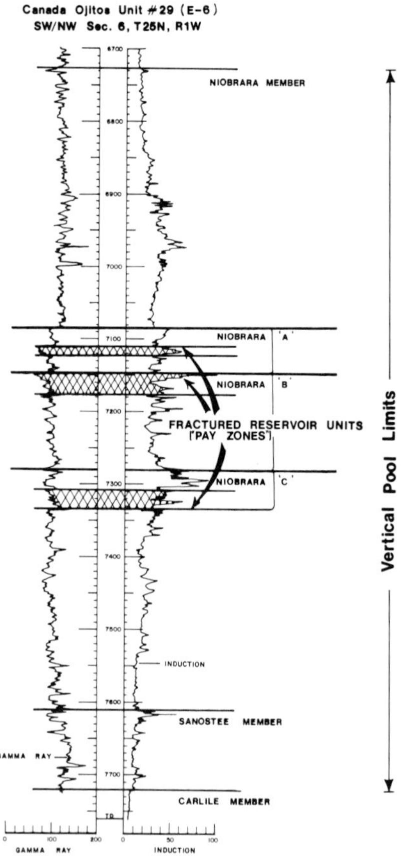

Figure 3. Type log showing units within the Niobrara. Fractured reservoir units, or "pay zones," are found in the "A," upper "B," and lower "C" submembers. The gross reservoir thickness is approximately 250 ft, while "net pay" is in the range of 30 to 100 ft.

drainage, was 600 to 800 bbl of oil/acre (NMOCC, 1988).

Given the lower dips, greater depths, greater volume of gas in solution, and higher shrinkage, "conventional" exploitation at West Puerto Chiquito would yield only solution gas drive recovery: No significant gravity drainage would occur here, despite its contribution in other fields. Furthermore, a reservoir of this type does not lend itself to efficient and economic development under competitive operations: Overdrilling and excessive production rates are a constant threat. Such conventional development would maximize current income but would leave large volumes of oil unrecovered, and the discounted present worth of the recoverable resource would be far less than its potential.

Therefore, a unique plan of development was implemented to achieve gravity drainage recoveries through pressure maintenance, wide spacing, and reservoir management. Reservoir management could only be accomplished through control of spacing and production—objectives that could not reasonably be achieved through conventional pool rules and spacing regulations under competitive operations.

In an unprecedented action, the New Mexico Oil Conservation Commission (NMOCC) approved Benson-Montin-Greer Drilling Corp.'s application to establish an oil pool covering four townships after only a few wells were drilled. To achieve adequate reservoir management, the operator formed a large federal exploratory unit (Canada Ojitos) covering most of the pool. Through a series of hearings before the NMOCC, and as reservoir information was developed, spacing was increased from 40 ac/well to 160, from 160 to 320, and finally from 320 to 640 ac. Development within the unit, however, proceeded on much wider spacing. The initial density of recovery wells approximated one well per four "downstructure" sections such that the drainage geometry for a particular wellbore was approximately 1 mi (1.6 km) in the strike direction and 4 mi (6.4 km) in the dip direction.

Development updip revealed a gas cap area of apparent low porosity and permeability. The Unit Agreement was modified to allow into participation noncommercial lands for gas injection purposes and the unit was expanded to include these lands. Low productivity wells (5 to 10 BOPD and 5 to 10 MCFG/day) were converted to useful gas injectors with injection rates of 1 to 3 million ft^3/day/well.

Productivity of recovery wells remained high because of the pressure maintenance and wide spacing. Efficient recoveries of large volumes of oil resulted—as much as 2.3 million bbl/well. Low capacity wells also maintained producing rates at proportionately high levels and accumulated relatively high recoveries: 120,000 bbl from wells with initial productivities of 25 BOPD and 250,000 bbl for wells with initial rates of 50 to 70 BOPD.

Expansion of the initially developed unit area was deterred in the 1970s by government-mandated oil price controls. Any new wells drilled in the unit would receive "old" oil prices while new wells outside a designated unit qualified for "new" oil prices. A special order sought from the Department of Energy permitting upper tier prices for wells drilled on expansion lands of this particular unit was entered in January 1981. Westward expansion of the unit was then undertaken through an initial five-well program.

Overdrilling offsetting the west boundary of the unit occurred in the 1980s and required the drilling of additional wells to protect against drainage. Most of these wells are unnecessary for the recovery of reserves in West Puerto Chiquito. As a consequence of drilling in the 1980s there was an increase in production rate and acceleration of depletion (Table 1). Spacing of wells in West Puerto Chiquito on the west boundary is approximately 640 ac/well.

New Fields

The Gavilan Mancos Pool, offsetting West Puerto Chiquito (Figure 2), was developed in the mid-1980s. Spacing was initially on temporary 320 ac/well but was later changed to 640 ac with an optional second well. Because of the close spacing and high allowables, central Gavilan was essentially oil depleted

Table 1. West Puerto Chiquito production. Note that from August 1968 through August 1987, all Canada Ojitos Unit produced gas was reinjected and 3.1 bcf of nonunit supplemental gas was acquired and injected. Gas has occasionally been marketed since 1 August 1987.

Year	Oil Production Barrels	Gas Production (Mcf)
1962	1,426	0
1963	31,601	7,268
1964	159,135	52,146
1965	238,942	80,158
1966	340,141	127,269
1967	347,789	189,291
1968	396,511	192,896
1969	519,922	230,108
1970	732,118	230,025
1971	769,573	290,146
1972	638,160	306,136
1973	511,566	487,041
1974	438,692	554,186
1975	321,875	525,000
1976	309,652	501,981
1977	315,158	519,816
1978	262,117	602,288
1979	239,031	550,477
1980	236,599	466,252
1981	237,877	535,511
1982	227,295	525,400
1983	222,455	495,013
1984	220,750	481,476
1985	379,405	449,841
1986	849,776	697,710
1987	851,195	1,499,595
1988	852,058	2,327,924
	10,650,819	12,924,954

five years after initial development (two years after substantial development). At that time and with few exceptions, all wells were in stripper oil status.

Tests of Other Zones

Pictured Cliffs Sandstone—Wells drilled to the Pictured Cliffs Sandstone (Figure 6) generally have high water saturations and produced low volumes of gas.

Mesa Verde—Cores of the Mesa Verde Formation indicate that it is probably water saturated.

Dakota Formation—Three wells have tested the Dakota Formation: one showed only water; a second approximately 100 MCFG/day with about 10 BOPD distillate and 10 BWPD; the third about 2 BOPD distillate and 2 BWPD and 40 to 50 MCFG/day.

DISCOVERY METHOD

The West Puerto Chiquito discovery well was drilled to test fractured reservoir development on a large synclinal flexure parallel to the Hogback monocline. Previous activity showed that commercial oil production could be developed in certain zones in the Niobrara in places where the monocline was tectonically fractured. Development to the west in the Lindrith Gallup–Dakota Pool, to the northeast in the East Puerto Chiquito Pool, and in an untested well drilled in 1952 3 mi (4.8 km) northwest of the discovery well indicated the Niobrara reservoir interval was present. In spite of the favorable geologic indications, the initial well and subsequent development wells were high risk, since economic success in the Niobrara play is dependent on wellbore communication with a high conductivity (tectonic) fracture system.

STRUCTURAL CHARACTERIZATION

The San Juan basin is located in the Colorado Plateau structural province. The basin is an asymmetric synclinal depression with a steep northern rim and a gently dipping southern flank. The basin is bounded by several major structural elements: to the east by the Archuleta anticlinorium and Nacimiento uplift, to the south and southwest by the Puerco fault zone and Chaco slope, to the southwest and west by the Zuni and Defiance uplifts, to the northwest by the Four Corners platform, and to the northeast by the San Juan Mountains. The central portion of the basin is bounded by the Hogback monocline on the west, north, and east sides (Figure 5).

Tectonic History

Although regional stratigraphic relations indicate present-day positive features such as the Archuleta anticlinorium, Nacimiento uplift, and Zuni uplift have an ancestral (pre-Laramide) component of structural development, Laramide tectonic forces operative in the Colorado Plateau are primarily responsible for the present configuration of the San Juan basin and its bounding structural elements. Subsidence began in Late Cretaceous time, prior to deposition of the Ojo Alamo Sandstone. The Archuleta anticlinorium and Hogback monocline on the east side of the basin were positive elements. During the Paleocene, the Ojo Alamo, Nacimiento, and Animas formations (Figure 6) were deposited in a depositional basin including the present-day San Juan and Chama basins and the Nacimiento and Archuleta uplifts (Baltz, 1967).

Laramide compressional forces operative in the Cordilleran structural province were transmitted into the Colorado Plateau, resulting in northeastward shift (Woodward and Callendar, 1977). Structural intensity peaked during late Paleocene–early Eocene: Subsidence of the central basin was at a maximum, the Nacimiento uplift was folded sharply, the Hogback monocline formed along the basin margins, and en echelon, northwest-trending folds developed on the eastern edge of the basin.

Deposition of the synorogenic San Jose Formation on the flanks of bounding uplifts preceded further northeastward shift of the Colorado Plateau. This post-San Jose (Eocene or Oligocene) deformational event resulted in wrench faulting along the Nacimiento and Gallina uplifts and uplift and tilting of all San Jose and older rocks along the eastern margin (Baltz, 1967). Late Cenozoic (Miocene or Pliocene) extensional faulting observed in the Puerco fault zone and along the Archuleta anticlinorium is thought to have reactivated earlier Laramide faulting and to be related to the development of the Rio Grande rift (Woodward and Callendar, 1977).

Regional Structure

West Puerto Chiquito field is located on the southeast basin flank, immediately north of the synclinal axis of the basin and along the west-dipping monocline. Crossing the monocline are several northwest-trending folds. Notable among these are the East Puerto Chiquito anticlinal nose, which is the locus for the East Puerto Chiquito field, and a prominent nose formed at the intersection of the Nacimiento and Gallina faults, which is the location of recent drilling activity in the southeastern portion of the West Puerto Chiquito field.

Field Structure

The structural geometry of the West Puerto Chiquito reservoir includes three genetically related structural elements: from east to west, the west-dipping monocline, a north-trending syncline, and a north-plunging anticlinal nose (Plate 1). Subsurface and outcrop data reveal no significant faulting or vertical displacement is present at reservoir level. Dips at the outcrop of the Niobrara, approximately 1 to 3 mi (1.6–4.8 km) east of the limits of the pool, range from 53 to 59°. Dips decrease rapidly to the west, varying from 6° in the gas injection (updip) portion of the reservoir to essentially flat in the synclinal trough marking the west boundary of the pool (Figure 7). The average dip in the reservoir is approximately 2° or 185 ft/mi. The prominent, north-plunging fold immediately west of the pool boundary—the "Gavilan nose"—exhibits low relief and dips in the range of 1° and is the locus for fractured Niobrara production at Gavilan field.

A map constructed on the top of the Niobrara reservoir interval reveals a clearly defined synclinal trough at the base of a monoclinal panel of dip (Plate 1). In general, the highest capacity wells are located at points of synclinal flexure (Gorham et al., 1979) along the monocline, and along major fracture trends.

Tectonic and Regional Fracture Orientation

Landsat, aerial photography, and geophysical methods provide evidence of large-scale fracture orientations at West Puerto Chiquito. The dominant (surface) fracture trends comprise a conjugate set oriented approximately northwest-southeast and northeast-southwest (Figure 8). Fracture orientations and overall "fabric" change dramatically across the (west northwest-east southeast) Lleguas fracture zone (northern portion of T25N), implying the presence of a rotational component. Several prominent northeast fracture trends are truncated by the Lleguas zone, and trends north of the zone undergo a pronounced northward shift in azimuth. Surface analysis indicates tectonic fracturing is present to some degree over all three structural elements—monocline, syncline, and anticlinal nose—localizing production in West Puerto Chiquito and Gavilan fields, a total area of approximately 150,000 ac. Electromagnetic surveys conducted in specific areas of the field indicate fractures are vertical to near-vertical in inclination.

The extrapolation of the surface fracture interpretation to reservoir depths involves difficult assumptions concerning timing of fracture development and mechanical response of rock units of varying composition and thickness. For example, the fractures cannot be throughgoing in any effective sense, since the vertical limits of production are confined to a few tens of feet. Pressure transient analyses reveal a reservoir geometry of "fracture blocks" bounded by high conductivity fractures (Appendix 3), which is confirmed by surface data and electromagnetic surveys. Therefore, Landsat-photo analysis is a useful exploratory tool.

STRATIGRAPHY

Regional Reservoir Considerations

The San Juan basin produces commercial volumes of oil and gas from structural and stratigraphic accumulations in reservoirs of Pennsylvanian, Jurassic, Tertiary, and, most important, Cretaceous age. A time-stratigraphic nomenclature chart for the San Juan basin is shown by Figure 6. The Pennsylvanian Paradox Formation produces oil and gas from cyclic sequences of restricted marine carbonates on the Four Corners platform in the northwest portion of the basin. Low-energy algal bank complexes in the shelf carbonate sequences form excellent stratigraphic traps. The Jurassic Entrada Sandstone produces oil from sand reservoirs of eolian origin occupying paleotopographic "highs" developed beneath the sealing organic limestones of the Todilto Formation.

The Upper Cretaceous formations represent a classic cyclical sequence typical of deposition in the Late Cretaceous western interior. The northwest-southeast depositional trends of the major units, and the existence of a prominent stratigraphic "rise" during the major transgressions and regressions, are the result of complex interplay between a shallow seaway to the northeast and a clastic sediment supply source to the southwest. Five major transgressions and regressions of the shoreline, and many minor ones, left a stratigraphic sequence that created ideal trapping mechanisms for large oil and gas reserves (Figure 4).

The Upper Cretaceous Dakota Formation represents the initial transgressive sand in the sequence depicted in Figure 4. Sandstone reservoirs in the Dakota produce gas and oil and are subdivided into four members. Basal units in the Dakota are generally fluvial in origin and therefore of erratic reservoir quality. Progressively younger sand units in the Dakota become more marine, with correspondingly better reservoir characteristics.

The Gallup sandstone, of Carlile age, produces oil on the Chaco slope, is the first regressive wedge in the basin, and is unique, having no equivalents elsewhere. The regressive Gallup sands are to be clearly distinguished from the transgressive, Niobrara-age sands ("transgressive Gallup," "Tocito," "Stray," or "Bisti") that are elongate, northwest-trending offshore bars. The bar sands, which are younger and seaward of the regressive Gallup, produce oil in several large fields in the central portion of the basin.

The Niobrara Member represents a major transgressive event that has equivalents throughout the western interior. Substantial dolomite contents,

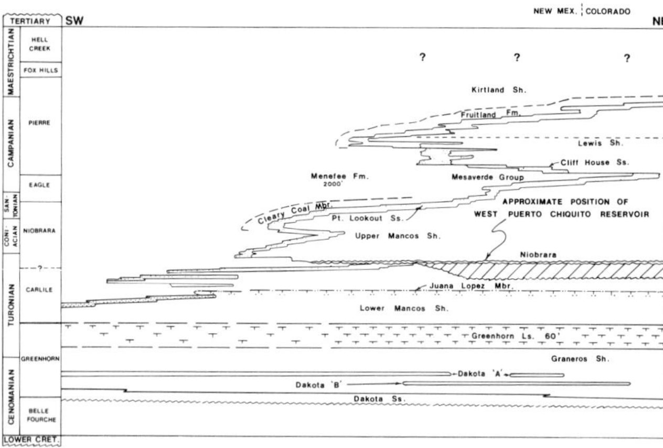

Figure 4. Upper Cretaceous time/stratigraphic section of the San Juan basin showing the stratigraphic position of the Niobrara reservoir interval at West Puerto Chiquito. (After Molenaar, 1977a.)

particularly in the highly laminated siltstone-carbonate sequences, give rise to the resistive character noted on logs (Figure 3) and impart the brittle nature responsible for fractured reservoir development on the basin flanks (London, 1972).

The Mesaverde Group consists of the Point Lookout, Menefee, and Cliffhouse formations. The Point Lookout is a regressive, coastal barrier sandstone that exhibits total northeast stratigraphic rise, over 130 mi (210 km), of approximately 1200 ft (365 m). The rise is responsible for the large, northwest-trending stratigraphic traps that produce gas throughout the central part of the basin. The Menefee is a lagoonal, back-barrier sequence of shale, coal, and fluvial sands that were deposited landward of the Point Lookout and Cliffhouse shorelines. The Cliffhouse is a transgressive sandstone developed during the overall Lewis transgression. The sands clearly intertongue with the marine Lewis shale to the northeast and the nonmarine Menefee to the southwest. A total stratigraphic rise, from northeast to southwest, of approximately 1300 ft (396 m) formed stratigraphic traps that are gas-productive through-out the basin.

The Upper Cretaceous Pictured Cliffs is a marine sandstone deposited in a variety of nearshore environments during the final regression of the Late Cretaceous sea. Prominent stratigraphic rise of approximately 1100 ft (335 m) is in large part responsible for the northwest-trending stratigraphic accumulations of gas in the central portion of the basin.

The Fruitland Formation is a sequence of nonmarine swamp and alluvial plain deposits laid down shoreward of the sea as it retreated from the basin for the last time. Fruitland Formation coalbeds and sandstones produce gas over a large percentage of the northern and central portions of the basin.

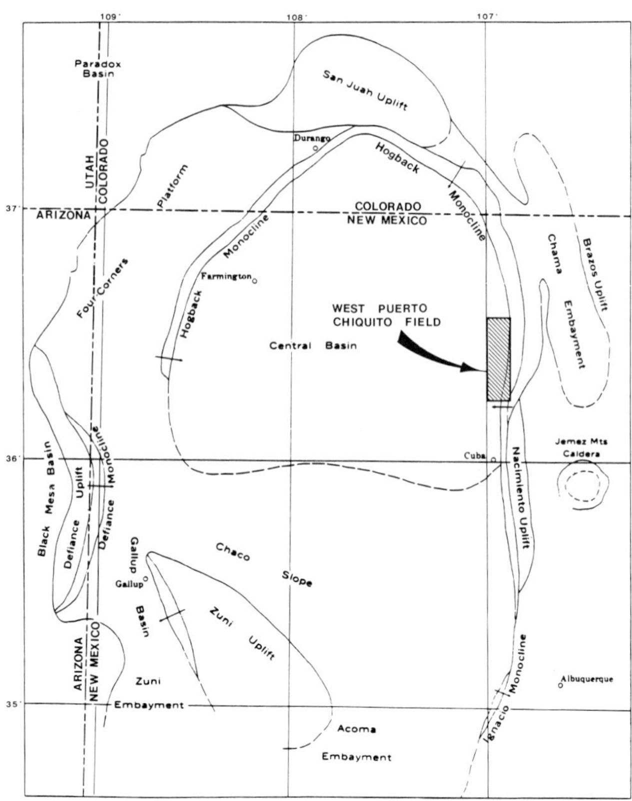

Figure 5. Map showing the location of the major tectonic features in the San Juan basin and surrounding areas.

Potential Source Rocks

Source-reservoir mechanisms have been postulated for the principal reservoirs in the San Juan basin. The Pennsylvanian Paradox carbonate reservoirs produce oils derived from the sapropelic, oil-generative shale units in the carbonate-evaporite sequence. Jurassic Entrada oils are thought to be sourced by organic limestones of the overlying Todilto Formation, which also provide the critical reservoir seal (Vincelette and Chittum, 1981).

Hydrocarbons produced from the upper Cretaceous Dakota sands are sourced by the intertonguing shales. Oil-generative marine shales in the Carlile and Niobrara members provide the source for oil produced in the regressive Gallup sand and younger basal Niobrara bar sands. Fractured reservoirs in the Niobrara shale section above the bar sands are sourced by indigenous organic-rich marine shales (see below).

Mesaverde sand reservoirs produce hydrocarbons sourced by shales of the Mancos and Lewis formations, as well as shale and coal units within the interval. Gas in the Pictured Cliffs and Fruitland reservoirs is derived from substantial volumes of methane expelled from coalbeds and shales in the Fruitland Formation.

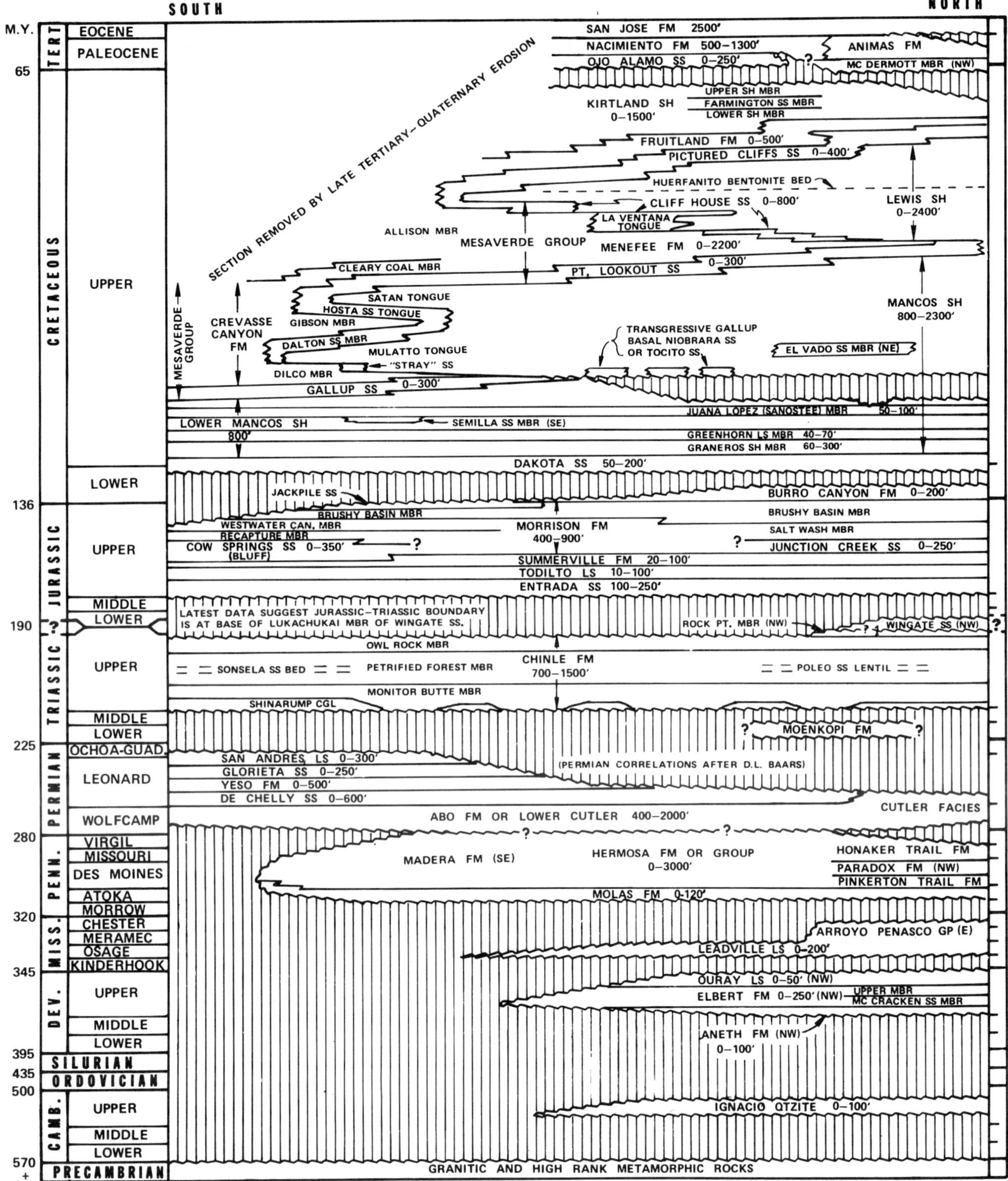

Figure 6. San Juan basin time-stratigraphic nomenclature chart. (From Molenaar, 1977b.)

TRAP

West Puerto Chiquito field is an areally extensive fracture trap of essentially indeterminate extent. Trapping of hydrocarbons at West Puerto Chiquito is provided by a loss of fracture permeability, both in the lateral and vertical directions. Although brittle reservoir units within the Niobrara interval are fractured to some extent over a large part of the basin flank, commercial production rates are dependent on

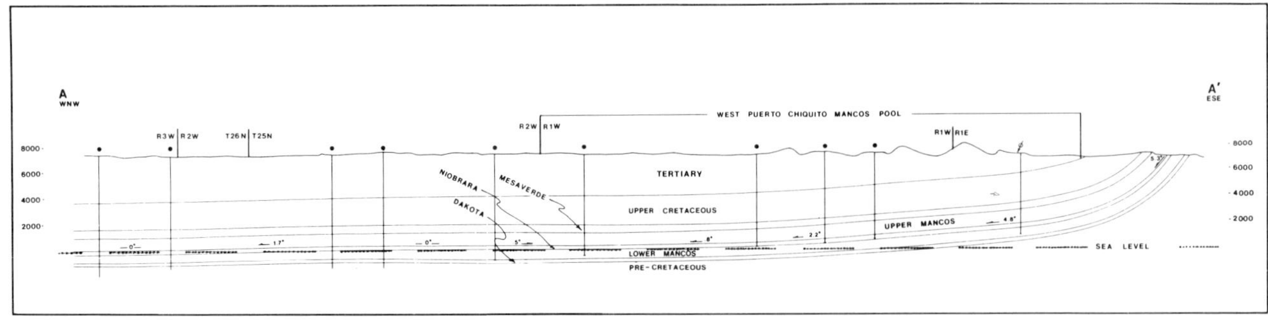

Figure 7. Structural cross section through West Puerto Chiquito field. Location of section is shown on Plate 1. No vertical exaggeration.

wellbore communication with a high-conductivity tectonic fracture system. The tectonic fracture system has been found to diminish in intensity and capacity away from the monocline (i.e., to the west), so that the "trap" is essentially a loss of commercial reservoir quality. Loss of commercial reservoir quality is known, through development and extensive reservoir study, to occur at approximately the current pool boundaries.

Vertical sealing of the reservoir is provided by massive shales of the upper and lower Mancos Shale above and below the Niobrara reservoir interval, respectively. These more ductile lithologies, though certainly fractured to some extent, are effectively "healed" and prevent the escape of significant hydrocarbons. Within the Niobrara reservoir interval, the brittle reservoir units are separated by more massive (and therefore ductile) lithologies. Vertical communication between reservoir units is limited to the area around some wellbores and along major fractures.

The West Puerto Chiquito "trap" exhibits no structural closure (Plate 1) and produces oil and gas over a vertical column of approximately 2100 ft (640 m). The reservoir has produced no water during its producing history. Trap integrity is attested to by the fact that the updip limits of the reservoir are within 1 to 3 mi (1.6–4.8 km) of the outcrop (Plate 1 and Figure 7) with no apparent hydrocarbon leakage or water incursion. Furthermore, elevation of the updip limit of the reservoir is approximately 1200 ft (366 m) structurally low to the oil-water contact in the East Puerto Chiquito field.

Although no communication has been observed between widely separated Niobrara reservoirs on the east flank during the history of production, there is nevertheless sufficient permeability in the fracture system to have equalized pressures on the east flank over geologic time (Figure 9). Thus, the original pressure observed in the West Puerto Chiquito reservoir, 1620 psig at a datum of +1195 ft (364 m), corresponds to a pressure gradient of approximately 0.33 psi/ft and reflects an "oil static" pressure differential from the datum elevation of the well to

an elevation approximately 900 ft below the outcrop, or approximately 6100 ft (1859 m) above sea level (Figure 10).

Wells in a newly developed area that show stabilized reservoir pressures less than the above gradient are indications that regional migration has taken place as a consequence of fluid withdrawal from the area.

Little or no vertical displacement has been noted along major fractures in the reservoir. The major component of movement is horizontal shear associated with the tectonic forces responsible for basin subsidence. Fractures constitute a conduit rather than a barrier for subsurface fluid flow. Minor displacement faulting (50 to 100 ft [15–30 m] of throw) has been observed in the Niobrara reservoir interval in the outcrop and subsurface (using seismic data) in the southeastern portion of the West Puerto Chiquito Pool (Plate 1).

Reservoir

Stratigraphy, Lithology, and Depths

The West Puerto Chiquito reservoir consists of a highly interconnected fracture network developed in discrete reservoir units in the Niobrara Member of the Mancos Shale. The Upper Cretaceous Niobrara Member represents a major transgressive event with equivalents throughout the western interior and was deposited on an open marine shelf under low energy conditions (Cole et al., 1989).

The Niobrara reservoir interval, in local nomenclature the "A," "B," and "C" units (Figure 3), consists of a laminated sequence of shale, siltstone, mudstone, and carbonate. X-ray diffraction, petrographic, and SEM analyses suggest the reservoir lithology is predominantly a calcareous silty claystone, with percentages of quartz and carbonate (predominantly calcite and dolomite cements) ranging from 30 to 49% of the bulk volume (Cole et al., 1989). Log data suggest a high degree of stratigraphic continuity for individual units in the Niobrara throughout the southeast portion of the San

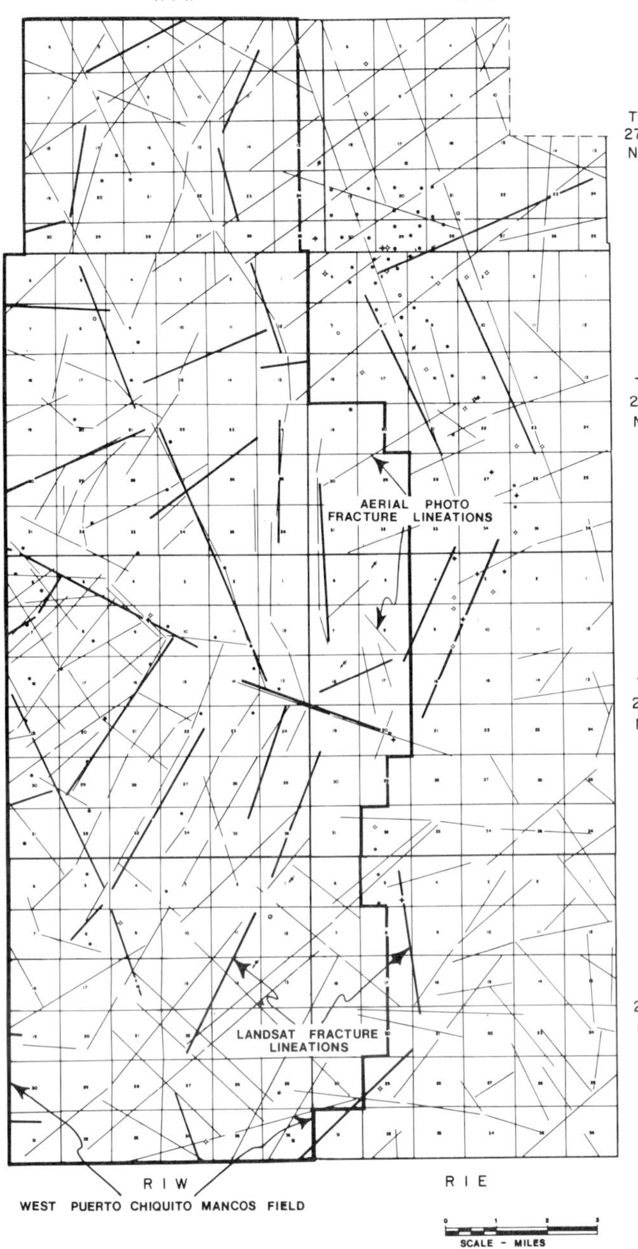

Figure 8. Map showing surface fracture interpretation of West Puerto Chiquito field. Dominant fracture "fabric" has orientations of northwest-southeast and northeast-southwest. Constructed from aerial plats and Landsat images.

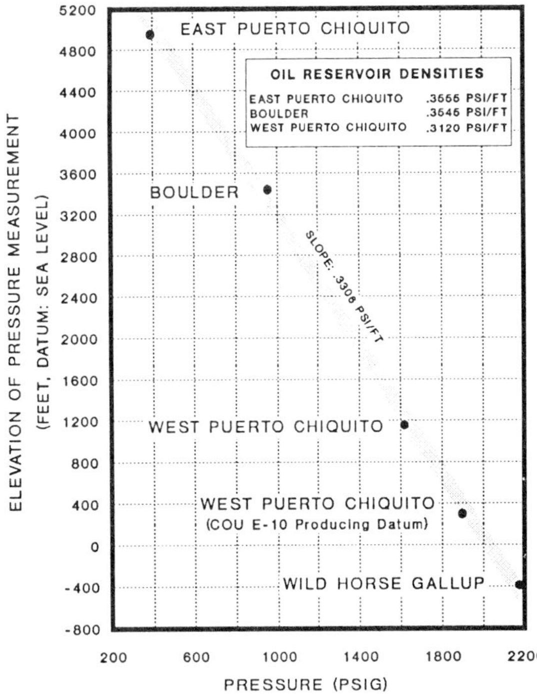

Figure 9. Plot showing initial pressures of Niobrara pools, southeast flank of San Juan basin, related by oil pressure gradient.

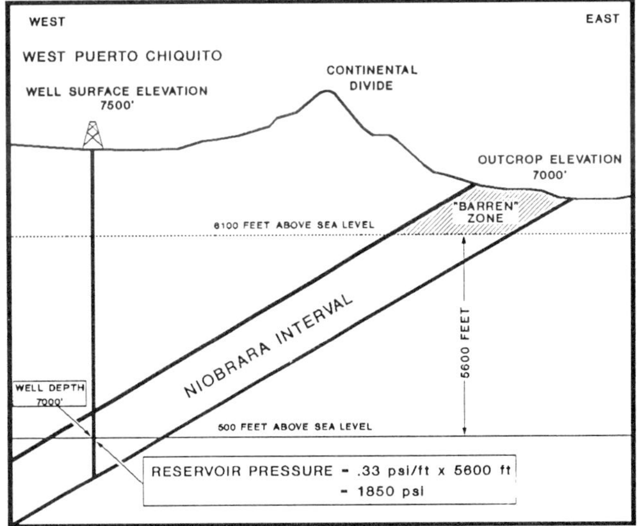

Figure 10. Sketch showing "oil static" equilibrium of pressures in the Niobrara from a "base-level" of 6100 ft above sea level. (See Figure 9.)

Juan basin. Detailed visual examination of cores, however, reveals pronounced horizontal and vertical heterogeneity, particularly in the highly laminated siltstone-carbonate sequences. Individual laminae are quite discontinuous on a scale of millimeters to centimeters, both vertically and horizontally. The laminated sequences range in thickness from 10 to 30 ft (3–10 m), and are characterized by the high resistivity response noted on logs of the Niobrara.

Certain compositional and textural aspects of the reservoir interval restrict the development of fractures to discrete zones. High dolomite contents and the thin-bedded nature of the laminated sequences result in a more brittle, fracture-prone rock unit. Examination of core shows the discrete "brittle zones" are encased in more massive shale sequences—"ductile zones." Preferentially fractured "brittle zones" have been identified in the lower "A," upper "B," and lower "C" units (Figure 3). All are laterally continuous within the pool area, have similar mechanical properties, and behave as reservoir response units. Essentially all of the hydrocarbons produced to date in West Puerto Chiquito and adjacent producing areas issue from fractures developed in these units.

The reservoir units occur in a gross interval of 250 to 300 ft (76–91 m) at depths ranging from approximately 5300 ft (1615 m) at the updip limits of the pool to 7600 ft (2316 m) at the base of the monocline.

The Niobrara "Matrix"

A key element in fractured reservoir characterization and performance prediction is the potential for matrix contribution to production. The interaction between matrix and fracture components is often the most poorly understood reservoir mechanism and yet is most critical for performance projections, particularly in a reservoir such as West Puerto Chiquito. Comprehensive analysis of core data from wells in West Puerto Chiquito and adjacent fields leads to the conclusion that "matrix porosity" is ineffective and will not contribute to production at any point in the reservoir history (Appendix 1). Therefore, effective porosity must necessarily be fracture porosity. The fractures at West Puerto Chiquito, in other words, provide essentially all of the storage capacity and permeability.

Thus, we differ with those who propose a reservoir system (NMOCC, 1987, 1988) that is controlled by dual porosity and/or imbibition—hypotheses that require the presence of a physical characteristic, effective matrix porosity, that our work indicates does not exist.

Reservoir Physical Properties

Fracture Development. Fractures contributing to reservoir fluid flow derive from two different processes: oil generation and tectonic forces. The process of oil generation involves the conversion of solid kerogen to liquid hydrocarbons and results in a net volume expansion and the development of microscopic "expulsion" fractures. Because microfractures derive from internal body forces, they are generally pervasive and randomly oriented, with extremely small aperture widths at reservoir depths. Fracture lengths, the degree of interconnection, and aperture widths are such that the permeability of the microfracture system, to the extent it could be measured, would most likely be isotropic and quite low. The degree of interconnection between microfractures and the tectonic fracture system will control the contribution of the microfractures to fluid flow and storage. Degree of interconnection of the microfracturing cannot be determined using petrographic techniques.

Tectonic fractures associated with Laramide and younger basin subsidence and east flank folding resulted in an extensive system of high conductivity fractures with orientations approximately northwest, northeast, and north. Spacing and aperture widths of the high conductivity fractures are unknown. Production and pressure information reveal exceptionally high permeabilities and a strong directional anisotropy in the high conductivity system. Petrographic and visual core analyses indicate the naturally occurring high conductivity fractures are vertical to near-vertical while the microfractures may occur in any orientation.

Proper characterization of any reservoir is dependent on the sampling of a representative reservoir volume. In a complex, heterogeneous fracture system such as exists at West Puerto Chiquito, the conventional methods of analysis using data from individual wells or groups of wells are inadequate. Core, log, and individual well pressure tests sample extremely limited reservoir volumes. Similarly, the averages derived from large amounts of such data acquired in multiple wells in a reservoir of this type will be intrinsically flawed. Pressure build-up tests, for example, sample reservoir areas measured in terms of tens of acres (and, in places, are further constrained by the bounds of the individual fracture block within which the well is completed) whereas interference tests reveal average properties of reservoir areas measured in thousands of acres. It is because of the reservoir geometry, that of a highly interconnected, high-conductivity, tectonic fracture system bounding blocks of lower conductivity fractures, that true average reservoir properties can only be determined through the use of interference testing (Appendix 3).

Pay Thickness. An accurate determination of "pay thickness," in this case fracture height, is difficult in a reservoir of this type. The average fracture height is never known independently. The porosity-thickness product, or unit pore volume, however, can be determined through the use of interference tests. Production logs and tests indicate virtually all significant hydrocarbon volumes issue from the lower "A," upper "B," and lower "C" zones (Figure 3). None of the reservoir units produce uniformly throughout the pool: The "C" unit has historically been the principal producing unit in the southern portion of the reservoir, and the "A" and "B" units are primary contributors in the northern portion of the reservoir. The average total "net pay thickness" would be in the range of 30 to 100 ft (9–30 m).

Porosity. Effective hydrocarbon pore volume for West Puerto Chiquito is wholly contained within fracture porosity (Appendices 1 and 2). Total effective hydrocarbon porosity is approximately 0.3 to 0.4% of bulk volume, with a maximum upper limit of 1%. While Nelson (1985) and others have developed methods of estimating reservoir pore volume of fractures from width and frequency data derived from cores and outcrops, others (Aguilera, 1980) suggest this method will not apply in all cases. The method certainly does not apply at West Puerto Chiquito. The only reliable method of estimating effective hydrocarbon pore volume is interference (or frac pulse) testing of producing wells. Such testing shows a pore volume (ϕh) of 0.25 to 0.30 porosity-feet, which for 30 to 100 ft (9–30 m) of net pay corresponds to a porosity of 0.3 to 1% (Appendix 2).

Hydrocarbon Characteristics

The oil produced from the West Puerto Chiquito Pool is a sweet, 39 to 40° API hydrocarbon. The solution GOR, from undersaturated fluid samples recovered from the Canada Ojitos Unit L-11 (NW/SW Sec. 11, T25N, R1W), a carefully conditioned well sampled early in the life of the reservoir, was 480 scf/bbl. West Puerto Chiquito fluid properties are discussed in Appendix 4.

Formation Compressibility and Relative Permeability

Formation compressibility and relative permeability data are important in performance predictions at West Puerto Chiquito. Although the relative permeabilities of oil and gas per se are not addressed here, some insight is offered as to the ratio of relative permeabilities, k_{rg}/k_{ro} (referred to herein as k_g/k_o). This is presented in Appendix 5. We also present in Appendix 5 analysis of available data concerning formation compressibility, C_f. Some of the salient points concerning formation compressibility in general are noted below:

1. Reservoir storage is contained in fractures of low total pore volume. Typically formation compressibility, C_f, increases with decrease in porosity. It is reasonable to expect compressibility of this formation to be higher than that normally found in oil reservoirs. The question, of course, is how much higher.
2. The value of C_f was found to approximate 15×10^{-6} ($\Delta v/v$/psi) from interference testing of Canada Ojitos Unit wells at pressures both above and below the bubble point (v = the rock volume).
3. Reported laboratory data from cores taken from nearby wells are not definitive, covering a wide range from 6×10^{-6}/psi to 150×10^{-6}/psi (Appendix 5). Moreover, compressibility from laboratory measurements of cores suffers from the same limitations inherent in all core analyses of the fractured Niobrara: Samples, individually or in the aggregate, are not representative of the reservoir.
4. Also countering the high compressibility concept is the fact that the reservoir is underpressured (less than 2000 psia at depths of 7500 ft [2286 m] or greater), indicating possible vertical support that protects from overburden pressures. Typically, the reservoir rock's horizontal stress is much less than the vertical (Nelson, 1985). Thus, it is

impossible to determine C_f by analogy with formations of higher porosity or directly from laboratory data. The conventional wisdom, that a low porosity formation has exceptionally high compressibility, may not apply here. Furthermore, shearing, nonparallel separation, and natural propping of fracture faces will reduce closure effects and prevent high formation compressibility from developing.

Reservoir Fluid Flow

Fluid flow occurs only in fractures, with very high rates of diffusion as a consequence of the high conductivity fracture system. Values of $k_o h$, a measure of reservoir conductivity, are typically 1 to 20 darcy-feet over large parts of the reservoir. Using sensitive pressure gauges, interference effects are routinely observed between some wells over a distance of 4 mi (6.4 km) in a period of 10 hours.

It is important to note that the "tight fracture" low-conductivity networks constitute *restrictions to flow* rather than barriers to flow, since wells completed in the tight fracture blocks routinely observe transients from neighboring wells and rates of pressure decline similar to area-wide figures. In fact, there are no areas of the reservoir that are not in communication to some degree, attesting to the pervasive and highly interconnected nature of the system.

Extensive interference testing and production data indicate a strong directional anisotropy in reservoir fluid flow. Permeabilities are preferentially higher in the north-south direction. Note that this preference direction is parallel to structural strike and perpendicular to the direction of basin subsidence (Plate 1).

Given the nature of the fracture system, regional migration is to be expected and in fact has been observed throughout the southeast flank of the basin. A notable example is the Bear Canyon Unit 4 to 6 mi (6.4–9.6 km) northeast of the central Gavilan area (Figure 2), where initial pressures were 1000 psi less than virgin pressure but within a few pounds of central Gavilan's pressure at that time.

Development Considerations

Variations in Recovery. Wells producing in fractured Niobrara pools on the east flank of the basin exhibit a range of cumulative oil recoveries over two orders of magnitude: per well recoveries of 15,000 to 20,000 bbl in certain areas, and 1,500,000 to 2,000,000 bbl in others. Recoveries on a per-acre basis show a narrower range, over one order of magnitude, of 80 to 800 bbl/ac. The apparent paradox is further highlighted by the range of pore volume in commercially productive areas: approximately 1500 to 3000 bbl/ac. The differences between areas are only partially accounted for by intensity of fracturing; differences are primarily attributable to drainage area of the wells and extent to which gravity drainage recovery is achieved. Since the reservoir pore volume and fluid flow derive from fractures, production is characterized by poor relative permeability (k_g/k_o) characteristics. However, this phenomenon, namely high early permeability to gas, results in good gravity segregation and therefore good gravity drainage characteristics. Some gravity drainage will occur throughout, but the recovery will be more significant in areas of greater structural dip and conductivity and when production is controlled to optimize this important depletion process.

Misleading Nature of Reservoir's Characteristics. Methods of development are dependent on interpreted values of reservoir properties. The physical properties of fractured reservoirs are difficult to determine. Furthermore, the properties are of lower quality than the production issuing from them implies.

Example 1: Unit (per-acre) Recoverable Reserves— The high productivity and large drainage areas of individual wells are misleading as to the per-acre volume of oil in place and recoverable reserves. This feature results from the high conductivity fracture systems found throughout large areas. During early development of a field adjoining West Puerto Chiquito (Gavilan, Figure 2), individual wells located on 320 ac spacing units accumulated production volumes of hundreds of thousands of barrels, implying large per-acre recoveries. In fact, however, the wells were simply draining large areas (a township or more).

As an example of this phenomenon, the first high capacity well in Gavilan, the Native Son #2 (Sec. 27, T25N, R2W), produced large volumes of oil with small pressure declines, suggesting large per-acre reserves. This well had produced 240,000 bbl more than its south offset, the Homestead Ranch #2 (Sec. 34, T25N, R2W) when this south offset went on permanent production in 1986. By 1989, both wells were essentially oil depleted. Despite the fact that the Homestead Ranch #2 had a higher productivity, its cumulative production at near depletion of about 170,000 bbl was still 240,000 bbl less than the previously completed Native Son #2. Although both wells had drained far greater areas than their spacing units, by the time the Homestead Ranch #2 went on permanent production, additional wells in the area had dramatically reduced the thousands of acres initially available to these wells. Thus, it was impossible for the Homestead Ranch #2 to achieve a recovery comparable to the first well.

Some have proposed that the difference in recoveries of examples such as this merely reflect the time at which the wells went on production: The longer the producing time, the greater the recovery. Such an explanation is untenable here, however, since both wells are approaching oil depletion at the same time.

Example 2: Stratified Reservoir—As another example of the reservoir's misleading character, the reservoir is stratified and not all zones deplete uniformly. Gravity segregation causes upper zones to have higher gas saturations than the lower ones. Numerous field examples of such segregation have been observed. An oil and gas zone with a high gas saturation will produce in a fashion similar to a gas zone, with volumes proportional to the difference of squares of pressures, whereas zones with high oil saturation will produce more in proportion to the difference of pressures. The consequence is lower gas-oil ratios at higher rates of production, falsely suggesting greater efficiency. In fact, however, high production rates favor only certain individual wells and exacerbate problems of migration and protection of correlative rights. In a reservoir sense, however, lower rates of production permit more efficient gravity drainage and greater ultimate recovery. Greater reservoir efficiency can be achieved, of course, by first producing the lower zones. This practice was utilized in West Puerto Chiquito (see the section titled *Canada Ojitos Unit Pressure Maintenance Project* under *Recovery Efficiency*, below). Under competitive conditions, however, this conservation method can be accomplished only under strict pool regulations.

Example 3: Declining Reservoir Productivity—A further example of the reservoir's misleading character is declining reservoir productivity that is not apparent. Once production in an area is such that a group of wells capable of producing at high rates is not draining reservoir substantially outside its developed area, and the wells are produced at capacity, exceptionally high decline rates will occur.

The relatively small volume of oil in place for a given productivity of this fractured reservoir (Appendix 2) causes extremely high production decline rates. Constant percentage ($P_o/P = e^{dt}$) decline rates in excess of 100% can be expected (see *Symbols and Basic Equations* preceding Appendices). However, if the production rates are limited by equipment capacity (or proration), the decline may not be evident from the wells' performance.

Wells produced by pumping will sometimes flow intermittently. When gas-oil ratios rise with depletion, a pumping well's production is augmented by an increasing amount of flow. As a consequence, a well's true productivity may be declining but its daily oil production stays relatively the same (or in some cases increases) until a critical point is reached, when oil production rates drop abruptly.

Examples of this occurred in 1987 when the New Mexico Oil Conservation Commission ordered for both West Puerto Chiquito and Gavilan a period of high-rate allowables: 1280 BOPD and 2000:1 GOR for 640-ac proration units. Some of the higher productivity wells were located in the boundary area of the two pools in T25N. Average developed well density in the immediate area of these wells was approximately one well to 640 ac. Average production rates were in the range of 300 to 400 BOPD. Though not apparent in the pumping wells limited by equipment capacity, reservoir productivity was declining rapidly as a result of the high rate of depletion. The production of the Howard 1-8 well (NE/4 Sec. 1, T25N, R1W) offsetting the Canada Ojitos Unit is a notable example of this phenomenon (Figure 11). Although the apparent decline rate of the Howard 1-8 is high, the true decline rate for this area is more accurately reflected by the production performance of the offsetting well, the Canada Ojitos Unit E-6 (Sec. 6, T25N, R1W), a well produced by continuous gas lift (Figure 12). Earlier interference tests showed the wells in this area to have a high degree of communication.

As shown on Figure 12, once all wells in the local area were on production at rates limited primarily by equipment capacity (mid-September 1987), the constant percentage rate of production decline ($P_o/P = e^{dt}$) was 300%/year for this well, despite at least some pressure maintenance support. This decline rate extrapolates, for example, from 300 BOPD to 15 BOPD in one year.

Confirming this implication, one year later when unrestricted allowables were ordered for a two-month period (winter 1988–1989), wells in the boundary area not receiving pressure support from other areas had production rates of 1 to 15 BOPD. In the interim, increasing production rates of north Gavilan wells had essentially stopped migration from that direction. Productivity of wells receiving pressure

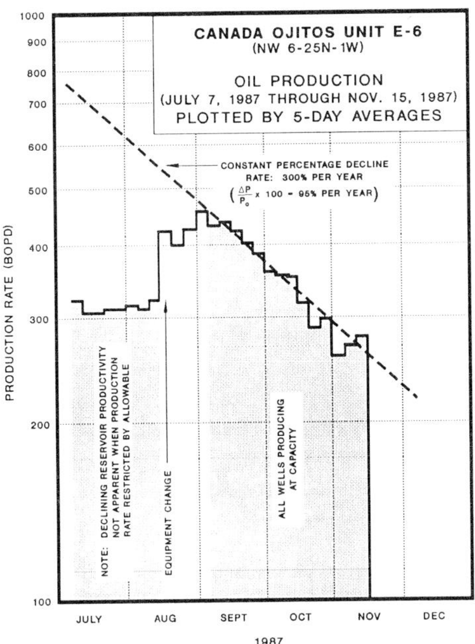

Figure 12. Production rate plot showing reservoir's true decline rate for the area of the well (Figure 11) on west boundary of the Canada Ojitos Unit in West Puerto Chiquito.

Analysis Type	Decline Rate, Percent per Year
$\Delta P/P$	300
$\Delta P/P_o$	95

maintenance support had declined, but not as dramatically as the others (Figure 13).

In summary, note that the high depletion rates depicted here are not to be considered exceptional: They are the norm for fractured Niobrara wells produced at high rates. Since well productivity varies as a function of its capacity (Kh), but the reservoir storage varies as only the cube root of Kh (Appendix 3 Figure 3-2), the higher the producing ability of wells, the greater will be the rate of depletion if produced at the high rates.

Forecasting Well Quality. Although a certain degree of communication exists throughout West Puerto Chiquito and surrounding areas, there are discrete subareas containing high conductivity fracture systems that are not directly connected to other high conductivity subareas. The exceptional conductivity of these fracture systems permits early evaluation of the development potential of a subarea. An example is southeast West Puerto Chiquito. At this writing (January 1989), activity has recently centered here where initial wells had high productivities. With the exception of a few wells that might receive pressure maintenance support from the unit, however, we anticipate there will be no future wells with the high ultimate recoveries experienced in West

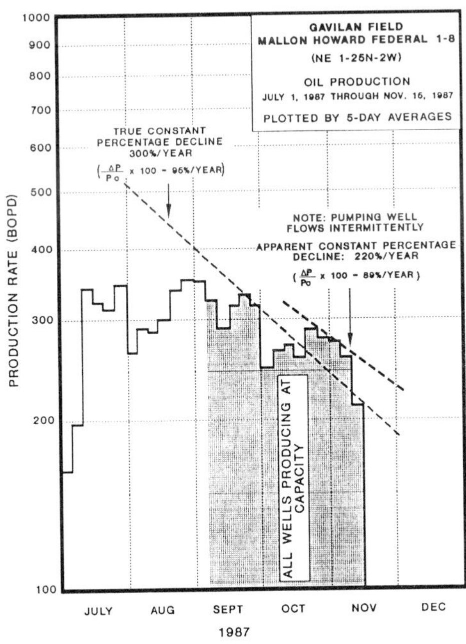

Figure 11. Production rate plot showing the masking of the Niobrara's exceptionally high decline rate when migration is reduced and production is limited by equipment capacity. *True* reservoir decline rate (see Figure 12) is greater than the apparent decline.

Analysis Type	Decline Rate, Percent per Year	
	Apparent	True
$\Delta P/P$	220	300
$\Delta P/P_o$	89	95

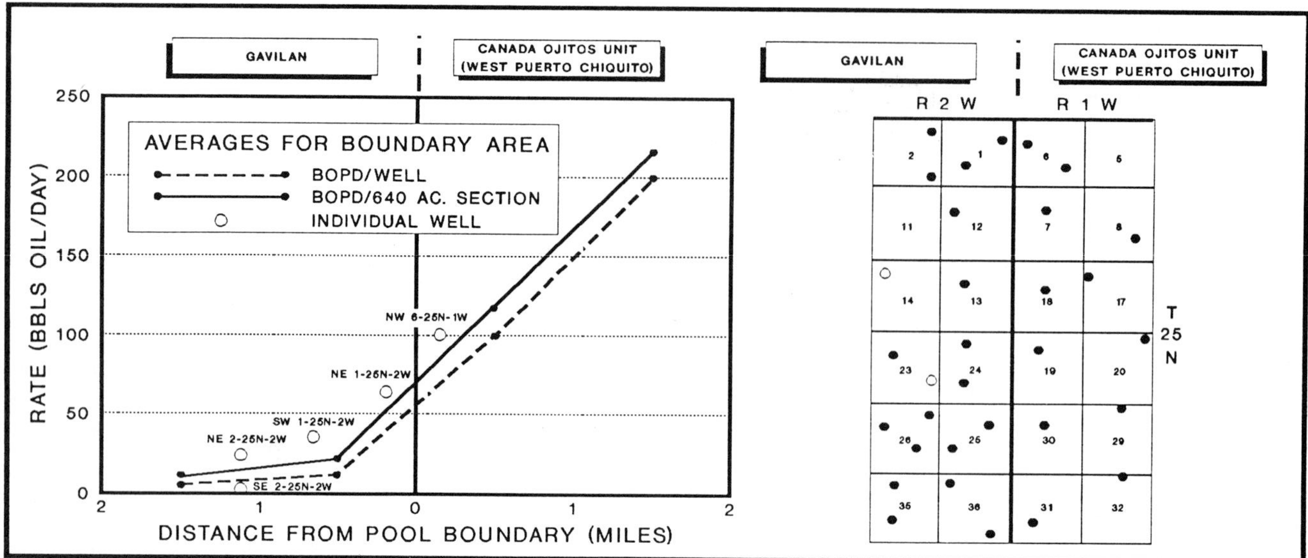

Figure 13. Map of the boundary area between West Puerto Chiquito (Canada Ojitos Unit) and Gavilan fields showing the location of producing wells. Plot of rate versus distance from the boundary shows amount of pressure maintenance support depends on well location. Averages plotted are for each of the four tiers (each tier is 1 mi east-west and 6 mi north-south).

Puerto Chiquito and in a few of the early Gavilan wells.

Two of the initial three wells in southeast West Puerto Chiquito produced at initial rates of 500 to 600 BOPD. Wells with equivalent initial rates in West Puerto Chiquito recovered up to 2.3 million bbl. Early wells of similar productivity in Gavilan produced hundreds of thousands of barrels. It is unlikely, however, that the analogy of high capacity wells and large ultimate cumulative recoveries will apply here. Rather, it is expected that future wells in southeast West Puerto Chiquito will realize ultimate recoveries of only 60,000 to 90,000 bbl/well if the existing 640 ac spacing is maintained. Closer spacing, of course, would result in proportionately lower per-well recoveries.

The method of analysis leading to this conclusion (Appendix 6) has general application to any newly drilled subarea that has a high conductivity fracture system.

Allowables and Producing Life. A development consideration for fractured Mancos formation wells is the allowable to be expected if a new pool is discovered. We anticipate that it will be similar to the allowable set in 1988 for the West Puerto Chiquito and Gavilan Pools of 800 BOPD and 2000:1 GOR for 640 ac spaced wells, an allowable inordinately high given the recoverable reserves.

Presumably the high allowable encourages exploration and development. In the fractured Niobrara, however, high allowables create problems of protection of correlative rights. Except for controls governing spacing, the pools are for practical purposes producing under the "law of capture" (see, for example, how high capacity wells can affect productivity of smaller wells—Appendix 6, Figure 6-5).

Expected producing life is a development consideration of some importance. In the absence of pressure maintenance and/or control of production to optimize gravity drainage, the depletion process will be solution gas drive, with ultimate recoveries in the range of 5 to 10% of oil in place, or 100 to 150 bbl/ac (60,000 to 100,000 bbl/well for 640 ac spacing).

To illustrate the effect of high allowables on the producing life of fractured Niobrara reservoirs, consider the development on 640 ac spacing of a subarea with 150 bbl of recoverable reserves per acre, simultaneous onset of production in all wells, and a 10 BOPD economic limit:

Per Well Average Initial Productivity (BOPD)	Producing Life (Years)
800	1.5
400	2.5
200	4.0
100	6.5

Controls Governing Well Spacing. Given the high conductivity throughout most of West Puerto Chiquito, oil recovery is essentially independent of well spacing. Of course, the greater the number of wells, the higher the total rate of reservoir withdrawal and the lower the ultimate recovery once the gravity drainage rate is exceeded.

In the Canada Ojitos Unit the initial density of recovery wells was approximately four sections per well with a drainage geometry of 1 mi (1.6 km) in the strike direction and 4 mi (6.4 km) in the dip direction. Pressure maintenance augmented the high transmissibility such that high productivities were maintained for a significant proportion of the depletion cycle and large per-well recoveries resulted. In contrast, when pressure and production are allowed to decline, additional wells are required to recover a given amount of oil in the same length of time. In summary for West Puerto Chiquito, its spacing since 1980 has been 640 ac per well. The drilled density is far less.

In contrast, the offsetting pool to the west, Gavilan, was initially spaced on 320 ac per well over the objection of operators requesting 160 ac spacing. During the latter stages of depletion, Gavilan spacing was changed to 640 ac with an option for a second well. A substantial part of southern Gavilan was developed on 320 ac spacing. The northeast part of the pool was developed later with a drilled density of approximately 1000 ac per well. At this writing (January 1989), lack of infill drilling suggests the operators are moving in the direction of wider spacing. In our opinion, the trend toward wider spacing will become generally accepted practice as additional production data become available.

Recovery Efficiency

As noted under *Development Considerations*, the large differences in per-well recoveries for oil wells producing from the Niobrara in the eastern San Juan basin are due primarily to the different drainage areas supporting the wells, along with the amount of gravity drainage realized.

Gravity drainage recovery, estimated at 55 to 60% of oil in place, is some ten times greater than the 5 to 6% expected from depletion by solution gas drive (Appendix 5). Given the reservoir geometry of a high-conductivity fracture network bounding lower-conductivity fracture blocks (Appendix 3), only parts of the reservoir can be depleted by gravity drainage. Except for certain areas of the reservoir with adequate structural dip (NMOCC, 1969), the lower conductivity fracture blocks will be depleted by solution gas drive.

That part of the reservoir occupied by high-conductivity fracture system(s) will provide some gravity drainage, even in areas of low dip angle, if production rates are not excessive. The high differential pressures that exist from the wellbore to the edge of the fracture blocks disappear in the surrounding fracture network, such that gravity segregation can take place here. In the high-conductivity system, vertical fractures apparently penetrate all producing zones, such that where the formation dip is low, the fracture height of 30 to 100 ft (9–30 m) can provide "dip" along portions of the fracture channels. In the low-conductivity fracture blocks, fractures are confined to the brittle zones and generally have insufficient conductivity to provide significant gravity drainage.

Obviously, the lower the per-acre withdrawal rate, the greater the opportunity for gravity drainage. In any newly developed area where high-conductivity fracture systems exist, production from the first few wells derives from an area encompassing a large number of undrilled spacing units. Thus, for the early part of the depletion cycle, the resulting low per-acre withdrawal and low differential pressures will facilitate gravity drainage. In areas of low dip angle, gravity drainage can yield a significant incremental recovery over that due to solution gas drive, albeit only a small proportion of the total gravity drainage potential.

Pressure maintenance greatly improves the opportunity to achieve gravity drainage recoveries (Muskat, 1945). Succinctly summarized by Muskat, "The effect of the gas injection is largely that of maintaining reservoir pressure while the gravity drainage is taking place."

Muskat notes further, "It seems likely that the case of complete gas segregation will in general require such low production rates, or pressure differentials over the pay, that it will seldom be practicable to operate a field in this manner." This constraint does not apply to the fractured Niobrara, however, where the reservoir storage is in fractures. Here the ratio of permeability to porosity is such as to make it uniquely suited to exploitation under gravity drainage depletion. As shown in Appendix 2 (Figure 2-1), for any given permeability, the porosity of a fracture system is at least an order of magnitude less than that of typical sandstone reservoirs. Consider a sandstone reservoir with permeability that would require 200 years to deplete at rates restricted to provide complete gravity segregation. The same permeability in a reservoir where fractures provide the storage would coincide with a volume of one-tenth as much oil in place. Accordingly, the same production rate would allow depletion in 20 years or less of the gravity drainage recoverable oil from the reservoir containing only fracture porosity.

Canada Ojitos Unit Pressure Maintenance Project. Significant gravity drainage recoveries have been achieved in the Canada Ojitos Unit. One method of identifying that gravity drainage took place in the Canada Ojitos Unit pressure maintenance project is through comparison of GORs versus accumulated recovery of certain wells (Figure 14). Since pressure decline for the histories shown was limited to a level approximately 100 psi below the bubble point, little solution gas drive recovery could have occurred. Continuous and extensive monitoring of pressures over three decades provides the facts with respect to reservoir pressure decline. Note the large proportion of ultimate recovery before significant increase in GOR. In addition, the COU E-10 well (Sec. 10, T25N, R1W) has the largest ultimate recovery and is also at the lowest structural datum (Plate 1). As the gas-oil contact moved downdip and GORs increased, the wells were shut in. One exception was the COU C-34 well (Sec. 34, T25N, R1W), which was maintained on production as an experiment to determine gravity drainage following gas breakthrough. The well produced 300,000 bbl at GORs in excess of 10,000 ft³/bbl. Since the solution gas drive process cannot operate without a pressure decline and pressure decline in the C-34 well was minimal, the recovery was necessarily a consequence of gravity drainage aided in small part by gas drive moving the oil thus accumulated to the wellbore. Comparison is made with the GOR history of the average of Gavilan's 40 best wells (Figure 14).

Although of relatively small amounts, some gas drive depletion has taken place in the gas cap area. If large volumes of gas continue to be "cycled," this process is expected to yield additional recovery from increasingly larger parts of the reservoir. That gas drive has taken place near the injection wells is evident from increasing permeability to gas reflected by injection pressures and volumes (NMOCC, 1989). These gas drive recoveries from the low capacity gas cap area, albeit small volumes compared to the total, are amounts that could not otherwise be recovered, as the gas cap area cannot be economically developed by recovery wells.

In order to produce the reservoir most efficiently (at minimum GOR), the lowest producing zones were opened to production first ("C" zone in the south part of the unit), with the intention of opening the upper zones in the latter stages of depletion when cycling operations were planned. Later upon testing the "A" and "B" zones where the "C" zone had originally been produced, productivity of gas was increased but not oil. Apparently the "A" and "B" zones were oil depleted by wells completed only in the "C" zone. Therefore, efficient recovery was realized without handling large volumes of gas.

There are large areas with low-conductivity fracture blocks, whose interconnecting high-conductivity fracture system is of a lower capacity than the better areas, as for example of some of the injection wells in the gas cap. Because of the orientation (along the strike) of some of the low capacity areas, they serve a useful purpose in dividing the reservoir into subareas permitting the pressure maintenance to be more effective and therefore increasing efficiency of recovery from the better areas. The dominance of north-south directional permeability causes minimum pressure differentials in a north-south direction (along strike) and significant pressure differentials (because of the intervening lower capacity systems) east-to-west (perpendicular to strike). The consequence is a series of areas that reflect "pressure plateaus." Insofar as the pressure maintenance project is concerned, this characteristic aids in efficient oil recoveries: Injected gas moves north-south along the strike and then diffuses downdip across the intervening lower conductivity system, first into one pressure plateau where it again moves north-south and then diffuses into the next. Note that if the preferential permeability were directed vice versa, there would be a greater tendency for the injected gas to channel directly downdip to recovery wells. Since the gas cap is in a low-conductivity portion of the reservoir with associated low storage volume, the gas cap pressures have increased with rising GORs as greater volumes of gas are produced and higher pressures are required to move the injected gas out of the gas cap (Figure 15).

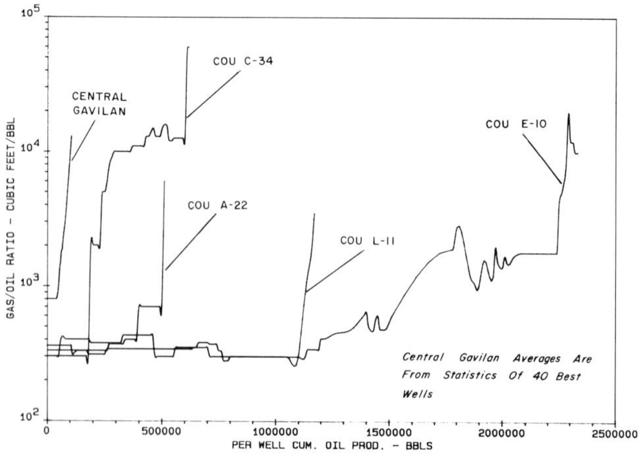

Figure 14. Plot of GOR versus cumulative oil production for four Canada Ojitos Unit wells. Note large recoveries prior to GOR increase, reflecting gravity drainage. Note well with largest recovery, COU E-10, Sec. 10, T25N, R1W, is at the lowest structural datum. Note large recovery of C-34 after gas breakthrough which, in view of minimal pressure decline, is mostly from gravity drainage. Comparison is made with average of Gavilan's 40 best wells.

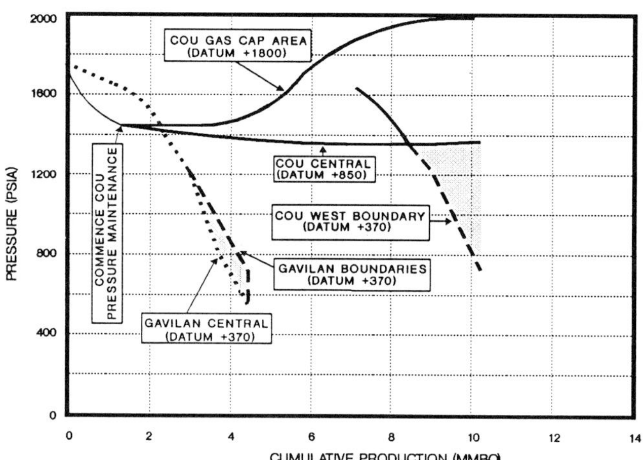

Figure 15. Plot of COU reservoir pressure versus cumulative production. Note the increase in gas cap pressures as larger volumes of gas are handled following GOR increases. Note the leveling of pressures associated with the onset and continued pressure maintenance in the Canada Ojitos Unit. Note also the rapid pressure declines in the Gavilan Pool and the concurrent effect on the Canada Ojitos Unit area adjacent to the pool boundary.

Finally, we note that management of the Niobrara through pressure maintenance and controlled production will yield efficient recoveries on wide spacing (NMOCC, 1969). This is evidenced by the remarkable performance of certain Canada Ojitos Unit wells. For example, the Canada Ojitos Unit E-10 well (Sec. 10, T25N, R1W) produced 2.3 million bbl of oil while the reservoir pressure declined approximately 100 psi. The depletion process clearly was not solution gas drive. The process of solution gas drive operating over that pressure decline would result in a recovery of 40 to 50 bbl/ac (equivalent to a drainage area of 45,000 to 60,000 ac). The higher estimated gravity drainage recovery for this area, 30 to 40% of oil in place, resulted in a recovery of 600 to 800 barrels per acre and *efficient* depletion of a reservoir area of 3000 to 4000 ac for this well.

In summary, a substantial area within the Canada Ojitos Unit reservoir is estimated to recover 30 to 40% of the oil in place (NMOCC, 1969, 1980, 1988). The amount attributable to recovery by gravity drainage will be 25 to 35%, five times the amount that would have resulted from "conventional" development on close spacing and at high production rates.

Field Comparison.   To illustrate the difference in recovery efficiency between fractured Niobrara pools developed under different operating conditions, comparison is made between West Puerto Chiquito and central Gavilan, an offsetting pool (Figure 2) developed under competitive conditions.

Close spacing and high allowables caused a high rate of depletion in central Gavilan that prevented the field-wide segregation of oil and gas that must occur to permit significant gravity drainage recovery (as it did in West Puerto Chiquito). High conductivity fracture systems are known to exist in Gavilan: Values of $K_o h$ of 10 to 20 darcy-feet have been measured through both interference and frac pulse testing (NMOCC, 1988). This conductivity along with Gavilan's low dip angles provided gravity drainage recoveries of 2 to 4% of oil in place which, when combined with the solution gas drive recovery of about 6%, provides a total recovery of 8 to 10% of oil in place, one-quarter to one-third of that realized for the Canada Ojitos Unit.

This estimate of the efficiency of Gavilan's recovery is arrived at as follows: Central Gavilan is estimated to produce approximately 5 million bbl of oil, 1 million bbl of which was recovered while pressures were above the maximum assumed bubble point of 1660 psi. Of the remaining 4 million bbl produced below bubble point pressure, approximately 1 million bbl resulted from regional migration, leaving 3 million bbl produced below the bubble point that originated under central Gavilan lands. Solution gas drive recovery is estimated at 6% of the oil in place (Appendix 5). Central Gavilan's initial oil in place is estimated at 30 to 40 million bbl (NMOCC, 1988), which therefore provides a solution gas drive recovery of 1.8 to 2.4 million bbl. The remaining 0.6 to 1.2 million bbl results from gravity drainage. Of the ultimate production (from pressure below the bubble point and after migration) this approximates one-sixth to one-third—a volume large compared to the total, but only 3 to 6% of the gravity drainage potential.

Further insight into the comparison of recoveries under gravity drainage and pressure maintenance versus development under competitive conditions comes from data of pressures and cumulative production. Because of cross-boundary migration, it is not possible to compare directly pressures and cumulative production of the two areas. A generalization of relative efficiencies of recovery, however, is evident from a comparison of plots of pressure versus cumulative production (Figure 15). The Canada Ojitos Unit initial pressure decline below bubble point was less than Gavilan's. Ultimate recovery, however, for the Canada Ojitos Unit is estimated to be greater.

As can be seen from the above review of the issues affecting recovery, it is impossible to acquire the information necessary to make precise analyses of efficiency. The foregoing, however, provides insight that can be used as a base for understanding recovery processes of the fractured Niobrara reservoir. The database will improve with time and production. When north Gavilan is depleted, for example, and its reservoir volume better known, a refined estimate can be made of migration prior to first production.

Source

The most likely source of the hydrocarbons produced at West Puerto Chiquito is the organic-rich marine shale of the Niobrara. Total organic carbon (TOC) contents reported from wells northwest of West Puerto Chiquito range from 1 to 3% (Cole et al., 1989). A reconstruction of the time and temperature history of the reservoir interval using the method of Lopatin (1971) gives a preliminary estimate of source rock maturity and the timing of generation and entrapment of hydrocarbons at West Puerto Chiquito (Figure 16). Of particular importance to the accuracy of the method are the burial and thermal histories of the Niobrara at the selected well, in this case the Canada Ojitos Unit 29 (E-6; SW NW Sec. 6, T25N, R1W). Successive periods of deposition, erosion, and uplift brought the Niobrara to its present burial depth of approximately 7100 to 7700 ft (2164–2347 m). Evidence exists, however, to place the unit at a maximum burial depth nearly 1000 ft (305 m) deeper during the Eocene and early Oligocene as a result of increased original thicknesses of Nacimiento and San Jose sediments (Baltz, 1967; Fassett and Hinds, 1971). Furthermore, although the present-day geothermal gradient is approximately 1.6°F/100 ft, higher heat flows associated with the development of the San Juan volcanic field most likely elevated the geothermal gradient during the early to mid-Tertiary (Bond, 1984; Reiter and Clarkson, 1983; Reiter and Mansure, 1983).

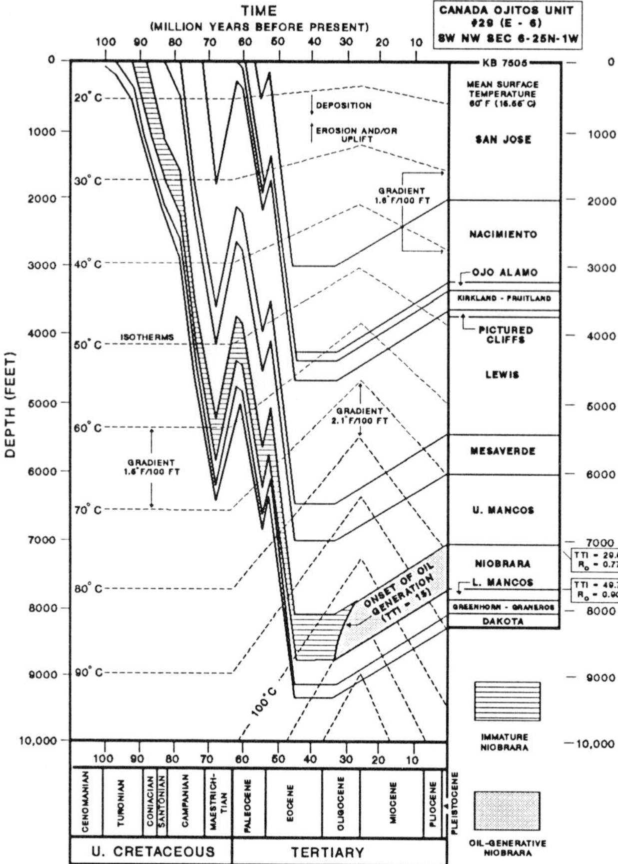

Figure 16. Burial and thermal history reconstruction at the Canada Ojitos Unit 29 well (E-6; SW NW Sec. 6, T25N, R1W) showing onset of oil generation in the Niobrara and estimate of present-day maturity level.

The thermal history of the Niobrara interval prior to the end of the Cretaceous (Maastrichtian) is assumed constant at 1.5°F/100 ft (Tissot and Welte, 1978), followed by an increase in the gradient to a maximum (at West Puerto Chiquito) of 2.1°F/100 ft at the end of the Oligocene, at which time volcanism and intrusive activity associated with the San Juan volcanic field had peaked (Steven, 1975). From the end of Oligocene to the present, gradients are assumed to decline at a constant rate to the present 1.6°F/100 ft.

In spite of the inherent uncertainty in the burial and thermal history data, the reconstruction is considered a reasonable preliminary estimate of the thermal maturity of the Niobrara at West Puerto Chiquito. As a further check on the accuracy of the method as applied here, the calculated thermal maturities (as measured by the "TTI," or "Time-Temperature Index") were compared to the maturities derived from vitrinite reflectance (R_o) data using the relationship established by Waples (1980).

Timing of Generation and Entrapment

Development of the Hogback monocline and associated tectonic fractures at West Puerto Chiquito occurred in late Paleocene–early Eocene, approximately 48 to 54 Ma (Baltz, 1967) and prior to hydrocarbon generation. The Niobrara interval entered the oil-generative phase, corresponding to a TTI of 15.0 (Waples, 1980), approximately 28 to 33 Ma during the mid- to late Oligocene (Figure 16). The presence of open fractures at the time of hydrocarbon generation may have facilitated the expulsion of oil from the shales to the fracture system, preventing the build-up of abnormal pressures frequently associated with generation in a closed system (e.g., the Bakken system in the Williston basin or the Green River–Wasatch system in the deep Uinta basin). Present-day maturities calculated for the Niobrara at West Puerto Chiquito range from TTI values of 29.5 at the top of the interval to 49.7 at the bottom, corresponding to vitrinite reflectance values ranging from 0.77 to 0.90%, respectively. Vitrinite reflectance values obtained from laboratory analyses of samples from the top and bottom of the Niobrara reservoir interval in the Canada Ojitos Unit #29 (E-6) well were 0.86 and 0.90%, respectively, in good agreement with values predicted by the Lopatin (1971) method. Both the burial history and laboratory methods indicate the Niobrara interval is thermally mature and probably still within the oil-generative phase at West Puerto Chiquito.

The East Puerto Chiquito "Problem"

Migration of hydrocarbons appears to have occurred to a limited extent in and adjacent to West Puerto Chiquito. Fractures in the Niobrara produced nearly 4 million bbl of oil at East Puerto Chiquito field on the northeast boundary of West Puerto Chiquito but at structural elevations ranging from 1200 to 5000 ft (366–1524 m) higher (Plate 1). Reconstruction of the burial and thermal history at East Puerto Chiquito indicate the Niobrara is below the threshold maturity level for oil generation. Hydrocarbons apparently migrated into East Puerto Chiquito from the deeper, oil-generative Niobrara to the west at some point prior to the late Tertiary (Miocene–Pliocene?) normal faulting that separated the two reservoirs.

EXPLORATION AND DEVELOPMENT CONCEPTS

Regional Play

Exploration for fractured Niobrara oil reservoirs has been pursued on the flanks of the San Juan basin, in northwest Colorado, and along the Front Range of eastern Colorado. Development of a Niobrara fracture play involves, at a minimum, recognition of the following fundamental criteria:

1. Presence of intense folding, faulting, and/or fracturing of the type normally associated with the flanks of foreland basins.

2. The concept that fracture systems form in networks with a specific spatial relationship to folds and generally over the entire structural feature.

3. The presence of brittle, fracture-prone lithologies, usually with high dolomite contents, isolated vertically by more ductile lithologies.

4. The presence of a fracture-prone texture, characterized by a thin-bedded, highly laminated sequence of brittle lithologies.

5. Sufficient source rock quality and thermal history to place the Niobrara interval within the generative phase during or after fracture formation.

6. Little or no later tectonism to destroy trap integrity.

Lessons Learned

Modern exploratory techniques, such as seismic data, could provide important data concerning fold geometry and extent but would be unlikely to locate major fracture trends with small vertical displacements. The application of seismic techniques, even at a point prior to the discovery of West Puerto Chiquito, would be unlikely to enhance the structural interpretation of the area due to the abundant subsurface and outcrop control. Seismic data in faulted portions of the reservoir, however, may assist in location of wells. Landsat and aerial photographic techniques appear to provide a cost-effective preliminary interpretation of the orientation and relative intensity of tectonic and regional fracturing in a prospective area. Certain electrical techniques are likely to provide data complementary to surface techniques, identifying subsurface locations of fracture trends to assist in well location.

Perhaps the most important lesson learned is that successful exploration and development of fractured Niobrara reservoirs can only be accomplished through use of large exploratory units. The Niobrara cannot be efficiently, economically, and equitably developed under competitive conditions. For example, West Puerto Chiquito wells separated by 40 ac spacing distances and draining from the same large reservoir area have shown differences in productivity of two orders of magnitude (the equivalent of 40 BOPD versus 4000 BOPD). It is therefore impossible in a practical sense to unitize fractured Niobrara reservoirs after discovery of a new field. Furthermore, competitive operations lead to adversarial positions and defeat the cooperative efforts needed for acquisition of test data, joint studies, and the free flow of information.

Had West Puerto Chiquito been developed under competitive conditions, for example, the result would have been more wells, less ultimate recovery, and a discounted net operating income of only a fraction of that realized.

REFERENCES CITED

Aguilera, R., 1980, Naturally fractured reservoirs: Petroleum Publishing Company, 703 p.

Arps, J. J., 1945, Analysis of decline curves: Trans. AIME 1945, v. 160, p. 228–247.

Baltz, E. H., 1967, Stratigraphy and regional tectonic implications of part of Upper Cretaceous and Tertiary rocks, east-central San Juan Basin, New Mexico: U.S. Geological Survey Professional Paper 552, 101 p.

Bond, W. A., 1984, Application of Lopatin's method to determine burial history, evolution of the geothermal gradient, and timing of hydrocarbon generation in Cretaceous source rocks in the San Juan Basin, northwestern New Mexico and Southwestern Colorado, in F. F. Meissner, et al., eds., Hydrocarbon source rocks of the greater Rocky Mountain region: RMAG, p. 433–447.

Bulnes, A. C., and R. U. Fitting, Jr., 1945, An introductory discussion of the reservoir performance of limestone formations: Trans. AIME 1945, v. 160, p. 179–201.

Cole, R. D., J. M. Allmaras, J. P. Zager, and G. E. Moore, 1989, Sedimentology, petrology, and X-ray mineralogy of the Coniacian-Santonian Niobrara shale, northeastern San Juan basin, New Mexico: American Association of Petroleum Geologists Section Meeting, Albuquerque, New Mexico.

Economides, M. J., and K. G. Nolte, 1987, Reservoir stimulation: Schlumberger Educational Services.

Fassett, J. E., and J. S. Hinds, 1971, Geology and fuel resources of the Fruitland Formation and Kirtland Shale of the San Juan Basin, New Mexico and Colorado: U.S. Geological Survey Professional Paper 676, 76 p.

Fatt, I., 1953, The effect of overburden pressure on relative permeability: Trans. AIME 1953, v. 198, p. 325–326. (Technical Notes Journal of Petroleum Technology, October 1953.)

Gorham, F. D., Jr., L. A. Woodward, J. F. Callendar, and A. R. Greer, 1979, Fractures in Cretaceous rocks from selected areas of San Juan basin, New Mexico—Exploration implications: American Association of Petroleum Geologists Bulletin, v. 63, n. 4.

Jones, F. O., 1975, A laboratory study of the effects of confining pressure on fracture flow and storage capacity in carbonate rocks: Journal of Petroleum Technology, January, 1975.

Keeling et al., 1964, 1969, Laboratory analyses of cores for relative permeability, El Paso Natural Gas Company 1964, Keeling engineering study for Gallegos Gallup Unit, Pan American Laboratories 1969, fractured Devonian chert and Ellenburger.

London, W. W., 1972, Dolomite in flexure-fractured petroleum reservoirs in New Mexico and Colorado: American Association of Petroleum Geologists Bulletin, v. 56, p. 815–821.

Lopatin, N. V., 1971, Temperature and geologic time as factors in coalification: Akademiya Nauk SSR Isvestiya Seriya Geologicheskikh, n. 3, p. 95–106.

Molenaar, C. M., 1977a, Stratigraphy and depositional history of Upper Cretaceous rocks of the San Juan Basin area, New Mexico and Colorado, with a note on economic resources: New Mexico Geological Society Guidebook, 28th Field Conference, p. 159–166.

Molenaar, C. M., 1977b, San Juan basin time-stratigraphic nomenclature chart: New Mexico Geological Society Guidebook, 28th Field Conference, p. xii.

Muskat, M., 1945, The production histories of oil-producing gas-drive reservoirs: Journal of Applied Physics, v. 16, p. 147–159.

Muskat, M., 1949, Physical principles of oil production: New York, McGraw-Hill Book Company, Inc., 922 p.

Nelson, R. A., 1985, Geologic analysis of naturally fractured reservoirs: Houston, Gulf Publishing Co., 410 p.

New Mexico Oil Conservation Commission, spacing hearing for the Boulder Pool, July 10, 1963.

New Mexico Oil Conservation Commission Case No. 3455, November 1966 and December 1969.

New Mexico Oil Conservation Commission Case No. 7075, November 1980.

New Mexico Oil Conservation Commission Case Nos. 7980, 8946, 8950, 9113, and 9114, March 1987.

New Mexico Oil Conservation Commission Case Nos. 7980, 8946, 8950, and 9111, June 1988.

New Mexico Oil Conservation Commission Case No. 9525, February 1989.

Ramey, J. H., Jr., 1964, Rapid methods for estimating reservoir compressibilities: Trans. AIME, v. 231, p. 447–454, Journal of Petroleum Technology, April 1964.

Reiter, M., and G. Clarkson, 1983, Relationships between heat flow, paleotemperatures, coalification and petroleum maturation in the San Juan Basin, northwestern New Mexico and southwestern Colorado: Geothermics, v. 12, p. 323–339.

Reiter, M., and A. J. Mansure, 1983, Geothermal studies in the San Juan basin and the Four Corners area of the Colorado Plateau: I. Terrestrial heat-flow mesurements: Tectonophysics, v. 91, p. 233–251.

Steven, T. A., 1975, Middle Tertiary volcanic field in the southern Rocky Mountains, *in* B. F. Curtis, ed., Cenozoic history of the southern Rocky Mountains: GSA Memoir 144, p. 75–94.

Thomas, R. D., and D. C. Ward, 1972, Effect of overburden pressures and water saturation on gas permeability of tight sandstone cores: U.S. Bureau of Mines, Journal of Petroleum Technology, February, 1972.

Tissot, B. P., and D. H. Welte, 1978, Petroleum formation and occurrence—a new approach to oil and gas exploration: New York, Springer-Verlag, 538 p.

Van Golf-Racht, T. D., 1982, Fundamentals of fractured reservoir engineering: Amsterdam, Elsevier, 710 p.

Vincelette, R. R., and W. E. Chittum, 1981, Exploration for oil accumulations in Entrada sandstone, San Juan basin, New Mexico: American Association of Petroleum Geologists Bulletin, v. 65, p. 2546–2570.

Waples, D. W., 1980, Time and temperature in petroleum formation, application of Lopatin's method to petroleum exploration: American Association of Petroleum Geologists Bulletin, v. 64, p. 916–926.

Woodward, L. A., and J. F. Callendar, 1977, Tectonic framework of the San Juan Basin, *in* San Juan Basin III: New Mexico Geological Society 28th Field Conference Guidebook, p. 209–212.

APPENDICES

Symbols and Basic Equations

Symbols and Basic Equations

$BOPD$ = barrels oil per day

STB = stock tank barrels

MCF/D = thousands of standard cubic feet per day

GOR = gas-oil ratio, cubic feet per barrel

C_f = formation compressibility

h = thickness

ϕ = porosity, fraction of pore volume

K, k = permeability

k_o = permeability to oil

k_g = permeability to gas

k_{ro} = relative permeability to oil

k_{rg} = relative permeability to gas

k_g/k_o = ratio of relative permeability

μ = viscosity as in Kh/μ

μ_o = viscosity of oil

μ_g = viscosity of gas

Symbols for constant percentage decline curve (also known as "constant rate," "exponential," "semi-log") (Arps, 1945; Muskat, 1949):

$$P = P_o\, e^{-dt} \text{ or } P_o/P = e^{dt} \tag{1}$$

where P_o = initial production rate, any units

P = production rate at time t, same units as for P_o

e = base natural logarithms

d, t = decline rate (fraction or decimal) and time in any consistent units

Also C = $(P_o - P)/d$

where C = cumulative recovery between production rates P_o and P

d = decline rate in units consistent with P_o and P

(Accordingly, a plot of P versus C is straight line on coordinate paper.)

Symbols for a variation of analysis of these decline curves sometimes used:

$$\Delta P/P_1 = D$$

Where ΔP = production rate decline from P_1 to P_2

D = decline rate in consistent units (fraction or decimal per month, per year, etc.)

Also C = $(P_1 - P_2)/(-(\ln 1 - D))$

Where C = cumulative recovery between production rates P_1 and P_2

D = decline rate in units consistent with P_1 and P_2

$\ln$ = natural logarithm

Appendix 1. Matrix Porosity Ineffective

The conclusion that matrix porosity is ineffective is based on the following observations from core data in West Puerto Chiquito and adjacent areas:

1. *Porosities and permeabilities are extremely low.*

Significant reductions occur in porosity and permeability measurements made at ambient conditions upon restoration to reservoir conditions (Nelson, 1985).

Analyses at ambient conditions of recent cores indicate average porosities in the range of 1.2 to 2.1% and (geometric) mean permeabilities of 0.015 to 0.024 md. At reservoir confining stress, these porosity values reduce to approximately 0.8 to 1.5%; and the corresponding permeabilities are estimated to reduce to 0.000046 md to 0.00011 md, respectively, using the method of Thomas and Ward (1972). Note that this method recognizes the inherent increase in formation compressibility with decreasing over-burden stress that occurs when cores are brought to the surface and tested at ambient conditions. Review of all available sources forces the conclusion that there are no known sandstone reservoirs in which a "matrix" of this quality contributes commercial oil volumes, with or without fracture enhancement.

Capillary pressures are likely to be very high, such that drainage of oil from tight pores is not likely to occur over the range of pressure encountered in depletion. More fundamental, however, is the question whether the microporosity *ever* contained hydrocarbons. Displacement of water by hydrocarbons from microporosity in the shale and siltstone during oil generation would be unlikely.

2. *Petrographic analyses indicate ineffective porosity.*

Microporosity in the siltstone laminae and carbonate laminae, low originally because of grain size, has been effectively destroyed by secondary diagenetic effects: Intergranular porosity is occluded by abundant carbonate cement and authigenic clay minerals throughout the reservoir interval. Photomicrographs and SEM analyses provide clear evidence that "matrix porosity" is essentially ineffective waterfilled microporosity.

3. *Visual examination indicates no oil saturation.*

Cores taken in the reservoir section in a well adjacent to West Puerto Chiquito were examined under ultraviolet light in the field immediately upon removal to the surface, and again in the lab. No saturation, staining, or fluorescence of the "matrix" was observed (Kurt Fagrelius, 1987, personal communication). However, fluorescence and staining were observed along fracture and bedding planes.

4. *Laboratory oil saturations are very low to nonexistent.*

Laboratory measurement and calculation of fluid saturations in reservoir lithologies of this type are fraught with uncertainty. For example, using conventional retort methods, oil saturations as high as 30 to 50% were reported for a recent core from a well adjacent to West Puerto Chiquito. The data from this analysis are unreliable because of the possibility of kerogen "cracking" and loss of bound water due to exposure to the elevated temperatures encountered in the retort process.

To achieve more reliable results, the Dean-Stark extraction method was used to analyze plugs from a subsequent well adjacent to West Puerto Chiquito field, resulting in low indicated oil saturation. This analysis revealed geometric mean oil saturation for 49 samples of 10.6%, highlighting the error in results associated with the application of conventional analysis methods to a rock of this character. Even this more reliable method is subject to significant analytical errors (see below), which when combined with such low oil saturations in a rock of extremely low permeability, argue for *no* effective oil saturation in the matrix.

5. *Water saturation supports minimal oil saturation.*

The average water saturation of cores analyzed by the Dean-Stark method at ambient conditions was 62.5%. At reservoir confining stress, the samples experienced porosity reductions in the range of 25 to 30%. Thus, a core of 90 to 100% water saturation in the reservoir becomes, at surface conditions, a core with water saturations in the 60 to 70% range. Furthermore, since connate waters are commonly gas saturated, the reduction to surface pressures may cause minor water expulsion due to solution gas drive, resulting in measured water saturations lower than actually exist in the reservoir.

In any case, pressure of the gas evolving from the formation water would likely be sufficient to prevent filtrate invasion of the microporosity. Thus, any water saturation measured must be formation water, is most likely too low, and is *not* representative of reservoir conditions. The customary assumption that at ambient conditions the difference between 100% of pore volume and that occupied by water represents hydrocarbon saturation is not valid here. The 60 to 65% water saturation measured at ambient conditions probably reflects actual water saturation of 100% at reservoir conditions.

6. *Low oil saturation not caused by flushing.*

Flushing of oil from a core sometimes occurs in the coring process, resulting in low measured oil saturation.

Such an explanation for the low oil saturation of the Niobrara cores is untenable: Rock of 0.02 md is not likely to be flushed, and oil staining and fluorescence were observed on some fractures and bedding planes (Kurt Fagrelius, 1987, personal communication). If flushing occurred, the first to be flushed would be the fractures; since the fractures were not flushed, it follows that the matrix of 0.02 md permeability was not flushed.

7. *Laboratory oil saturations are of questionable accuracy.*

The Dean-Stark procedure, while more reliable than the retort method, does not measure oil volumes directly; rather, the oil saturation is calculated by weight difference using a measured water volume (weight) recovered. Given the low porosities and apparent low oil saturations in an extremely low sample pore volume, the weight difference attributed to oil saturation is vanishingly small and subject to large measurement errors. Whatever the actual oil volume measured in the cores, it almost certainly originated in the fractures, which themselves occupy an extremely small proportion of the total sample volume analyzed.

Appendix 2. Fracture Porosity: The Reservoir Storage

The effective hydrocarbon pore volume in the West Puerto Chiquito field is wholly contained in fractures. This fundamental conclusion derives from both assessment of core analyses (Appendix 1), which effectively preclude any "matrix" contribution, and from extensive interference and production testing, which provide direct evidence that the storage volume, as well as the permeability, most likely is of fracture porosity.

To understand the storage system in West Puerto Chiquito, we begin with a comparison of porosity (ϕ) and permeability (K) for matrix (sandstone) and fracture systems, and then relate well test information to these relationships.

The permeability versus porosity relationships for 2200 sandstone specimens (Bulnes and Fitting, 1945) and a fracture system with fractures parallel to the direction of flow are presented in Figure 2-1 (NMOCC, 1966). Note that for any given fracture density (or spacing) and a given permeability, the fractures are of equal aperture width. The graph provides an overview of the relation of fracture width and spacing to fracture porosity and permeability.

This particular fracture flow system is defined analytically for fracture spacing in centimeters (Jones, 1975) as:

$$\phi f = 4.93 \times 10^{-3} \left(\frac{K}{S^2}\right)^{1/3} \qquad (1)$$

ϕf = fracture porosity (%)

K = flow test permeability (darcys)

S = fracture spacing or average distance between fractures (centimeters)

If fracture "density" (as opposed to "spacing") is given in fractures per foot, F (NMOCC, 1965), the equation is:

$$\phi f = 5.05 \times 10^{-4} (KF^2)^{1/3} \qquad (2)$$

To understand fracture flow it is instructive to examine the relation of fracture spacing (or density) and aperture width to permeability from equation (1) or (2) or Figure 2-1.

For any given permeability, the porosity becomes a function of fracture spacing: As spacing decreases (density increases), the porosity increases. Furthermore, for any given permeability, the porosity is essentially independent of aperture width—a concept not easy to grasp at first introduction. Note that aperture width does not appear (directly) in equations (1) or (2). Porosity is simply a function of permeability multiplied by the $^2/_3$ power of fracture density, a concept useful in estimating maximum fracture porosity (see below).

For fracture systems with nonuniform aperture widths, the permeability increases as the cube of the porosity. For a given permeability, therefore, increasing aperture width results in decreasing density (and fracture porosity). Thus, the maximum ratio of porosity to permeability will occur for fractures of uniform width rather than for mixed widths.

Clearly, in a naturally fractured reservoir, fractures are not uniformly parallel to the direction of the flow, and therefore, porosity is expected to be greater than that for an idealized system of parallel flow. It has been suggested (Van Golf-Racht, 1982) that for a fracture system in which the fractures are not parallel to the direction of flow but at random distribution, the porosity would be increased by the value of the cube root of pi squared, an increase in porosity values (for a given permeability) over those shown by equations (1) and (2) and Figure 2-1 of 35%.

In natural (nonideal) systems, fractures have flow restrictions and are not perfectly interconnected such that the porosity will be higher for any given permeability than that shown by the (ideal) parallel flow system.

Whatever the true relation of porosity to permeability for a system, changes in fracture aperture width—as with changes in overburden pressure—will cause porosity to vary as the cube root of the permeability (Jones, 1975). Furthermore, in comparing one area of a reservoir with another, if the fracture density is the same, but the fracture width is different, then the ratio of porosities for the two areas will be proportional to the cube root of the ratio of the permeabilities.

In defining an approximate upper limit of ratio of porosity to permeability for fracture systems, we

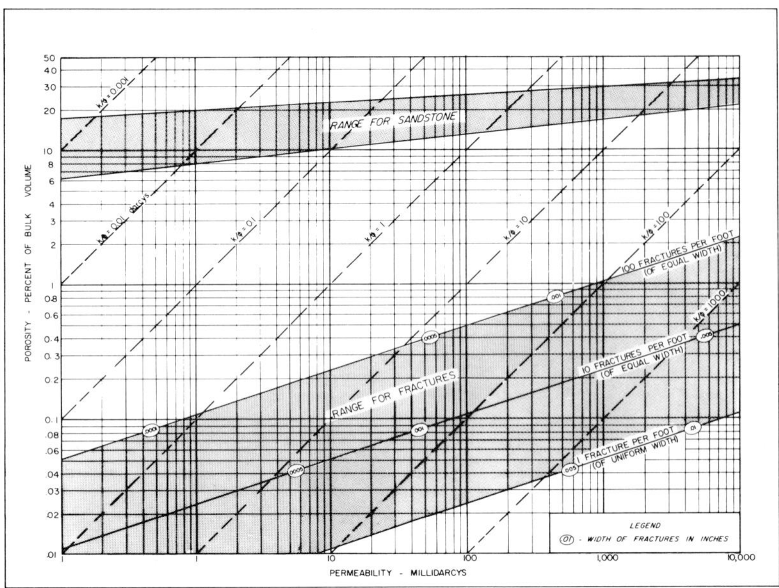

Figure 2-1. Relation of porosity to permeability for typical sandstone reservoirs compared with that for a fracture system in which the fractures are parallel to the direction of flow.

have arbitrarily selected a relation that provides porosities equivalent to approximately twice that shown for the parallel system of 100 fractures/ft (NMOCC, 1969) and is the upper line on the field identified as "most fracture systems" on Figure 2-2. Figure 2-2 is an expansion of Figure 2-1 except that the above maximum limits line and a lower (not limiting) line have been added in lieu of the basic parallel fracture system of Figure 2-1. The purpose of Figure 2-2 is to aid in distinguishing fracture porosity from typical sandstone matrix porosity through production and interference testing. Central to this effort is identification of a maximum upper porosity limit for fracture systems of a given permeability.

The selected upper limit is equivalent to a system of fractures parallel to flow of 282 fractures/ft (24 fractures/in.). It is also equivalent to randomly distributed fractures of 180 fractures/ft (15/in.) using Van Golf-Racht's adjustment. It is difficult to conceive of a fracture system of any description having a porosity greater than that shown by a system of parallel fractures with a density of 24 fractures/in. (or randomly distributed fractures of 15 fractures/in.). For the system of parallel flow and 282 fractures/ft, the fracture widths for 10, 100, and 1000 md are, respectively: 0.0002 in., 0.0004 in., and 0.0009 in. (0.005, 0.01 and 0.02 mm).

The lower limit of ratio of porosity to permeability for a fracture system approximates the lower range of values determined for West Puerto Chiquito and Boulder fields, and may represent reasonable values for other commercial fractured reservoirs. It is not, however, intended to be a lower limit in the same context that the upper line is probably close to a true maximum upper limit for any fractured reservoir.

The curves of Figure 2-2 have been expanded both directions from the range of Figure 2-1 in order to examine the porosity-permeability relationships in "tight" reservoir rock and the hypothetical relations of the curves at a pore volume of 100% of the bulk volume.

The convergence of the upper and lower bounding lines for sandstone samples at 100% of pore volume results from the selection of the approximate boundaries of the Bulnes and Fitting (1945) sample population. There appears to be some analytical justification for this to occur because at pore volumes of 100% of bulk volume permeabilities of all systems of whatever nature will have to coincide.

On the other hand, the junction at 100% pore volume of the lower line for fracture systems is purely coincidence, and although of a certain amount of interest, it has no analytical significance. This line represents a fracture density of about 28 fractures/ft for a parallel flow system or about 18 fractures/ft for randomly distributed fractures; and at a permeability of 10^7 darcys the porosity of such fracture systems is projected to occupy 100% of the bulk volume (with the fractures apparently separated by infinitely small layers).

It is apparent from Figure 2-2, except for sandstone with unusual characteristics, if the production and interference testing shows total porosity values which fall below the sandstone relation for the corresponding permeability then at least some fracturing is present. Also if the total porosity falls above the maximum line for fractures of the corresponding permeability, there has to be some matrix contribution. Further, if the total porosity falls below the maximum fracture line, then the fracture porosity is probably a large part of the total porosity—and may even be 100% of it.

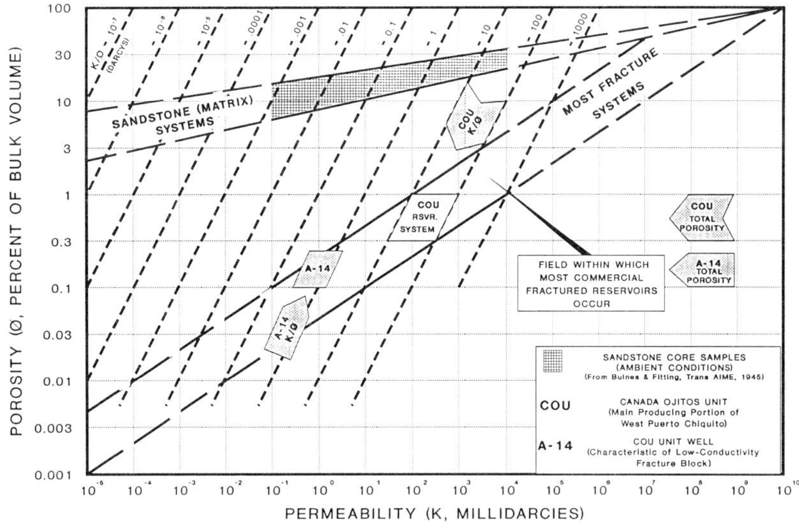

Figure 2-2. Data of Figure 2-1 expanded with fracture system replaced with upper (limiting) and lower (nonlimiting) definition of field for "most fracture systems." Reservoir properties from interference and production testing for Canada Ojitos Unit (COU) areas within West Puerto Chiquito are identified.

To locate the West Puerto Chiquito characteristics, we note that it is impossible from interference and production testing (or by any other method for that matter) to determine directly the values of porosity or permeability. Such testing, however, will reveal values for kh/μ, k_oh, ϕh, and k/ϕ. If k_{ro} is known or can be closely approximated as, for example, in tests at pressures above or slightly below the bubble point, then Kh can be determined. By "sampling" several thousand acres under such conditions, the bounds of the K/ϕ directing arrow for the "COU reservoir" field of Figure 2-2 were determined.

The bounds for the porosity field cannot be directly calculated but can be approximated as follows:

From the same interference testing, the average unit pore volume approximates 0.25 porosity-feet. Identifying the porosity requires estimating the pay thickness. It is impossible to measure exactly the pay thickness in West Puerto Chiquito wells. It is probable, however, that the pay thickness is confined to the identified resistive zones on electric logs. A reasonable assessment of this suggests a range of 30 to 100 ft (9–30 m) for pay thickness.

These thicknesses and 0.25 porosity-feet pore volume translate to a porosity of 0.3% to 1.0%. These bounds are plotted on Figure 2-2 identified as "COU total porosity." The "COU total porosity" relation intersects that for K/ϕ at the field identified as "COU reservoir system." By the same method Kh can be reduced to K and the field identified from ϕ and K. From visual inspection of Figure 2-2, however, the effect of varying thicknesses, and resultant ϕ, is more readily seen by interconnection of K/ϕ lines with those of ϕ.

The location of the "COU reservoir system" field lying in the graph's "most fracture systems" area means that it is possible—we believe probable—that all of the porosity is fracture porosity. For it to be otherwise, i.e., containing a matrix porosity, the porosity of the matrix after allowing a minimal amount for fractures, would have to be substantially less than 1%. We think this unlikely (see Appendix 1).

Note that for a typical fractured dual porosity (sandstone matrix) system, essentially all of the permeability will be from fractures. In this case the share of the total porosity that is from fractures plotted against (total) permeability will fall in the field "most fracture systems"; whereas the same permeability compared with total porosity will fall high above (in the sandstone field). Valhall field (Nelson, 1985), for example, produces from a highly porous chalk. Well test analyses here indicated a total permeability of 66 to 100 md and corresponding fracture porosity of 0.3 to 0.4% (Nelson, 1985), which falls in the same field as the "COU reservoir system," whereas Valhall total porosity falls far above this field. In contrast, the independent determination of total porosity at West Puerto Chiquito places it entirely in the field for fracture systems.

The "average" determined for the Canada Ojitos Unit results from the influence of the combined properties of both the low capacity fracture blocks and the high capacity fracture system. It is virtually impossible to determine the values of the high capacity system alone. In some cases, however, reasonably accurate determinations can be made of the characteristics of an individual fracture block, as, for example, the block containing the A-14 injection well (Sec. 14, T25N, R1W) described in Appendix 3 and plotted as Figure 2-2.

Note here the extremely low porosity (0.1 to 0.2%) of the A-14 fracture block and the fact that the permeability and total porosity fall within the field for fracture systems. As before, if independent knowledge showed the total porosity for the fracture block to be significantly greater than that for fracture porosity then it would indicate some matrix porosity. Clearly this is not the case here.

The A-14 fracture block porosity is an order of magnitude lower than the "COU reservoir system." This significantly lower porosity, typical of fracture systems, is markedly different from that for matrix porosity systems. Changes in permeability from one area to another in a matrix reservoir typically show relatively small porosity changes. If, for example, the West Puerto Chiquito reservoir contained effective matrix porosity, the change in porosity for a permeability change similar to the difference between the COU average value and that of the A-14 fracture block would be by a factor of less than two rather than the factor of ten shown by the tests. Furthermore, a matrix porosity of 0.1 to 0.2% is outside the bounds of any known producible matrix porosity reservoir.

In summary, this analysis supports the conclusion reached from interpretations of core data that the reservoir contains no effective matrix porosity.

Appendix 3. Geometry of the Niobrara Fractured Shale Oil-Producing Reservoir of West Puerto Chiquito

The main producing reservoir of the West Puerto Chiquito Pool lies within the Canada Ojitos Unit. It comprises a system of lower capacity fracture blocks joined by an interconnecting, high-capacity fracture system. In the main part of the reservoir, the size of the individual fracture blocks is in the range of 20 to 200 ac. The less productive parts of the reservoir may have fracture blocks of larger sizes.

That this is the geometry of the system is supported by the fact that many individual well build-up, drawdown, or pressure fall-off tests show capacities (Kh) in the range of 0.02 to 0.5 darcy-feet; whereas the overall average reservoir capacity is from one to three orders of magnitude higher than that of the individual blocks. The overall average physical properties can only be determined from interference and frac pulse tests. A frac pulse test is analysis of the pressure pulse generated by hydraulic fracture treatment (NMOCC, 1988).

Although some have suggested that the reservoir might be of dual porosity type, the production and pressure data do not support this premise. If the reservoir were of the typical dual porosity type (matrix porosity laced with fractures), then the reservoir average transmissibility determined from interference tests would be in the same general range as that shown by testing of individual wells. The difference—up to three orders of magnitude—indicates the individual well fracture blocks are linked together by a high capacity fracture system with exceptionally high transmissibility in order to bring the overall average up by such a degree.

Individual well pressure tests reflect a flow regime covering small areas, generally 100 ac or less. The only reservoir geometry that can be satisfied by pressure and production testing is the fracture block system described above and shown in idealized form on Figure 3-1. Note that this basic geometry is corroborated by Landsat and aerial photography (Figure 8). Although most of the reservoir area is occupied by the lower capacity fracture blocks, at least half of the in-place reservoir volume lies in the high-capacity fracture system bounding the fracture blocks. In some areas, as much as 80 to 85% of the total recoverable oil comes from the high-capacity

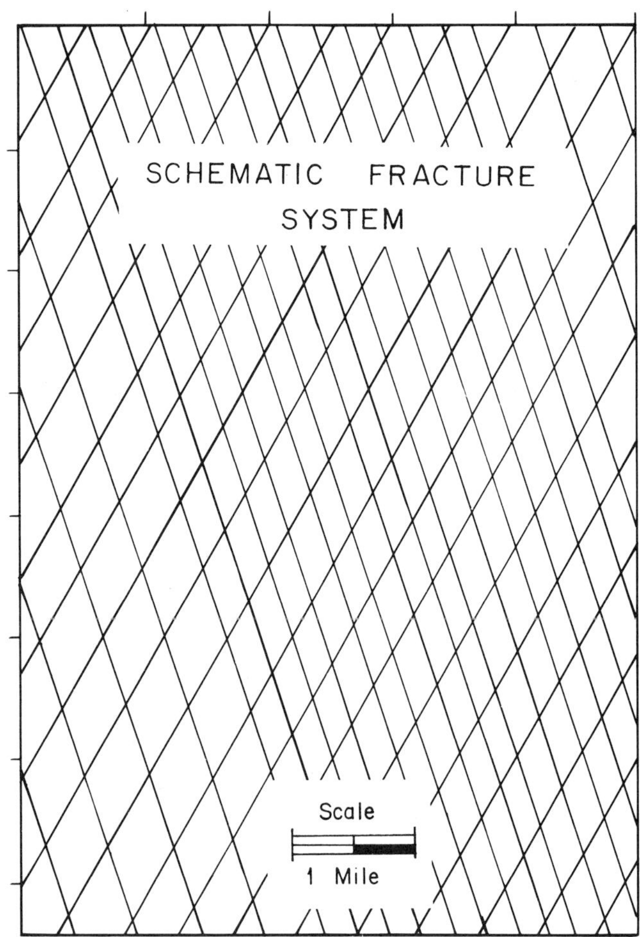

Figure 3-1. Sketch showing idealized fracture network of low capacity fracture blocks bounded by interconnecting high capacity fracture system.

system, depending upon the degree of operation of the gravity displacement producing mechanism.

It is impossible, of course, to determine the exact shape of a fracture block. However, it is not essential to the understanding of the reservoir flow system and its bearing on reservoir management to know the exact shapes or sizes of the fracture blocks. What is important is to recognize that the reservoir comprises a network of low-capacity fracture blocks linked by a high-capacity fracture system.

A well produced from a low-capacity fracture block has a relatively low transmissibility compared to that of the high-capacity fracture system such that the flow regime within the block is *essentially* that of "constant pressure at the boundary" for any time in its depletion history. For those parts of the reservoir where pressure is maintained by gas injection, the system is *exactly* that of constant pressure at the boundary. This is a fundamental distinction to make in analyzing data to determine reservoir properties.

The reservoir overall (the low-capacity fracture blocks and high-capacity fracture system taken together) average physical properties—Kh/μ, K_oh, and ϕh—can be determined from interference and frac pulse tests. Complete understanding of the reservoir and its behavior, however, requires a knowledge of these properties for the individual fracture blocks as well, particularly with respect to the type of porosity contained therein.

We make the observation that the unit pore volumes of the fracture blocks vary with their permeabilities, but in general these volumes are small compared to the overall reservoir average pore volume. For the individual fracture block, the value of Kh/μ, K_oh and K_gh can be approximated from build-up and drawdown or pressure fall-off tests. The value of unit pore volume, ϕh, for an individual fracture block, however, is more difficult to determine.

The calculation for an individual fracture block of its total pore volume, estimated unit pore volume (ϕh, in terms of stock tank barrels per acre, STB/ac, of hydrocarbon pore space) and its area are

presented here. Estimates of Kh and porosity are included, as well. In making such analyses it is preferable to use tests in which there is a minimum influence of wellbore storage and afterflow. The best wells for such testing, therefore, are gas injection wells. Furthermore, to minimize the effects of stratification, it is preferable to use wells with only one zone open to the wellbore. One such test is shown by the type curve, Figure 3-2, of a pressure fall-off test made in 1987 on the Canada Ojitos Unit A-14 injection well (Sec. 14, T25N, R1W) (NMOCC, 1989). The system is that of constant pressure at the boundary as determined from the leveling of pressures at t_{DX_f} values above 2 (if it were a closed system the plotted points would fall to the left of the infinite line, or if a boundary had not been reached, the points would follow the infinite line) (see SPE Paper 6015, Transactions AIME 1978, v. 265, p. 139).

The fact that the measured points depart from the theoretical line at values between t_{DX_f} of 0.4 to 1 may indicate that the fracture block is not of exactly uniform dimensions. In addition it is unlikely that the well is exactly centered on the induced fracture. This phenomenon would affect the shape of the curve to some extent, but since the conductivity of the fracture is so much higher than that of the surrounding rock, there would not be a significant difference if the well were located at one end of the fracture rather than in the middle.

The same reasoning applies regarding the location of the fracture within the fracture block. The leveling of the pressures will occur as a function of the shorter distances to the edge of the block such that any estimate of the block's volume or size will be on the minimum side.

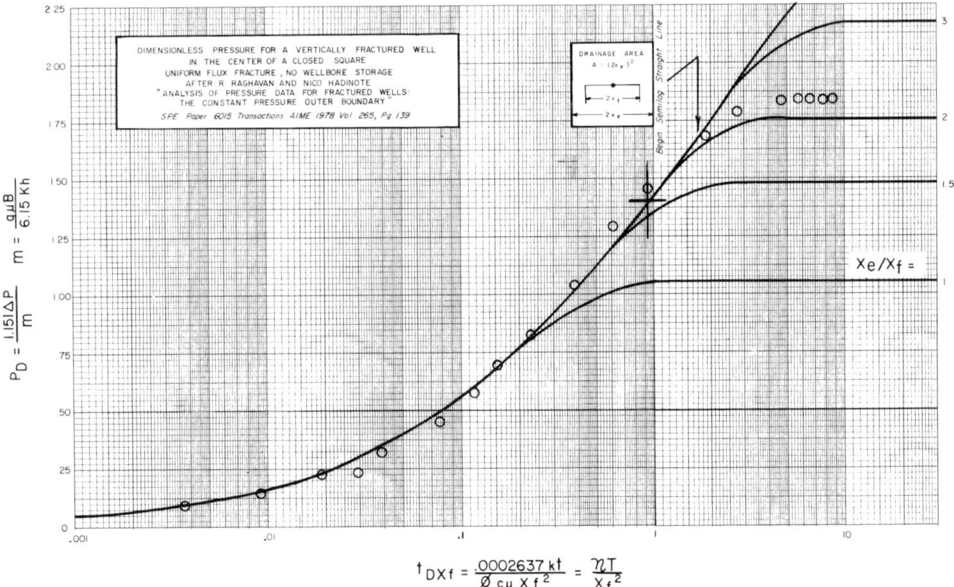

Figure 3-2. Plot of measured points of pressure fall-off test COU injection well A-14 (Sec. 14, T25N, R1W) in COU within the West Puerto Chiquito field. Plotted points are against background type curve for "constant pressure at the boundary."

Once a curve match has been obtained, then the total pore volume can be calculated. In the above example, $K_g h/\mu$ is 0.46 darcy-feet/cp and for C_T equal to 370×10^{-6} and the ratio of t to t_{DX_f} equal to $1/0.94$, the total pore volume in the fracture block is calculated to be 28,000 bbl of hydrocarbon pore space (equivalent to 22,000 STB).

Although it is impossible to determine fracture length or unit pore volume from this information alone, it is possible to determine how the value of one is dependent on the other. Furthermore, the relation of the total fracture length, $2x_f$, to unit pore volume can be described. Figure 3-3 shows this limit of definition in determining fracture length or unit pore volume. If one knows the fracture length, the unit pore volume can be determined, or vice versa. In the absence of independent information as to one or the other, however, this relation—one variable dependent on the other—is the limit of definition that can result from analysis only of a build-up or fall-off test.

To make an independent determination of the length of the induced fracture, we analyze the frac treatment causing the induced fracture. Although aided by analysis of the frac treatment, the unknown quantities are such that it is impossible to construct a mathematically exact solution that will provide unit pore volume. Given the fact that maximum average pore volume is probably less than 2000 bbl per acre (from interference tests in West Puerto Chiquito), it is possible to combine information from the two sets of data to arrive at an approximate solution.

With respect to analysis of the frac treatment, we note that in other wells tested by radioactive tracer surveys following frac treatment, and by production testing with zones individually fractured, the results show that there is very little build-up of frac height during frac treatments using low viscosity fluids in the Niobrara. Rather, the frac appears to be confined

to, if not the perforated interval, the particular zone (A, B, or C). For the subject well the perforated interval was 30 ft (9 m), and it is believed that the total frac treated interval could not exceed 50 ft (15 m).

The main variable in the instant case in estimating the length of the induced fracture is the leak-off of the frac fluid, or frac efficiency. Using frac efficiency as the variable, the fracture lengths were computed by the KGD model (Economides and Nolte, 1987) and the results for 30 ft (9 m) and 50 ft (15 m) fracture lengths displayed graphically (Figure 3-4).

These curves show the limit of definition in estimating frac length (one variable dependent on the other). Amount of leak-off, and efficiency, is next to impossible to determine. The relation does, however, provide an entirely independent method of estimating frac length. In calculating the curves by the KGD model, an oil viscosity of 3 cp, Young's Modulus of 5.6×10^{-6}/psi, frac treatment rate of 67½ BPM, and total volume of 3600 bbl were used.

With the knowledge that overall reservoir unit pore volume will probably not exceed the equivalent of 2000 STB/ac, this is used as the highest point on the ordinate of the plot in Figure 3-3. Using this information from Figure 3-3 and combining it with that of Figure 3-4 provides a method of estimating frac length as shown by reproducing the curves of Figures 3-3 and 3-4 on Figure 3-5. Although the maximum ordinate for the pressure fall-off test is only an estimate for the reservoir, it is a reasonable estimate. The result is not particularly sensitive to this estimated figure: If the correct figure were 1000 instead of 2000 bbl/ac, the consequence would be a reduction in pore volume of 50 to 100 bbl/ac.

From Figure 3-5, we determine the total frac length to be in the range of 700 to 800 ft (213–244 m), and the corresponding hydrocarbon pore volume to be 300 to 400 STB/ac. From this and the fracture block's

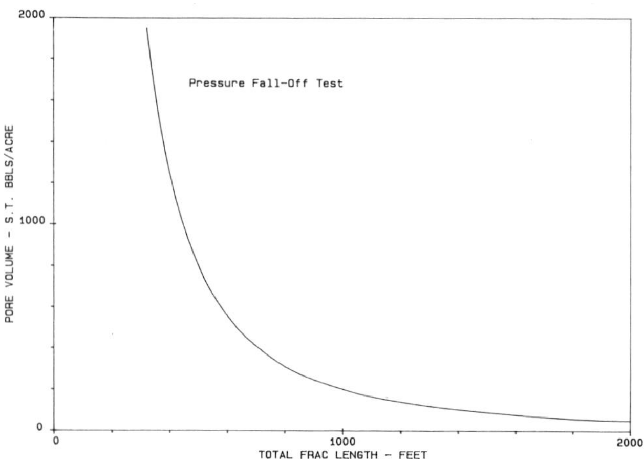

Figure 3-3. Relation of unit pore volume (stock tank barrels per acre) to fracture length of hydraulically induced fracture of the COU A-14 well determined from pressure fall-off test shown in Figure 3-2.

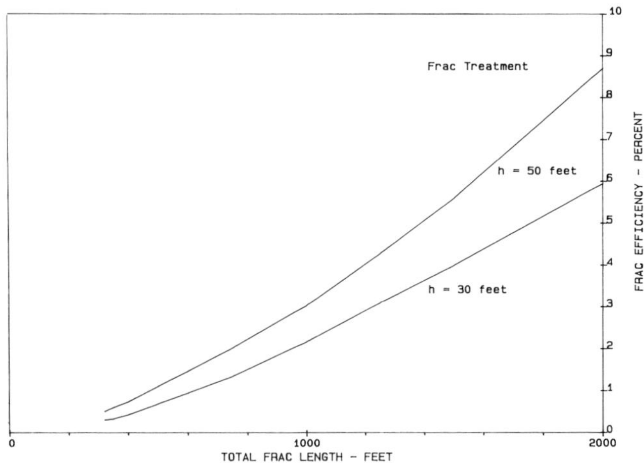

Figure 3-4. Relation of frac efficiency to total induced fracture length for hydraulic fracture treatment of the COU A-14 injection well calculated by the KGD and PKN methods.

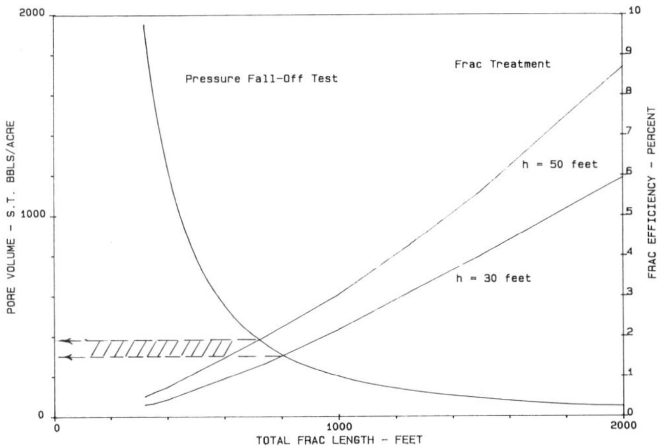

Figure 3-5. Combination of Figures 3-3 and 3-4 using arbitrarily selected maximum ordinate of 2000 stock tank barrels per acre for location of data of pressure fall-off test. Interpreted frac length 700 to 800 ft. Pore volume of well's low capacity fracture block approximately 300 to 400 barrels per acre using KGD method of hydraulic fracture analysis.

total pore volume, the minimum area occupied by the fracture block can be calculated; in this case, 55 to 75 ac.

Fracture lengths versus the frac efficiency were also calculated by the PKN method (Economides and Nolte, 1987). Using this method and a similar combination of data, the unit pore volumes are indicated to be in the range of 200 to 270 STB/ac with a fracture block size of 80 to 110 ac.

The 30 ft frac "height" line intersects the other at a fracture length of 800 ft (244 m) (Figure 3-5). For this fracture length, the fracture width is 0.19 in. This amounts to an effective sand volume supporting the propped fracture of ±44% of the 147,000 lb of frac sand used.

For the 50 ft (15 m) frac height line intersection at 730 ft fracture length, the average fracture width is 0.16 in., which indicates 58% of the sand volume is effective in propping the fracture.

As gas continues to be injected in this well, permeability to gas is increasing. The maximum value so far shown is a $K_g h$ of 0.015 darcy-feet. Since permeability to gas has increased approximately sevenfold since initial injection, it is reasonable to believe that K_{rg} is approaching a value of 0.5. Kh for this block would be 0.025 to 0.035 darcy-feet using a range of 0.4 to 0.6 for K. As noted above, the corresponding unit hydrocarbon pore volume is 200 to 400 STB/ac. From this $\phi h = 0.033$ to 0.067 and the resulting K/ϕ values range from 0.4 to 1.1. The corresponding average porosity values for the estimated 30 to 50 ft (9–15 m) of pay (and the 200 to 400 STB/ac determined above) range from 0.1 to 0.2%. These values (covering the ranges for both KGD and PKN models) are plotted in Figure 2-2 of Appendix 2 (example identified therein as A-14).

These porosity and corresponding permeability values for the A-14 fracture block converted to hydrocarbon pore volume (bbl/acre) are compared with those obtained in interference tests sampling large areas of the reservoir by displaying its field on a graph of capacity (Kh) versus unit pore volume in STB/ac (Figure 3-6).

On Figure 3-6, the basic X, Y, and Z lines derive from the porosity and permeability relations for the field identified as "some fracture systems" on Figure 2-2 of Appendix 2. The "probable upper limit for fracture porosity" is identified as the "A" relation, the lower part of the field as the "C" relation, and halfway in between, the "B" relation. Using these values and the number of feet of pay as shown on Figure 3-6, lines X, Y, and Z have been computed.

Figure 3-6 was prepared in 1969 (NMOCC, 1969) as a basis to work from in comparing unit pore volume as it might be dependent on Kh. Most of the test data were acquired by the operator of the Canada Ojitos Unit. Kh data for Boulder were from tests made by Standard of Texas (NMOCC, 1963). Location of the X, Y, and Z lines of Figure 3-6 is based on the premise that the thicknesses of the "pay zones," in this case the fractured reservoir units, are relatively uniform throughout the reservoir. Increases in pore volume between areas are caused by greater curvature (or bending stress) applied to the reservoir units, resulting in increased aperture width but relatively similar fracture density throughout. Thus, the ratio of oil in place of two areas varies as the cube root of the ratio of their respective capacities (Kh). Clearly we cannot expect this to be the case throughout; but it is surprising the number of tests that fall in the projected ranges.

The character of the fracture network makes unreliable such analyses as Horner plots where extrapolated to estimate reservoir pressure. One can be assured only that the reservoir pressure is as high as the last pressure measured. Extrapolation can be reliable only in the sense that it marks the maximum possible. Because of this infirmity, the Canada Ojitos Unit operator uses pressure fall-off tests following fracture treatment to estimate maximum reservoir pressure indicated by a well.

Numerous examples corroborate the existence of "tight" fracture blocks and nearby high capacity fracture systems. The Canada Ojitos Unit E-10, for example, was drilled through the pay zones with air. The flow stream analysis showed no hydrocarbons above that of the background overlying shales, and the well produced no oil or gas "natural." After frac treatment, however, it produced at high rates and has accumulated 2.3 million bbl. Clearly, the character of the formation in the bore-hole was not representative of the reservoir.

In conclusion, the reservoir geometry underscores the difficulty of attempting to analyze the reservoir through core analyses. The bulk of the recoverable reserves are probably located in open fractures of the high capacity fracture system. Reservoir property

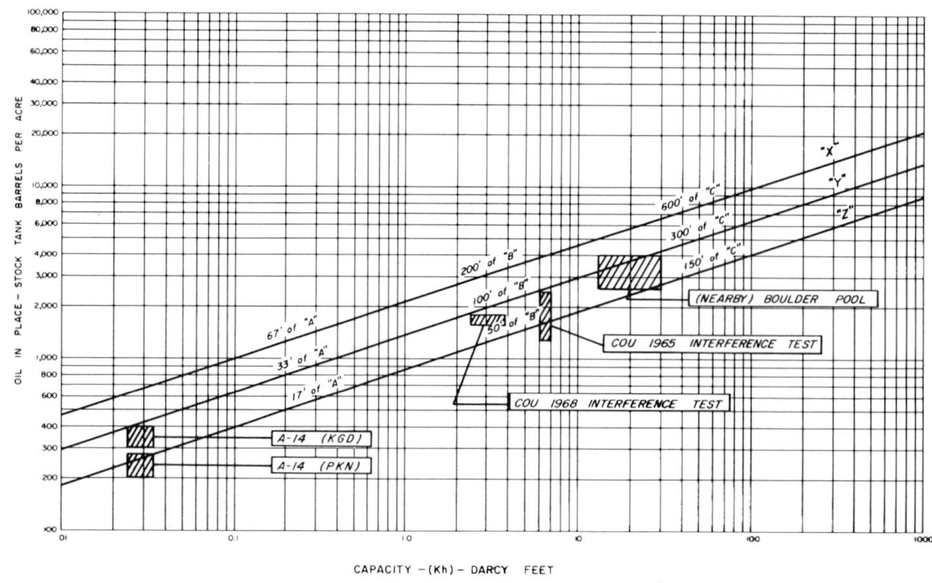

Figure 3-6. Plot of pore volume (stock tank barrels per acre) as dependent on capacity, *Kh*, determined from interference and pressure testing.

determination through core analyses of the low capacity "tight" block must be considered suspect. The properties of the high capacity system, from which the bulk of the reserves will come, are even more difficult to determine. Average characteristics reflected by interference and frac pulse testing, however, yield data useful in development planning and reservoir management.

Appendix 4. West Puerto Chiquito Fluid Property Data

Although in most solution gas drive reservoirs the process involves a mix of flash and liberation processes, in making our analyses we have used differential liberation fluid property data (Figure 4-1) since this more nearly represents the entire depletion process here. In support of this position, we offer the following observations.

With early free gas movement in a fractured solution gas drive reservoir, the free gas is removed from the system as soon as it reaches the wellbore. With early high GORs, this commences with first production, and the reservoir process is clearly one approaching strict differential liberation. Although liberated gas as it moves through the reservoir to the wellbore stays in contact with reservoir oil until it reaches the wellbore, so does gas liberated in the laboratory stay in contact with some oil for the length of the pressure "step" during the laboratory analysis. The laboratory simulation of the reservoir process may not be perfect, but values obtained from it are probably as accurate as the values determined for the other factors influencing a reservoir analysis. Most of the production in West Puerto Chiquito has been the consequence of gravity displacement: Here the liberated gas moves—not to the wellbore—but

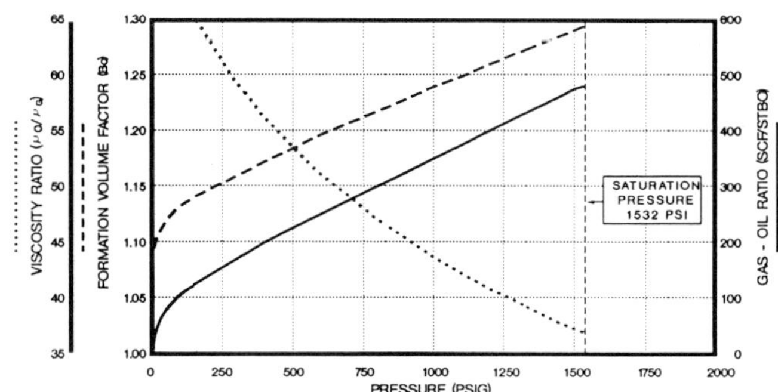

Figure 4-1. Fluid property data from COU K-13 well, Sec. 13, T25N, R1W, in West Puerto Chiquito field.

updip to form a gas cap, and the result is the same, differential liberation.

Since part of the recovery in West Puerto Chiquito will be by solution gas drive and most of the recovery in nearby fields will be by solution gas drive, we now review in more detail the influence of both differential and flash liberation characteristics for solution gas drive.

In estimating the portion of the overall depletion process that is influenced by differential and flash characteristics, it is instructive to review the mathematical relation of the various factors influencing solution gas drive depletion. This is most clearly accomplished by inspection of Muskat's method summarized in Appendix 5 (Muskat, 1945). The first step of equation (1) in Appendix 5 defines change in reservoir oil saturation with pressure decline. This step, as noted above, is clearly independent of flash data. Simply expressed, the gas and oil arrive at the wellbore by the differential depletion process, what happens thereafter will not have a retroactive effect on the reservoir's oil saturation.

The only effect of flash data lies in the determination of its share of the process which relates the volume of produced oil to the volume it occupied in the reservoir. Note that this produced volume is small (5 to 10% of oil-in-place for solution gas drive) compared to the remaining reservoir volume controlled by differential data. Furthermore, note that the fraction of the total stream that reaches the stock tank as oil depends on the "path" the mixture takes from the bottom hole of the well bore up the production string(s) to the stock tank. The paths taken by the production in the different wells may be different; but field-wide for the various operators the overall effect of the flash process modifying reservoir performance calculated by differential data will be small. In summary, it amounts only to the differences in amount of gas dissolved in oil at the stock tank for the different paths taken by the oil up the production strings. These different paths are defined by the production method used.

In our analyses, we use differential data all the way through. Not only is the reservoir process differential; but also we note that a substantial amount of a field's production as it moves from the bottom of the wellbore to the surface is differential liberation:

1. A large volume of oil has been produced in West Puerto Chiquito with submersible hydraulic pumps. Here as the produced oil moves to the surface, it is in continual contact with the power oil, and as gas comes out of solution from the produced oil, it immediately goes into solution in the power oil in a strictly differential liberation process. On flashing into the separator, the gas still in solution in the power oil prevents it from contacting the produced oil; so here again the process is still largely differential liberation.
2. Sometimes pumping wells are produced such that the gas is separated from the oil in the bottom of the well and moves up the annulus separate from the oil. Here the process is differential liberation until the two streams are combined in the separator.
3. In some of the wells that flow by gas lift by heads, the gas and oil are separated in the bottom of the wellbore and stay separated all the way to the stock tank—a form of the differential liberation process.

In summary, fluid sample data determined by the differential liberation process is appropriate for reservoir analyses at West Puerto Chiquito.

Appendix 5. Relative Permeability and Formation Compressibility

Fluid withdrawal from a closed reservoir causes pressures to decline and net overburden pressures to increase, resulting in a reduction in pore volume. Change in overburden pressure causing a reduction of as much as 50% of the pore space has been determined not to affect the ratio of relative permeabilities (gas to oil) in a sandstone (Fatt, 1953). In the absence of information to the contrary, we think it probable that the same will hold for a fractured reservoir.

Although this characteristic does not change, as pressure declines formation compressibility (C_f) causes a reduction in pore volume and, as a consequence, reduces the free gas saturation. Accordingly, the corresponding relative permeability ratio (k_g/k_o) will be smaller than that had the pore volume not been reduced. If the formation compressibility were of the same order of magnitude as system compressibility, it would have a marked effect on reservoir performance.

At West Puerto Chiquito and vicinity, neither the value of C_f nor that of k_g/k_o is precisely known. Since each property operates to influence the same factors controlling reservoir performance, we examine these two characteristics together.

Reported Formation Compressibilities

Three separate analyses for formation compressibility have been reported for West Puerto Chiquito and the offsetting pool, Gavilan:

1. Comparison of interference tests in West Puerto Chiquito at pressures above and below the bubble point.
2. Brine squeeze test of core samples from a Gavilan field well.
3. Special test to estimate fracture compressibility from cores of another Gavilan field well.

Remarks with respect to these tests follow.

1. An interference test that "sampled" several thousand acres in the Canada Ojitos Unit in West Puerto Chiquito in 1965 (NMOCC, 1966) when the oil was undersaturated and the formation compressibility was significant (with respect to system compressibility) showed reservoir volume of oil in place ranging from 1000 to 2500 STB/ac for the assumed range of formation compressibilities of 26×10^{-6}/psi to 6×10^{-6}/psi. Another interference test run in 1968 (NMOCC, 1969) in part of the area covered by the 1965 test showed stock tank oil-in-place volume approximating 1700 bbl/ac. This test was run when pressures were below the bubble point and the formation compressibility was believed to be relatively insignificant. A pore volume of 1700 bbl/ac shown by the second interference test approximates the average estimated from the first test. The formation compressibility would then be the average of 6×10^{-6}/psi and 26×10^{-6}/psi; or approximately 15 to 16×10^{-6}/psi.
2. Brine squeeze tests were run using cores from the Mobil B-38 well, Sec. 4, T24N, R2W (personal communication, Mobil to Gavilan Engineering Committee, 1986). From the plot of volume change versus applied pressure, the calculated compressibilities were determined to range from 6×10^{-6}/psi to 16×10^{-6}/psi.
3. Information from the special tests directed at estimating fracture compressibility from cores of the Mallon Davis Federal 3-15, Sec. 3, T25N, R2W, was reported for three samples (NMOCC, 1987). They were approximately 50×10^{-6}/psi, 100×10^{-6}/psi, and 150×10^{-6}/psi, a spread of 100×10^{-6}/psi. The laboratory qualified its results because of the difficulty in supporting the samples in such a fashion as to properly simulate reservoir conditions.

We believe the best information as to value of formation compressibility is the comparison of the interference tests. The brine squeeze tests, while yielding values approximating those of the interference tests, are subject to the same limitations inherent in all core analyses of this reservoir: Core samples are not representative of the producing reservoir (Appendices 2 and 3).

It is to be expected that compressibility will vary somewhat from one area to another. It is unlikely, however, that it will vary over the extreme ranges indicated in item 3 above. Even so, the study was expanded to include analyses involving the reported high compressibilities in order to cover completely the effect of such phenomena should they in fact exist.

"Critical Formation Compressibility"

For an overview of the effect of formation compressibility on solution gas drive performance, we inspect Muskat's basic formula (Muskat, 1945), equation (1) below. Note that the left-hand member of equation (1) is actually value *per unit pore volume* and has the same dimensions as formation compressibility. Note, also, early in the depletion cycle when free gas saturation and k_g/k_o are low, that $\Delta S_o/\Delta P$ approaches the value of lambda. Lambda is the dominant member of the two terms defining saturated oil compressibility (Ramey, 1964), which for the subject reservoirs is 200 to 300×10^{-6} (Figure 5-1). This is the same order of magnitude as the high values (100 to 150×10^{-6}/psi) of formation compressibility indicated by the tests of item 3 above. Thus, without analyses, but simply from inspection of the formula, it is clear that if formation compressibilities were in fact this high, they would modify significantly the otherwise normal reservoir performance.

For instance, if the initial value of $\Delta S_o/\Delta P$ were the same as that of formation compressibility, then the volume change of the oil saturation would be balanced by the pore volume change due to the formation compressibility and little free gas would evolve as oil is withdrawn. Gas oil ratios would decline initially and never reach high values with depletion. The fractures would contract such that the porosity would be significantly reduced from its original value. If the initial GORs were low, a very large part of the initial oil in place would have been expelled at depletion.

We refer here to the formation compressibility that initially balances $\Delta S_o/\Delta P$ as the "critical" formation compressibility. If, initially, formation compressibilities are above "critical," reservoir pressures will not decline on withdrawal of oil, and the solution gas drive process cannot take place. We have not included production behavior for such a situation, as we think

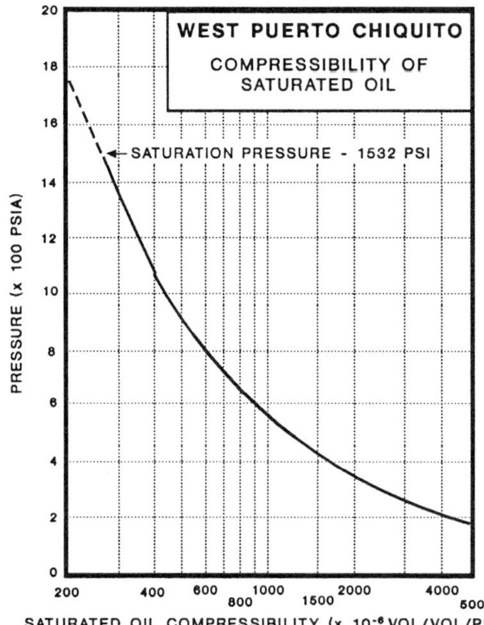

Figure 5-1. Compressibility of saturated oil from West Puerto Chiquito fluid property data.

it unlikely that it would actually occur. We make this note, however, since for one of the k_g/k_o curves studied herein, the critical formation compressibility approximates 100×10^{-6}/psi—well within the range of the 50 to 150×10^{-6}/psi reported for the special compressibility tests.

To illustrate production behavior where the value of formation compressibility approximates, but is slightly less than, "critical," as well as sensitivity of the production histories to different values of C_f, we present GOR histories utilizing three relative permeability ratio curves (curves I, II, and III of Figure 5-2). These three curves lie in the range indicated by laboratory measurements of relative permeability of fractured formations (Keeling et al., 1964, 1969). Curves I and II will "bracket" the characteristics on the high and low sides; and curve III lies in between these extremes.

Only GOR histories are displayed, since this provides a means of comparison with field performance from a minimum of data; namely, oil and gas production. In all of our determinations of sensitivity to formation compressibility, certain precautions were taken to insure that the differences in curves were indeed the consequence of changing C_f and not caused by mathematical inaccuracies. Straight line curves (saturation varies as the logarithm of k_g/k_o) were chosen since they permit precise calculation of k_g/k_o for any given saturation. Also the Muskat method, with formation compressibility recognized

(equations 5, 6, and 7 below), was used. Once composite functions have been determined for the average pressure of each step, there remain only two unknowns: oil (or free gas) saturation and the corresponding relative permeability ratio. Although in developing production histories of typical reservoirs it is possible to obtain acceptable results with this method by the direct calculation at each step from a reasonable projection of the average saturation for each preceding step, greater precision is reached by converging the saturation and relative permeability ratio values to exact matches. An integral part of the computer program used performs this convergence. Approximately 40 precise points were used to define each curve up to a free gas saturation of 20%. The program identifies and then makes a straight line interpolation between the relevant points to determine the convergence. Thus, although the k_g/k_o relations used do not exactly follow the curves, being instead a series of points on the curve with interconnecting straight lines, they are precisely the same for each set of C_f comparisons.

Fluid property data for West Puerto Chiquito (Appendix 4) were used along with 10% connate water saturation and an abandonment pressure of 125 psia. With these parameters the "critical formation compressibility" approximates the values in Table 5-1. (Higher water saturations would result in lower values for critical formation compressibilities.

Table 5-1

For curve (of Figure 5-2)	Initial k_g/k_o	Approximate "Critical" Formation Compressibility
I	0.1	100×10^{-6}/psi
II	0.001	300×10^{-6}/psi
III	0.01	230×10^{-6}/psi

Plots of GOR versus oil recovery in percent of oil in place are shown on Figure 5-3 for these three relative permeability curves and associated "critical" formation compressibility. Note the relatively high recoveries for curves II and III and the overall shape of the GOR curves: an initial decrease with only a slight increase over the depletion history. The curve shapes bear little resemblance to those of typical solution gas drive reservoirs.

Overview: Sensitivity of Oil Recovery to Formation Compressibility

An example of variation in oil recovery with C_f is demonstrated using relative permeability curve II of Figure 5-2. Here k_g/k_o and all parameters other than C_f are held constant. The results are displayed on Figure 5-4 where the GOR histories are plotted against cumulative recoveries in percent of oil in place for the several formation compressibilities shown.

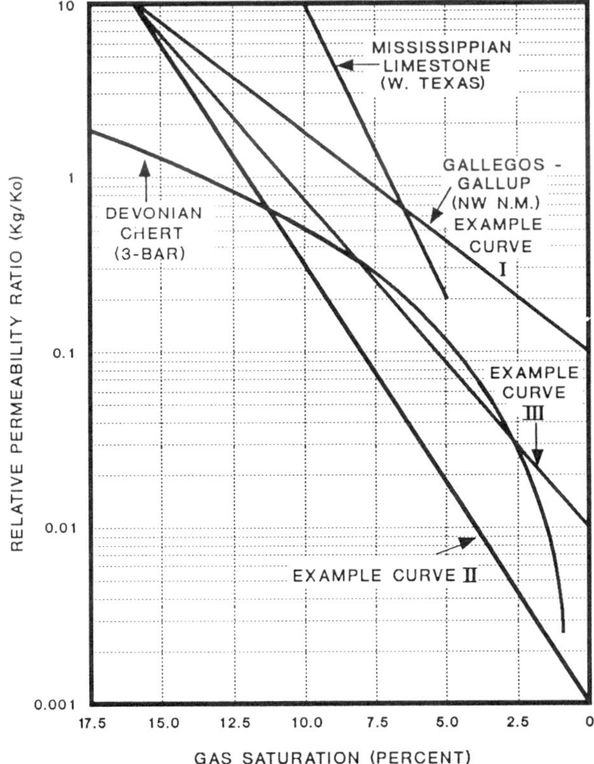

Figure 5-2. Ratio of relative permeability characteristics for some fractured reservoirs and example curves used in analyses.

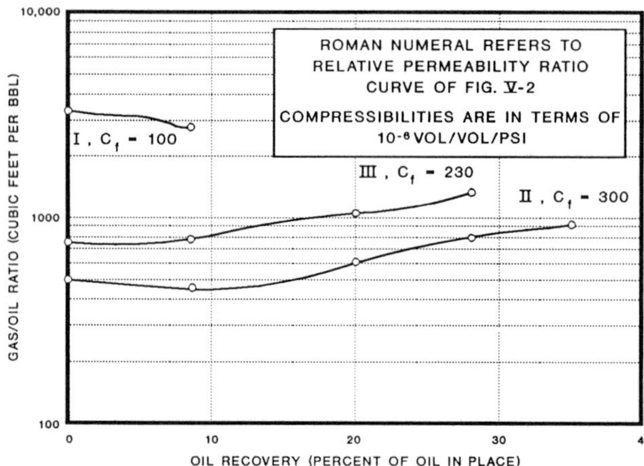

Figure 5-3. Solution gas drive GOR histories using "critical" formation compressibility. Note high recoveries for k_g/k_o curves II and III and that shape of all GOR curves bears no resemblance to that of typical solution gas drive reservoirs.

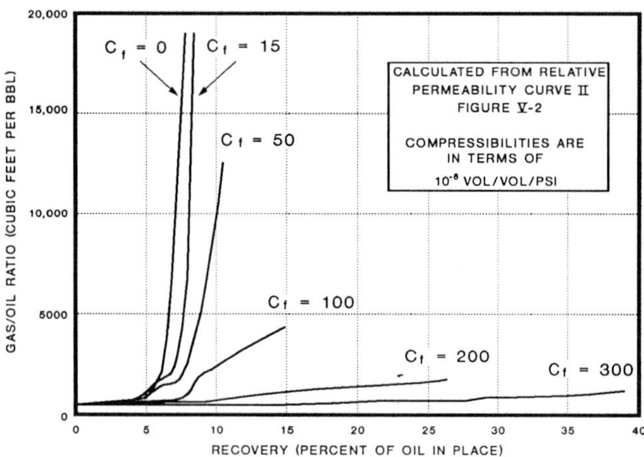

Figure 5-4. Solution gas drive recovery using relative permeability curve II and various formation compressibilities. Note extreme range of recoveries for the different compressibilities.

Here the ultimate recovery is identified by the end point of each GOR line. They show a range of recovery of five to one (confirming the observation that high C_f's would indeed be significant) for C_f ranging from the "critical" to zero (Table 5-2).

Estimate of Relative Permeability and Formation Compressibility through Comparison with Field Data

For purposes of comparison with calculated performance using the k_g/k_o curves and several values of C_f, we use the field performance of central Gavilan offsetting West Puerto Chiquito. Note that West

Table 5-2

C_f (10⁻⁶/psi)	Ultimate Recovery of Oil in Place
0	8.8
15	9.2
50	11.0
100	15.7
200	28.0
300	41.6

Puerto Chiquito's pressure maintenance and relatively large amounts of gravity drainage preclude accurate determination of k_g/k_o from its production data. Although central Gavilan (Figure 2) has received substantial gravity drainage and support from migration, its dominant recovery mechanism, at least for the latter part of the depletion cycle, is solution gas drive. Analysis of solution gas drive provides k_g/k_o data.

In estimating relative permeability characteristics from central Gavilan production data, we note the following:

1. Because of the high conductivity of the fracture systems, significant gravity drainage has occurred (see "efficiency" above), even in areas of low dip. The result is lower GORs initially and apparently lower relative permeability ratio (k_g/k_o) than the reservoir's true character. In the latter stages of depletion, however, an effect of the earlier gravity drainage is to cause lower oil saturations than would be the case for pure solution gas drive. Therefore, the GORs and apparent relative permeability ratios will appear higher than had the process been pure solution gas drive. Gravity drainage, in other words, renders the determination of oil saturation associated with a given relative permeability ratio quite difficult.

2. Different areas of the reservoir will have different degrees of fracturing and may cause actual relative permeability ratios to be different in the different areas. The principal effect of the varying degrees of fracturing, however, will be reflected in the gravity drainage effects rather than difference in relative permeability ratios.

3. Even if reservoir pressures are reasonably well known and k_g/k_o can be calculated from GORs and estimated reservoir pressures, determination of the corresponding true oil saturation is more difficult, and sometimes the procedure cannot be completed. In Gavilan, where indefinite amounts of gravity drainage and regional migration have occurred, it is impossible to determine precisely the concurrent oil saturation. Regional migration precludes estimation of free gas saturation from the elementary relation of oil produced below the bubble point and an estimate of oil in place, along with formation volume factors. To enlarge the study area to include lands providing regional

migration in the analyses brings in complications more difficult to deal with: The Bear Canyon Unit (Figure 2) is downdip, with a lower well density, more efficient gravity drainage (a depletion process significantly different from Gavilan's), and migration from the east, where pressure maintenance support is involved. In spite of the problems of dealing with production above the bubble point, gravity drainage, and migration, the GOR history of Gavilan nevertheless provides some insight with respect to ratio of relative permeability and formation compressibility.

Since central Gavilan is essentially oil depleted (Figure 5-5), it is possible to make a reasonable projection of (economic) ultimate recovery: 5 million bbl. Any economically recoverable oil beyond that amount will be from stripper wells supported by gas sales or from "protected" wells still receiving gravity drainage. With an estimate of central Gavilan's ultimate recovery, it is possible to relate its GOR history to recovery in percent of ultimate. This provides one method of estimating C_f and k_g/k_o from a minimum of required field data; namely, oil and gas production.

Central Gavilan encompasses approximately 24,000 ac and has produced largely by solution gas drive from both high and low capacity wells with a high degree of communication. Although by no means a "perfect laboratory" for determining C_f and k_g/k_o, the overall production with appropriate adjustments should reveal nearly average formation properties.

k_g/k_o characteristics influence, and can be determined from, reservoir performance at pressures below the bubble point. Analysts have not agreed on Gavilan's bubble point pressure, but the highest reported (NMOCC, 1987) is 1660 psi. Approximately 1 million bbl were produced prior to the reservoir reaching this pressure.

Statistics for central Gavilan's GOR history for production below this bubble point pressure are set out in Table 5-3. A plot of GOR versus ultimate recovery is shown in Figure 5-6. The first 65% of ultimate recovery reflects a smoothed curve of the published data (NMOCC, 1988) for production below the bubble point. Data for the rest of the curves come from monthly reports to the New Mexico Oil Conservation Commission.

The lower curve extending from this base ("all wells," Figure 5-6) derives from reported data for all wells in central Gavilan, while the upper curve ("all wells except two," Figure 5-6) excludes two wells (Hill Trust #1, Sec. 5, and High Adventure #1, Sec. 8; both in T25N, R2W). These two wells are still receiving substantial gravity drainage and do not reflect the relative permeability characteristics shown by the other approximately 60 wells. If the production of these wells were to be included in determining average GOR (and k_g/k_o values), the result would be values lower than the true average reservoir characteristics.

Similarly, other wells initially receiving gravity drainage and indicating k_g/k_o values lower than the true reservoir character will, in the later stages, exhibit higher gas saturations and k_g/k_o values than the true reservoir average. As a consequence, the curve for "all wells less 2" (Figure 5-6) shows higher GORs than would be the case in the absence of gravity drainage. We have therefore selected the GOR history between these two extremes ("average" curve of Figure 5-6) as more nearly representative of the actual relative permeability character for this part of the depletion cycle.

GOR histories for various formation compressibilities were calculated for the (straight line) k_g/k_o curves I, II, and III (Figure 5-2) and compared to the central Gavilan GOR history on Figures 5-7, 5-8, and 5-9. Clearly, the calculated curve forms for compressibilities exceeding 15×10^{-6}/psi bear no resemblance to the curve of field performance, forcing the preliminary conclusion that the value of C_f is probably low.

GOR histories using relative permeability ratio curves I and II (Figures 5-7 and 5-8) "bracket" the field data on both the high and low sides. As k_g/k_o characteristics move in the direction of a curve "match" (from curve II to curve III on Figures 5-8 and 5-9), the curves with higher C_f's show greater divergence at the latter part of the depletion cycle. Clearly, as convergence is approached between calculated and observed curves, values of C_f decrease, making it unlikely C_f will be greater than 15×10^{-6}.

Finally, use of k_g/k_o curve IV (Figure 5-10) with $C_f = 15 \times 10^{-6}$/psi results in an approximate match with field performance (Figure 5-11). Use of higher C_f values (not plotted) show greater divergence from field performance as the match improves. Moreover, for the reasons noted below, the true k_g/k_o relation is probably one that would shift the curves farther left (past a "match"), reducing even further the

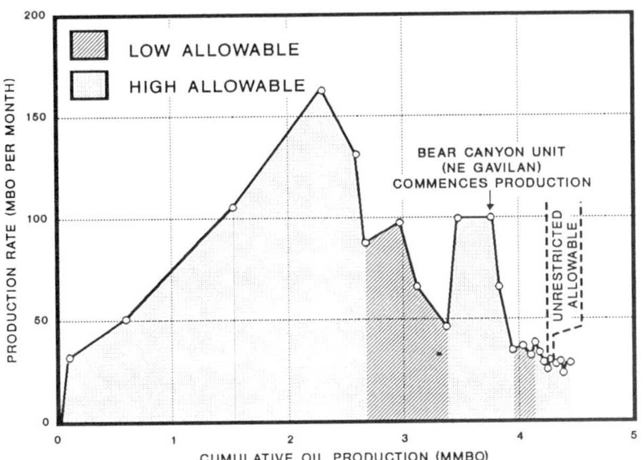

Figure 5-5. Plot of production rate versus cumulative recovery for central Gavilan. Periods of high and low allowables are identified along with one period of unrestricted allowable. Economic ultimate recovery of 5 million bbl estimated from this plot.

Table 5-3. Central Gavilan gas-oil ratios and cumulative production

		Prod. below Bubble Point		Gas-Oil Ratios *		
	Cumulative	Cumulative	Ultimate	Central	Central Gavilan Less	
Date	Oil Prod.	Less 1000	Recovery	Gavilan	Two Wells **	Average
	(M Bbl)	(M Bbl)	(%)	(cf/bbl)	(cf/bbl)	(cf/bbl)
	0			550	550	550
07/85	1,000	0	0	850	850	850
12/85	1,500	500	12.5	1,100	1,100	1,100
06/86	2,000	1,000	25.0	1,400	1,400	1,400
08/86	2,500	1,500	37.5	2,000	2,000	2,000
01/87	3,000	2,000	50.0	3,150	3,150	3,150
08/87	3,500	2,500	62.5	4,600	4,600	4,600
05/88	4,050	3,050	76.3	8,000	9,600	8,800
07/88	4,100	3,100	77.5	9,000	11,600	10,300
10/88	4,200	3,200	80.0	10,800	15,200	13,000
01/89	4,300	3,300	82.5	13,000	18,800	16,000
		(4,000) ***	(100.0) ***			

*   GORs are smoothed averages.
** Hill Trust #1 and High Adventure #1, Sec. 5 and 8, T25N, R2W.
*** Ultimate recovery for production below bubble point estimated at 4000 M bbl.

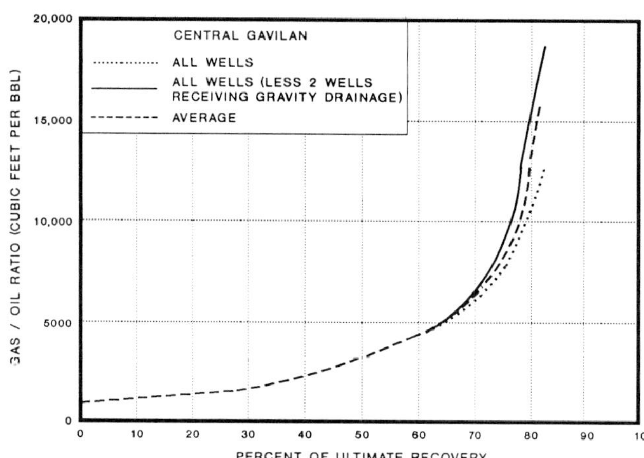

Figure 5-6. GOR history of central Gavilan from field data. Average curve is used in making comparisons of calculated histories with field performance. Percent of ultimate recovery is based on cumulative recovery of production below maximum assumed bubble point pressure of 1660 psi.

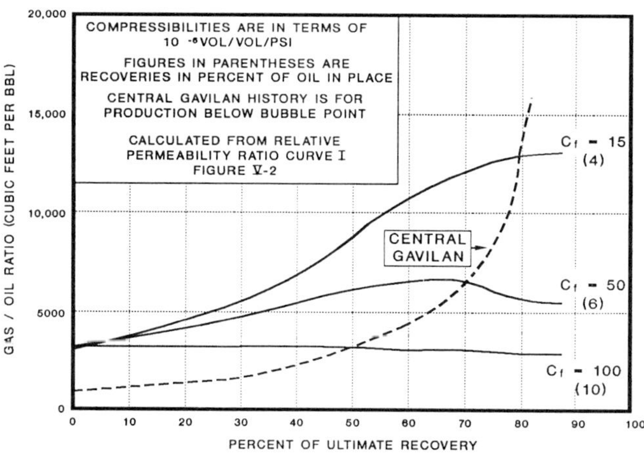

Figure 5-7. Calculated solution gas drive recoveries using k_g/k_o curve I of Figure 5-2 and various formation compressibilities. Histories are GOR versus ultimate recovery. Note failure of calculated curves to "match" field performance.

possibility that high formation compressibilities exist.

Gavilan's greater depths and higher temperatures would result in a higher bubble point than that determined for West Puerto Chiquito oil. Solution gas drive recoveries, however, are not sensitive to small differences in initial bubble point pressures; recognition of a different bubble point would not materially affect such analyses. Furthermore, a higher bubble point than that used would result in lower values of the composite function lambda and cause even greater divergence between the calculated curves using high formation compressibilities and the field data.

The foregoing analysis, while not completely definitive, clearly eliminates formation compressibilities in the range of 100 to 150 × 10⁻⁶/psi. In fact, it is unlikely that C_f is higher than 15×10^{-6}/psi.

94

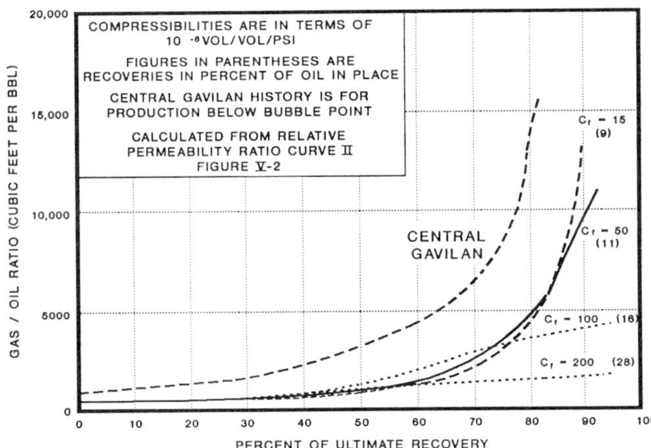

Figure 5-8. Calculated solution gas drive recoveries using k_g/k_o curve II of Figure 5-2 and various formation compressibilities. Note these curves fall below that of field performance. Note divergence from field data of curves for $C_f = 100$ and 200×10^{-6}.

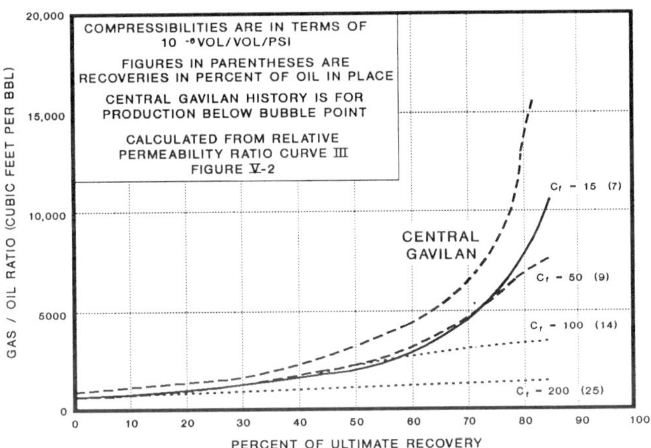

Figure 5-9. Calculated solution gas drive recoveries using k_g/k_o curve III of Figure 5-2 and various formation compressibilities. Note low value of maximum GORs for curves with C_f of 50×10^{-6} and greater.

Data of this analysis, although adequate to eliminate high values of C_f as being appropriate, are not sufficiently precise to identify formation compressibility in the comparatively narrow ranges up to 15 or 20×10^{-6}/psi. For this definition we rely on the inherently more accurate results obtained from the interference test data.

Probable Range of "True" k_g/k_o Characteristics

An approximate match of GOR and percent ultimate recovery results from relative permeability curve IV and C_f of 15×10^{-6}/psi (Figures 5-10 and 5-11). No adjustment has been made to compensate

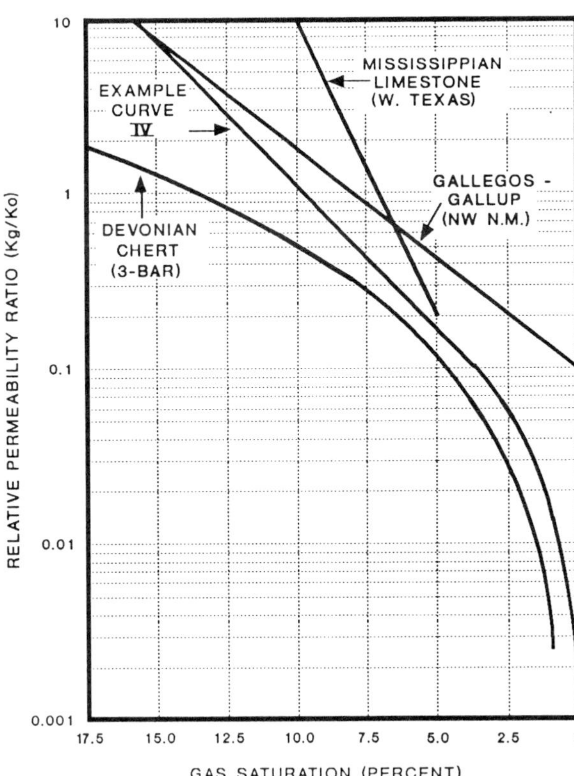

Figure 5-10. k_g/k_o example curve IV compared with those of some fractured formations.

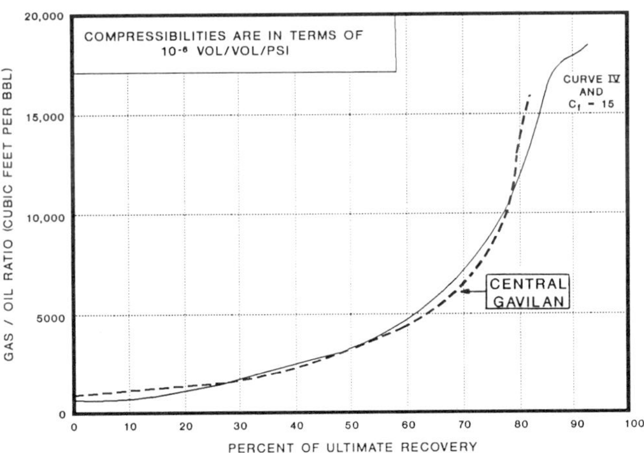

Figure 5-11. Calculated solution gas drive recoveries using k_g/k_o curve IV (Figure 5-10) and using formation compressibility of 15×10^{-6}. Despite apparent match authors believe true representation of k_g/k_o, in the absence of migration and gravity drainage, would be shifted to the left.

for regional migration or gravity drainage effects early in the depletion cycle. Regional migration would have no effect on the shape of the curve if it were present *throughout* the depletion cycle. The effect in central Gavilan, however, where regional migration was substantial initially but decreased with time,

95

is similar to that of gravity drainage when GORs are plotted against ultimate recovery: lower initial GORs than for no migration; then as migration is reduced, rising more steeply than had no migration been present.

In view of the above, curve IV, despite its apparent "match" with field performance, is somewhat optimistic for the "true" k_g/k_o relation. Curve IV represents limiting lower values of k_g/k_o; any adjustment for gravity drainage and migration would shift the k_g/k_o relation higher on the plot of Figure 5-10, at least for the early part of the curve. The latter part of the k_g/k_o curve (at higher gas saturations) would most likely "flatten" such that the probable range of "true" k_g/k_o values would assume a form within the shaded area shown on Figure 5-12.

The effect on ultimate solution gas drive recoveries of the different curves is shown by example:

At 175 psi abandonment pressure and $C_f = 15 \times 10^{-6}$/psi, curve IV yields a solution gas drive recovery of 5.94% of oil in place, while the curve represented by the extended shading results in a recovery of 5.65%. Apparently the k_g/k_o values for gas saturations greater than 10% have a minor effect on recovery. At a free gas saturation of 10%, the solution gas drive recovery using curve IV is 90% of ultimate recovery (recovery using the extended shading curve is 80% of its ultimate).

In summary, we conclude that formation compressibility is that shown by the interference tests: approximately 15×10^{-6}. The exact shape of the k_g/k_o relation will turn on the indefinite amount of migration and gravity drainage of central Gavilan, with minimum values represented by curve IV (Figures 5-11 and 5-12). It is probable that true k_g/k_o fits in the shaded area of Figure 5-12. Whatever its exact shape, it appears that the resulting solution gas drive recovery will not exceed 6% of oil in place.

Muskat Method of Calculating Solution Gas Drive Performance

The Muskat method (Muskat, 1945), after substituting S for ρ, and R_s for S, can be described, in incremental form, as follows:

$$\Delta S_o / \Delta P = \frac{S_o \lambda + (1 - S_o - S_w) \varepsilon + S_o \eta (\psi)}{1 + (\mu_o/\mu_g)} \qquad (1)$$

$$\text{Gas Oil Ratio, } R = \alpha \psi + R_s \qquad (2)$$

Oil Recovery (fraction of initial pore space) =
$$(S_{oi}/B_{oi}) - (S_o/B_o) \qquad (3)$$

Oil Recovery (fraction of initial oil in place) =
$$(S_{oi}/B_{oi} - S_o/B_o)/(S_{oi}/B_{oi}) \qquad (4)$$

where $\lambda(p) = (1/B_o\gamma) (dR_s/dp)$

$$\varepsilon(p) = (1/\gamma)(d\gamma/dp)$$

$$\eta(p) = (1/B_o)(\mu_o/\mu_g)(dB_o/dp)$$

$$\alpha = (\gamma B_o)(\mu_o/\mu_g)$$

S_o = oil saturation, fraction of pore space

S_w = water saturation, fraction of pore space

μ_o/μ_g = ratio of viscosity of oil to gas

ψ = relative permeability ratio

B_o = reservoir barrels of oil per stock tank barrel

γ = standard cubic feet of gas per reservoir barrel of oil

R_s = gas in solution in the oil at the subject pressure, cubic feet per barrel

i means initial conditions

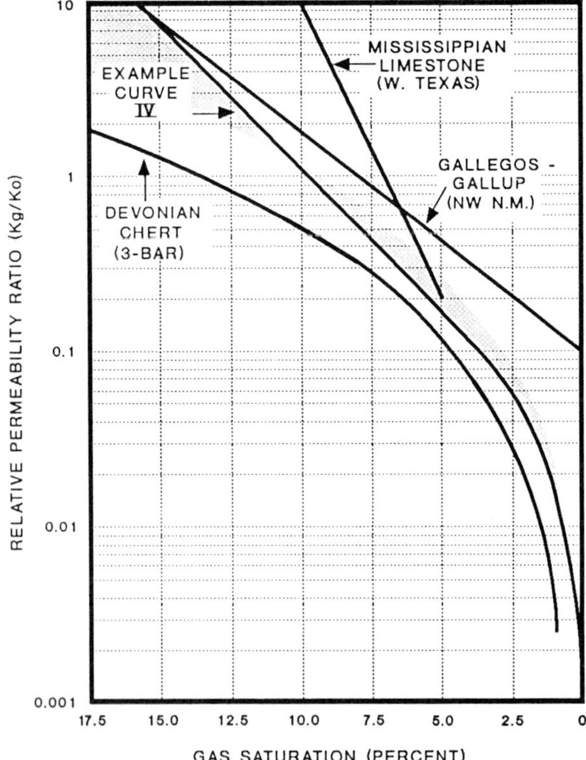

Figure 5-12. k_g/k_o curve IV showing adjustment (shaded area) in which true k_g/k_o relation should lie (after accounting for migration and gravity drainage).

(The derivatives dR_s/dp, $d\gamma/dP$, and dB_o/dP in the composite functions λ, ε, and η are determined by graphical, or numerical, integration.)

Recognition of Formation Compressibility Using the Muskat Method

Muskat treated formation compressibility as negligible. Recognition of formation compressibility is accomplished as follows:

Although the left-hand term of equation (1) shows $\Delta S_o/\Delta P$, actually it is change in oil saturation *per pore volume* per pressure differential, and as such, it has the same dimensions as formation compressibility, which in English units is $\Delta V/V/$psi. Accordingly, equation (1) becomes:

$$\Delta S_o/\Delta P = \frac{S_o\,\lambda + (1 - S_o - S_w)\,\varepsilon + S_o\,(\psi) - C_f}{1 + (\mu_o/\mu_g)(\psi)} \qquad (5)$$

where C_f = formation compressibility, $\dfrac{1}{\Delta}\left(\dfrac{\Delta V}{\Delta P}\right)$

And since at the end of a pressure step, the pore volume has shrunk by an amount equal to formation compressibility times pressure change, $C_f \times (P_i - P_2)$, then the volume of oil remaining at the end of a pressure step (P_2) has likewise been decreased by formation compressibility in the same proportion, so equation (3) above becomes:

Oil Recovery, fraction of initial pore space:

$$S_{oi}/B_{oi} - (S_o/B_o)\,((1 - C_f\,(P_i - P_2))) \qquad (6)$$

where P_i = initial pressure

P_2 = pressure at end of a ΔP step

And accordingly, equation (4) above becomes:

Oil Recovery, fraction of initial oil in place:

$$\frac{S_{oi}/B_{oi} - (S_o/B_o)\,((1 - C_f\,(P_i - P_2))}{S_{oi}/B_{oi}} \qquad (7)$$

Appendix 6. Example of Development Forecast through Recognition of High Conductivity in a Reservoir Subarea (Recent Development South Part of West Puerto Chiquito)

The reservoir geometry that exists in the main West Puerto Chiquito reservoir (Appendix 3) has been found in nearby "subareas" of the field. The subareas may be in communication with each other through low permeability fractures, but the internal high capacity fracture network of one subarea may not be in close communication with other subareas. Within a subarea the interconnected fracture system can be revealed by interference, frac pulse, and occasionally shut-in pressure testing.

The pressure behavior of an individual well will often be defined by flow systems of "constant pressure at the boundary" (Appendix 3). Some will show influence of a nearby fault or fracture of high capacity. Typically, engineers do not attempt to analyze the "late time" portion of a pressure build-up. In the fractured Niobrara, however, with properly conducted tests, analysis of the late-time period of pressure build-up can be distinctive in assessing such reservoir character. Sensitive pressure gauges are desirable in acquiring late-time data.

An example of pressure build-up revealing presence of a high capacity fault or fracture in southeast West Puerto Chiquito (NMOCC, 1989) occurs in the Amoco Schmitz Anticline No. 1 well, Sec. 25, T24N, R1W from tests conducted in September 1988 (Figures 6-1, 6-2, and 6-3). Figures 6-2 and 6-3 are expanded scale sections of Figure 6-1. Significant are the pressures from the 60th to the 116th hours of shut-in shown on the greatly expanded plat of Figure 6-3 and comparison with those just preceding. Although the pressures were not taken with a sensitive instrument, it is possible to average the points of equal pressure (scanner reads several points

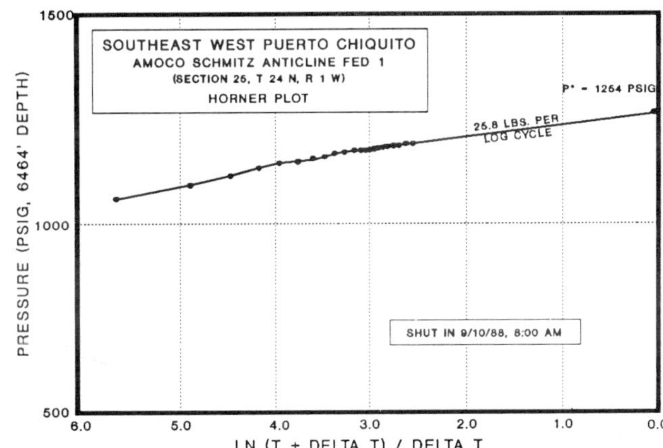

Figure 6-1. Plot of pressure buildup test of a well in southeast West Puerto Chiquito, a reservoir "subarea" first produced in 1985.

identically) and determine, as shown in Figure 6-3, that the pressures fall in a straight line, not at all like the "rounding" of pressures that occurs in a closed reservoir. Rather, the well test data indicate the presence of a linear high capacity fault or fracture.

Note that a straight line sealing fault will cause the slope of a pressure build-up curve to double. It can also be shown by the same type of analysis (method of images) that a linear fault or fracture of infinite capacity will cause the slope to decrease by one-half.

For the Schmitz well the slope of the build-up curve decreases by somewhat less than half (40.5 to 25.8

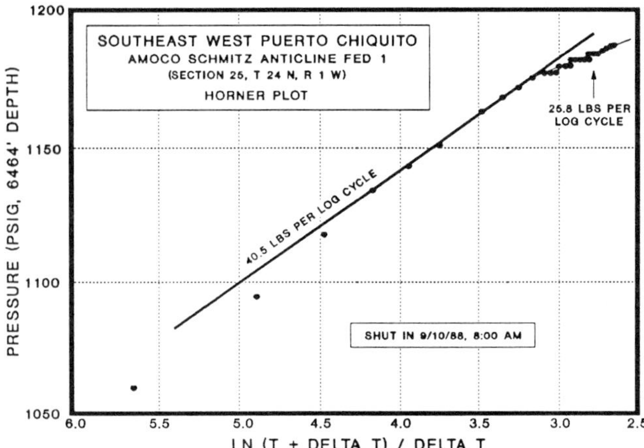

Figure 6-2. Plot of portion of pressure build-up (Figure 6-1) expanded to show that part of curve with a slope of 40.5 lb per log cycle.

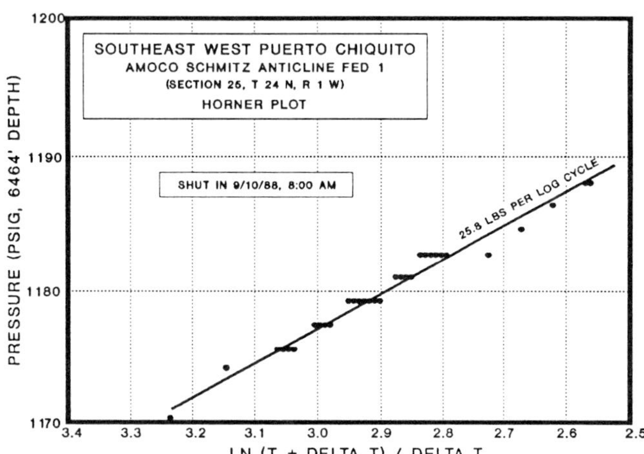

Figure 6-3. Plot of portion of build-up of Figure 6-1 greatly expanded to show "late-time" straight line slope of approximately 26 lb per log cycle, evidencing presence of high capacity fracture in well's immediate drainage area.

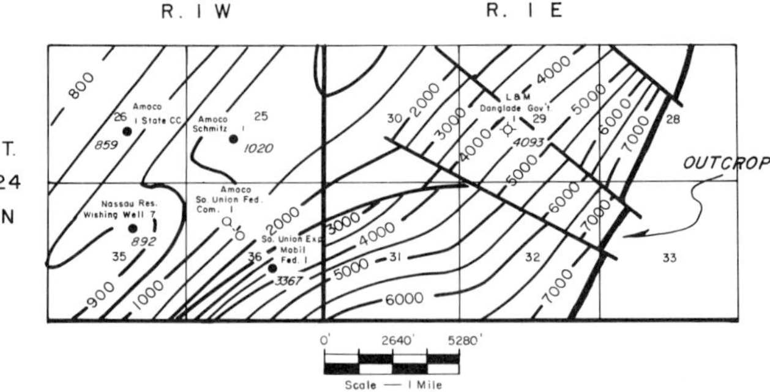

Figure 6-4. Plat showing structural contours and surface identified faults part of southeast West Puerto Chiquito "subarea" with northwest-southeast orientation.

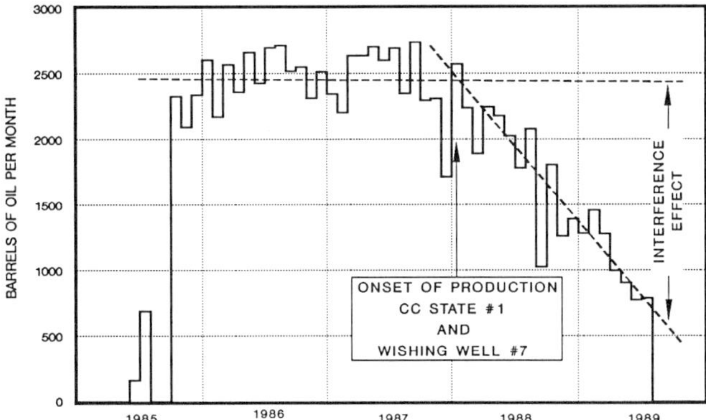

Figure 6-5. Production rate plot showing interference effect of new wells producing at high rates on stabilized production of existing wells.

psi/log cycle), indicating a high capacity fault or fracture (but not one of infinite capacity). Surface fault orientations (Figure 6-4) suggest that the possible extension of the high capacity fracture system observed in the Schmitz well would extend to the vicinity of the second well drilled in this area, the Amoco C.C. State No. 1 (Sec. 26, T24N, R1W). Confirmation that an interconnected fracture system exists between the wells was observed in the production interference effects on the Schmitz well when the C.C. State went on production in February 1988 (Figure 6-5).

Thus, although no interference testing was available in southeast West Puerto Chiquito, the existence of a high capacity fracture system was established using:

1. The late-time straight line slope of pressure build-up.
2. Production interference effects.
3. Equalized pressure of high capacity and low capacity wells.

Knowledge of the existence of the interconnecting fracture system permits forecast of the prospects of extension drilling as follows:

From pressure decline and production volumes for approximately four months, the indicated subarea reservoir volume is 2,500,000 STB. Pressure build-up showed a value of $kh/\mu = 1$ for the better wells.

In the absence of interference (or frac pulse) testing, pore volume per acre cannot be determined directly, nor can diffusivity, which depends on it. By assuming a series of reservoir per-acre volumes, calculating the diffusivity constant from these and Kh/μ, however, a plot can be made of reservoir unit pore volume versus two indicated areas: the area of the total reservoir oil volume; and the area that could be reached by steady state conditions (Figure 6-6).

The curve showing area that could be reached by steady state conditions in 120 days for $kh/\mu = 1$ approximates the curve for the reservoir volume area (lower two lines of Figure 6-6). If the subarea internal fracture system's diffusivity were the same as that of individual wells ($kh/\mu = 1$), then there would be no indication that the subarea reservoir was limited in size; extension drilling might provide similar wells.

If the subarea contains a high capacity fracture system, however, then the subarea's diffusivity and the area that could be reached by steady state conditions will be greater. An example is the upper curve in Figure 6-6, which results from a 3 to 1 ratio of reservoir Kh/μ to individual well Kh/μ. In West Puerto Chiquito and Gavilan, the Kh/μ for the high capacity fracture system is many multiples of that for individual wells. The same is probably true here.

Thus, for any assumed values of pore space per acre, the area that could be reached by steady state conditions is far greater than that indicated by reservoir volume. The surrounding area, therefore, must be of significantly lower permeability, and extension wells located therein will be limited in area drained to that of their own tracts. Recoveries, therefore, are expected to average 60,000 to 100,000 bbl per well. Further, since the permeability is known only to be significantly lower than for the existing wells, extension wells could very well be marginally economic, or nonpaying.

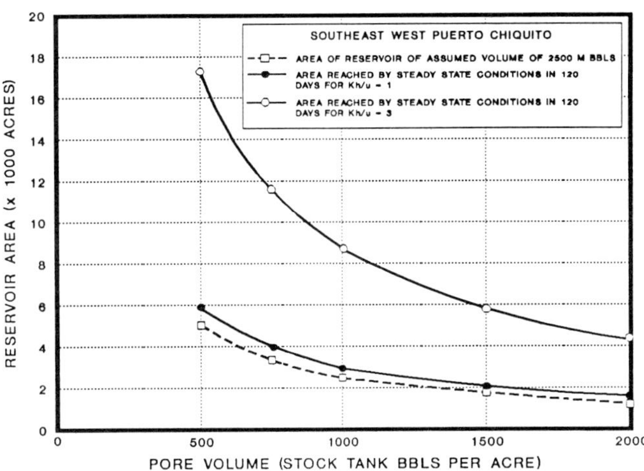

Figure 6-6. Plot showing reservoir area covering volume of 2.5 million bbl as dependent on per acre volume along with the concurrent area that could be reached by steady state conditions for the transmissibilities shown.

Another "tool" that may be of some use in evaluating a new subarea is Figure 3-6 of Appendix 3. Until information is available identifying existence of a high capacity fracture system, we suggest caution in assuming that the reservoir of a new area is of better quality than Figure 3-6 would imply.

Appendix 7. Field Description

Field name .... *West Puerto Chiquito field*

Ultimate recoverable reserves .... *15–20 million bbl (depending on future depletion method)*

Field location:

    Country .... *U.S.A.*

    State .... *New Mexico*

    Basin/Province .... *San Juan basin*

Field discovery:

    Year first pay discovered .... *Fractured Niobrara Member of U. Cretaceous Mancos Shale   1962*

Discovery well name and general location:

    First pay .... *No. K-13 Canada Ojitos Unit*

Discovery well operator .... *Bolack-Greer*

IP:

    First pay .... *15 BOPD and 6 MCF/D*

All other zones with shows of oil and gas in the field:

Age	Formation	Type of Show
Upper Cretaceous	*Mesa Verde*	*Gas*
	Pictured Cliffs	*Gas*
Lower Cretaceous	*Dakota*	*Gas*

Geologic concept leading to discovery and method or methods used to delineate prospect:
Postulated Niobrara fracture trap formed by Laramide development of basin-bounding monocline.

Structure:

    Province/basin type .... *Colorado Plateau/Structural; Klemme IIB, Bally 222*

    Tectonic history
    Pre-Laramide faulting and uplift associated with development of Ancestral Archuleta anticlinorium rejuvenated during Eocene with vertical uplift and lateral shift, which created the Hogback monocline and a series of north-trending folds.

    Regional structure
    West-dipping flank of Hogback monocline and adjacent north-plunging structural "nose."

    Local structure
    5° homoclinal dip to west flattening to 0° at base to form synclinal flexure on west side.

Trap:

    Trap type(s)
    Fracture system isolated from outcrop by calcite vein-filling, and without connection to water.

Basin stratigraphy (major stratigraphic intervals from surface to deepest penetration in field):

Chronostratigraphy	Formation	Depth to Top in ft*
Tertiary	*Nacimiento, San Jose*	*Surface*
Upper Cretaceous	*Ojo Alamo*	*3000*
	Kirtland-Fruitland-PC	*3300*
	Lewis	*3700*
	Mesa Verde: Cliff House, Menefee,	
	Pt. Lookout	*5200*
	Mancos	*6000*
	Niobrara	*7000*
	Sanostee, Carlile, Greenhorn,	
	Graneros, Dakota	*7500*
	Entrada	*8300*

**Varies from 2000 ft shallower to 500 ft deeper than schedule depending on structural position and topography.*

Reservoir characteristics:

Number of reservoirs *1*
Formations .... *Niobrara Member of Mancos Shale; A, B, C lithologic units*
Ages .... *Upper Cretaceous*
Depths to tops of reservoirs .... *5000–7500 ft depending on structural position and topography*
Gross thickness (top to bottom of producing interval) .... *250 ft*
Net thickness—total thickness of producing zones
    Average .... *50–100 ft*
    Maximum .... *100–150 ft*
Lithology .... *Highly laminated shale, siltstone, and minor limestone and dolomite sequence*
Porosity type .... *Fracture*
Average porosity *0.5–1.0%*
Average transmissibility .... *Varies 1–50 darcy-feet*

Seals:

Upper
    Formation, fault, or other feature .... *Massive shale (Mancos) and loss of fractures*
    Lithology .... *Shale*
Lateral
    Formation, fault, or other feature .... *Niobrara—loss of fractures*
    Lithology .... *Shale*

Source:

Formation and age .... *Niobrara (Upper Cretaceous)*
Lithology .... *Shale*
Average total organic carbon (TOC) *1–3%*
Maximum TOC .... *NA*
Kerogen type (I, II, or III) .... *NA*
Vitrinite reflectance (maturation) .... $R_o = 0.77$–0.90%
Time of hydrocarbon expulsion .... *Tertiary*
Present depth to top of source .... *5000–7000 ft*
Thickness .... *300 ft*
Potential yield .... *NA*

Appendix 8. Production Data

Field name .... *West Puerto Chiquito field*

Field size:

    Proved acres *80,000*
    Number of wells all years *40*
    Current number of wells *40*
    Well spacing .... *640 ac*
    Ultimate recoverable .... *15–25 million bbl*
    Cumulative production .... *10 million bbl*
    Annual production .... *0.5 to 1.0 million bbl*
    Present decline rate .... ***
        Initial decline rate .... ***
        Overall decline rate .... ***
    **Rates not meaningful because of pressure maintenance project, continuous development, gravity drainage.*
    Annual water production .... *Nil*
    In place, total reserves .... *50–100 million bbl*
    In place, per acre foot .... *NA*

Primary recovery .... *NA; pressure maintenance started early in life*
Secondary recovery .... *NA; see above*
Cumulative water production .... *Nil*

Drilling and casing practices:

Amount of surface casing set .... *500 ft*
Casing program
Early wells: 7⅝-in. intermediate through Mesa Verde, set from 100 ft to 500 ft above A zone, 5½-in. liner through pay
Later wells: 5½-in. intermediate from surface through pay

Drilling mud .... *Fresh water, viscosity 35 seconds shallow, 50–70 deep; water loss 4.8 though producing zones*
Bit program .... *Recent wells: 12¼-in. surface hole, 8¾-in. through Lewis shale, 7⅞-in. to TD*
High pressure zones .... *None*

Completion practices:

Interval(s) perforated .... *Primarily A, B, and C zones*
Well treatment .... *Sand frac*

Formation evaluation:

Logging suites .... *Induction-gamma ray, neutron density, some frac logs*
Testing practices .... *Open-hole: none; all testing done after running production casing*
Mud logging techniques .... *Sample logging, observation of lost circulation zones, occasionally gas monitoring*

Oil characteristics:

Type .... *NA*
API gravity .... *38–40°*
Base .... *Paraffin*
Initial GOR .... *480 ft³/bbl (gas in solution)*
Sulfur, wt% *0*
Viscosity, SUS .... *0.62 cp at initial reservoir conditions*
Pour point .... *NA (low)*
Gas-oil distillate .... *NA*

Field characteristics:

Average elevation .... *Surface 7000–8000 ft*
Initial pressure .... *1620 psig at datum +1195*
Present pressure .... *2000 psi near injection wells, 1000 psi most remote wells*
Pressure gradient .... *0.313 psi/ft (initial reservoir conditions)*
Temperature .... *155–170°F*
Geothermal gradient .... *0.015–0.018°F/ft*
Drive .... *Solution gas*
Oil column thickness .... *Indeterminate*
Oil-water contact .... *None*
Connate water .... *Indeterminate*
Water salinity, TDS .... *NA*
Resistivity of water .... *NA*
Bulk volume water (%) .... *NA*

Transportation method and market for oil and gas:
90% moves through pipelines, remainder trucked.

Dukhan Field—Qatar
Arabian Platform

Qatar General Petroleum Corporation (QGPC)
Doha, Qatar

Amoco Qatar Petroleum Company
Doha, Qatar

FIELD CLASSIFICATION

BASIN: Arabian
BASIN TYPE: Foredeep
RESERVOIR ROCK TYPE: Limestone and
   Dolomite
RESERVOIR AGE: Jurassic
PETROLEUM TYPE: Oil
TRAP TYPE: Anticline over Salt Diapir

RESERVOIR ENVIRONMENT OF DEPOSITION: Shallow Marine, Tidal Flat, and Sabkha

LOCATION

The Dukhan field is located on the western flank of the north–south-trending Qatar Peninsula in the southwestern Arabian Gulf (Figure 1). The peninsula is the topographic expression of the geologically long-lasting Qatar-Fars arch. Dukhan is a giant oil and gas field about 43.5 mi (70 km) long and about 2.5 to 3.7 mi (4–6 km) wide (Figure 2). The field is a single "Arabian type" fold, explained in the *Structure* section, with one major and several minor culminations along strike (Sugden, 1962). Initial recoverable reserves from its four reservoirs are 4.57 billion bbl of oil and 10.8 tcf of gas. Dukhan is located between the world's largest oil field (Ghawar in Saudi Arabia) and one of the world's largest gas fields (North Dome in offshore Qatar). The peninsula itself is arid and of low relief. Maximum elevation of about 295 ft (90 m) is reached along the Dukhan anticline.

HISTORY

Pre-Discovery

A reconnaissance geological survey of Qatar in 1933 recognized the Dukhan anticline on the basis of surface geology. The structure was mapped in greater detail by a later follow-up surface survey in 1937–1938, subsequent to the first oil concession in Qatar (in 1935), granted to the Anglo-Persian Oil Company. In 1936, Petroleum Development (Qatar) Ltd. was organized as the operating company for the concession.

Discovery

The first exploration well drilled on the Dukhan anticline was begun in 1938 on the basis of previous field mapping and was the discovery well. This well, the Petroleum Development (Qatar) Dukhan No. 1 (25°25′16.2″N; 050°47′00.6″E), was completed in January 1940. The well was operated by Qatar Petroleum Co. Partners included:

Royal Dutch Shell	23.75%
British Petroleum Company	23.75%
Cie Francaise des Petroles	23.75%
Standard of New Jersey	11.87%
Mobil Corporation	11.87%
Partex (Gulbenkian Foundation)	5.00%

The well struck oil in what was then called the Jurassic Zekrit Formation, now known as the Arab C of the Jurassic Qatar Formation (Figure 3). Total depth of 5685 ft was reached in 459 days and was near the depth limit of the rig. Initial production was 4480 BOPD of 39° API oil at a wellhead pressure of 367 psi. The Dukhan No. 2, completed 10 mi to the south of No. 1 in March 1941, produced 5000 BOPD of 36° API oil from the Arab C. Arab D was first encountered in this well and was found to be gas-bearing in what turned out to be a major associated gas cap. The Dukhan No. 3 was completed in edge water 2.5 mi (4 km) east of No. 1 in May 1942, defining the lateral limit of both Qatar Formation Arab zone accumulations.

By 1944, the field was already considered a giant, with an estimated 500 MMBO reserves. Operations and exploration were ceased at that point until 1947

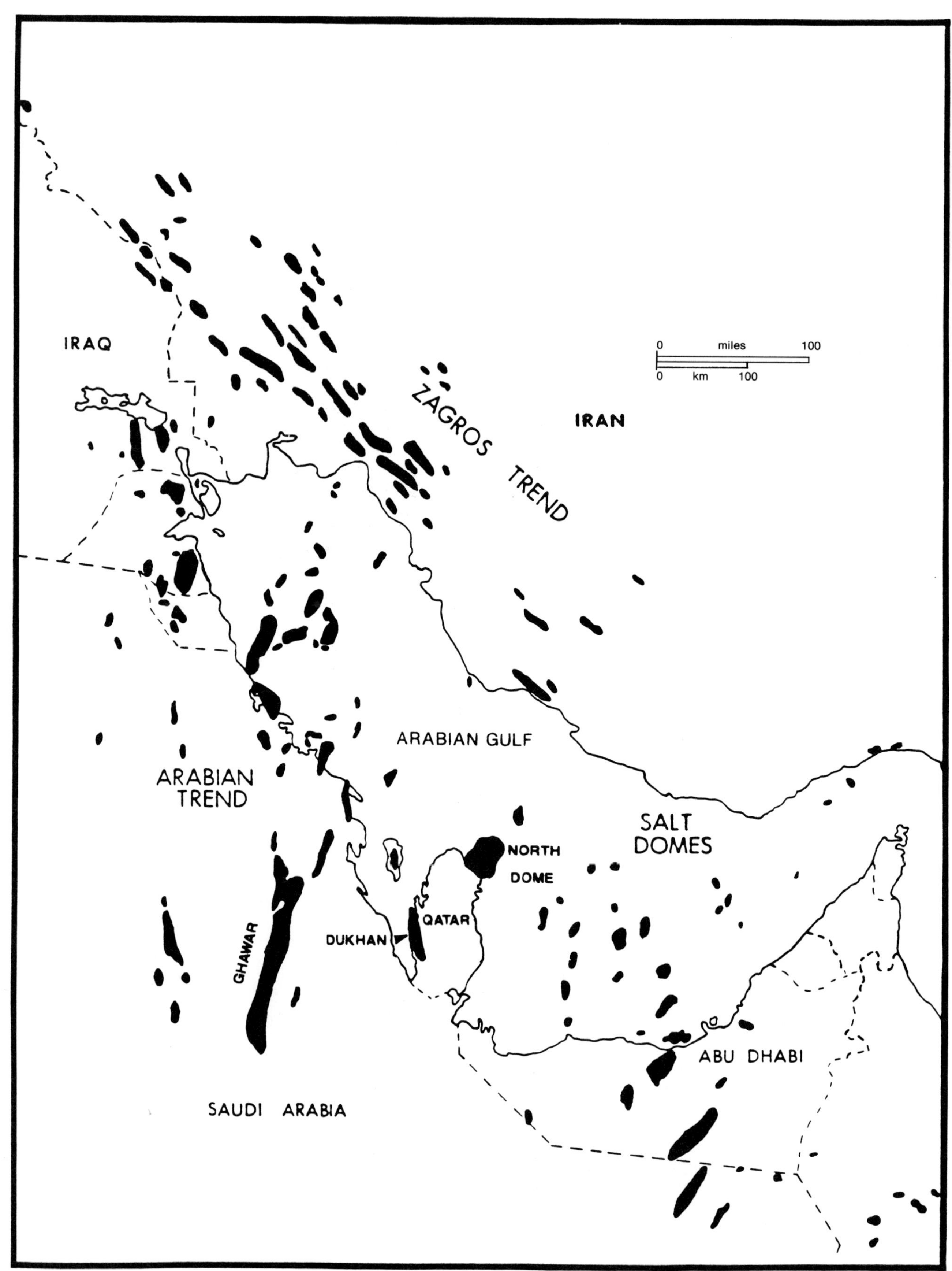

Figure 1. Index map showing major oil fields in the Arabian Gulf area and location of Dukhan field.

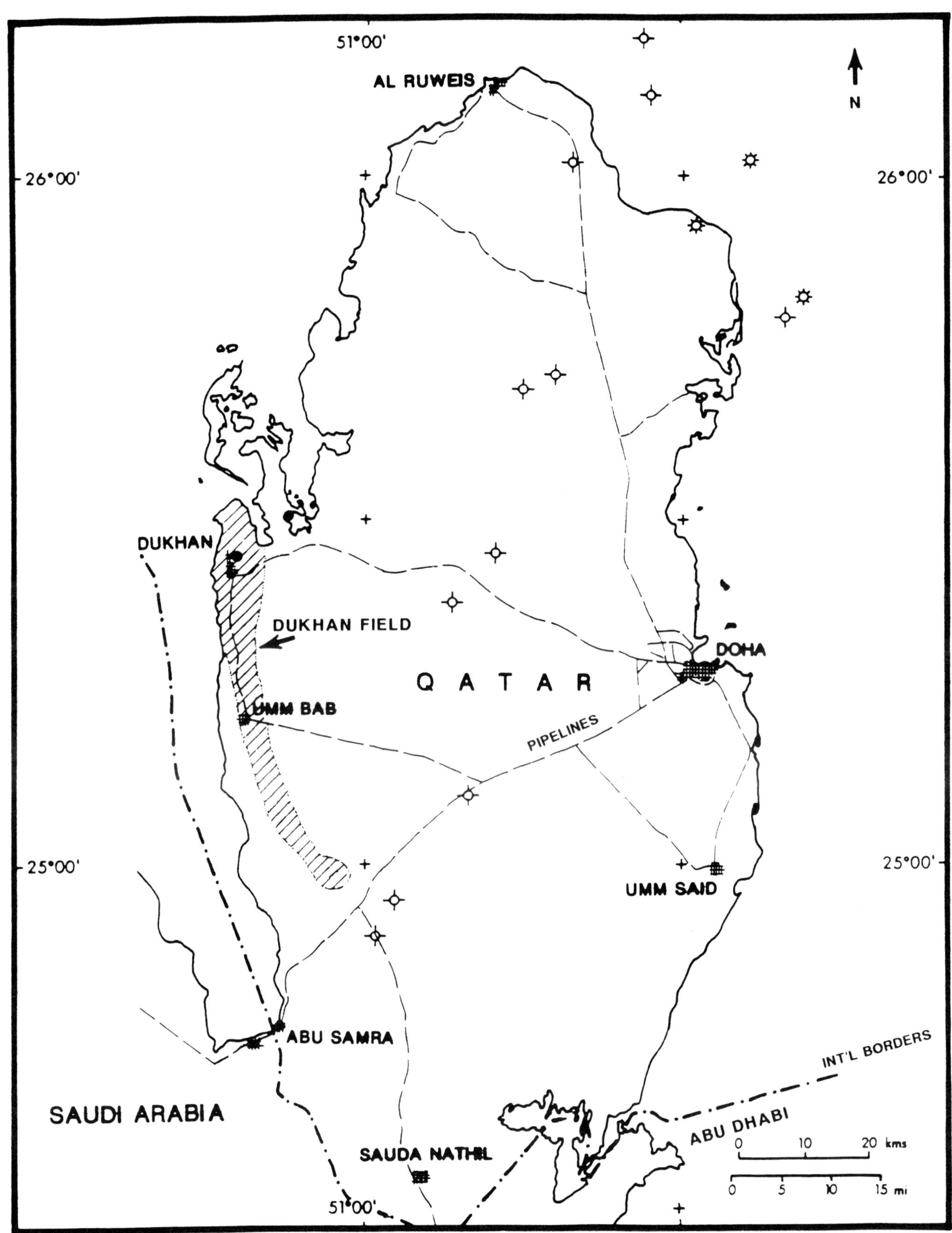

Figure 2. Index map of Qatar displaying the overall shape of Dukhan field.

Era	Period	Series	Stage/Age	Group	Formation	Member	Marker	Reservoir
CENOZOIC			QUATERNARY		SUPERFICIAL DEPOSIT			
CENOZOIC	TERTIARY	MIOCENE-PLIOCENE			HOFUF			
CENOZOIC	TERTIARY	MIOCENE-PLIOCENE			DAM			
CENOZOIC	TERTIARY	PALEOCENE-EOCENE	LUTETIAN		DAMMAM			
CENOZOIC	TERTIARY	PALEOCENE-EOCENE	YPRESIAN		RUS			
CENOZOIC	TERTIARY	PALEOCENE-EOCENE	THANETIAN MONTIAN?		UMM ER RADHUMA			
MESOZOIC	CRETACEOUS	UPPER	MAASTRICHTIAN	ARUMA	SIMSIMA			
MESOZOIC	CRETACEOUS	UPPER	CAMPANIAN	ARUMA	FIQA			
MESOZOIC	CRETACEOUS	UPPER	SANTONIAN	ARUMA	HALUL			
MESOZOIC	CRETACEOUS	UPPER	CONIACIAN	ARUMA	LAFFAN			
MESOZOIC	CRETACEOUS	UPPER	(TURONIAN)	WASIA	MISHRIF			MISHRIF
MESOZOIC	CRETACEOUS	UPPER	CENOMANIAN	WASIA	KHATIYAH	UPPER		
MESOZOIC	CRETACEOUS	UPPER	CENOMANIAN	WASIA	KHATIYAH	MIDDLE		
MESOZOIC	CRETACEOUS	UPPER	CENOMANIAN	WASIA	KHATIYAH	LOWER		LOWER
MESOZOIC	CRETACEOUS	LOWER	ALBIAN	WASIA	MAUDDUD	UPPER LIMESTONE		UPPER LIMESTONE
MESOZOIC	CRETACEOUS	LOWER	ALBIAN	WASIA	MAUDDUD	LOWER MARL		
MESOZOIC	CRETACEOUS	LOWER	ALBIAN	WASIA	NAHR UMR			LOWER SAND
MESOZOIC	CRETACEOUS	LOWER	APTIAN	THAMAMA	SHUAIBA	UPPER		A; B UPPER; C
MESOZOIC	CRETACEOUS	LOWER	APTIAN	THAMAMA	SHUAIBA	LOWER		D LOWER
MESOZOIC	CRETACEOUS	LOWER	BARREMIAN HAUTERIVIAN	THAMAMA	HAWAR			
MESOZOIC	CRETACEOUS	LOWER	BARREMIAN HAUTERIVIAN	THAMAMA	KHARAIB	B — LIMESTONE		B-LIMESTONE
MESOZOIC	CRETACEOUS	LOWER	BARREMIAN HAUTERIVIAN	THAMAMA	KHARAIB	MIDDLE MARL		
MESOZOIC	CRETACEOUS	LOWER	BARREMIAN HAUTERIVIAN	THAMAMA	KHARAIB	C — LIMESTONE		C-LIMESTONE
MESOZOIC	CRETACEOUS	LOWER		THAMAMA	LEKHWAIR	FAHIA		
MESOZOIC	CRETACEOUS	LOWER		THAMAMA	LEKHWAIR	ZAKUM		ZAKUM
MESOZOIC	CRETACEOUS	LOWER	(HAUTERIVIAN)	THAMAMA	YAMAMA			YAMAMA
MESOZOIC	CRETACEOUS	LOWER	VALANGINIAN BERRIASIAN	THAMAMA	SULAIY			SULAIY
MESOZOIC	JURASSIC	UPPER			HITH			
MESOZOIC	JURASSIC	UPPER	TITHONIAN		QATAR	ARAB A		ARAB A
MESOZOIC	JURASSIC	UPPER	TITHONIAN		QATAR	UPPER ANHYDRITE		
MESOZOIC	JURASSIC	UPPER	TITHONIAN		QATAR	ARAB B		ARAB B
MESOZOIC	JURASSIC	UPPER	TITHONIAN		QATAR	MIDDLE ANHYDRITE		
MESOZOIC	JURASSIC	UPPER	TITHONIAN		QATAR	ARAB C		ARAB C
MESOZOIC	JURASSIC	UPPER	TITHONIAN		QATAR	LOWER ANHYDRITE		
MESOZOIC	JURASSIC	UPPER	KIMMERIDGIAN		JUBAILA	FAHAHIL		ARAB D
MESOZOIC	JURASSIC	UPPER	KIMMERIDGIAN		JUBAILA	MARLY LIMESTONE		
MESOZOIC	JURASSIC	UPPER	(?OXFORDIAN)		HANIFA			
MESOZOIC	JURASSIC	LOWER-MIDDLE	CALLOVIAN		ARAEJ	UPPER		A; B; C
MESOZOIC	JURASSIC	LOWER-MIDDLE	CALLOVIAN		ARAEJ	UWAINAT		UWAINAT
MESOZOIC	JURASSIC	LOWER-MIDDLE	BATHONIAN		ARAEJ	LOWER		LOWER
MESOZOIC	JURASSIC	LOWER-MIDDLE	BAJOCIAN HETTANGIAN		IZHARA		KEY BED	IZHARA
MESOZOIC	TRIASSIC	MIDDLE-UPPER	NORIAN?		HAMLAH			HAMLAH
MESOZOIC	TRIASSIC	MIDDLE-UPPER	CARNIAN LADINIAN		GULAILAH		KHAIL ANHYDRITE	
MESOZOIC	TRIASSIC	LOWER			SUDAIR			
MESOZOIC	TRIASSIC	LOWER			KHUFF	K1 — K2		K1
MESOZOIC	TRIASSIC	LOWER			KHUFF	K1 — K2		K2
PALEOZOIC	PERMIAN	UPPER			KHUFF	K — 3		K3
PALEOZOIC	PERMIAN	UPPER			KHUFF	UPPER ANHYDRITE		
PALEOZOIC	PERMIAN	UPPER			KHUFF	K4		K4
PALEOZOIC	PERMIAN	UPPER			KHUFF	MEDIAN ANHYDRITE		
PALEOZOIC	PERMIAN	UPPER			KHUFF	K5		K5
PALEOZOIC	PERMIAN	LOWER			HAUSHI			
PALEOZOIC	DEVONIAN (?CARB)				TAWIL			
PALEOZOIC	SILURIAN	LOWER-UPPER			SHARAWRA	UPPER SANDY		
PALEOZOIC	SILURIAN	LOWER-UPPER			SHARAWRA	LOWER MARLY		
PALEOZOIC	ORDOVICIAN	MIDDLE-UPPER			TABUK			

Figure 3. Generalized stratigraphic nomenclature for Qatar. (From QGPC.)

tested in 1950 by Dukhan No. 12 located on the east flank, flowing 3500 BOPD, 40° API, on a 1-in. choke with 750 psig.

Post-Discovery

Later drilling in the field after World War II led to discovery of two additional producing horizons within the field: the Uwainat member of the Middle Jurassic Araej Formation (oil with gas cap), and the Permo-Triassic Khuff Formation (dry gas). Gas was discovered in the Khuff in 1959 by the first deep test, Dukhan No. 65. Production of gas from the Khuff began in 1976 (Brindley et al., 1984). As of December 1989, the field had 160 producing wells (138 oil, 22 gas). The total number of wells drilled to date has reached at least 383. A dumpflood water injection scheme has been in operation in the Arab C and Arab D limestone reservoirs since 1970. In a dumpflood, water is flooded into the depleted zone by gravity force alone from a higher perforated water zone in the same hole.

Of the four producing reservoirs in Dukhan, the Arab D produces the greatest volume of hydrocarbons and would, by itself, be considered a giant field with 6.82 billion bbl of original oil in place. Initial in-place oil and gas in the four reservoirs are calculated to be 11.3 billion bbl of oil and 19.6 tcf of gas. As of 1 January 1990, cumulative production was 2.57 billion bbl of oil and 4.08 tcf of gas, and remaining reserves were 2.00 billion bbl of oil, including gas liquids, and 6.7 tcf of gas. Annual production in 1989 was 86.9 MMBO and 185 BCFG.

DISCOVERY METHOD

The Dukhan field was discovered in 1940 based on field mapping in the 1930s: reconnaissance mapping in 1933, and detailed mapping in 1937–1938. The long, gentle anticline is well expressed at the surface. Modern seismic sections show that the subsurface structure corresponds closely to surface structure. Dukhan, like most of the giant Arabian platform onshore fields, was discovered in the "heyday" of field mapping in the oil industry. For these fields, little additional delineation was necessary prior to drilling. Methods would not be different today for structures like the Dukhan anticline. No oil seeps are known in the area of Dukhan field, which reflects the efficiency of the anhydrite seals of the Hith and Arab formations, particularly the 60 ft (18 m) thick unit between Arab B and C reservoirs (Figure 3). Whereas Arab C and D reservoirs are full to spillpoint, Arab A and B reservoirs hold only minor accumulations.

due to World War II. In 1948, Dukhan No. 6 encountered a probable gas/oil contact (GOC) in the Arab D based on traces of oil in DST #2 and oil-stained core, but no oil tests were possible due to a lost core barrel. The Arab D oil column was first

STRUCTURE

Tectonic History

The Arabian platform has been an area of relative structural stability since "Pan African" times (the last of the Precambrian orogenies greater than 550 million years ago). In the Infracambrian (a stratigraphically defined period coincident with the later portions of the Pan African event), two salt basins developed in the area of the present Arabian Gulf. These two basins (Eastern and Western basins) were separated by the north-south ancestral Qatar-Fars arch. The lower Paleozoic in the area was dominated by clastic deposition, giving way in the Permian to slow, continuous platform-carbonate and evaporite deposition, ceasing in the Miocene with emergence. A significant period of Triassic rifting (starting in the Permian) far to the southeast and northeast of this area formed the Tethys Ocean basin. This event seems to have had little effect in the immediate area of Qatar, however.

Intercontinental collision began in the Cretaceous at the sites of earlier rifting, northeast of Qatar, as the Tethys Ocean basin was consumed along the Zagros suture. The collision resulted in a post-Oligocene northerly tilt of the Qatar-Fars arch in the northern offshore area of Qatar as the proto-Arabian Gulf began to form as a foreland sag. Continued convergence along the Zagros suture resulted in the formation of the Zagros Fold and Thrust belt and the equivalent Border Fold belt on the Arabian plate side of the convergent margin. This Miocene–Pliocene compressional event was important in the southeastern Gulf region, such as in the Musandam Peninsula and to the northeast in Iran, but had only minor effects in the Dukhan field area. Minor northward tilting of Dukhan may have been responsible for late filling of the southernmost culmination, Diyab, explaining anomalous oil properties in that section. The Qatar Peninsula was an area of remarkable tectonic stability since Precambrian time. It was generally inboard from most major Phanerozoic convergent and divergent plate edge tectonic events.

Regional Structure

The area of the southwestern Arabian Gulf has experienced two mild structural events in the Phanerozoic: Jurassic to Miocene salt diapirism, and Miocene–Pliocene Zagros compression. According to Sugden (1962), Infracambrian salt was buried deep enough in the area to begin diapiric rise in the Jurassic (and possibly as far back as the Permian). The diapirism took two forms: salt domes to the east of Qatar and salt anticlines to the west. The salt domes occur sporadically throughout the Gulf area, many of which in offshore Qatar and Abu Dhabi have associated oil fields, such as Yaz, and many with present-day reefs above them.

Other diapirs may have risen over small displacement faults in the basement. These faults are probably Precambrian in age and were reactivated just prior to initiation of salt motion. The result is a long salt ridge with a "forced folded" anticline above it, much like that seen in the Paradox basin of Utah in the U.S.A. The position, orientation, and length of the resulting structure is controlled by the small-displacement basement fault, while the amplitude of the structure is controlled by the magnitude of salt rise. These forced folds are prevalent in the area and have been called "Arabian Folds" or "Arabian Plains-Type Folds," (Sugden, 1962). Most have a roughly north-south orientation (see Figure 1). Several of the world's largest oil fields occur in this area on such structures (Ghawar, for example). Continuous growth history, based on isopach studies and gravity data, support the conclusion of a component of salt diapirism in the creation of these structures.

Local Structure

The Dukhan anticline can be mapped at the surface in Tertiary age rocks. Its surface trace is shown in Figure 4. The structure is a long and gentle north–south-striking anticline with an overall teardrop shape in map view (Figures 5 and 6). Surface limb dips vary from $2°$ to $10°$. While gentle in curvature, the anticline is asymmetric, varying in both degree and vergence direction along the fold. In the northern portions of the fold, asymmetry is dominantly toward the western flank (Figures 7 and 8). Four culminations exist at the Arab reservoir level along strike (see Figure 6). This broad fold configuration at the Arab level gives way to an axial graben at the deeper Permian Khuff level as a result of early extension above the rising salt ridge (see Figure 5). Both levels exhibit a pronounced inflection or change in strike to the southeast in the southern portion of the field. The possibility of left-lateral strike-slip faulting is being studied.

For years, the anticline's origin was a matter of debate. Currently, however, its origin as primarily a salt anticline is generally accepted. A schematic section showing a diapiric Dukhan onshore and diapiric salt domes offshore is shown in Figure 9. The concept of salt movement as the origin of the structure is supported by: (1) characteristic multiple culminations with time; (2) a pronounced gravity low associated with the northern structure; (3) the position of the anticline on the easternmost updip edge of the western Infracambrian salt basin; and (4) the increase in amplitude of the structure with depth. Growth in amplitude was continuous and steady until the Miocene, which is another evidence of a salt movement origin (Sugden, 1962).

Structural closure at the Arab D level is at least 1600 ft (488 m) at the apex of the structure and 2100 ft (641 m) at the Khuff reservoir level (see Figure 7). Seismic response in the area is generally good

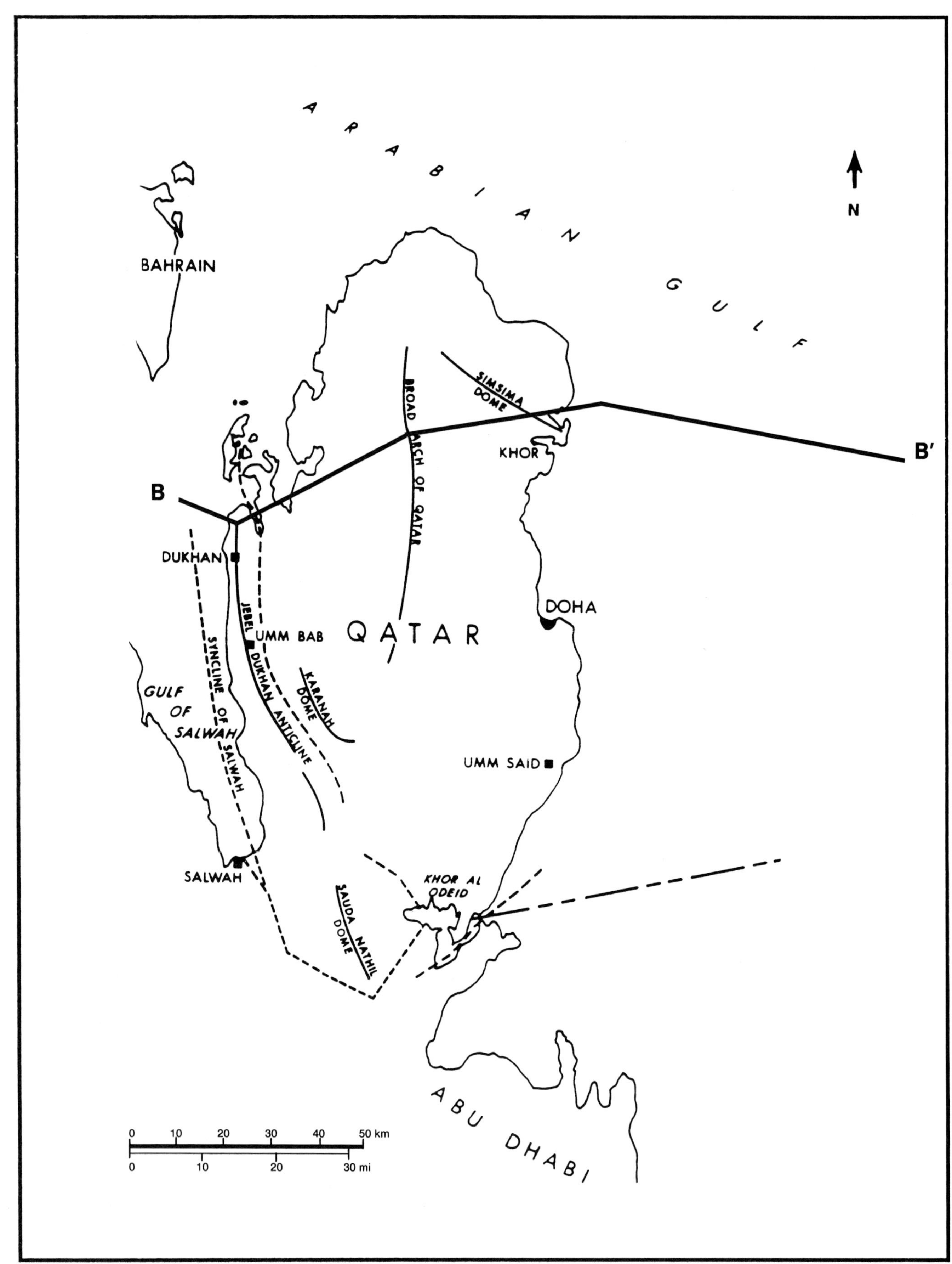

Figure 4. Major structural features of the Qatar arch.

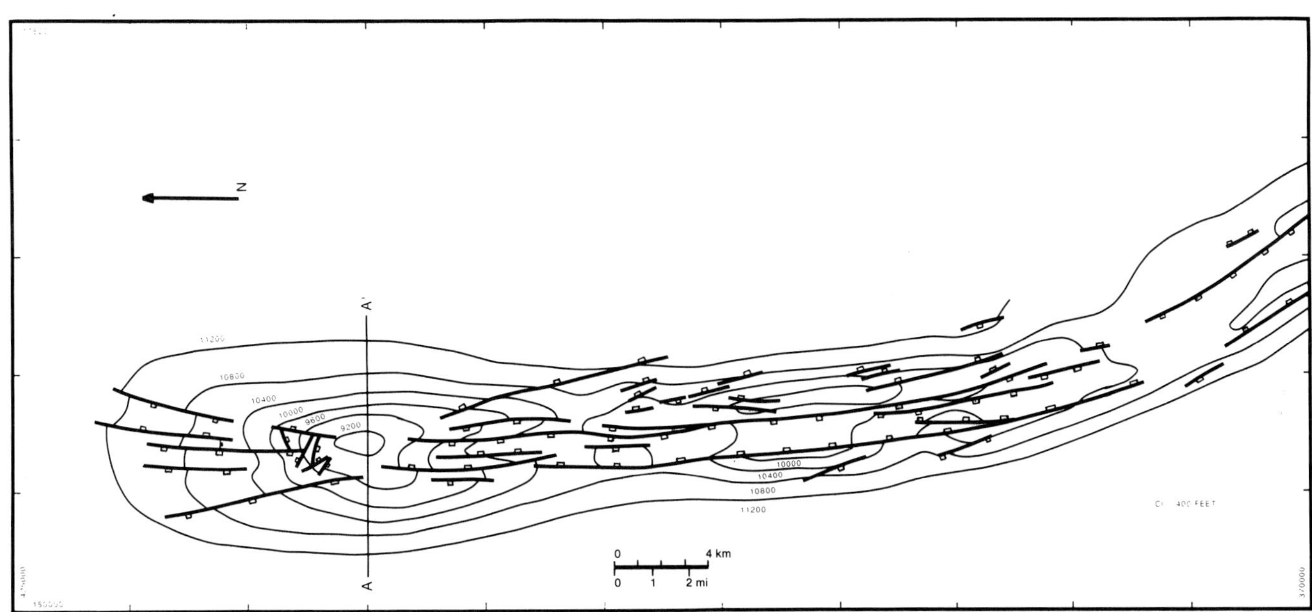

Figure 5. Structural contour map at top of the Khuff Formation for Dukhan field. (From QGPC.)

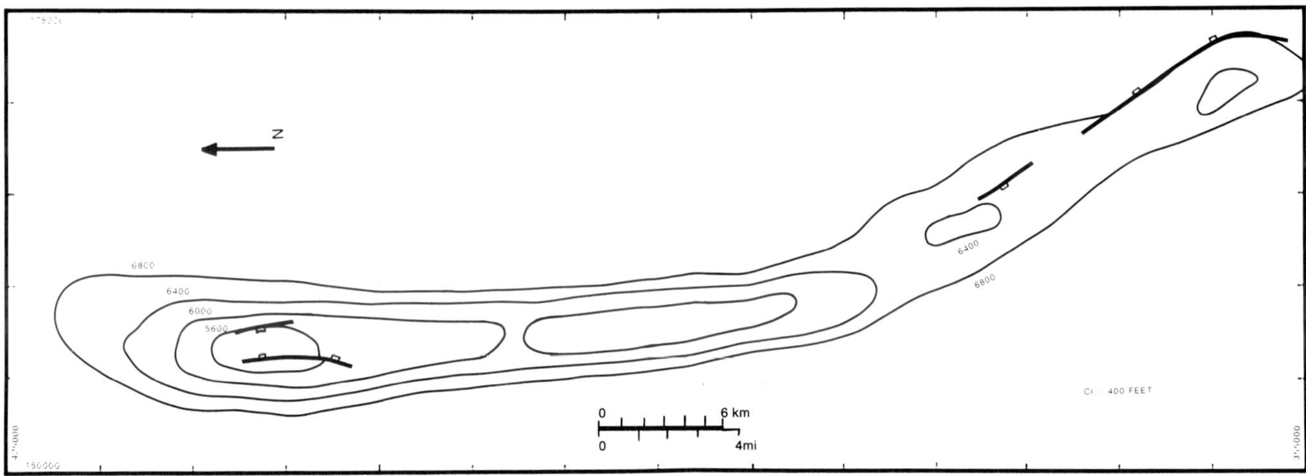

Figure 6. Structural contour map at top of the Arab D (Arab-4) limestone for Dukhan field. (From QGPC.)

and displays the steeper west limb of the structure, Figure 8.

STRATIGRAPHY

The generalized stratigraphic column for Qatar is shown in Figure 10. In general, the lower Paleozoic section of Qatar is dominated by clastic deposits. The deepest well on Dukhan, the Dukhan DKG-27, with a TD of 16,310 ft (4975 m) in pre-Tabuk, has penetrated strata only as old as the Ordovician (see

Appendix 1). Pre-Khuff clastic reservoirs are of very poor quality, but may be gas-bearing.

Beginning with deposition of the Permian Khuff, carbonate deposition dominated the area of Qatar. Ramp-type interbedded limestones, dolomites, and evaporites of predominantly sabkha origin make up the majority of the strata from Permian to Miocene. The Miocene to Recent is represented by mixed carbonate and clastic deposition, reflecting varying sediment sources, some associated with Zagros terrane.

The Dukhan reservoirs are dominantly peloidal to oolitic or bioclastic limestones and sucrosic to

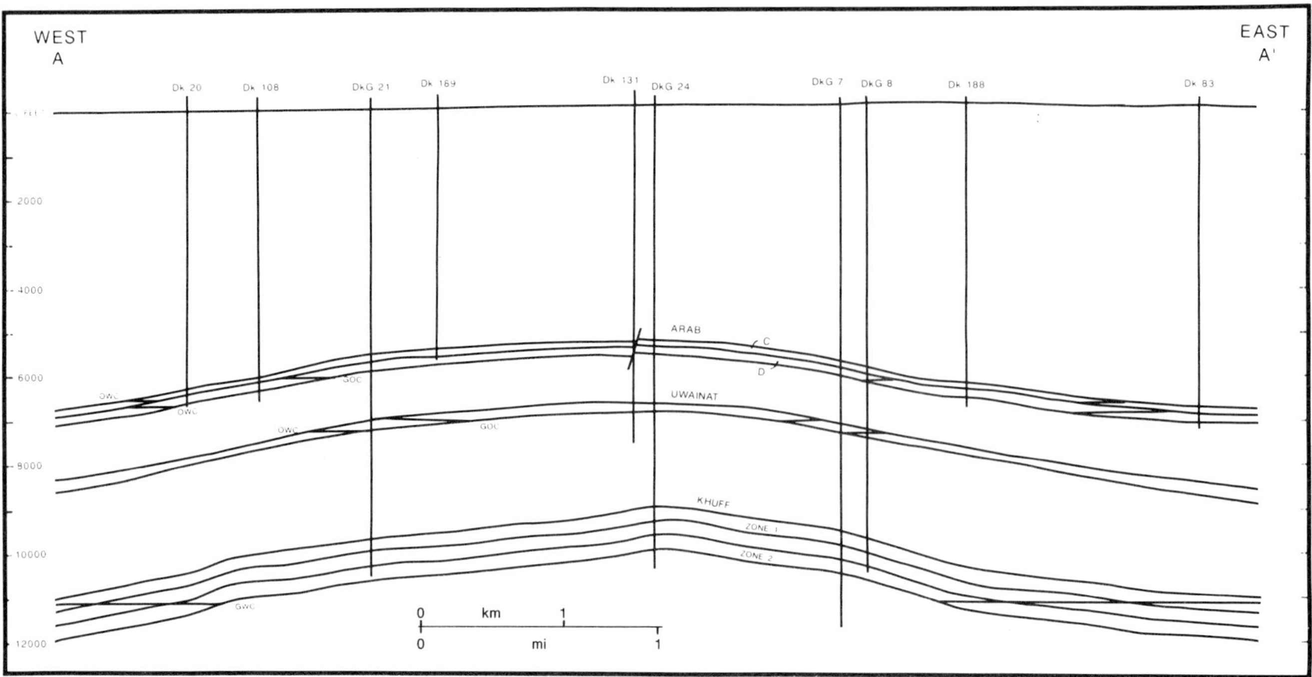

Figure 7. Structural cross section (A–A′) across Dukhan field. (From QGPC.) Location on Figure 5.

microcrystalline dolomites. The Upper Jurassic Hanifa Formation underlies the Arab reservoirs and is a kerogen-rich lime mudstone; TOC up to 6% has been observed.

TRAP

Dukhan is an anticlinal trap. Interbedded anhydrites within the reservoirs and the Jurassic Hith anhydrite formation comprise the vertical and lateral seals of the field. Structural closure is 1600 ft (488 m) at the Arab level and 2100 ft (641 m) at the Khuff. Oil-water contacts occur at –6520 ft (1989 m), –6780 ft (2068 m), and –7250 ft (2211 m) in the Arab C, Arab D, and Araej (Uwainat), respectively. The gas-water contact is at –11,050 ft (3370 m) in the Khuff.

An interesting feature of Dukhan is the presence of bitumen mats (often referred to as "tar mats") within the Jurassic reservoirs, principally in the Arab D and Uwainat. These bitumen mats may represent paleo oil-water contacts now domed upwards, indicating growth or rotation of the structure after a phase of migration into the structure. Similar tar mats have been documented in other offshore Arabian Gulf fields (Pinnington, 1981).

The very effective seals of the Arab reservoirs in Dukhan are thick anhydrite beds. The Uwainat and Khuff are sealed by tight argillaceous limestones and shales. Daniel (1954) postulated that it was the gentle folding with subsequently mild fracturing that kept

the anhydrites from rupturing those seals locally. Perhaps a tighter structural closure would have created a less effective seal in Dukhan.

Reservoir Rock

The reservoirs of Dukhan field are all carbonate rocks. Most are grainstones with intergranular porosity. The main producing zones in the Arab and Uwainat reservoirs are dominated by grainstones that have a common evidence of leaching and vadose diagenesis. Dolomite with intercrystalline porosity also contributes an important component of the Arab reservoirs, while the Uwainat is essentially pure limestone. The Khuff reservoir is mainly dolomite, and the relatively thin producing intervals are dominated by vuggy, intercrystalline, and relict mouldic porosity. The Arab reservoirs in Dukhan are similar in characteristics and depositional environments to those of nearby giant fields which, in combination, contain a large portion of the world's known oil reserves.

The Jurassic Arab reservoirs and seals in Dukhan are interpreted as upward-shoaling shelf or ramp-grainstone deposits that terminate in supratidal sabkhas. These shoaling environments produced grainstones and algal carbonates that were often quickly dolomitized in the near-surface environment. The rocks of the four main shoaling cycles (Arab A through D) are interbedded with thick anhydritic deposits. Porosities and permeabilities for the Arab

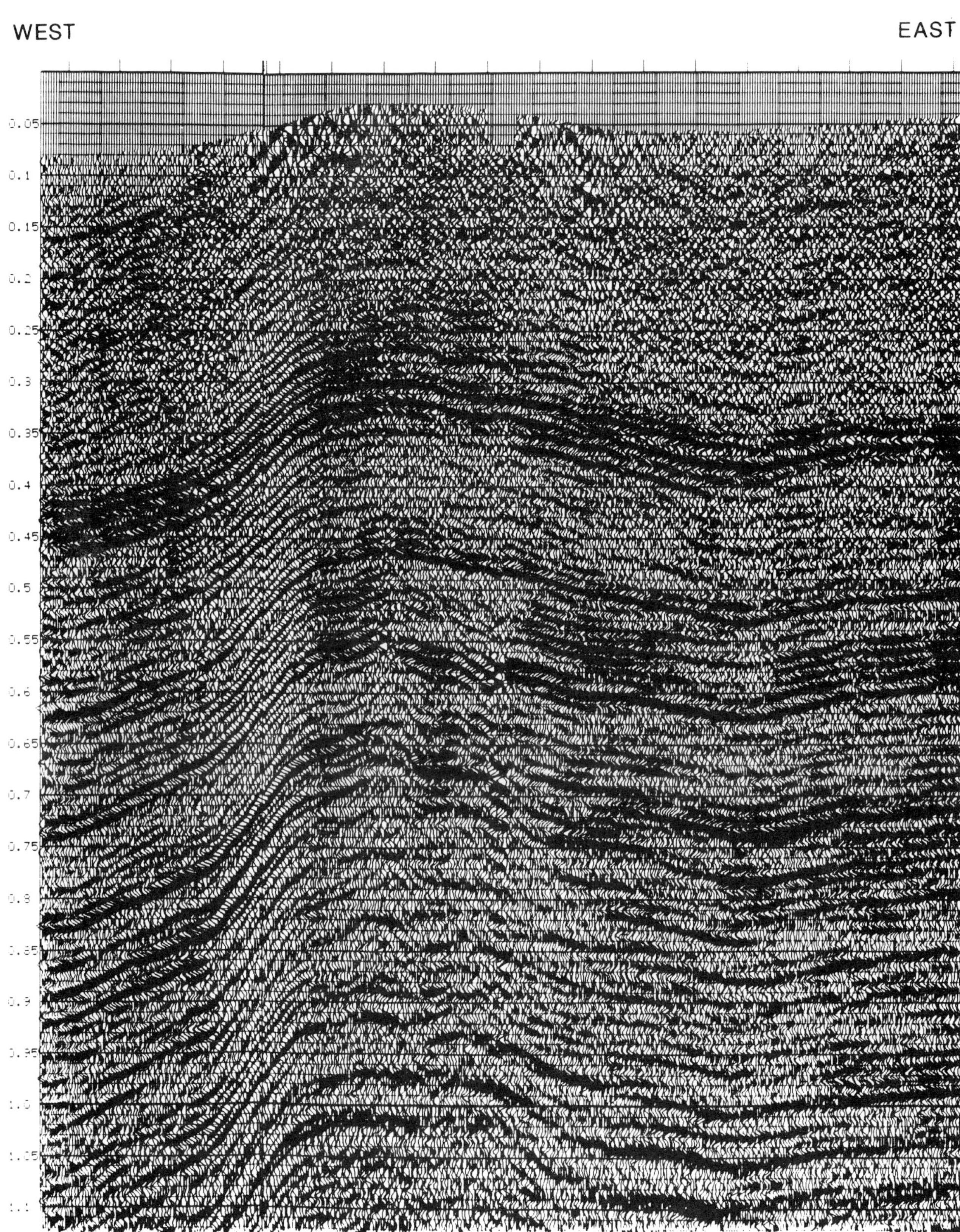

Figure 8. Typical seismic section across the northern portion of the Dukhan anticline. (Courtesy of Amoco Production Company.)

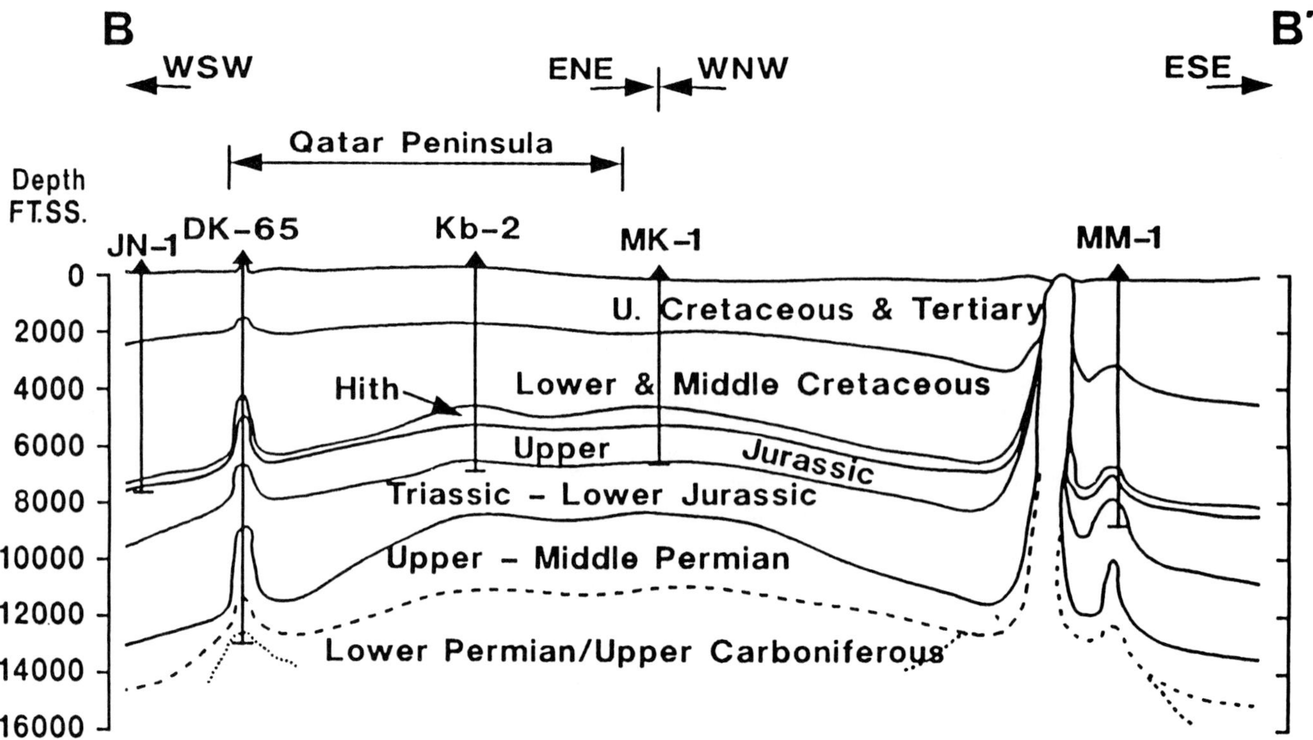

Figure 9. Schematic cross section (B-B′) across the Qatar Arch. (Courtesy of P. Bettis, after Pinnington, 1981.) General line of section is shown as B-B′ on Figure 4.

C and D units average 18% with 30 md and 20% with 70 md (Figures 11 and 12), respectively.

Depositional characteristics of the Uwainat member of the Araej Formation and Khuff reservoirs are similar but with larger bioclastic input. The Uwainat averages 18% porosity and 15 md permeability (Figure 13) but contains extreme vertical variations in permeability from 0.01 md to several darcys. The Khuff averages 5% porosity and 30 md permeability. Due to its thickness, finer-grained nature, and deeper burial, the Permian Khuff is more fractured than the other reservoirs in the field, but evidence of fracture production is rare. Fracture contributions to the Arab reservoirs are of minor importance in Dukhan. While primary sedimentary facies are the dominant control on production in the Dukhan reservoirs, the reservoir properties of the producing intervals have been significantly enhanced by diagenesis, particularly with respect to vadose leaching and dolomitization.

Source

The source rocks for the oil in the Arab reservoirs of Dukhan field are most likely to occur in the Late Jurassic Hanifa and possibly the lower Jubaila formations (Frei, 1985). The Hanifa source rock varies in thickness across Qatar, reaching its greatest thickness of 50 to 90 ft (15-27 m) in the Dukhan area. According to Frei (1985), based upon core from the upper Hanifa, the organic matter in the Hanifa consists of type II kerogen, predominantly sapropelic and partially bacterially transformed, with liptodetrinites and undetermined algae as main macerals. He further states that the unit is a good-to-excellent oil source with TOC averaging 2.1 wt.% (ranging 1.2-5.1 wt.%) and a sulfur content of 8.9% of extracted hydrocarbons. Gas chromatograms for the Hanifa and lower Jubaila are given in Figure 14.

The Hanifa occurs at a depth of 6200 to 7500 ft (1891-2288 m) over the Dukhan structure. Interpreted vitrinite reflectance has been reported as 0.62% at 5740 ft (1750 m) subsea (Frei, 1985). Potential yield may be 50-100 L/m^3.

Gas in the Khuff Formation may have been sourced from the Silurian lower Sharawara Formation.

EXPLORATION CONCEPTS

Dukhan field, like many of its giant neighbors, was discovered by surface mapping of low relief structures. In this area of the Arabian Gulf, a good knowledge of the relatively monotonous early

AGE			AMOCO FORMATION	LITHOLOGY (onshore | offshore)	DESCRIPTION	SOURCE	RESERVOIR PRIMARY	RESERVOIR SECONDARY	SEAL
CENOZOIC		QUATAR	HOFUF		quartz and calcareous sands and gravels				
CENOZOIC		MIOCENE / PLIOCENE	DAM		sandstone, clay, marls, "chalky" lime-mudstones; local dolomite and gypsum				
CENOZOIC		EOCENE	DAMMAM		dolomite, limestone, subordinate shale				
CENOZOIC		EOCENE	RUS		dolomite, dolomitic limestone with gypsum and anhydrite layers				
CENOZOIC		PALEOCENE	UMM ER RADHUMA		dolomitic limestone, locally porous vugular dolomite				
MESOZOIC	CRETACEOUS	UPPER	SIMSIMA		dolomite, porous bioclastic limestone; limestone, marly at base				
MESOZOIC	CRETACEOUS	UPPER	FIQA		"chalky" bioclastic limestone				
MESOZOIC	CRETACEOUS	UPPER	HALUL		shale, grey-green SEISMIC HORIZON "L"				
MESOZOIC	CRETACEOUS	UPPER	LAFFAN					○	✔
MESOZOIC	CRETACEOUS	UPPER	* MISHRIF/RUMAILA		limestone, crystalline				
MESOZOIC	CRETACEOUS	UPPER	AMADI — AMADI SLT / AMADI LST		"chalky" limestone, marly with shale at base				
MESOZOIC	CRETACEOUS		MAUDDUD		lime-mudstones and marls				✔
MESOZOIC	CRETACEOUS		NAHR UMR		shale, with lenses of sandstone SEISMIC HORIZON "SH"			○○	✔
MESOZOIC	CRETACEOUS	LOWER	SHUAIBA		"chalky" bioclastic limestone, intercalated marls			○○	✔
MESOZOIC	CRETACEOUS	LOWER	HAWAR		gray/green shale and marl				
MESOZOIC	CRETACEOUS	LOWER	KHARAIB (B / C)		"chalky" pelletal limestone, argillaceous at base			○○	
MESOZOIC	CRETACEOUS	LOWER	* LEKHWAIR					○	
MESOZOIC	CRETACEOUS	LOWER	YAMAMA		"chalky" pelletal limestones, mudstones				
MESOZOIC	CRETACEOUS	LOWER	SULAIY		argillaceous lime mudstone SEISMIC HORIZON "H"				✔
MESOZOIC	JURASSIC	UPPER	HITH		anhydrite, thin dolomitic streaks				
MESOZOIC	JURASSIC	UPPER	ARAB (A / B / C)		pelletal limestone or sucrosic dolomite separated by anhydrite layers		○○		
MESOZOIC	JURASSIC	UPPER	ARAB D		bioclastic pelletal limestones				
MESOZOIC	JURASSIC	UPPER	JUBAILA — MARLY LST		locally dolomitic, argillaceous at base				
MESOZOIC	JURASSIC	UPPER	HANIFA		marl, bituminous limestone	○			
MESOZOIC	JURASSIC	LOWER & MIDDLE	UPPER ARAEJ		lime mudstone, porous at top		○○○		
MESOZOIC	JURASSIC	LOWER & MIDDLE	UWAINAT		"chalky" limestone with thin porous bioclastic limestones				
MESOZOIC	JURASSIC	LOWER & MIDDLE	LOWER ARAEJ		lime mudstone				
MESOZOIC	JURASSIC	LOWER & MIDDLE	IZHARA — KEYBED		lime mudstones, dolomitic streaks; anhydrite; argillaceous lime mudstone, shale				
MESOZOIC	TRIASSIC	MIDDLE & UPPER	* HAMLA		finely crystalline dolomite, marl; anhydrite streaks				
MESOZOIC	TRIASSIC	MIDDLE & UPPER	GULAILAH		anhydrite with thin microcrystalline dolomite streaks SEISMIC HORIZON "S"				
MESOZOIC	TRIASSIC	LOWER	SUDAIR		purple, red, green argillaceous siltstones; micro-crystalline dolomites				✔
PALEOZOIC	PERMIAN	UPPER	KHUFF		tight micro-crystalline dolomite, occasionally sucrosic, locally undolomitized bioclastic/oolitic limestones; anhydrite streaks; median anhydrite; micro-crystalline dolomite with sucrosic streaks SEISMIC HORIZON "BK"		○○ ○		✔
PALEOZOIC	PERMIAN	LOWER	HAUSHI		shale, grey-green quartzitic sandstone				
PALEOZOIC	CARB? / SIL.-DEV.		TAWIL		quartzitic sandstone and siltstones				
PALEOZOIC	SILURIAN		SHARAWRA (UPPER SANDY / LOWER MARLY)			○			
PALEOZOIC	ORD		TABUK						

* Formation truncated over North Dome area

Figure 10. Generalized petroleum geology stratigraphic section for Qatar. (Courtesy of Amoco Production Company.)

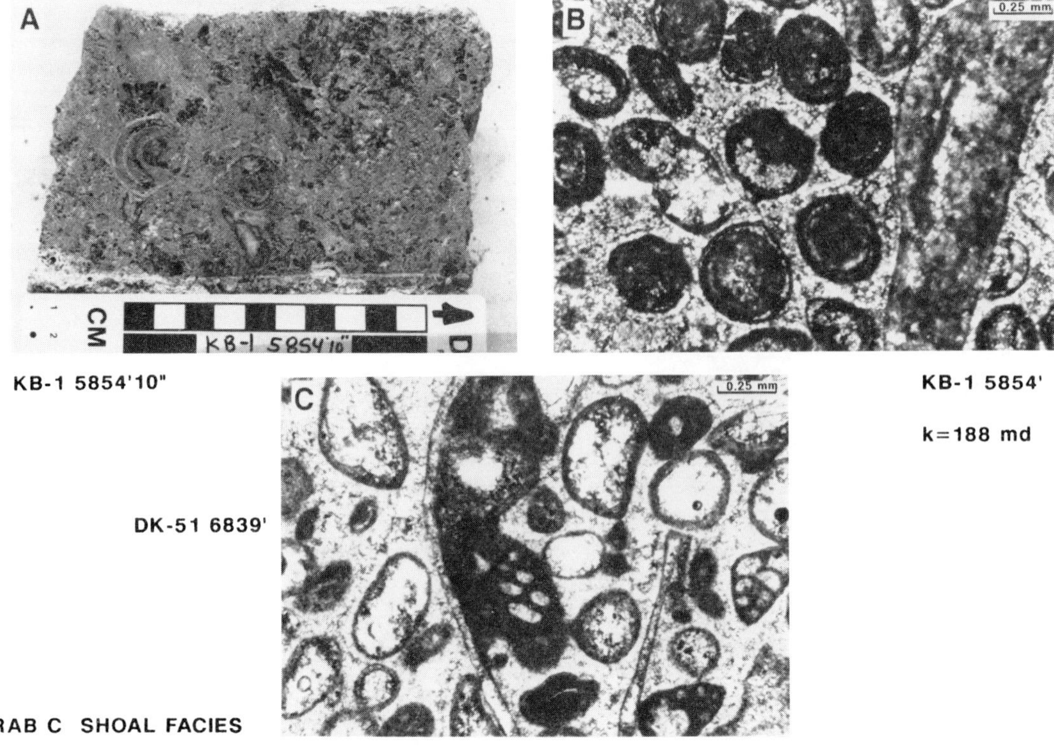

Figure 11. Petrologic characteristics of the Arab C (Arab-3) reservoirs in Dukhan field. (Courtesy of E. Biller.)

Figure 12. Petrologic characteristics of the Arab D (Arab-4) reservoirs in Dukhan field. (Courtesy of E. Biller.)

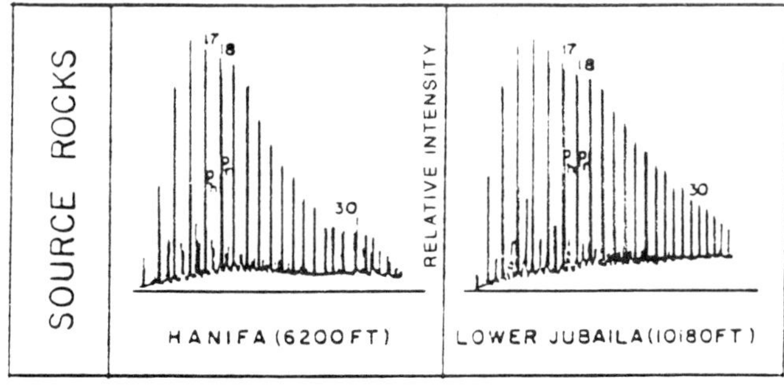

Figure 13 (upper photograph) carries these labels:

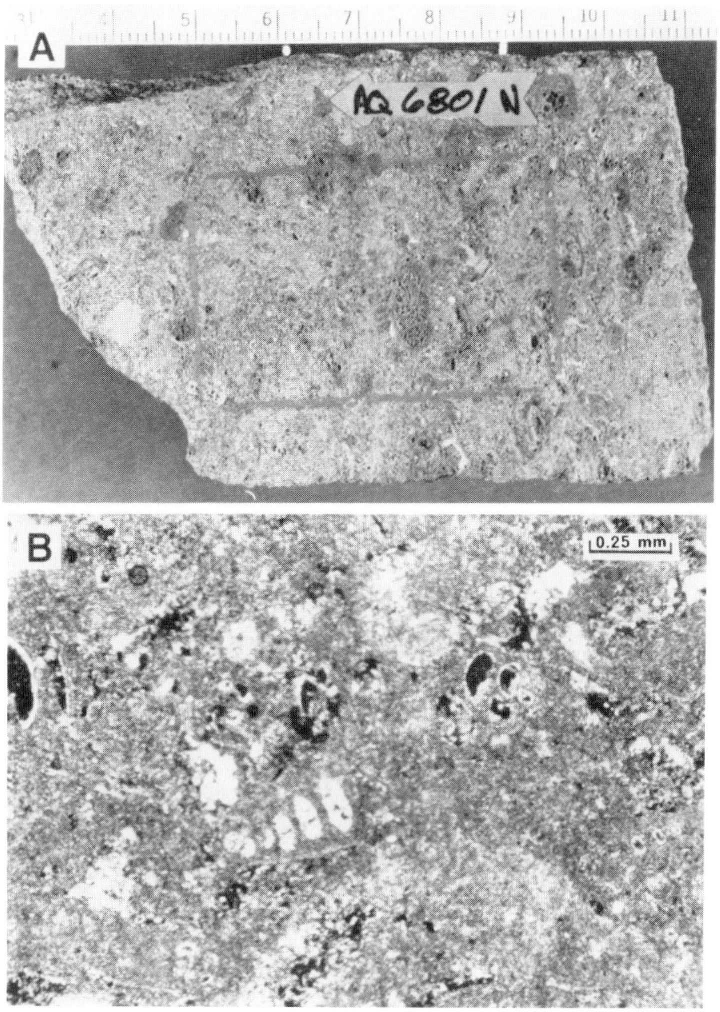

Figure 13. Petrologic characteristics of the Araej reservoirs in Dukhan field. (Courtesy of E. Biller.)

Figure 14. Gas chromatograms for Hanifa and lower Jubaila source rocks of Qatar. (From Frei, 1985.)

Tertiary carbonate package is necessary to effectively correlate units in the field and thus map closure at the surface. Of prime importance in the Jurassic reservoirs of Dukhan and other nearby fields is the depositional environment and diagenesis of the reservoir units. The reservoirs are shelf-type carbonates interbedded with and overlain by evaporates often dominated by sabkha environments, leaching, and early dolomitization. The intimate interbedding of excellent reservoir and seals and the large, broad nature of the structure make Dukhan an excellent world-class field.

ACKNOWLEDGMENTS

We wish to thank the Qatar General Petroleum Company and Amoco Production Company for permission to publish this manuscript, and Colleen Dowse for providing word processing. The manuscript was cooperatively compiled and edited by R. A. Nelson, W. H. Pierce, and C. F. Harpke of Amoco and C. D. Letteney, M. A. Sharif, M. E. Conefrey, J. S. Gomes, and M. E. Zahran of the Qatar General Petroleum Corporation.

REFERENCES CITED

Brindley, A. W., A. J. A. N. Mohammad, and A. A. M. Al-Dulaimi, 1984, Dukhan Khuff gas reservoir performance: Proceedings of 8th Doha Tech. Symp., November 26–28, 1984, Doha, Qatar, p. 111–116.

Daniel, E. J., 1954, Fractured reservoirs of Middle East: American Association of Petroleum Geologists Bulletin, v. 38, n. 5, p. 774–815.

Frei, H. P., 1985, Mesozoic source rocks and oil accumulations in Qatar: Proc. OAPEC Sem. Source and Habitat of Petrol. in Arab Countries, Kuwait, 1985, p. 115–124.

Pinnington, D. J., ed., 1981, Schlumberger Well Evaluation Conference UAE/Qatar, Chapter 1—Geology, p. 1–32.

Sugden, W., 1962, Structural analysis and geometrical prediction for change of form with depth, of some Arabian plains-type folds: American Association of Petroleum Geologists Bulletin, v. 46, n. 12, p. 2213–2228.

<h1 style="text-align:center">Appendix 1. Field Description</h1>

Field name .... *Dukhan field*

Ultimate recoverable reserves .... *Primary 4.57 billion BO and 10.8 tcf gas*

Field location:
    Country .... *Qatar*
    Basin/Province .... *Arabian platform*

Field discovery:
    Year first pay discovered .... *Jurassic C and D 1940*
    Year second pay discovered .... *Uwainat post-1945*
    Year third pay discovered .... *Khuff 1959*

Discovery well name and general location:
    First pay .... *Petroleum Development (Qatar) Dukhan No. 1 25°25′16.2″N; 050°47′00.6″E*

Discovery well operator .... *Qatar Petroleum*

IP:
    First pay .... *4480 BOPD*

Geologic concept leading to discovery and method or methods used to delineate prospect

Surface geology, recognized in 1933 and mapped in 1937–1938.

Structure:
    Province/basin type .... *Bally 221, Klemme IICa*

    Tectonic history
    Stable platform from Precambrian to middle Cretaceous. Regional tilting to northeast at closure of Tethys and development of the Arabian Gulf. Light Miocene deformation in response to Zagros deformation to northeast.

    Regional structure
    On western flank of long-lasting Qatar-Fars arch and eastern edge of Infracambrian salt basin. Anticline caused by salt diapirism localized over basement fault of low displacement.

    Local structure
    An "Arab Fold." North-south anticline with gentle limb dips (10–15° max). Position and strike controlled by low displacement basement fault. Fold amplitude generated by salt diapirism above basement fault. 1200 ft (366 m) closure.

Trap:
    Trap type(s) .... *Anticlinal trap*

Basin stratigraphy (major stratigraphic intervals from surface to deepest penetration in field):
Well Dkg 27

Khuff gas producer	*Location*	*156,982*	*RT/KB:*	*152 ft*
(deviated)		*411,926*	*GL:*	*120 ft*

Chronostratigraphy	Formation	True Vertical Depth Subsea in ft (m)	Isochore Thickness in ft (m)
Upper Cretaceous	*Simsima S-1*	*1018 (310)*	*507 (155)*
	Halul	*1525 (465)*	*37 (11)*
	Laffan	*1562 (476)*	*33 (10)*
	Mishrif	*1595 (486)*	*71 (22)*
	Ahmadi	*1666 (508)*	*565 (172)*
Lower Cretaceous	*Mauddud*	*2231 (680)*	*168 (51)*
	Nahr Umr	*2399 (732)*	*431 (131)*

DUKHAN

117

Chronostratigraphy	Formation	True Vertical Depth Subsea in ft (m)	Isochore Thickness in ft (m)
	Shuaiba	*2830 (863)*	*371 (113)*
	Hawar	*3201 (976)*	*49 (15)*
	Kharaib	*3250 (991)*	*234 (71)*
	Lakhwair	*3484 (1063)*	*477 (145)*
	Yamama	*3961 (1208)*	*260 (79)*
	Sulaiy	*4221 (1287)*	*486 (148)*
Upper Jurassic	*Hith*	*4707 (1436)*	*388 (118)*
	Qatar	*5095 (1554)*	*278 (85)*
	Arab-A	*5095 (1554)*	*35 (11)*
	u. anhydrite	*5130 (1565)*	*39 (12)*
	Arab-B	*5169 (1576)*	*14 (4)*
	m. anhydrite	*5183 (1581)*	*48 (15)*
	Arab-C	*5231 (1595)*	*80 (24)*
	l. anhydrite	*5311 (1620)*	*62 (19)*
	Jubaila	*5373 (1639)*	*859 (262)*
	Arab-D res.	*5373 (1639)*	*231 (70)*
	base Arab-D res.	*5604 (1709)*	
	Hanifa	*6232 (1901)*	*325 (99)*
Middle–Lower Jurassic	*Hanifa res.*	*6232 (1901)*	*9 (3)*
	base Hanifa res.	*6241 (1904)*	
	Araej	*6557 (2000)*	*620 (190)*
	Uwainat res.	*6655 (2030)*	*178 (54)*
	base Uwainat res.	*6833 (2084)*	
	l. Araej res.	*6908 (2107)*	*29 (9)*
	base l. Araej res.	*6937 (2116)*	
	Izhara	*7177 (2189)*	*434 (132)*
	u. Izhara	*7177 (2189)*	*72 (22)*
	base u. Izhara res.	*7249 (2266)*	
	l. Izhara res.	*7429 (2266)*	*46 (14)*
	base l. Izhara res.	*7475 (2280)*	
Upper–Middle Triassic	*Hamlah*	*7611 (2321)*	*229 (70)*
	Hamlah Dolomite	*7674 (2340)*	*166 (51)*
	Gulailah	*7840 (2391)*	*840 (256)*
Lower Triassic	*Sudair*	*8680 (2647)*	*641 (196)*
Lower Triassic-Upper Permian	*Khuff*	*9321 (2843)*	*1660 (506)*
Lower Permian	*Haushi*	*10,981 (3349)*	*844 (257)*
Carboniferous(?)/Devonian	*Tawil*	*11,825 (3607)*	*1474 (450)*
	porous Tawil	*12,631 (3852)*	*315 (96)*
	base porous Tawil	*12,946 (3948)*	
Silurian	*Sharawara*	*13,299 (4056)*	*2108 (643)*
Ordovician	*Tabuk*	*15,407 (4699)*	*511 (156)*
	pre-Tabuk	*15,918 (4855)*	
	T.D. (D.D.)	*16,008 (4882)*	

Reservoir characteristics:

Number of reservoirs 4

Formations .... *Arab C, Arab D, Araej, Khuff (also read Arab 3 and 4 for Arab C and D)*

Ages .... *Upper Jurassic, Middle Jurassic, middle–Upper Permian*

Depths to tops of reservoirs .... *Arab C, 5600 ft (1708 m); Arab D, 5740 ft (1751 m); Araej, 7050 ft (2150 m); Khuff, 9500 ft (2898 m)*

Gross thickness (top to bottom of producing interval) ..... *Arab C, 80 ft (24.4 m); Arab D, 185 ft (56.4 m);*
Araej, 120 ft (36.6 m); Khuff, 1700 ft (518.5 m)

Net thickness—total thickness of producing zones
   Average .... *Arab C, 15 ft (4.6 m); Arab D, 120 ft (36.6 m); Araej, 75 ft (22.9 m); Khuff ?*

Lithology
Arab C and D: oolite-peloidal grainstone crystalline dolomite
Araej: chalky limestone with thin porous bioclastic limestones
Khuff: microcrystalline dolomite, bioclastic/oolite limestone, anhydrite streaks, often silty

Porosity type .... *Arab C and D, intergranular and intercrystalline; Araej, intergranular;*
Khuff, intercrystalline, moldic, intergranular, and fracture

Average porosity .... *Arab C, 18%; Arab D, 20%; Araej, 18%*
Average permeability .... *Arab C, 30 md; Arab D, 70 md; Araej, 15 md; Khuff, 30 md*

Seals:

  Upper
     Formation, fault, or other feature .... *Hith and Arab anhydrites*
         Lithology .... *Anhydrite*
  Lateral
     Formation, fault, or other feature .... *Closure*

Source:

  Formation and age .... *Upper Jurassic Hanifa Formation and lower Jubaila*
  Lithology .... *Kerogen and bitumen-rich lime mudstone*
  Average total organic carbon (TOC) .... *2.1% (1.2–5.1%)*
(10 samples from upper Hanifa and lower Jubaila)
  Maximum TOC *5.1%*
  Kerogen type (I, II, or III) .... *II (sapropelic organic matter)*
  Vitrinite reflectance (maturation) .... $R_o = 0.62$ *at 1750 ft (534 m)*
  Time of hydrocarbon expulsion .... *NA*
  Present depth to top of source .... *6153 ft (1877 m) on structure*
  Thickness .... *300–400 ft (92–122 m) (?)*
  Potential yield .... *50–100 L/m³; 10–60 BBO*

DUKHAN

Appendix 2. Production Data

Field name .... *Dukhan field*

Field size:

  Proved acres .... *80,000 ac (32,400 ha)*
  Number of wells all years *383*
  Current number of wells .... *160 producing*
  Well spacing .... *NA*
  Ultimate recoverable .... *4.57 BBO; 10.8 tcf*
  Cumulative production (1/90) .... *2.57 BBO; 5.08 tcf*
  Annual production (1989) .... *86.9 MMBO; 186 bcf*
  Present decline rate .... *NA*
    Initial decline rate .... *NA*
    Overall decline rate .... *NA*
  Annual water production .... *NA*
  In place, total reserves .... *11.3 BBO; 19.6 tcf*
  Primary recovery .... *4.57 BBO; 10.8 tcf*

Secondary recovery .... *NA*
Enhanced recovery .... *NA*
Cumulative water production .... *NA*

Drilling and casing practices: .... *NA*

Completion practices: .... *NA*

Formation evaluation: .... *NA*

Oil characteristics:

API gravity .... *Arab C, 37°; Arab D, 42°; Araej, 42.5°*
Base .... *Paraffin*
Initial GOR .... *Arab C, 735; Arab D, 1070; Araej, 1060*
Sulfur, wt% .... *Arab C, 1.8; Arab D, 1.1; Araej, 0.8*
Viscosity, SUS .... *Arab C, 0.6 est.; Arab D, 0.4 est.; Araej, 0.4 est.*
Gas-oil distillate *17%*

Field characteristics:

Average elevation .... *66 ft (20 m)*
Initial pressure
Arab D, 0.51 psi at 6250 ft (22,500 kPa at 1906 m)
Araej, 3520 psi at 7050 ft (24,625 kPa at 2150 m)
Khuff, 6090 psi at 10,000 ft (42,600 kPa at 3050 m)

Pressure gradient
Arab D, 0.51 psi/ft (11.8 kPa/m)
Araej, 0.50 psi/ft (11.5 kPa/m)
Khuff, 0.56 psi/ft (12.8 kPa/m)

Temperature .... *Arab C, 194°F (90°C); Arab D, 207°F (97°C);Araej, 219°F (104°C)*
Geothermal gradient .... *Arab C, 0.035°F/ft (0.064°C/m); Arab D, 0.033°F (0.060°C/m);*
Araej, 0.031°F (0.056°C/m)
Drive .... *Arab C, solution gas; Arab D, gas/water; Araej, gas/water; Khuff, gas*
Oil column thickness .... *Arab C, 920 ft (281 m); Arab D, 960 ft (293 m); Araej, 196 ft (60 m);*
Khuff, 1700 ft (518 m)
Oil-water contact .... *Arab C, 6520 ft (1989 m); Arab D, 6700 ft (2044 m); Araej, 7246 ft (2210 m);*
Khuff, 11,100 ft (3386 m)
Connate water .... *Arab C, 20%; Araej, 30%*
Water salinity, TDS .... *NA*
Resistivity of water .... *NA*
Bulk volume water (%) .... *NA*

Transportation method and market for oil and gas:
Oil pipeline to Umm Said

Sendji Field—People's Republic of Congo
Congo Basin

S. BAUDOUY
C. LEGORJUS
Elf-Congo
Pointe-Noire, République Populaire du Congo

FIELD CLASSIFICATION

BASIN: Congo
BASIN TYPE: Passive Margin
RESERVOIR ROCK TYPE: Sandstone and
  Dolomite

RESERVOIR AGE: Cretaceous
PETROLEUM TYPE: Oil
TRAP TYPE: Salt Dome

RESERVOIR ENVIRONMENT OF DEPOSITION: Nearshore Marine
TRAP DESCRIPTION: Salt dome bisected by a graben; minor stratigraphic influence

LOCATION

The Sendji field is located in the Congo basin at the southern end of the Gulf of Guinea, about 40 km (25 mi) offshore from Pointe-Noire and Pointe-Indienne of the People's Republic of Congo (Figure 1). Congo has a 150 km (93 mi) seacoast, trending northwest to southeast. The field is near Cabinda (Angola) in water depths of about 95 m (312 ft) on the Pointe-Noire Grands Fonds Exploration Permit, which was granted in 1968 to an association of Elf-Congo, 65%, and Agip Recherches Congo, 35%, for which Elf-Congo is operator.

The majority of crude oil production of the People's Republic of Congo, exceeding 6 million MT (43 million bbl) per year, is from six offshore fields that include the Sendji fields as well as Loango, Yanga, Tchibouela, Emeraude, and Likouala fields (Figure 2). Estimated total recoverable oil reserves exceed 17 million MT (120 million bbl) from the Sendji field.

HISTORY

The first seismic survey covering the Congo offshore was carried out in 1959. Despite poor quality, the seismic records indicated the southwestwardly continuation of salt-related structures known to parallel the Gabon coast along the margin of the Congo basin.

In 1968, drilling by Elf-Congo resulted in the discovery in May 1969 of the Emeraude field on a northwest–southeast-trending salt structure. The discovery initiated a more intense exploration effort in the Congo basin. Albian–Cenomanian salt-tectonic turtle structures were revealed by seismic surveys carried out in 1969, 1971, and 1972.

Northward, Agip discovered the Loango field in 1971, and to the south, Elf-Congo made the discovery of the Likouala field on an Albian turtle structure in 1972. These discoveries led to the drilling of the similar structure and discovery of the Sendji field.

The discovery well, Sendji Marine 1 (SEM-1), penetrated the post-salt formations and the pre-salt section (Figure 3), reaching total depth of 2887 m (9472 ft) in the Djeno Sandstone. The well encountered a 210 m (690 ft) gross oil column in the upper part of the Albian Sendji Carbonate. Evaluations at the time indicated that the field could not be economically exploited.

In the following years improving world oil prices and more definitive information about the structure and the possible extent of the field provided by the additional seismic surveys over the Sendji field area indicated that exploitation would be commercially feasible. As a result, the 50-year Yanga-Sendji claim was granted to Elf-Congo on 1 December 1979.

Following a lull from 1976 to 1977, exploration drilling in the Congo began again in 1978 and led to new offshore successes for Elf-Congo: Yanga and Tchendo in 1979, Nkossa and Boatou in 1983, Tchibouela in 1984, Loussima and Kombi in 1985, and Tchibeli in 1986. Agip discovered Zatchi in 1981.

Appraisal of the Sendji structure began with the drilling in the wells between 1979 and 1982. These

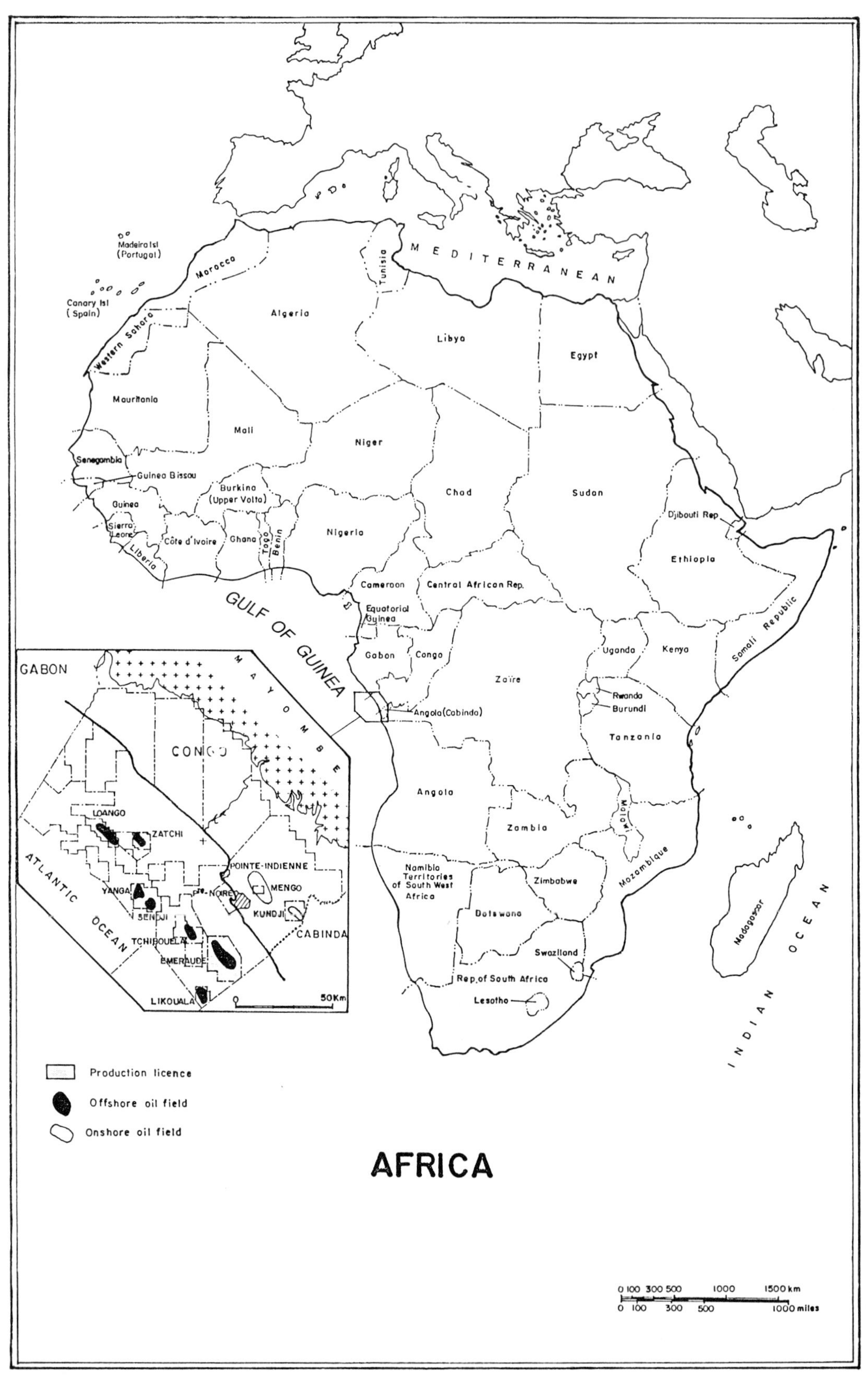

Figure 1. Location map. The enlargement shows the major northwest-southeast alignment of the Congolese coastal basin parallel to the present-day shoreline. (Mayombe is basement crystallines.)

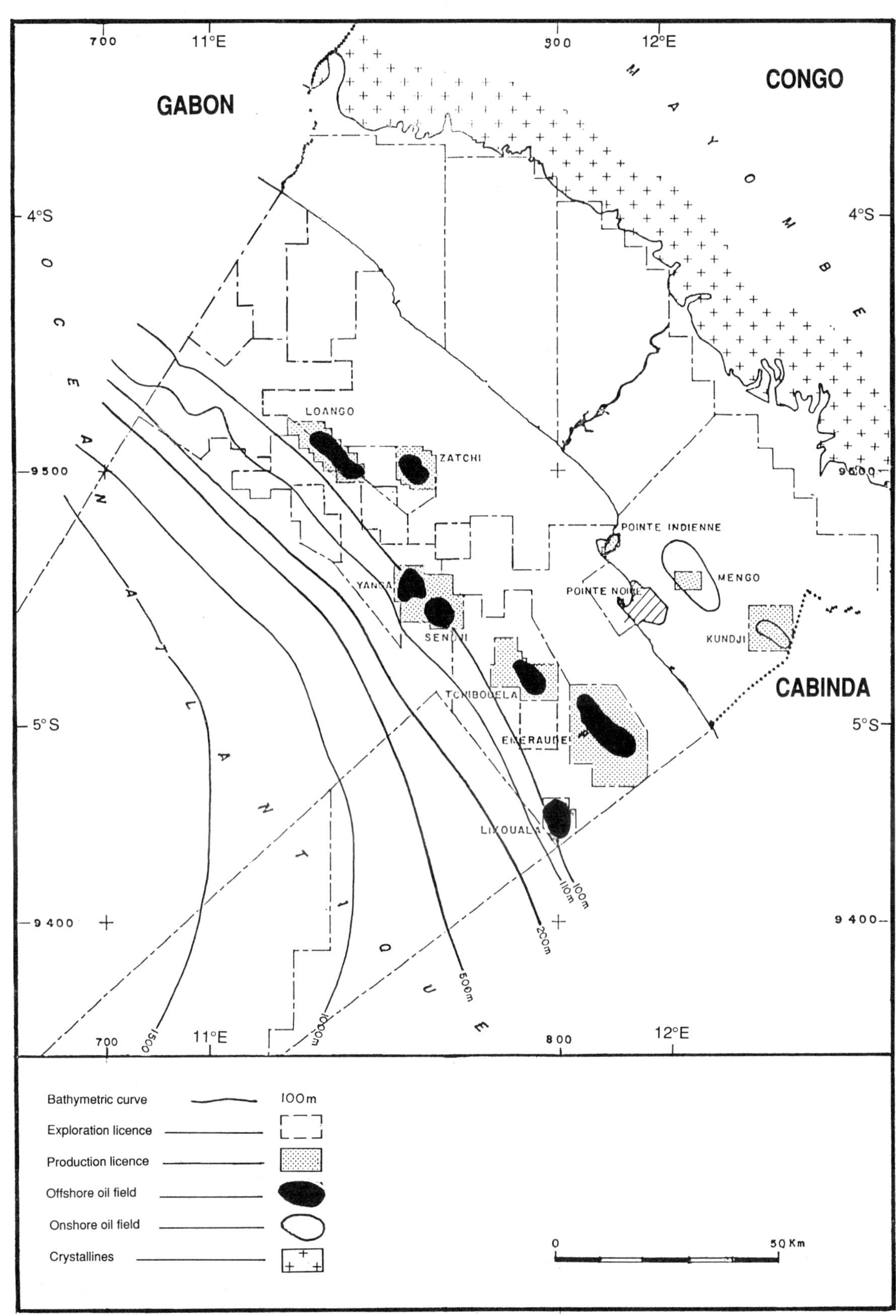

Figure 2. Bathymetry of the Congolese coastal basin. The 110 m isobath is the outer limit of developed fields in the offshore Congo.

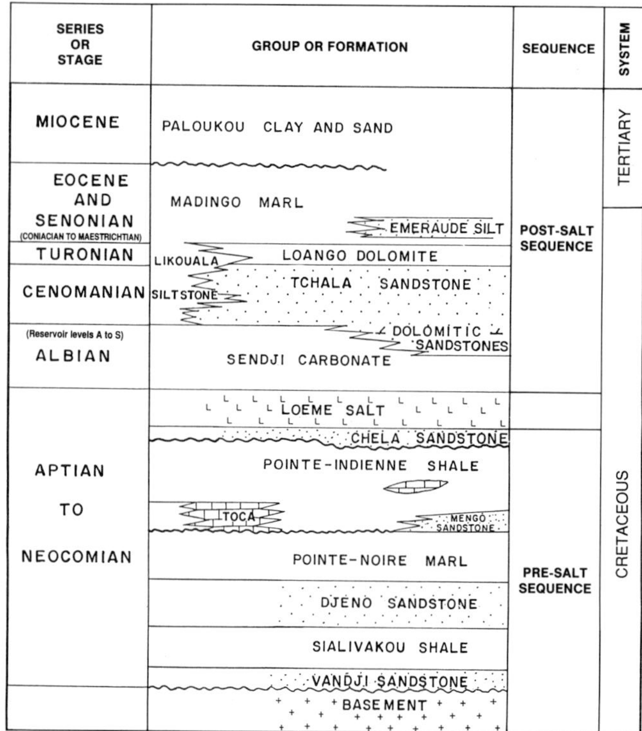

Figure 3. Lithostratigraphy of the Congolese coastal basin.

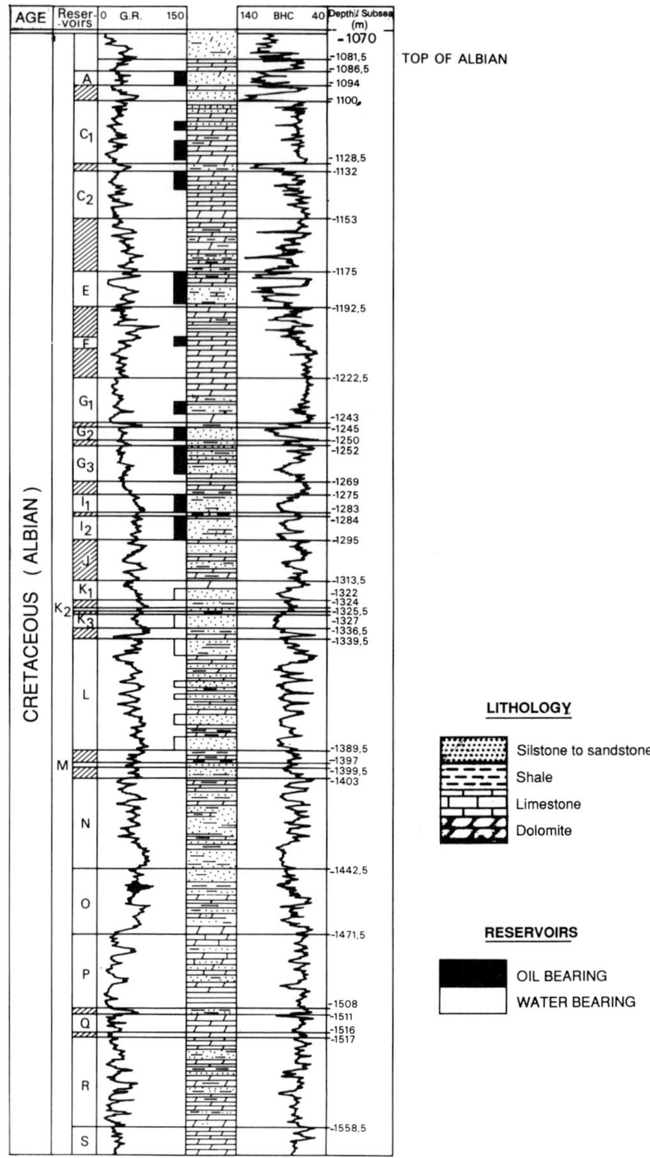

Figure 4. Sendji Marine 1 (SEM-1) gamma ray-sonic log reservoir column showing: log characteristics of reservoirs and seals and reservoir nomenclature; thickness of various zones; and simplified lithology.

wells, designated SEM-2, SEM-3, SEM-4, and SEM-5, were drilled on the structural flanks to evaluate the reservoirs. This evaluation was to include the determination of oil-water contacts in the different reservoirs and the study of facies changes as well as other reservoir properties. The locations of these four wells are indicated on the various geologic maps of the field (Figures 8, 9, 10, 11, and 15, but most readily seen on the small-scale structure map on Figure 7).

Development drilling of the Sendji field began in 1982 following the drilling of four appraisal wells. The data obtained from the discovery and appraisal wells permitted informal classification of the sequence of Sendji Carbonate zones (Figure 4) and identification of nine of these zones (or combinations) as being of reservoir quality and containing entrapped commercial quantities of oil. The nine reservoirs fall in a depth interval between 1050 and 1550 m (3450 and 5100 ft) below sea level.

Estimated recoverable oil reserves from each of the reservoirs (totaling 17.38 million MT or 124.6 million bbl) together with initial well potentials, the number of wells producing from each reservoir, and the number of injection wells (as of 1 January 1988) are shown in Table 1. Total oil in place is estimated to be 73.57 million MT (513.2 million bbl). Recovery factors of the reservoirs vary but average over 24%.

The development program was continually modified as data obtained during drilling were incorporated into the geological interpretation of the structure and reservoirs. The main development phase began in October 1982 with the drilling of SEF-101 from platform SEF-1. All wells drilled in the next two years during the initial development phase were from four platforms, designated SEF-1 through -3 and SEF-P (Figure 8). A supplementary development phase was proposed in 1984 and completed in 1986 from a fifth platform (SEF-4), emplaced in the northeast part of the field. Additional wells were drilled from the various platforms in 1987–1988.

To develop the Sendji field, 76 wells (50 producers and 26 injectors) were drilled from five platforms through 1988 (Table 1 and Figures 8–11).

Table 1. Albian Sendji Carbonate reservoirs, showing for each reservoir the estimated recoverable petroleum reserves, average initial production potentials of wells, and the number of wells completed as producing and as injection wells. These reservoirs, all in the upper part of the Sendji Carbonate, are shown by the lithologic/gamma ray/BHC sonic log stratigraphic column in Figure 4.

Sendji Carbonate Reservoir	Estimated Recoverable Reserves 10^6 MT (10^6 bbl)	Avg/Well Initial Daily Production Potential 10^2 m³ (10^3 bbl)	Number of Wells as of 1-1-88	
			Producing	Injection
A	2.8 (20.08)	3.1 (1.95)	6	5
E	3.5 (25.10)	2.3 (1.45)	9	4
G2 with Upper G3	5.8 (41.59)	1.8 (1.13)	15	7
Other G3	3.6 (25.81)	1.8 (1.13)	11	7
I K L	0.91 (6.52) 0.20 (1.43) 0.16 (1.15)	1.2 (0.75)	7	3
P	0.31 (2.22)	3.0 (1.89)	1	0
S	0.10 (0.72)	4.0 (2.52)	1	0
Totals	17.38 (124.62)	—	50	26

DISCOVERY METHOD

An initial seismic survey in 1959 of an area that included the field was of poor quality but was sufficient for recognition of the continuation from offshore Gabon of a belt of salt tectonic structures paralleling the coast and extending southeastward along the Congo basin. Subsequent seismic surveys in the area in 1969, 1971, 1972, and 1980 enabled a better definition of the structures, among these the Sendji structure.

Criteria leading to the drilling of the Sendji structure were:

- Recent discoveries of Loango and Likouala fields in the detrital facies at the base of the Cenomanian and the top of Albian.
- Known structures along a northwest-southeast trend.
- Size of the structure shown by seismic interpretation.
- Paleogeographic location.

STRUCTURE

Tectonic History

The Congolese coastal basin (Figure 5) is one of a series of basins belonging to the Atlantic passive margin. Its history can be divided into two main sequences:

1. Before oceanic opening (Neocomian to Aptian), fault basins of the rift type filled with lacustrine or brackish deposits.
2. After opening (Albian to Tertiary), prograding platforms were affected by salt tectonics.

These two sequences were separated by the Aptian salt episode corresponding to the formation of a huge basin spreading along nearly 2000 km (1240 mi) from South Angola north to the Cameroon shores. Deposition and movements of the salt constituted a major event in the petroleum geology of the basin. The salt is essential as a seal for the pre-opening sequence and also as the origin of structural deformation of the overlying formations. It is the boundary between the petroleum sequences above and the pre-salt source rock sequences beneath.

Regional Structure

The rift-type basins antedating the oceanic opening are characterized structurally by a succession of

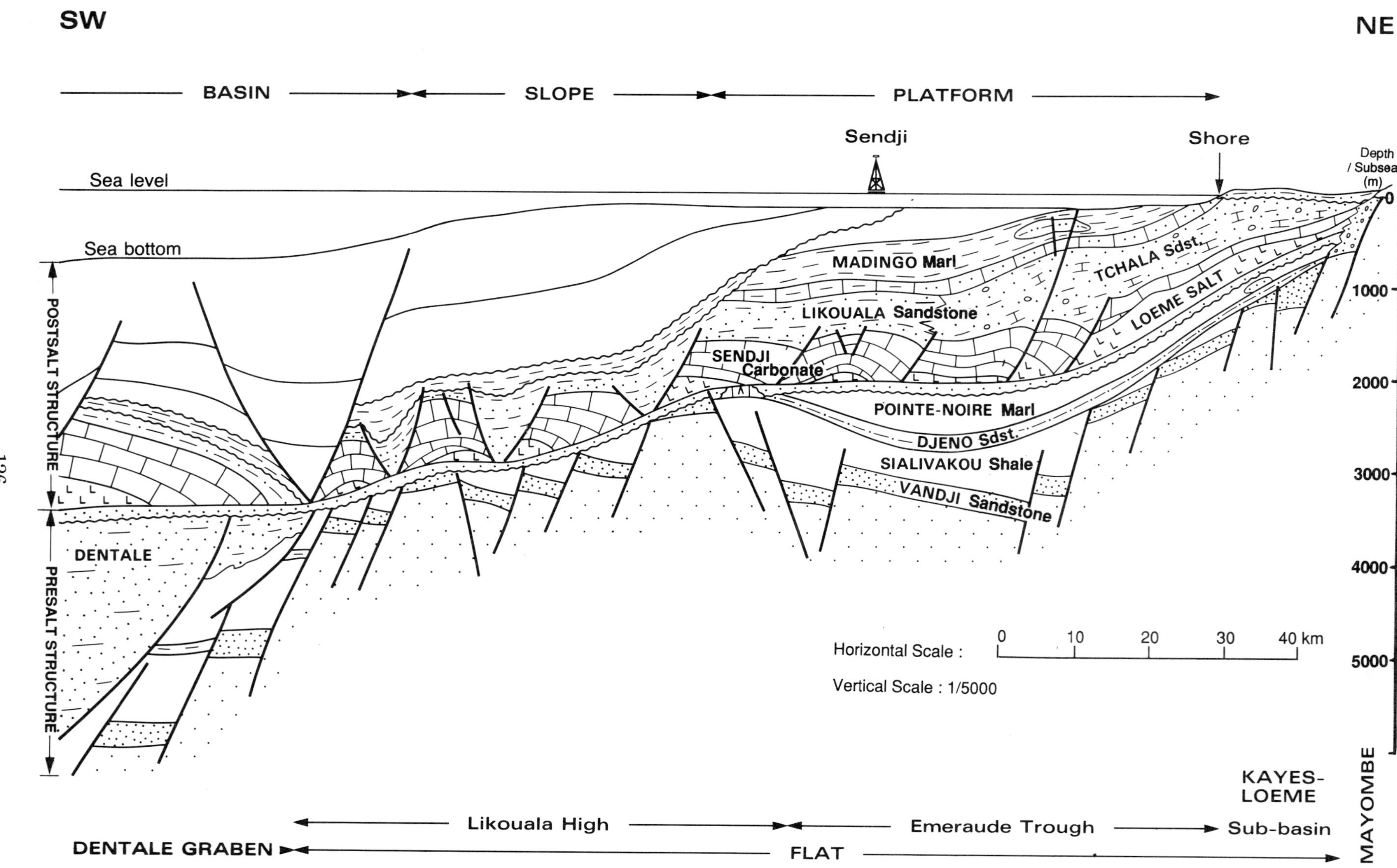

Figure 5. Schematic structural cross section of the Congolese coastal basin showing pre-salt and post-salt sequences.

126

horsts and half-grabens (Figure 5). They are filled with lacustrine to brackish deposits of Early Cretaceous age. The sealing of this system is ensured regionally by Aptian Loeme Salt, which is an effective barrier to migration except locally, where magnitude and type of halokinesis created discontinuities in the salt beds and opened the way to migration. Traps are mainly rotated fault blocks and anticlinal structures.

Post-opening formations have been affected by salt tectonics. Reservoir levels can be found throughout the stratigraphic column with sediments of marine to marginal-littoral origin. They consist of detrital or carbonate formations of Albian to Cenomanian age. Sealing is ensured by shales of different stratigraphic levels. Trapping depends on the presence of these shales.

Traps are of tectonic origin and are controlled by salt movements. They include simple closures with intumescence of the salt or fault-growth related structures and more complex closures of the turtle-back type or salt diapirs.

Local Structure

Structural analyses based on a seismic cross section (Figure 6) and structural maps and cross section (Figures 5, 7, 8, 9, 10, and 11) of the Sendji field reveal the asymmetry of the structural axis with increasing depth. On the northern flank an east–west-trending graben, shown on Figures 9, 10, and 11, was a synsedimentary feature during the upper Albian (Figure 15).

A Neogene (Miocene) channel just overlying the structure is indicated by seismic interpretation (Figure 6).

The formation of the structures with which the Sendji field is associated was linked with movements of the underlying salt. The dome is a turtle structure over a residual salt pillow capable of later mobilization. Salt movements lasted from Albian to late early Senonian. The seismic marker at the top of the Turonian shows a closure on the 80E65 seismic section (Figure 6). During upper Albian, an east-west graben was created. Its maximum downthrow is approximately 40 m (130 ft). As a result of synsedimentary growth, some reservoir zones are much thicker in the graben. The faults located north and south of the graben are minor faults.

Tectonic history plays an important role in the efficiency of the trapping mechanism. Seal thickness of some lower reservoirs is less than the fault throws. This leads to hydrocarbon leakage and decreases the reservoir area to an area corresponding in size to the last closed structural contour at the base of the seal on the opposite side of an affecting fault. The upper reservoirs do not leak hydrocarbons as the result of faults because the thickness of the beds acting as seals (zone F) or of tight beds (zone G1) is greater than the fault throws.

STRATIGRAPHY

Post-Salt Series: Reservoirs and Seal Rocks

The lithostratigraphy of the Congolese coast basin (Figure 3) shows almost continuous sedimentation from Early Cretaceous to Tertiary. Post-salt reservoirs are numerous and are present in all formations.

Albian (Sendji Carbonate Formation)

Petroleum reservoirs occur throughout the Albian section, and they are oil producing in the Loango and especially the Yanga and Sendji fields. They are made up of:

- High energy carbonates (lumachelle or oolitic limestone), generally dolomitic.
- Sands and clean sandstones deposited in tidal channels and as offshore bars or beaches.
- Silts of prelittoral origin.

Cenomanian (Tchala Sandstone and Likouala Siltstone Formations)

The chemical and detrital deposits of the Albian are succeeded by a detrital regressive period resulting from the rejuvenation of the continent. Several environments within the Cenomanian can be differentiated. The Tchala sandstones indicate a continental to marginal littoral environment. They are excellent reservoirs in the Loango, Yanga, Tchibouela, and Likouala fields. The Likouala Siltstone, made up of fine sandstones and siltstones, indicates a marginal littoral environment. Its reservoir quality is inferior to the Tchala sandstones. The Likouala Siltstone is the main reservoir of the Likouala field.

Turonian (Loango Dolomite Formation)

A general transgression started during the Turonian. Favorable sandstone reservoirs are at the base of the Turonian, with porosities ranging from 10 to 30%, averaging near 30%. The upper reservoirs, present in the Tchibouela and Emeraude fields, consist of shelly limestone.

Senonian (Emeraude Siltstone Formation)

The Emeraude Siltstone is located in the far south of the Congo coastal basin and constitutes the reservoir of the Emeraude field. It consists of poorly consolidated silts, shaly silts, and fine sandstones interbedded with shale clays and thin, shelly limestones. The latter are fractured and provide pathways for fluid migration. Despite the shaliness of the reservoir, porosities of 17 to 30% are recorded.

Pre-Salt Series: Potential Source Rocks

The potential source rocks for the post-salt oil fields in the Congo coastal basin are to be found in the

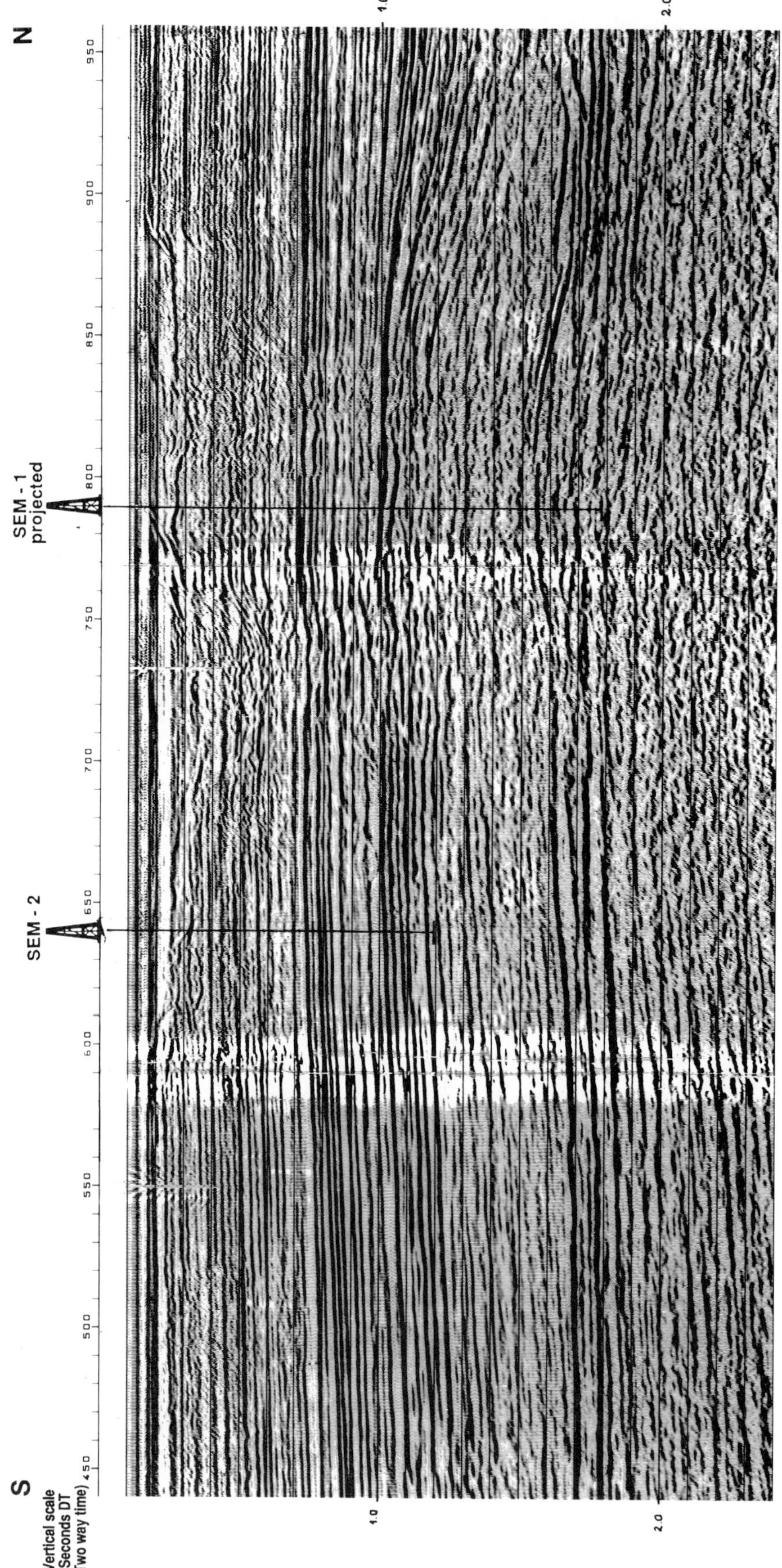

Horizontal scale :   0   .2   .4   .6   .8   1km

Figure 6A. Uninterpreted seismic line 80E65 cross section.

128

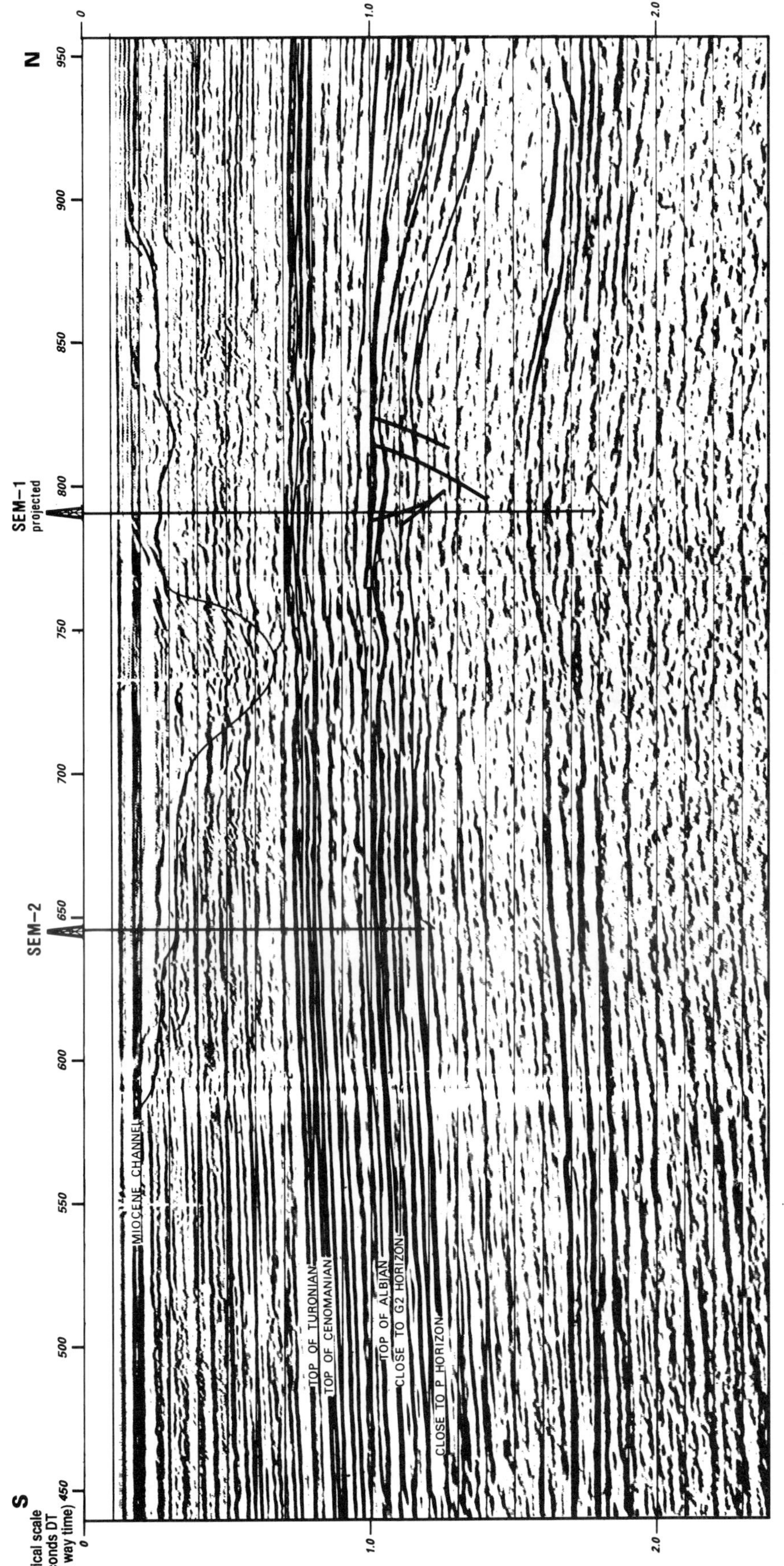

Horizontal scale : 0 .2 .4 .6 .8 1km

Figure 6B. Interpreted seismic cross section (line 80E65). Location of section shown on Figure 9.

129

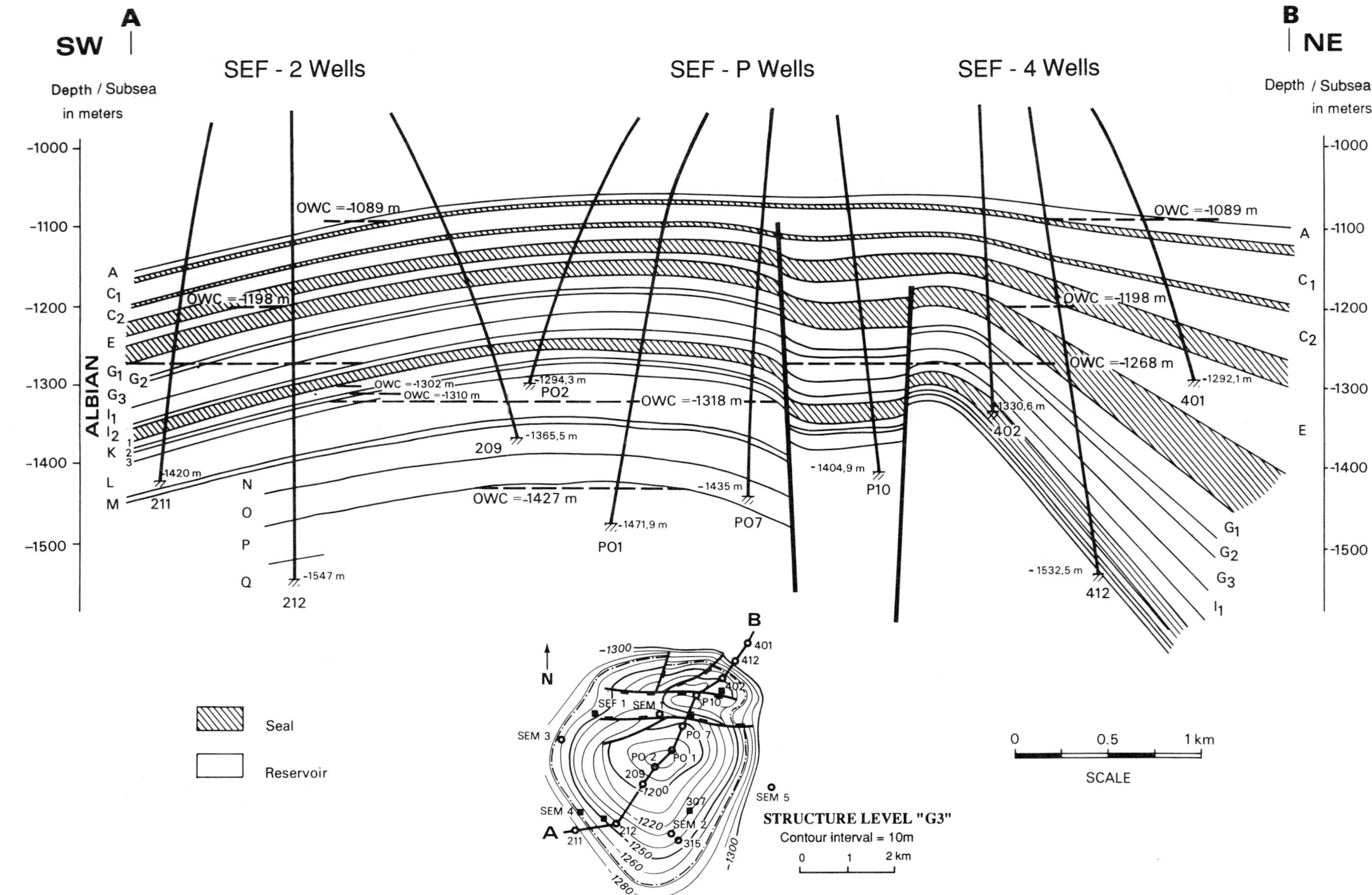

Figure 7. Geological cross section through the reservoirs. Zones K1, K2, and K3 have three different oil-water contacts.

130

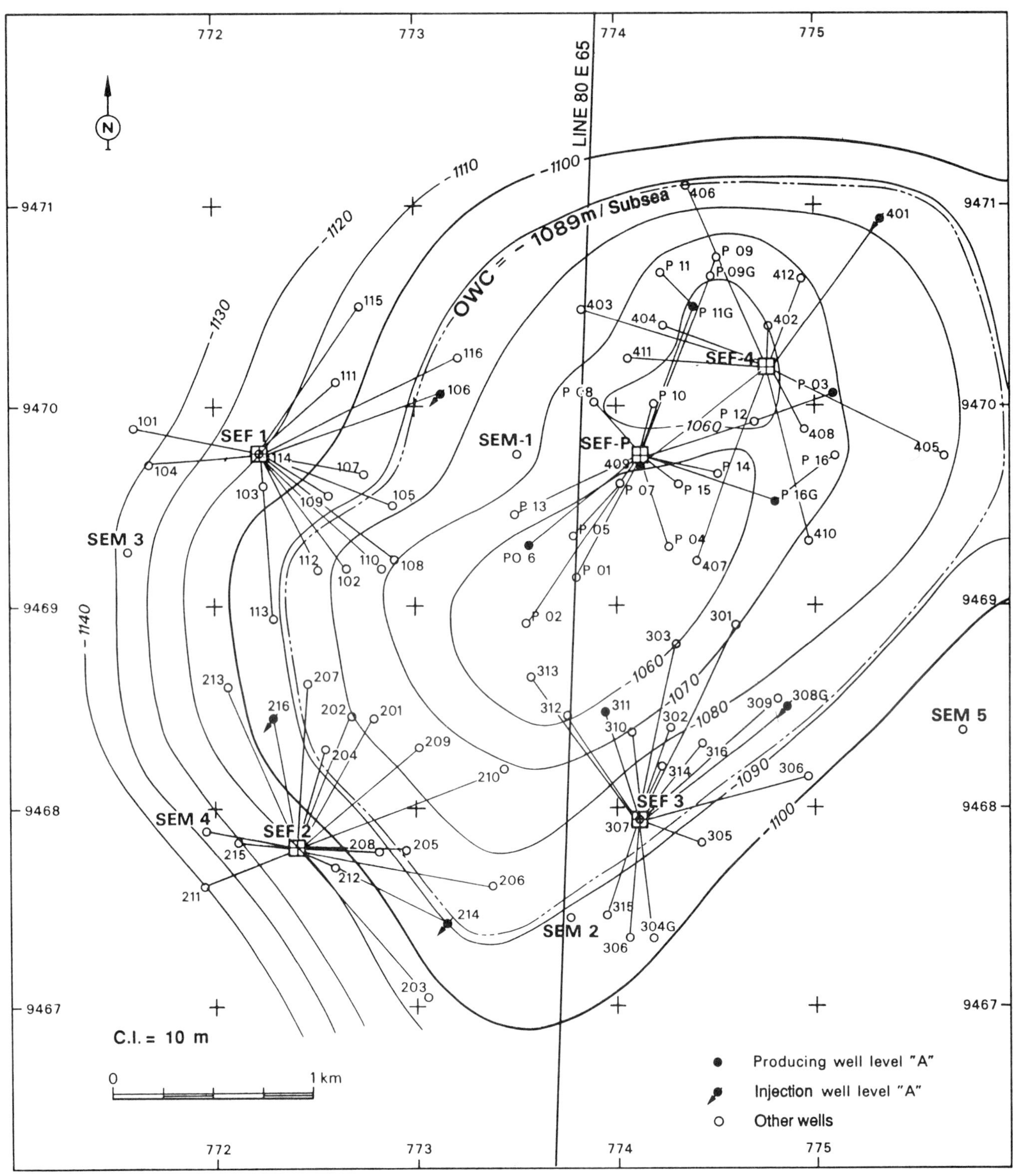

Figure 8. Structure map, top of zone A. Depth in meters subsea. SEM-1 to SEM-4 are evaluation wells. SEF-P and SEF-1 to SEF-4 are production platforms.

131

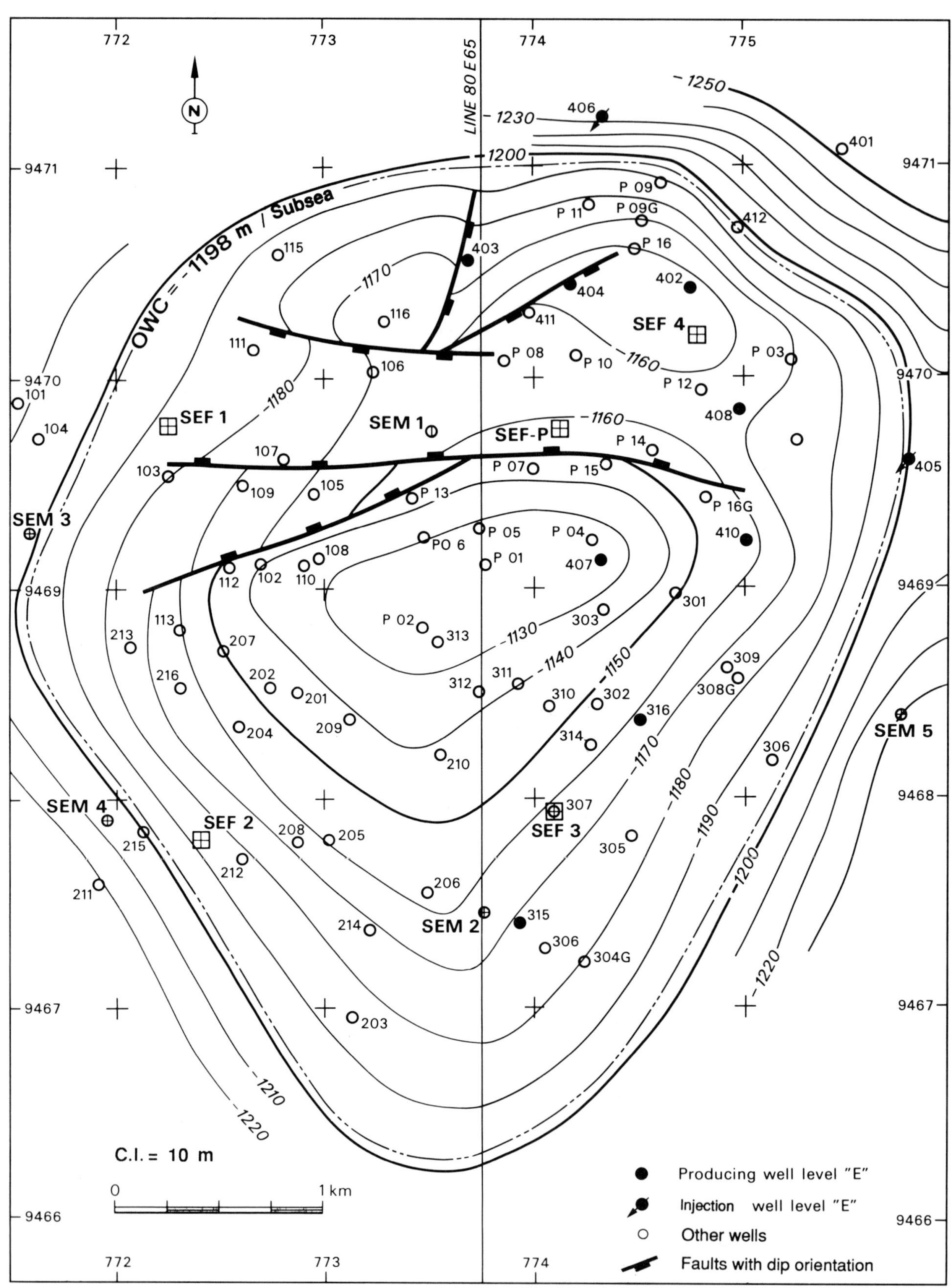

Figure 9. Structure map, top of zone E. Depth in meters subsea.

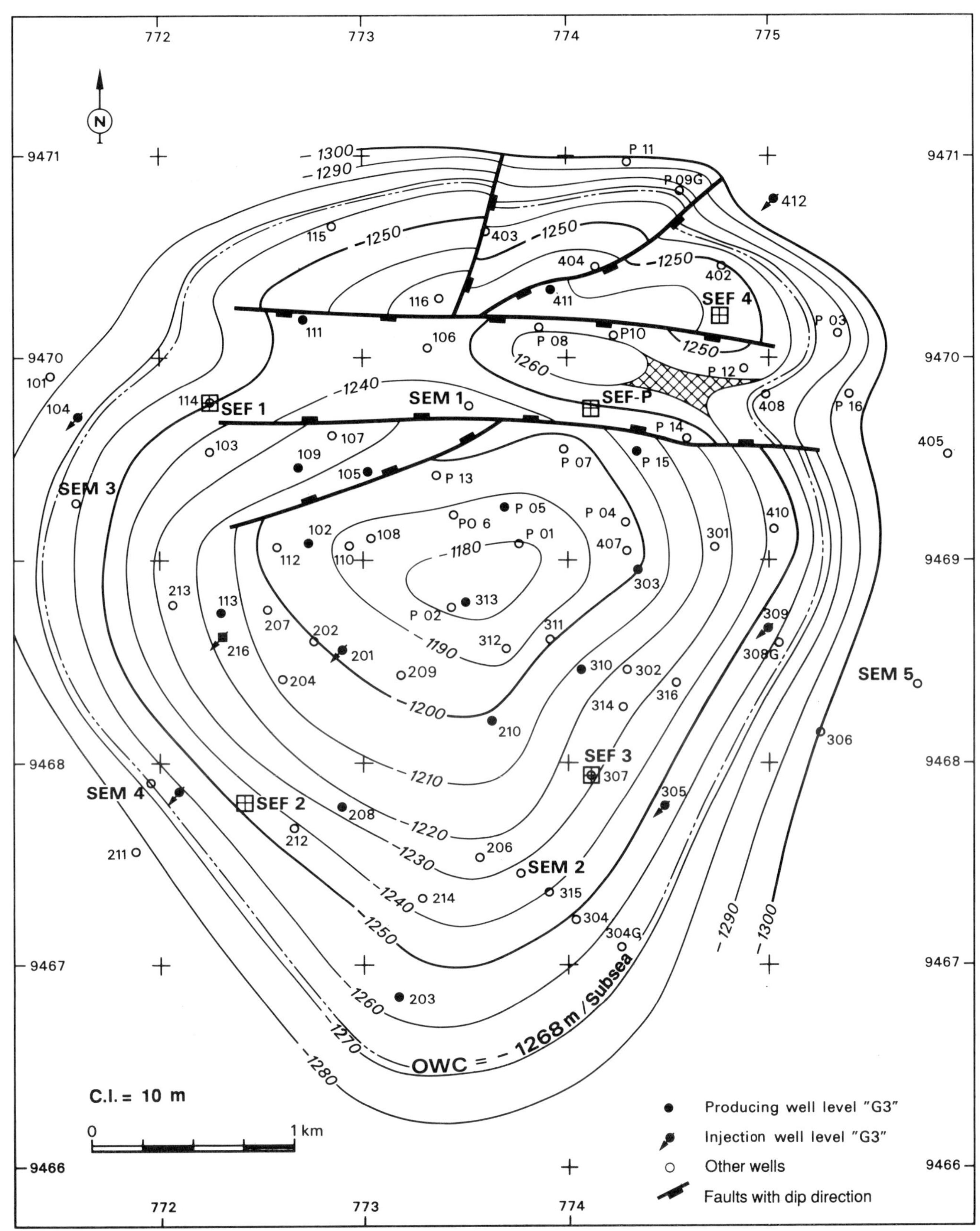

Figure 10. Structure map, top of zone G3. Depth in meters subsea.

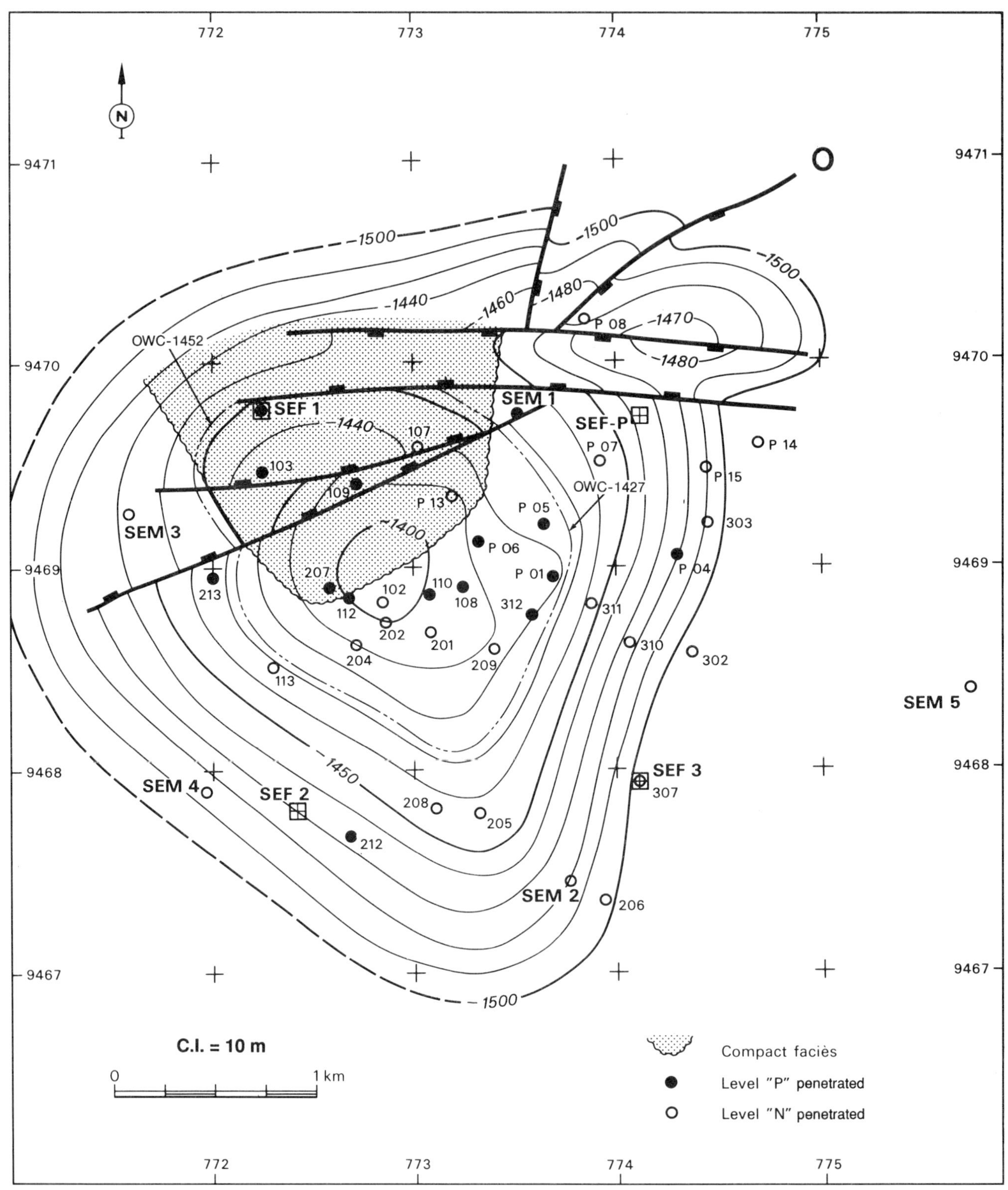

Figure 11. Structure map, top of zone P. A compact, tight facies, mostly segregated by faults, occupies the northwestern part of the structure and has a deeper oil-water contact.

pre-salt series. Geochemical analyses of oils show isotope ratios characteristic of oil originating from lacustrine shales. Although potential source rocks exist in the post-salt formations, they are not found to be thermally mature. Oil-prone lacustrine shales occur widely within the pre-salt sequence. They are found in the following formations.

Pointe-Noire Marl Formation

The most important source rock for oil found in the Congo coastal basin is in the Marne Noires Formation, a dolomitic, organic-rich black shale that is over 500 m (1640 ft) thick.

Pointe Indienne Shale Formation

Low in organic content, the Argiles de Pointe-Indienne Formation is considered to be a relatively poor source rock.

Djeno Sandstone Formation

The Djeno formation, containing dominantly clastic sediments with high shale content moderately rich in organic matter, is considered a moderate to good source rock.

Sialivakou Shale Formation

The source rock potential of the Sialivakou formation is essentially limited to offshore Congo. Onshore the shallow lacustrine marls have low organic content and are poor source rocks.

TRAP

Trap Type

The Sendji field structure is multilayered with several pays. Different oil-water contacts have been identified (Figure 7). Oil-water contacts (OWC) of all producing zones are known. The only oil zones in which OWC contacts are unknown are the C1 and C2 levels, which are undeveloped.

The traps result from halokinesis and are structurally nearly identical for all producing levels. They consist of simple domal anticlines.

Basal Cenomanian Reservoir

At the level of the basal Cenomanian reservoir, the structure is very flat and contains heavy oil (15.13° API). It has not been developed. The maximum pay thickness is 14.3 m (46.9 ft).

Albian Reservoirs

Zone A

The structure map of zone A (Figure 8) shows a southwest–northeast-elongated closure. The bounding faults of the graben, seen in the lower levels, die out before reaching upward to this level, where they are replaced by a saddle centered in the SEF-P platform area. The OWC corresponds closely to the 1089 m (3572 ft) subsea contour closure. As may be seen in Figure 8, there is a saddle between the Sendji structure and the salt dome to the east. The pay zone thickness totals 33.6 m (110 ft) and has a surface area of 8.7 km^2 (3.4 mi^2).

Zone C

Zone C has not been developed. Its southwest-northeast structural orientation is similar to the structure mapped at zone A. The OWC contact is unknown. Interpretation of the results of the SEF-4 platform wells is that a permeability barrier creates stratigraphic closure of zones C1 and C2 in the northeastern part of the field. To the northeast of this barrier, the oil-bearing reservoirs in zone C1 are isolated sandy lenses (Figure 17).

Zone E

Sealing of zone E (Figure 9) is provided by the overlying shaly E-C interlayer, while the dolomites and shales of zone F seal the underlying zones G1, G2, G3, and I. The whole of this sequence corresponds to the period when the graben was being filled with sediments, as shown by the reduced fault throws. The structure at the top of zone E maintains a southwest-northeast orientation. The OWC contact at 1198 m (3929 ft) subsea corresponds approximately to the lowest closing contour. The pay zone thickness equals 74 m (243 ft) and has an area of 14.2 km^2 (5.5 mi^2).

Zones G1, G2, G3, and I

The structural configuration of the field is quite similar at each of the zones G1, G2, G3 (Figure 10), and I. A common oil-water contact at 1268 m (4159 ft) subsea shows the seals between these levels were not efficient at the time of hydrocarbon accumulation. The spill point is determined by the lowest contour closure of the uppermost reservoir zone; other zones are not wholly filled and are probably connected by fault planes bounding the graben. These paths allow oil from zones I1, I2, G3, and G2 to migrate to zone G1, where the migration is stopped by the shales and compact dolomites of zone F, which is in fact thicker than the fault throws (Figures 4 and 7). The pay zones have thicknesses and surface areas of 110 m (361 ft) and 20.15 km^2 (7.8 mi^2) for zone G1; 95 m (312 ft) and 14.2 km^2 (5.5 mi^2) for zone G2; 88 m (289 ft) and 11.25 km^2 (4.3 mi^2) for zone G3; and 100 m (328 ft) and 8.4 km^2 (3.2 mi^2) for zone I.

Zone K

Zone K contains three secondary reservoirs, K1, K2, and K3 (Figures 4 and 7), each of which has its own OWC (respectively −1302 m [−4271 ft], −1310 m [−4297 ft], and −1318 m [−4323 ft]). The extent of the trap is limited by the southern bounding fault of the graben. The pay zone thickness for zone K is 48 m (157 ft) with a surface area of 4.3 km^2 (1.6 mi^2).

Zone L

Zones K and L have a common OWC at –1318 m (–4323 ft); they are both sealed by the overlying shaly and dolomitic zone J, and they are in communication via the faults, with maximum throws of 40 m (131 ft) bounding the graben. The pay zone thickness of zone L is 38 m (125 ft) for a surface area of 3 km² (1.2 mi²)

Zone P

Between zone L and zone P, zones M, N, and O have not been developed because of their poor production potentials. Trapping in zone P results from simple closure and facies change (Figure 11). In the center of the field, the main part of the reservoir has an OWC at –1427 m (–4681 ft); in the northwest part of the structure at this level, a compact, tight facies, partially segregated by faults, has an OWC at –1452 m (–4763 ft). The pay zone has a thickness of 27 m (89 ft) and an area of 1.25 km² (0.5 mi²).

Zone S

Only five wells have penetrated zone S, and its areal extent is uncertain. It is sealed by a thick bed of shaly dolomite. The OWC is not well defined; it is found at –1507 m (–4943 ft) at SEF-112 and at –1522 m (–4992 ft) at SEF-207. If one considers the OWC to be at –1522 m (–4992 ft), the pay thickness is 42 m (138 ft), with an area of 1.7 km² (0.6 mi²).

Reservoirs

There are both detrital and carbonate reservoir facies of the upper Albian in the Congolese coastal basin (Figure 3). In the Sendji field reservoirs consist of siltstone or fine- to medium-grained sandstones deposited in a littoral to sublittoral environment and of sandy, shelly, or locally oolitic dolomites deposited in an inner neritic to littoral high-energy environment. Apart from the P and S zones, only detrital reservoirs have been commercially productive. The multilayered oil-productive upper part of the Albian Sendji Carbonate, including seals and nonproductive zones, is an interval more than 500 m (1640 ft) thick. Gas:oil ratios and bubble points of the oils in these reservoirs are summarized by Table 2.

The quantity and the quality of the Sendji field reservoirs are determined by two kinds of factors. One is the kind of processes resulting in the paleostructures, including synsedimentary tectonic movements in the northern part of the structure that generated an east-west graben with fault blocks and steep dips; the graben is characterized by thick deposits of coarse detritals with increased net pay, while elsewhere around the structure, the thickness and net pay of each zone increases from the top of the structure toward the flanks. The other kind of factor is comprised of the elements of the paleogeography, including a progressive deterioration of the littoral deposits to low-energy platform deposits in the southwest where marine influence is more important (subtidal deposits); net pay decreases at the same time as facies change from inner-shelf detritals (sands and siltstone) and carbonates (sandy dolomites, vuggy dolomites) in the northeast to outer-shelf detritals (shaly siltstone and shales) and carbonates (dolomicrites and shaly and silty dolomites).

Basal Cenomanian Reservoir

The basal Cenomanian reservoir consists of coarse-grained sands and sandstone, locally dolomitic, deposited in a continental fluvial environment. Porosities range from 24 to 28%. The oil contained in this reservoir is heavy at 15.3° API and highly viscous at 3460 cp at 50°C (122°F).

Albian Reservoirs

Zone A—Zone A, the upper reservoir of the Albian Sendji Carbonate, consists of sands and very fine to medium-grained sandstones (Figure 12), dominant in the southwestern part of the structure, grading to shale and dolomitic siltstone, dominant toward the northeastern part of the structure. The thickness of zone A is constant over most of the structure, between 6 and 10 m (20 and 33 ft), and increases gradually in the northeastern part to 19 m (62 ft) at SEF-401. Net pay increases in the same direction.

A northwest–southeast-trending paleoshore influences the lithology, with thicker, coarser littoral beach deposits passing to finer-grained sediments toward the outer shelf. Porosities are in the 20 to 30% range. Permeability values computed from DSTs show that reservoir quality increases from the southwest to the northeast. Permeability is less than 400 md at SEF-214 and SEF-216, but is over 2000 md at SEFP-06, P-16G, P-03, and P-11G. The oil contained in zone A has a gravity of 29.30° API and viscosity is 14.6 cp at field conditions.

The pressure regime of zone A, as in all the other Sendji Carbonate reservoirs, is hydrostatic (initial shut-in pressure measured at SEM-1 at –1075 m [–3526 ft] was 114.4 bars [1659 psi]). The maximum flow was 1200 m³ (7550 bbl) oil per day. Production decline is 25% per year. Ultimate recovery enhanced by gas injection is expected to be 35.4% of the 8.9 million MT (64 million bbl) of the oil in place or 3.15 million MT (22.5 million bbl) (Table 1). Today zone A is being produced from six wells, with five water injection wells.

Zone C—The C1 and C2 subzones of zone C contain sandy or vuggy dolomite and dolomitic sandstone with a few layers of shale and anhydrite. They are separated by a bed of dolomitic shale. The depositional environment was intertidal to subtidal. A facies change toward the northeast creates a permeability barrier thereby controlling the stratigraphic closure of subzones C1 and C2, thus explaining why the oil column is greater than the structural closure.

Table 2. Gas:oil ratios and oil bubble points in Albian Sendji Carbonate zones A through S.

Sendji Carbonate Zone	Initial Pressure bars (psi)	Reference Depth—MSL m (ft)	Gas:Oil Ratio		Bubble Point bars (psi)
			m³/m³	SCF/STB	
A	114.7 (1663)	1095 (3525)	14.9	85	10.0 (145)
C	118.4 (1717)	1125 (3690)	2.4	14	10.0 (145)
E	122.0 (1769)	1150 (3775)	14.9	85	39.9 (579)
G2 with upper G3	127.5 (1849)	1200 (3935)	11.4	65	33.4 (483)
Lower G3	127.5 (1849)	1200 (3935)	11.4	65	33.4 (483)
I, K, L	127.5 (1849)	1200 (3935)	12.0	68	41.2 (597)
P	149.5 (2168)	1400 (4595)	17.3	88	36.0 (522)
S	159.5 (2313)	1500 (4920)	17.3	98	40.0 (580)

Porosities range from 19 to 30% for subzone C1 and between 17 and 24% for C2. Permeability varies greatly, the most reliable figures being obtained from DST results of SEM-1 (K = 329 md for subzones C1 and C2) and more recently from SEF-207 (K = 244 md for the subzones). Oil is heavy and viscous; gravity is 23.3° API and viscosity is 150 cp at field conditions.

This zone, produced from a single well by gas lift at a rate of 30 m³ (215 bbl) of oil per day, was abandoned in November 1984. The recoverable reserves for zone C amount to 570,000 MT (4.1 million bbl) of oil of the 54.3 million MT (389 million bbl) of oil in place. The high oil viscosity precludes a water injection system, and the reserves are calculated from monophasic expansion alone.

Zone E—Zone E consists of dolomites, siltstones, and shales, dominant in the southwest of the field, grading to fine- to medium-grained sandstones and sands, dominant in the northeast (Figures 12 and 13). The results of the SEF-4 wells in the northeastern part of the structure confirm the thickening of sandy layers of zone E in the graben (Figure 15). The filling of the graben during the deposition of zone E included a large quantity of coarse clastics. With the exception of the graben area, where detrital deposits are thick, zone E is similar to zone A in its development. The existence of a northwest–southeast-trending paleoshore is shown by the progressive changes of facies from an inner-lagoonal, very fine micritic facies in the south, to littoral deposits with coarser sands in the northeast. This evolution in depositional environments over the structure is reflected in the differences in reservoir quality. In the south, porosities range from 18 to 27% and permeabilities average 500 to 600 md, while in the northeast, porosities range from 20 to 30% with permeabilities increasing to between 600 and 900 md.

Oil is 26.4° API and viscosity is 12 cp at field conditions. Maximum flow reached by zone E was 1200 m³ (7550 bbl) of oil per day. The yearly decline is 15%. With a recovery factor of 28.6%, recoverable reserves are estimated to be 3.5 million MT (25.1 million bbl) of oil. There are eight producing wells and two water injection wells for production of the reserves of zone E (Table 1).

Zone G1—Recent production tests for zone G1 have given mainly poor results (DST No. 1 on SEF-102 showed PI of 0.1 bbl/d/psi, and DST No. 1 on SEF-315 showed a PI of 0.5 bbl/d/psi), and although included in original field development planning, this zone has not been developed to date. The G1 zone consists mainly of compact dolomite interbedded with vuggy or sandy dolomite, dolomitic sandstones, and dolomitic siltstones. Anhydrite, occurring mainly as nodules, is widespread in zone G1. These deposits are characteristic of a low-energy, restricted, inner lagoon or littoral sabkha-type environment.

The reservoir facies are detrital, occurring in lenses of limited areal extent that are not easily correlated between wells. Porosities range from 17 to 22%, while permeabilities are generally less than 10 md. In sandy layers, permeabilities can reach 300 md. Oil is 25.6° API and viscosity is 12 cp at field conditions. Oil in place is estimated to be 12 million MT (86 million bbl) of oil. No reserves have been attributed to the G1 level because of the limited extent of the reservoirs.

Zone G2—Zone G2 consists of a regressive sequence starting at the base with low-energy subtidal siltstone grading upward to intertidal beach sands and terminating in the G1–G2 interlayer with tidal flat or sabkha-type dolomites, dolomitic shale, and nodular anhydrite (Figure 13). The thickness of zone G2, including the pay zones, varies little over the field. Thickness depends essentially on the paleostructure, thinning over the top of the structure to less than 6 m (20 ft) and thickening to 9 m (30 ft) on the flanks of the structure. Unlike zones A and E, there is no noticeable influence of a paleoshore. Porosities are in the range of 25 to 30%. Permeabilities are relatively high, with values from 300 to 1400 md.

Hydrocarbons from zone G1 migrated along fault planes of the central graben and partly filled the G2 level. A common OWC occurs at –1268 m (–4160 ft). Oil is 25.6° API and viscosity is 12 cp at field

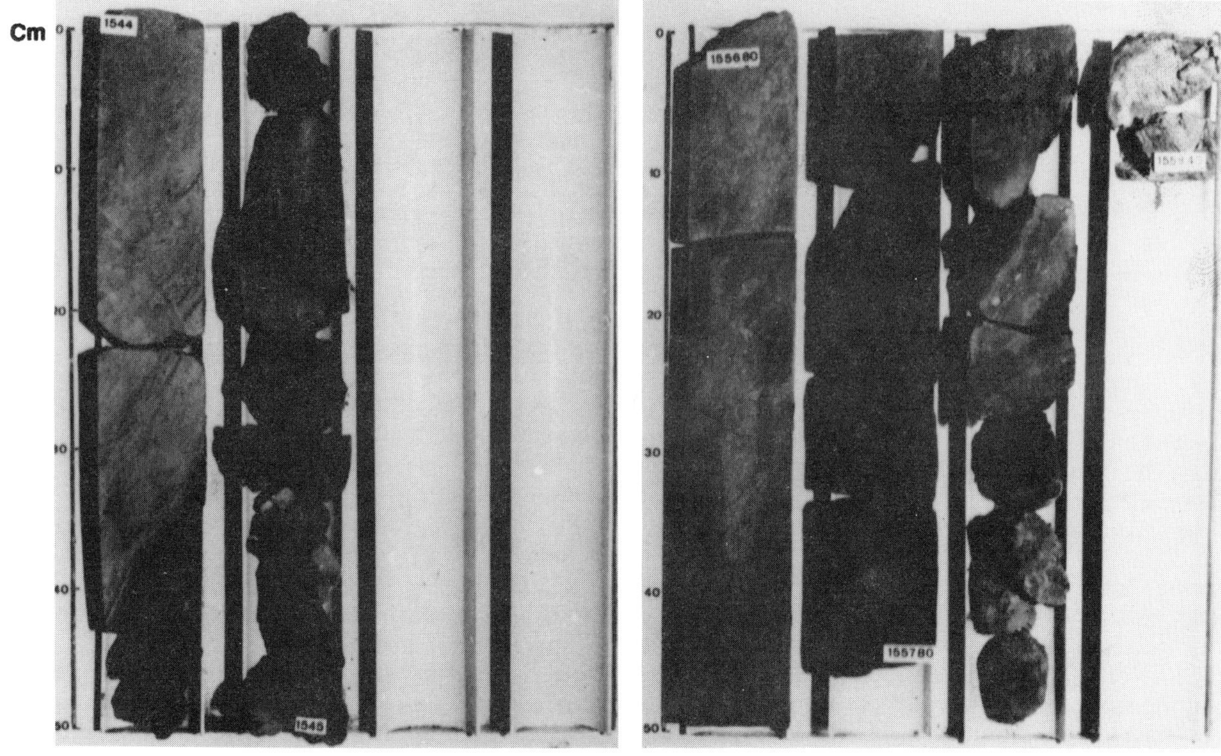

ZONE A

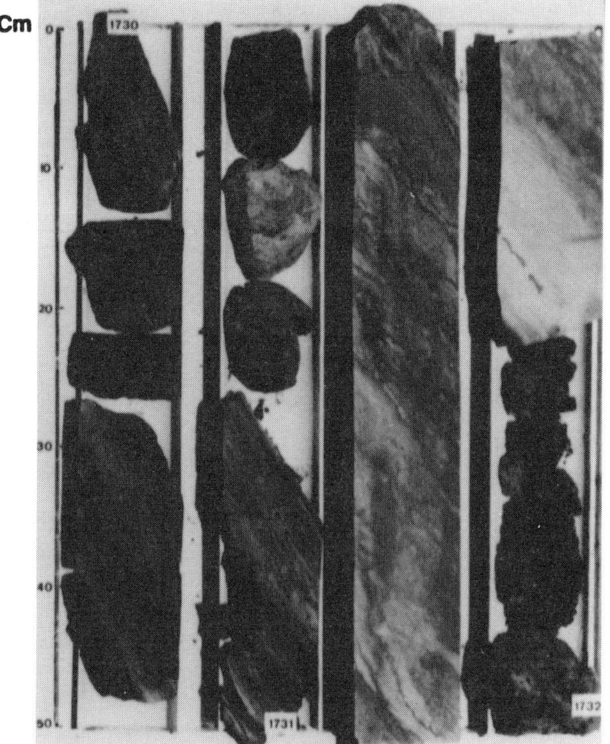

ZONE E

Figure 12. Core photographs, zones A and E (SEF-P11 well).

138

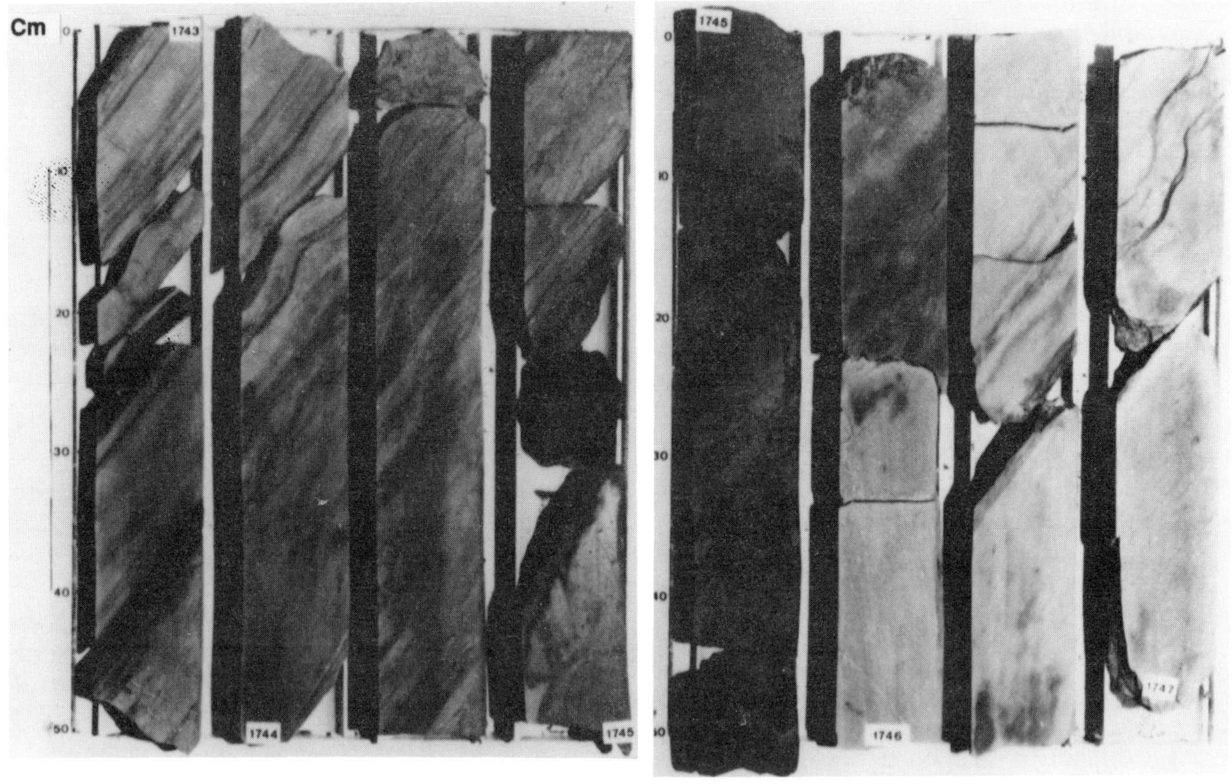

ZONE E

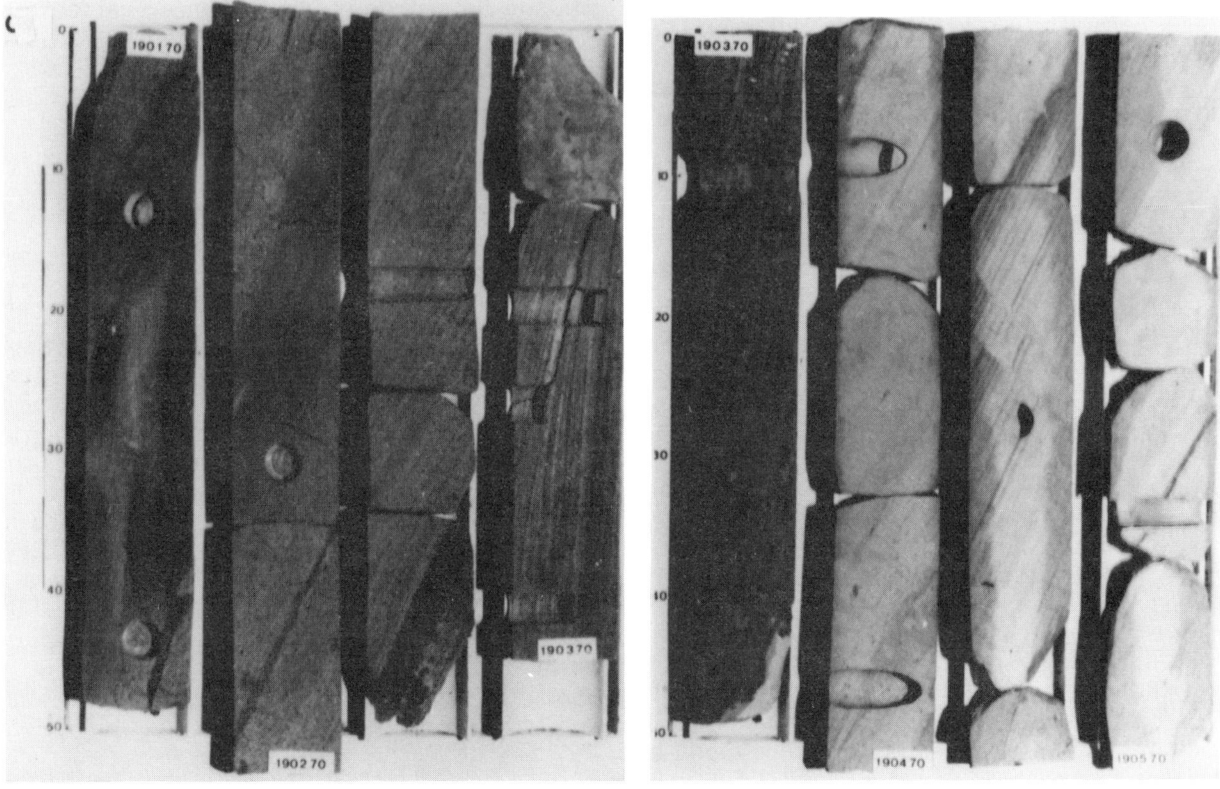

ZONE G2

Figure 13. Core photographs, zones E and G2 (SEF-P11 well).

conditions. Maximum flow reached was 1600 m³ (10,000 bbl) per day of oil; the annual decline is 10%.

Zone G2 and upper zone G3 are produced simultaneously and their combined recoverable reserves, the highest of the zones of the Sendji field, total 5.8 million MT (41.6 million bbl) of oil, with a recovery factor of 42.5%. To date, production comes from 15 producing wells (seven completed simultaneously in zone G2 and upper zone G3) and seven injector wells, one of which lies in a central position (SEF-P-02).

Zone G3—Zone G3 is comprised of two different sedimentological units. A detrital zone consisting of siltstone and very fine grained sandstone at the base of the G3 is particularly well developed along a northwest-southeast axis passing through SEF-P and SEM-5. The producing interval of G3 corresponds to this zone, which is called "lower G3."

The upper G3 corresponds to a sandy interval overlying a tight carbonate sequence. It has been interpreted as a meandering channel developed from north to southeast. This channel is produced simultaneously with zone G2.

Lower G3 sedimentation occurred in the form of an offshore bar paralleling the paleoshore. Relatively high-energy carbonates of bioclastic dolomite, packstone, and grainstone were deposited on both sides of the offshore bars. Thickening of the deposits toward the flanks of the structure suggests an influence of the paleostructure on depositional pattern. Average porosities range from 20 to 25%. Average permeability is 340 md and increases northeastwardly.

Oil is 25.6° API and viscosity is 12 cp at field conditions. Maximum flow was 1200 m³ (7550 bbl) of oil per day. Production decline is 20% per year. The G2 and G3 zones are connected via the graben faults, thus the structure is not filled to the G3 zone spill point. The recovery factor is 26.9%, and recoverable reserves are estimated to be 3.6 million MT (25.8 million bbl) of oil. The 11 producing wells and the seven injectors are all located south of the graben.

Zone I1—Zone I1 is a homogeneous reservoir throughout the field. It is essentially comprised of sublittoral very fine sandstones and siltstone. Interbedded dolomites, influenced mainly by paleostructure, thicken toward the southwest and the northeast over the flanks of the structure. Pay zone thicknesses, controlled primarily by shoreline proximity, are more than 10 m (33 ft) along a centrally located northwest-southeast trend, thickening toward the southeast. Porosities range from 20 to 24%. Permeabilities range from 5 md (dolomites) to 400–500 md (siltstone and sandstone). For zone I as a whole, the oil is 29.3° API. Viscosity is 8.9 cp at field conditions. Recoverable reserves are estimated to be 910,000 MT (6.52 million bbl) of oil. The recovery factor is 10%.

Zone I2—Reservoir facies of zone I2 are sandstones and siltstones deposited in a sublittoral to intertidal environment, separated into 2 or 3 layers by beds of intertidal dolomite. The western area (SEF-109, SEF-107, and SEM-1 area) is isolated by tight facies and faults. In this area a second OWC at –2388 m (–4226 ft) is seen as well as the common OWC of zones G1 to I2 at –1288 m (–4226 ft).

Average porosities range from 17 to 24%. Permeability values range from 5 to 200 md with an average of 70 md.

Zone K—The K1 and K2 units have porosities in the 17 to 26% range, but permeability is generally less than 5 md. The K3 unit, with porosities between 20 and 28%, has slightly better permeability, with values around 50 md being frequent.

Zone K oil is 29° API. Viscosity is 9 cp at field conditions. Recoverable reserves are estimated to be 200,000 MT (1.43 million bbl) of oil with a recovery factor of 3%.

Zone L—Zone L, like zone K, is composed of regressive sequences, including silty reservoirs, interbedded with shaly dolomites that act as seals. An increase in thickness is observed from west to east without any lateral facies variation. The low-energy, subtidal depositional environment with good lateral continuity is the same as in zone K. Porosities range from 18 to 23%. Permeabilities are generally under 5 md, with only a few measured values ranging from 5 to 200 md. Oil in the zone is 29° API with a viscosity of 11.5 cp at field conditions. Recoverable reserves of the zone are 160,000 MT (1.15 million bbl) of oil with a recovery factor of 3%.

Zones I, K, and L are produced simultaneously from seven wells (three from zone I, alone, and four commingled). The four injectors are completed only in zone I.

Zone P—Zone P constitutes the upper part of a carbonate sequence that is middle Albian in age. It is made up of three types of facies. An area of compact dolomitic facies, which creates a stratigraphic trap and which prevents oil from zone P from migrating through fault planes to higher levels, occurs in the north-northwest part of the field. Limestone facies, which are compact and tight in the SEF-207 area, are pay zones in the SEF-P-01, SEF-P-06, SEF-102, SEF-108, SEF-110, and SEF-312 region where they are interbedded with silty sandstone facies. Interbedded siltstone, dolomite, and sandstone facies are found in the southern part of the field.

Zone P depositional environments are partly controlled by the paleostructure. The tight dolomitic facies in the northern region develop over a structural high and are diagenetic in origin. Around the flanks of the structure are high-energy intertidal shoal deposits that grade into subtidal silts and carbonates further offshore.

Average porosities range from 19 to 23%. Permeabilities obtained from SEF-108 DST data range from 300 to 400 md. Oil in the zone is 31.1° API. Viscosity is 3.8 cp at field conditions. Maximum oil flow from the one well in production was 270 m^3 (1700 bbl) per day. Yearly decline is 50%. Reserves amount to 310,000 MT (2.22 million bbl) of oil with a recovery factor of 27%.

Zone S—Little is known about zone S since it has been penetrated by only five wells. Zone S reservoir facies are vugular dolomite with thin interbeds of sandstone. Porosities range from 20 to 24%. The single datum point obtained from SEF-207 gave a permeability value of 264 md.

Oil in zone S is 31° API, and its viscosity is 4 cp at field conditions. Maximum flow from the one well, SEF-207, the only well producing from zone S, has a potential flow of 170 m^3 (1070 bbl) of oil per day. With gas drive, recoverable reserves are estimated to be 100,000 MT (720,000 bbl) of oil with a recovery factor of 10%.

Fault System

During development of the Sendji field, the structural interpretation of the field has changed. The first seismic interpretations of Sendji field described a turtle structure (Figure 14). New information obtained during field development showed faults that affected the Albian series. Electric log correlations have shown depositional gaps within some levels. Depths can vary greatly between neighboring wells but still can be similar in more distant wells. The OWC is common for zones G1, G2, G3, and I, which implies either a fault system or inefficient seals.

Evaluation of seismic and structure maps led to the hypothesis of a fault trough in the northern part of the field that was initiated during deposition of zone G2 and that remained active until the deposition of zone C1 (Figures 6 and 15). Pressure measurements in producing zones have confirmed the presence of these faults.

The Sendji field is interpreted to be a turtle structure bisected from east to west by a graben. The graben is bounded by major east-west faults and by some secondary faults trending in a southwest-northeast direction.

Source Rocks

Potential source rocks of the Congolese coastal basin are numerous both in the Neocomian to Aptian pre-salt series and the Albian to Tertiary post-salt series, but because of the low maturity stage reached by post-salt series, these rocks have not contributed to the oil pool. Post-salt reservoirs have all been fed by pre-salt source rocks.

The pre-salt series, primarily the Pointe-Noire Marl of Barremian age, generated the oil of the Sendji field.

The environment of deposition was lacustrine, and the hydrocarbon generation potential is very high, with type I–II organic matter and an average TOC of 1 to 5%, locally reaching 20 to 30% (Figure 16).

Two types of source rocks are distinguished in the Sendji field area. The Sialivakou Marl and lower Djeno formation contain a type II–III organic matter with a TOC of 1%. The upper Djeno formation and the Pointe-Noire Marl, which contain type II to I–II saprolitic organic matter with a TOC of 1 to 5%, locally reach 20%.

In the source rock area of the Sendji reservoirs, vitrinite reflectance is around 1 at depths of approximately 3000 m (9840 ft). The oil of the Sendji field is the least mature in the Congolese coastal basin; it is also slightly biodegraded. The API gravities and viscosities of oils recovered from different zones in the SEM-1 well are shown by Table 3.

Important physical variations are evident in these oils having a common level of maturity and common origin in the Sendji field area. The differences in the oils were acquired after trapping by segregation in different traps as the result of variations in seal efficiencies. Migration paths of oil from pre-salt to post-salt formations are numerous and depended upon halokinetic movements that created windows in the salt formation and the faults and fractures that became the pathways.

EXPLORATION CONCEPTS AND CONCLUSIONS

The oil of the Sendji structure is trapped in reservoirs of the upper Albian, and therefore, this formation is of major interest in petroleum exploration of the Congolese coastal basin. These upper Albian reservoirs belong to the Sendji Carbonate formation. This formation, comprised of 75% carbonates and 25% quartz clastics, does not correspond to a classical sedimentary model. The entire Albian sequence can contain reservoir rocks. The multilayers are separated by numerous seals of different origins and natures. Lateral variability of facies is wide and a detailed scheme of distribution is difficult to develop. Oil occurrence depends upon the presence of an efficient seal extending over the reservoir. Stratigraphic traps could exist in these formations, as demonstrated by the facies changes in the C1 and C2 levels of the Sendji field (Figure 17), although the known traps of Sendji are predominantly structural in nature.

With such multilayered reservoirs containing various proportions of sandstone and carbonate, the geologist and reservoir engineer should cooperate to develop as detailed a geological model as possible before establishing the development scheme. The number and position of producer and injector wells

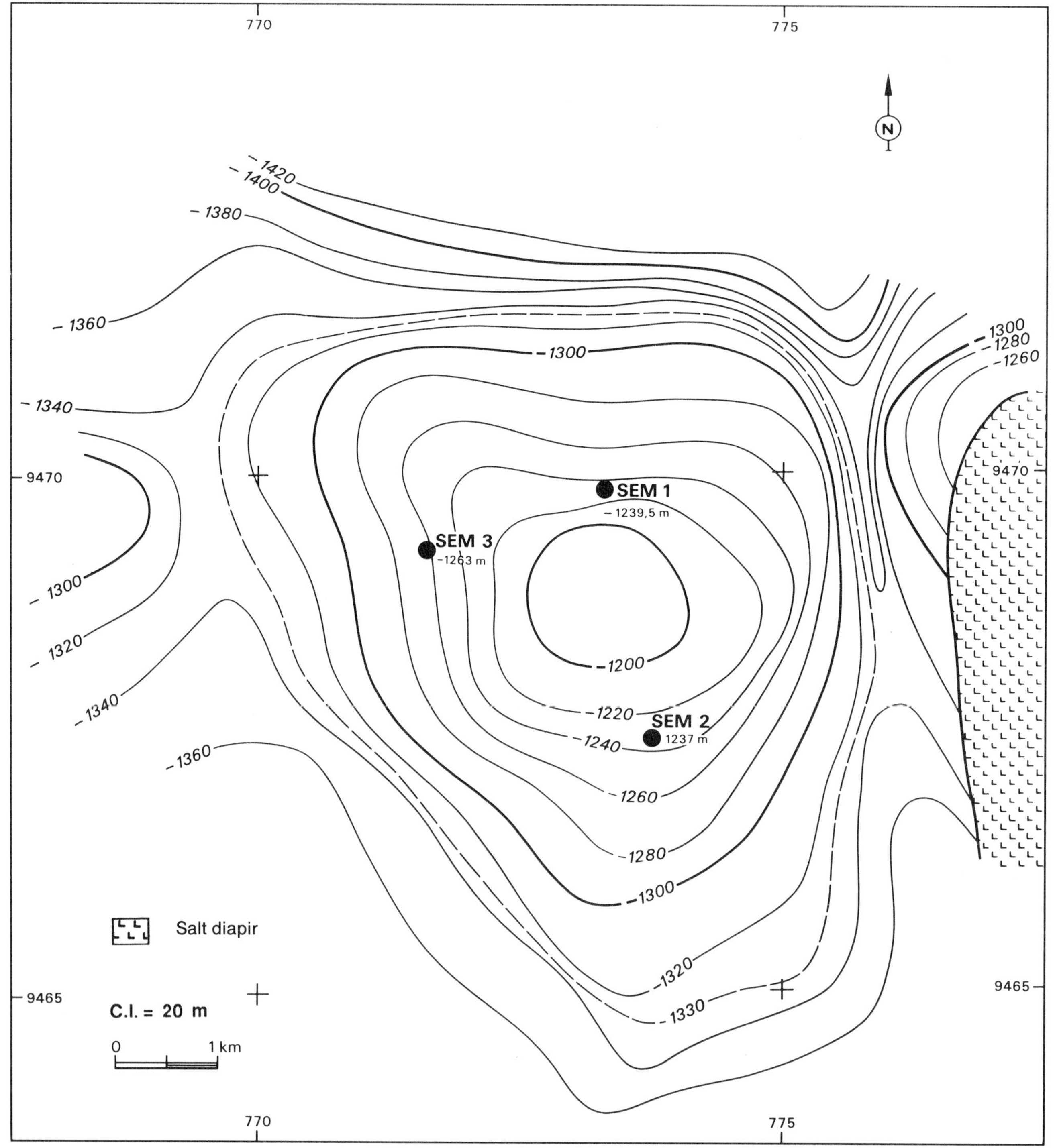

Figure 14. Structure map, top of zone G3 (1980 map). The east-west graben (Figures 9, 10, and 11) was recognized only after the drilling of development wells from SEF-P and SEF-1.

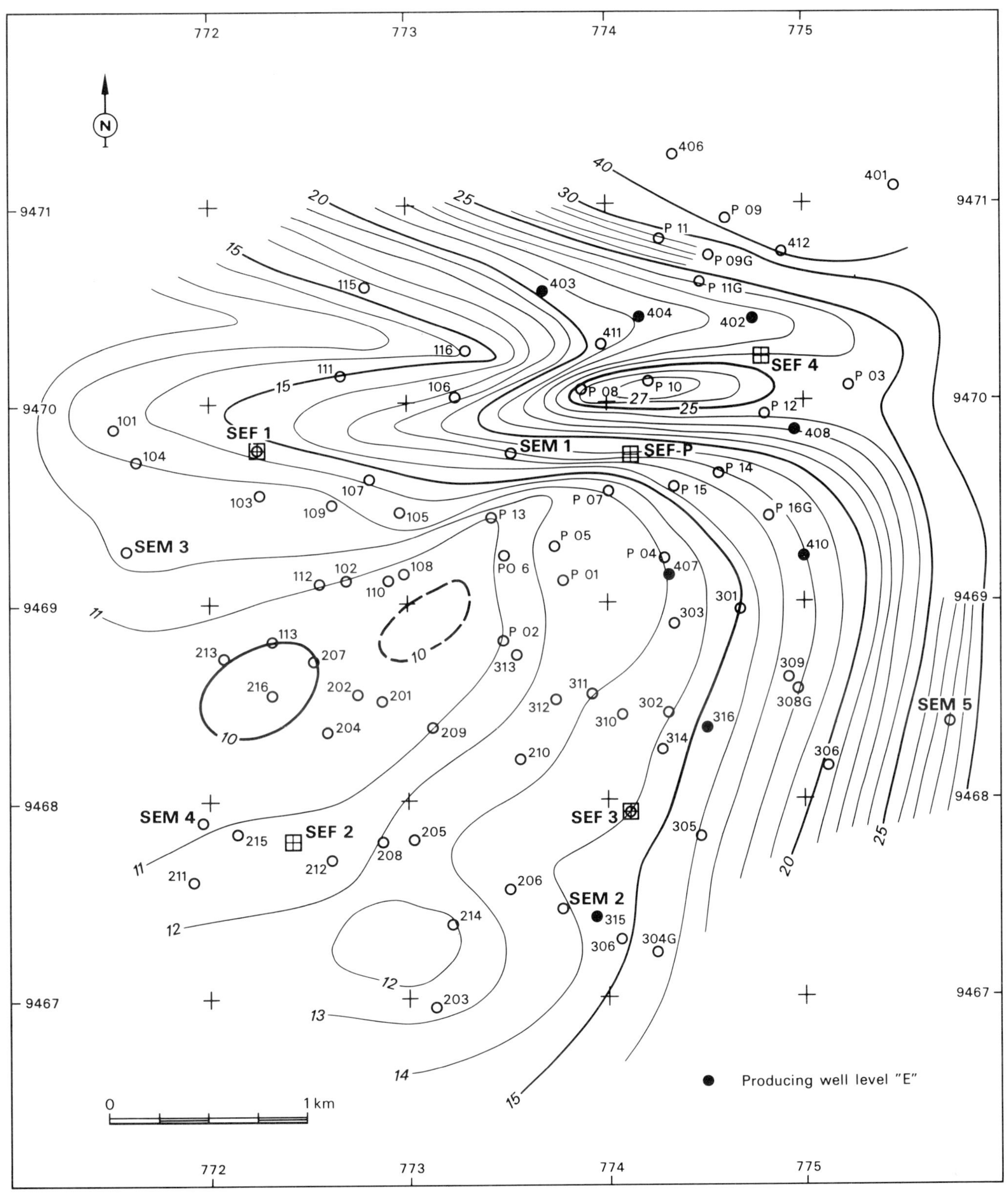

Figure 15. Isopach map of zone A. Contour interval, 1 m.

143

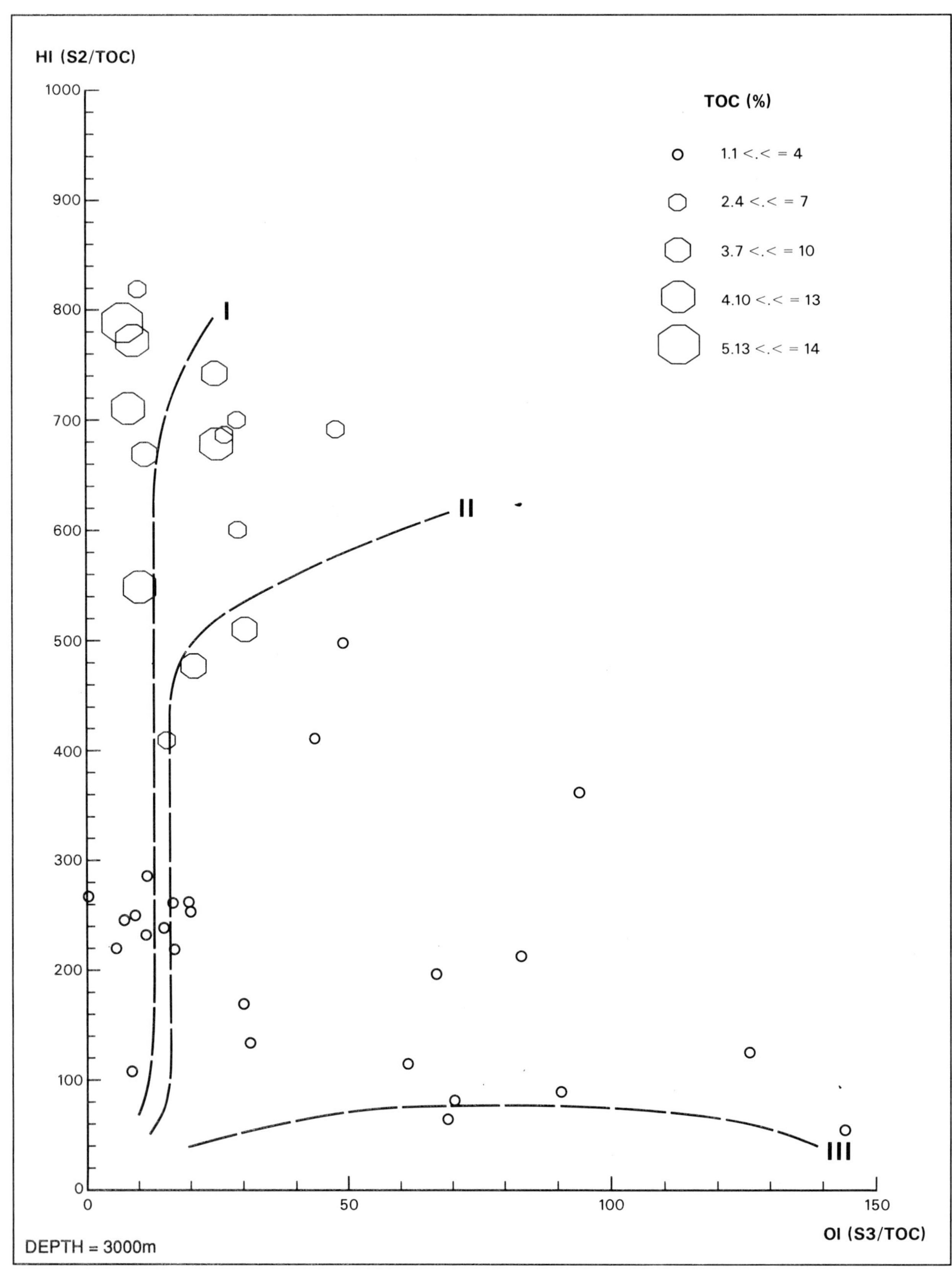

Figure 16. Hydrogen index versus oxygen index in the Pointe-Noire Marl source rock.

Table 3. API gravities and viscosities of oils from different Sendji Carbonate zones in the SEM-1 well.

Sendji Carbonate Zone	API° Gravity	Viscosity at 15°C (cp)
A	29.3	87
C	23.3	628
G2, G3	25.6	132

depend upon thickness of the formation and facies changes and several development phases have been necessary to develop the potential of Sendji field. Because of the thickness of layers, averaging only 20 to 25 m (66 to 82 ft), seismic resolution is not sufficient to characterize the heterogeneous nature of the different reservoirs.

With reserves of 17.38 million MT (124.6 million bbl) of oil and a yearly production in 1988 of 1.59 million MT (11.4 million bbl), Sendji was until December 1988, when the Tchibouela field came into production, the most important producing field in the Congo. Despite the complexity of the reservoirs having various lithologies, there is no major problem in the production of the Sendji field.

Fifty producing wells and 26 injectors are distributed on five platforms: SEF-1, SEF-2, SEF-3, SEF-4, and SEF-P. SEF-P also supports the equipment for oil production and transmission of Yanga field oil to the onshore terminal of Djeno, south of Pointe-Noire.

ACKNOWLEDGMENTS

I wish to thank Elf-Congo and Agip Recherches Congo for the opportunity to publish this study of the Sendji field in the Congolese coastal basin.

This work is based on unpublished information and particularly on an Elf-Congo internal report made by J. Voriot at the end of the first development phase in 1985.

I am particularly grateful to J. M. Masset, who greatly improved the English text. I thank B. Pialot, who supervised drafting in the Elf-Aquitaine drafting office in Paris.

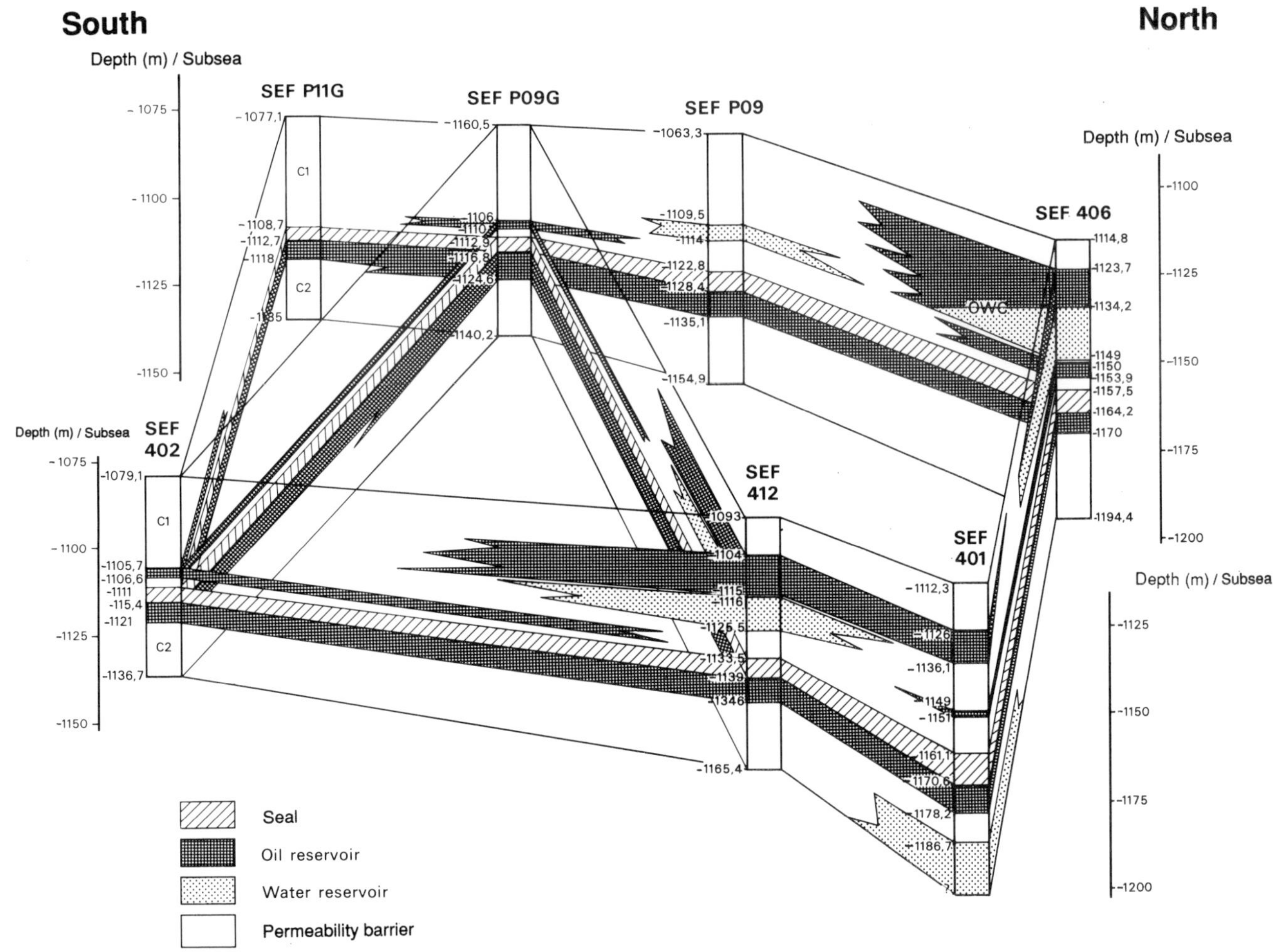

Figure 17. Fence diagram showing zone C reservoir facies distribution. Zone C1 is made up of sand lenses of limited areal extent. The northern flank has thicker, more extensive lenses. Zone C2 is more uniform in thickness and more areally extensive.

Field name .... *Sendji field*

Ultimate recoverable reserves .... *17.38 million MT (124.6 million bbl; 19.8 million m³) of oil*

Field location:

    Country .... *Congo*

    State .... *Offshore*

    Basin/Province .... *Gulf of Guinea, Congo coastal basin*

Field discovery:

    Year first pay discovered .... *Sendji contains 7 upper Albian Sendji Carbonate pays; the five upper pays were discovered in 1973 (zones A, E, G2 + G3 upper, G3 lower, and I + K + L)*

    Year lower two pays discovered .... *Zones P and S, 1983*

Discovery well name and general location:

    First pay .... *Sendji Marine 1 (SEM-1); 40 km west of Pointe-Noire*

    Second pay .... *SEF-108 (P)*

    Third pay .... *SEF-112 (S)*

Discovery well operator .... *Elf-Congo*

    Second pay .... *Elf-Congo*

    Third pay .... *Elf-Congo*

IP in barrels per day and/or cubic feet or cubic meters per day:

    First pay .... *A, 7466 bbl/day (1187 m³/d)*
    E, 6806 bbl/day (1082 m³/d)
    G2 + G3 upper, 9202 bbl/day (1463 m³/d)

    Second pay .... *G3 lower, 7806 bbl/day (1241 m³/d)*
    I + K + L, 2876 bbl/day (457 m³/d)

    Third pay .... *P, 736 bbl/day (117 m³/d)*
    S, 692 bbl/day (110 m³/d)

All other zones with shows of oil and gas in the field:

Age	Formation	Type of Show
Cenomanian	*Tchala Sandstone*	*Heavy oil*
Albian	*Sendji Carbonate (zones C, G1)*	*Oil*

Geologic concept leading to discovery and method or methods used to delineate prospect, e.g., surface geology, subsurface geology, seeps, magnetic data, gravity data, seismic data, seismic refraction, nontechnical:

Structures originated from salt movement. Seismic has been used to recognize and map the structures.

Structure:

    Province/basin type .... *Congo; Bally 1141; Klemme III c*

    Tectonic history

    The Congolese coastal basin is related to the formation of oceanic crust. It corresponds to a rifted passive margin. Two periods are clearly defined: (1) before the opening: earlier rift system; (2) after the opening: prograding platform overlying the rift system. These two periods are separated by an important salt deposit.

    Regional structure

    The post-opening basins have structures created by salt movements; Sendji field is a turtle structure.

    Local structure

    Anticlinal dome bisected by an east-west graben.

Trap:

    Trap type(s) .... *Anticlinal trap with several pays (simple closure); minor stratigraphic influences*

Basin stratigraphy (major stratigraphic intervals from surface to deepest penetration in field):

Chronostratigraphy	Formation	Depth to Top in ft (m)
Miocene	Paloukou clay and sand	360 (110)
Eocene to Senonian	Madingo Marl	1245 (380)
Turonian	Loango Dolomite	2425 (740)
Cenomanian	Tchala Sandstone	2690 (820)
Albian	Sendji Carbonate	3445 (1050)
Aptian	Loeme Salt	7450 (2270)

Reservoir characteristics:

Number of reservoirs 7
Formations .... Sendji Carbonate
Ages .... Albian
Depths to tops of reservoirs
A, 1055 m (3460 ft); E, 1125 m (3690 ft); G2 + G3 upper, 1174 m (3850 ft); G3 lower, 1180 m (3870 ft); I + K + L, 1208 m (3965 ft); P, 1400 m (4595 ft); S, 1480 m (4855 ft)

Gross thickness (top to bottom of producing interval) .... 450 m (1475 ft)
Net thickness—total thickness of producing zones
    Average .... 110 m (360 ft)
    Maximum .... 180 m (590 ft)
Lithology .... Calcareous dolomite, fine- to medium-grained sandstone and shaly siltstone
Porosity type .... Matrix porosity
Average porosity 25%
Average permeability .... 700 md

Seals:

Upper
    Formation, fault, or other feature .... Sendji Carbonate
    Lithology .... Shale, compact dolomite
Lateral
    Formation, fault, or other feature .... Downdip OWC

Source:

Formation and age .... Barremian Pointe-Noire Marl
Lithology .... Marl to silty marl
Average total organic carbon (TOC) 1–5%
Maximum TOC 20–30%
Kerogen type (I, II, or III) .... I–II
Vitrinite reflectance (maturation) .... $R_o = 1$
Time of hydrocarbon expulsion .... Recent
Present depth to top of source .... 3000 m (9850 ft)
Thickness .... 70 m (230 ft)
Potential yield .... NA

Appendix 2. Production Data

Field name .... Sendji field

Field size:

    Proved acres .... Zone A, 9 km² (2200 ac)
    Zone E, 14.2 km² (3500 ac)
    Zones G2 + G3 upper, 14.2 km² (3500 ac)
    Zone G3 lower, 11.3 km² (2800 ac)

Zones I + K + L, 6.0 km² (1480 ac)
Zone P, 2.0 km² (500 ac)
Zone S, 2.0 km² (500 ac)

Number of wells all years .... *Zone A, 6 prod., 5 inj.*
Zone E, 9 prod., 4 inj.
Zones G2 + G3 upper, 15 prod., 7 inj.
Zone G3 lower, 11 prod., 7 inj.
Zones I + K + L, 7 prod., 3 inj.
Zone P, 1 prod.
Zone S, 1 prod.
Total number of wells all zones, 50 prod; 27 inj., total 76 wells

Current number of wells .... *As above*

Well spacing .... *Zone A, 750 m; zone E, 600 m; zones G2 + G3 upper, 750 m;*
zone G3, 600 m; zones I + K + L, 500 m

Ultimate recoverable oil .... *Zone A, 3.58 × 10⁶ m³ (3.15 × 10⁶ MT) (22.5 × 10⁶ bbl)*
Zone E, 3.99 × 10⁶ m³ (3.5 × 10⁶ MT) (25.10 × 10⁶ bbl)
Zones G2 + G3 upper, 6.61 × 10⁶ m³ (5.8 × 10⁶ MT) (41.59 × 10⁶ bbl)
Zone G3 lower, 4.10 × 10⁶ m³ (3.6 × 10⁶ MT) (25.81 × 10⁶ bbl)
Zones I + K + L, 1.45 × 10⁶ m³ (1.27 × 10⁶ MT) (9.10 × 10⁶ bbl)
Zone P, 0.35 × 10⁶ m³ (0.31 × 10⁶ MT) (2.22 × 10⁶ bbl)
Zone S, 0.11 × 10⁶ m³ (0.10 × 10⁶ MT) (0.72 × 10⁶ bbl)
All zones, 19.81 × 10⁶ m³ (17.38 × 10⁶ MT) (124.62 × 10⁶ bbl)

Cumulative production (through 1988) .... *7.765 × 10⁶ m³ (6.81 × 10⁶ MT) (48.84 × 10⁶ bbl)*
Zone A, 1.664 × 10⁶ m³ (1.450 × 10⁶ MT) (10.47 × 10⁶ bbl)
Zone E, 0.791 × 10⁶ m³ (0.694 × 10⁶ MT) (4.98 × 10⁶ bbl)
Zones G2 + G3 upper, 2.470 × 10⁶ m³ (2.167 × 10⁶ MT) (15.54 × 10⁶ bbl)
Zone G3 lower, 1.660 × 10⁶ m³ (1.456 × 10⁶ MT) (10.44 × 10⁶ bbl)
Zones I + K + L, 0.800 × 10⁶ m³ (0.702 × 10⁶ MT) (5.03 × 10⁶ bbl)
Zone P, 0.288 × 10⁶ m³ (0.253 × 10⁶ MT) (1.81 × 10⁶ bbl)
Zone S, 0.092 × 10⁶ m³ (0.081 × 10⁶ MT) (0.58 × 10⁶ bbl)

Annual production (1988) .... *1.591 × 10⁶ m³ (1.397 × 10⁶ MT) (10.00 × 10⁶ bbl)*
Zone A, 0.330 × 10⁶ m³ (0.290 × 10⁶ MT) (2.08 × 10⁶ bbl)
Zone E, 0.305 × 10⁶ m³ (0.268 × 10⁶ MT) (1.92 × 10⁶ bbl)
Zones G2 + G3 upper, 0.505 × 10⁶ m³ (0.443 × 10⁶ MT) (3.18 × 10⁶ bbl)
Zone G3 lower, 0.267 × 10⁶ m³ (0.234 × 10⁶ MT) (1.68 × 10⁶ bbl)
Zones I + K + L, 0.134 × 10⁶ m³ (0.118 × 10⁶ MT) (0.84 × 10⁶ bbl)
Zone P, 0.030 × 10⁶ m³ (0.026 × 10⁶ MT) (0.18 × 10⁶ bbl)
Zone S, 0.020 × 10⁶ m³ (0.018 × 10⁶ MT) (0.12 × 10⁶ bbl)

Present decline rate ...... *Zone A, 25%; zone E, 15%; zones G2 + G3 upper, 15%; zone G3 lower, 20%;*
zones I + K + L, 20%; zone P, 50%; zone S, 60%

Initial decline rate .... *As above, except zone P 20%*

Annual water production .... *249 × 10³ m³ (1.566 × 10⁶ bbl)*
Zone A, 53 × 10³ m³ (333 × 10³ bbl)
Zone E, 1 × 10³ m³ (6 × 10³ bbl)
Zones G2 + G3 upper, 30 × 10³ m³ (189 × 10³ bbl)
Zone G3 lower, 137 × 10³ m³ (862 × 10³ bbl)
Zones I + K + L, 2 × 10³ m³ (13 × 10³ bbl)
Zone P, 21 × 10³ m³ (132 × 10³ bbl)
Zone S, 5 × 10³ m³ (31 × 10³ bbl)

In place, total oil reserves .... *81.38 × 10⁶ m³ (71.57 × 10⁶ MT) (513.2 × 10⁶ bbl)*
Zone A, 10.16 × 10⁶ m³ (8.91 × 10⁶ MT) (63.89 × 10⁶ bbl)
Zone E, 14.00 × 10⁶ m³ (12.28 × 10⁶ MT) (88.1 × 10⁶ bbl)
Zones G2 + G3 upper, 15.55 × 10⁶ m³ (13.64 × 10⁶ MT) (97.8 × 10⁶ bbl)
Zone G3 lower, 15.15 × 10⁶ m³ (13.29 × 10⁶ MT) (95.3 × 10⁶ bbl)
Zones I + K + L, 23.90 × 10⁶ m³ (20.97 × 10⁶ MT) (150.3 × 10⁶ bbl)
Zone P, 1.40 × 10⁶ m³ (1.23 × 10⁶ MT) (8.8 × 10⁶ bbl)
Zone S, 1.28 × 10⁶ m³ (1.12 × 10⁶ MT) (8.0 × 10⁶ bbl)

In place, per acre-foot .... *NA*

Primary and secondary recovery ... *Water injection for pressure maintenance and water sweep in use for major zones since beginning of production; oil recoveries as shown above in Ultimate Recoverable*

Enhanced recovery

Cumulative water production .... *361 × 10³ m³ (2.27 × 10⁶ bbl)*

Drilling and casing practices:

Amount of surface casing set

Casing program .... *16-in., 250 m (820 ft); 10¾-in., 800 m (2625 ft); 7⅝-in., 2100 m (6890 ft)*

Drilling mud .... *Bentonite + Biopolymère*

Bit program .... *14¾-in., 9⅝-in.*

High pressure zones .... *No*

Completion practices:

Interval(s) perforated .... *Varies*

Well treatment .... *Gravel pack behind casing*

Formation evaluation:

Logging suites ...... *SP, induction, sonic, gamma ray, compensated neutron, lithodensity, high resolution dipmeter, stratigraphic dipmeter*

Testing practices .... *DSI, RFT (repeat formation tester)*

Mud logging techniques .... *Sampling, ROP, CaCO₃, chromatograph, mud characteristics, depth control, drilling parameters control*

Oil characteristics:

Type .... *Naphthenic-paraffinic*

API gravity .... *29.5°*

Base .... *NA*

Initial GOR .... *15 m³/m³ (84 ft³/bbl)*

Sulfur, wt% *0*

Viscosity, SUS .... *12 cp*

Pour point .... *18°C (64°F)*

Gas-oil distillate .... *NA*

Field characteristics:

Average elevation .... *Sea level*

Initial pressure .... *127.5 bar (1850 psi)*

Present pressure .... *110.0 bar (1600 psi)*

Pressure gradient .... *0.11 bar/m (0.486 psi/ft)*

Temperature .... *65°C (140°F)*

Geothermal gradient .... *50°C/km (2.7°F/100 ft)*

Drive .... *Water*

Oil column thickness .... *450 m (1475 ft)*

Oil-water contact .... *Various reservoirs*

Connate water *25%*

Water salinity, TDS .... *230 g/L*

Resistivity of water *0.02*

Bulk volume water (%) .... *NA (varies)*

Transportation method and market for oil and gas:

Pipeline 50 km (31 mi) from Sendji field to Djeno terminal; loading system on tankers—floating buoy

Ruston Field—U.S.A.
Gulf Coast Basin, Louisiana

L. A. HERRMANN
Louisiana Tech University
Ruston, Louisiana

J. A. LOTT
Arkla Exploration Company
Shreveport, Louisiana

R. E. DAVENPORT
Louisiana Tech University
Ruston, Louisiana

FIELD CLASSIFICATION

BASIN: Gulf Coast
BASIN TYPE: Passive Margin
RESERVOIR ROCK TYPE: Sandstone
TRAP TYPE: Salt Pillow
RESERVOIR ENVIRONMENT OF DEPOSITION: Deltaic and Nearshore Marine
TRAP DESCRIPTION: Dome over salt pillow with multiple producing zones; some zones
  stratigraphically controlled

RESERVOIR AGE: Cretaceous and Jurassic
PETROLEUM TYPE: Gas

LOCATION

The Ruston gas field lies just north of Ruston in Lincoln Parish, Louisiana, 70 mi (113 km) east of Shreveport and 35 mi (56 km) west of Monroe in north-central Louisiana (Figures 1 and 2). The field covers a surface area of 64.5 mi^2 (sections) or approximately 41,280 ac (167 km^2) (Figure 3). It lies on the north flank of the Gulf Coast geosyncline within a Hosston–Cotton Valley producing trend extending from Caddo and DeSoto Parishes on the west to Ouachita and Caldwell Parishes on the east (Figure 1). Stratigraphic pinch-out or a combination of pinch-out and structural closure are the principal trapping mechanisms within this Lower Cretaceous–Upper Jurassic trend. The Ruston field contains both structural and stratigraphic traps.

Ruston is in a region known as the "piney woods," an area of rolling hills largely covered with pine forests, occasionally interspersed with cattle farms and small row-crop farms. The region is within the West Gulf Coastal Plain Physiographic Province as defined by the U.S. Geological Survey. Elevations are moderate, ranging from about 200 ft (61 m) to slightly more than 300 ft (91 m) above mean sea level. The highest elevation in the area is on top of Mt. Driskill some 15 mi (24 km) southwest of Ruston with an elevation of 535 ft (163 m) above mean sea level. The climate is moderate with temperatures seldom going below freezing or above 100°F (38°C).

Transportation to the area is provided by three main highways (U.S. 167, U.S. 80, and Interstate 20), and a number of state and parish roads. There is one railway serving the area, an east-west line through Ruston, operated by the Midsouth Rail Corporation of Jackson, Mississippi.

Eight companies were producing natural gas and a small amount of oil from the Ruston field as of December 1987, including Arkla Energy Resources, Arkla Exploration Company, Goodrich Oil Company, Lyons Petroleum, Inc., Murphy Oil U.S.A., Inc., Sonat Exploration Company, TXO Production Company, and Wessely Energy Company. Gas from the Ruston field is gathered by Arkansas Louisiana Gas Company, Lear Gas (formerly Producers Gas Company), Delhi Pipeline Company, Kerr-McGee Corporation, Louisiana Gas Purchasing Corp., and Mississippi River Transmission Co. Gas liquids are processed by Arkla Gas Company in North Ruston and by Kerr-McGee Corporation in Dubach, Louisiana. A small amount of oil is being produced from the Hosston and Sligo formations (Figure 4) at the present time, and it is being trucked to a refinery by Enron Oil Trading and Transportation Co. and Scurlock Oil Company.

Arkansas Louisiana Gas Company operates a gas storage unit within the Ruston field, utilizing the depleted James sandstone reservoir to store gas for use during periods of peak demand. Mississippi River Transportation Company, owned by ARKLA Inc.,

151

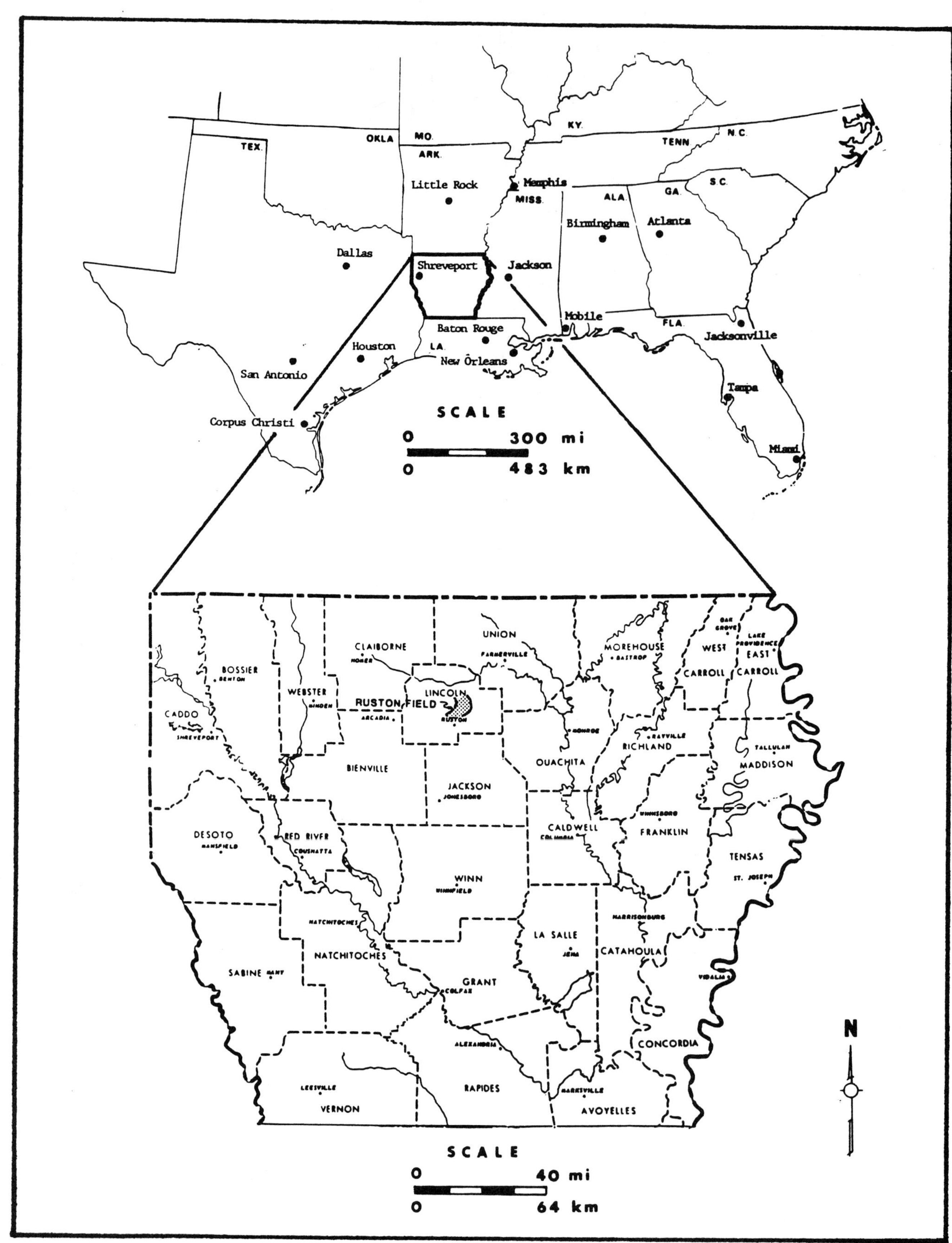

Figure 1. Location of Ruston field in Lincoln Parish, North Louisiana.

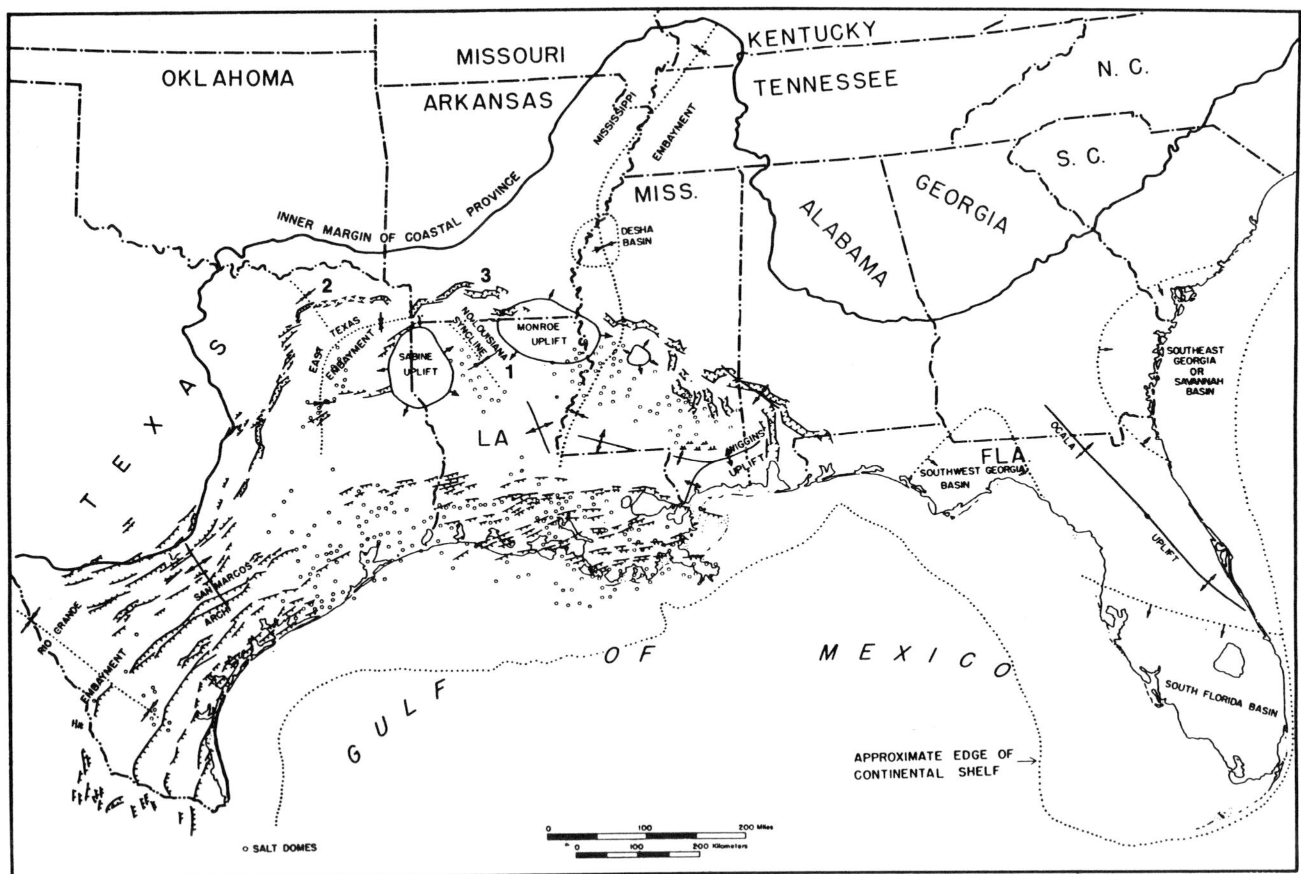

Figure 2. Natural gas fields in Lincoln Parish, Louisiana. The Ruston field is in the center of the parish, several miles north of the city of Ruston. (From Louisiana Geological Survey, 1980.)

operates a similar unit in the Unionville field just north of the Ruston field, utilizing the Vaughn sandstone of the Cotton Valley Group.

Ultimate hydrocarbon recovery from the Ruston field is estimated to be about 1,600,000 bbl of oil, 1,732,000 bbl of condensate, and 614 bcf (17.2 billion m^3) of natural gas, primarily from Hosston and Cotton Valley reservoirs. Production to January 1987 totaled 1,516,701 bbl of oil, 1,252,000 bbl of condensate, and 544 bcf (15.2 billion m^3) of natural gas.

HISTORY

Pre-Discovery

There are many prolific Hosston and Cotton Valley fields located in a belt that trends roughly east-west across North Louisiana. This belt includes most of the Cotton Valley strandline fields (Pate, 1963) and many of the Hosston fields of the region. The earliest fields with Hosston or Cotton Valley production discovered in this trend were Caddo–Pine Island (1905), Homer (1919), Cotton Valley (1922), Dixie (1929), Sugar Creek (1930), and Lisbon (1936) (Wheeler, 1963). During the 1930s and 1940s, exploration advanced rapidly eastward.

Late in 1930, Mr. A. E. Oldham, an employee of the Arkansas Louisiana Gas Company, made a surface reconnaissance survey covering an area of several townships north of the town of Ruston. He mapped a glauconitic sand member of the Cook Mountain Formation of Eocene age that he called the Minden beds. His map (Figure 5) reveals a surface domal structure, and based on this feature, Arkansas Louisiana Gas Company leased approximately 8500 ac (3443 ha) in the area (Walker, 1953). Surface drainage, traced from the Ruston 15 minute quadrangle and superimposed on Oldham's map (Figure 6), shows a radial pattern that corroborates the existence of a surface dome.

A seismograph survey was made in the area by Arkansas Louisiana Gas Company in 1934 to verify the results of the earlier surface mapping. A seismic map on the base of the Ferry Lake Anhydrite (Figure 7) shows that the crest of a subsurface structure lies about 2 mi (3.2 km) north of the surface feature. The seismic structure is compared with the surface structure on top of Minden beds in Figure 8.

Arkansas Louisiana Gas Company made another seismic survey in 1936, but before the survey was completed, a test well, interestingly, the Arkansas Louisiana Gas Company No. 1 Causey in the NE NW of Sec. 32, T19N, R2W, was drilled *halfway* between the surface structure and the original seismic

153

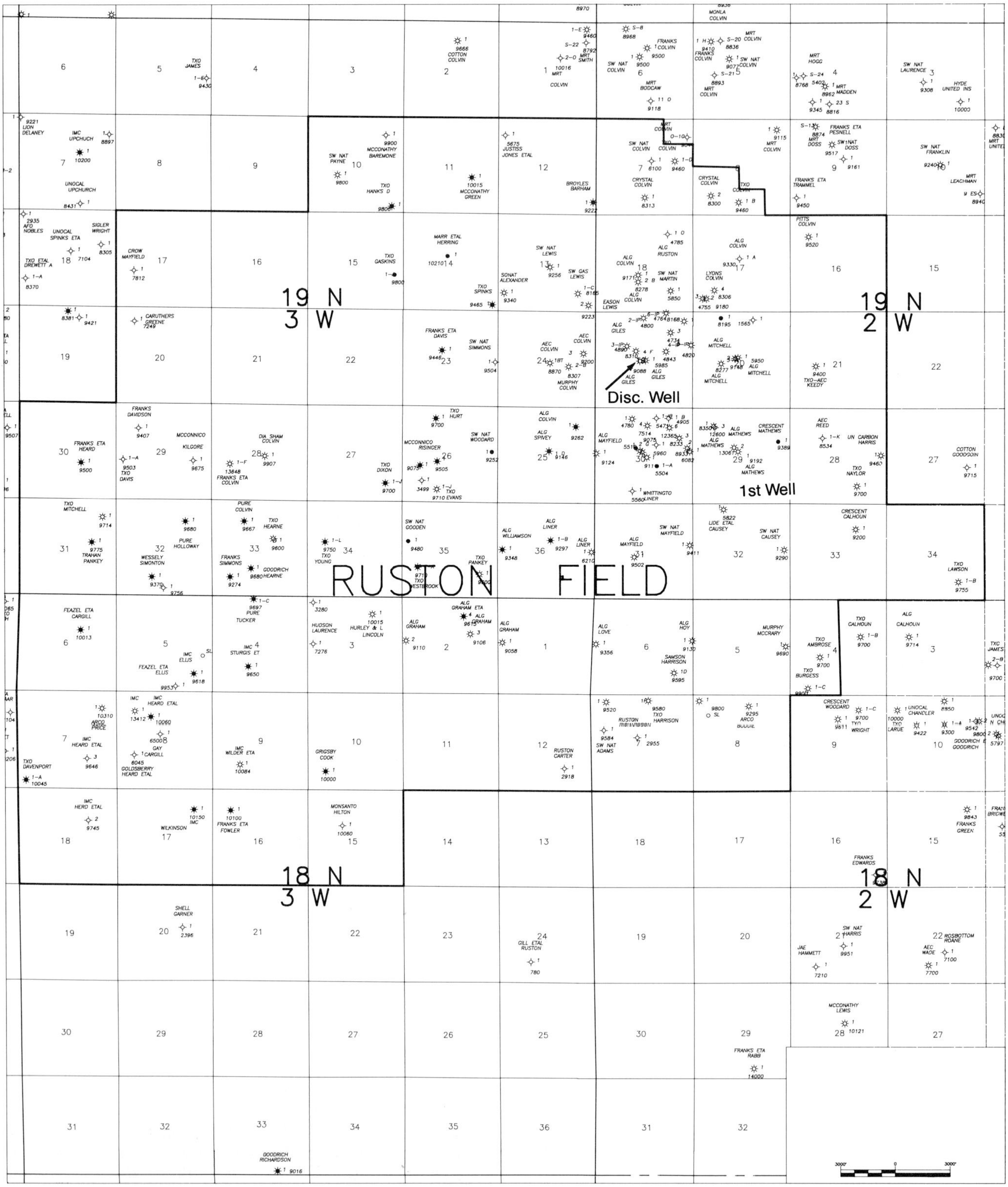

Figure 3. Ruston field outline, Lincoln Parish, Louisiana.

154

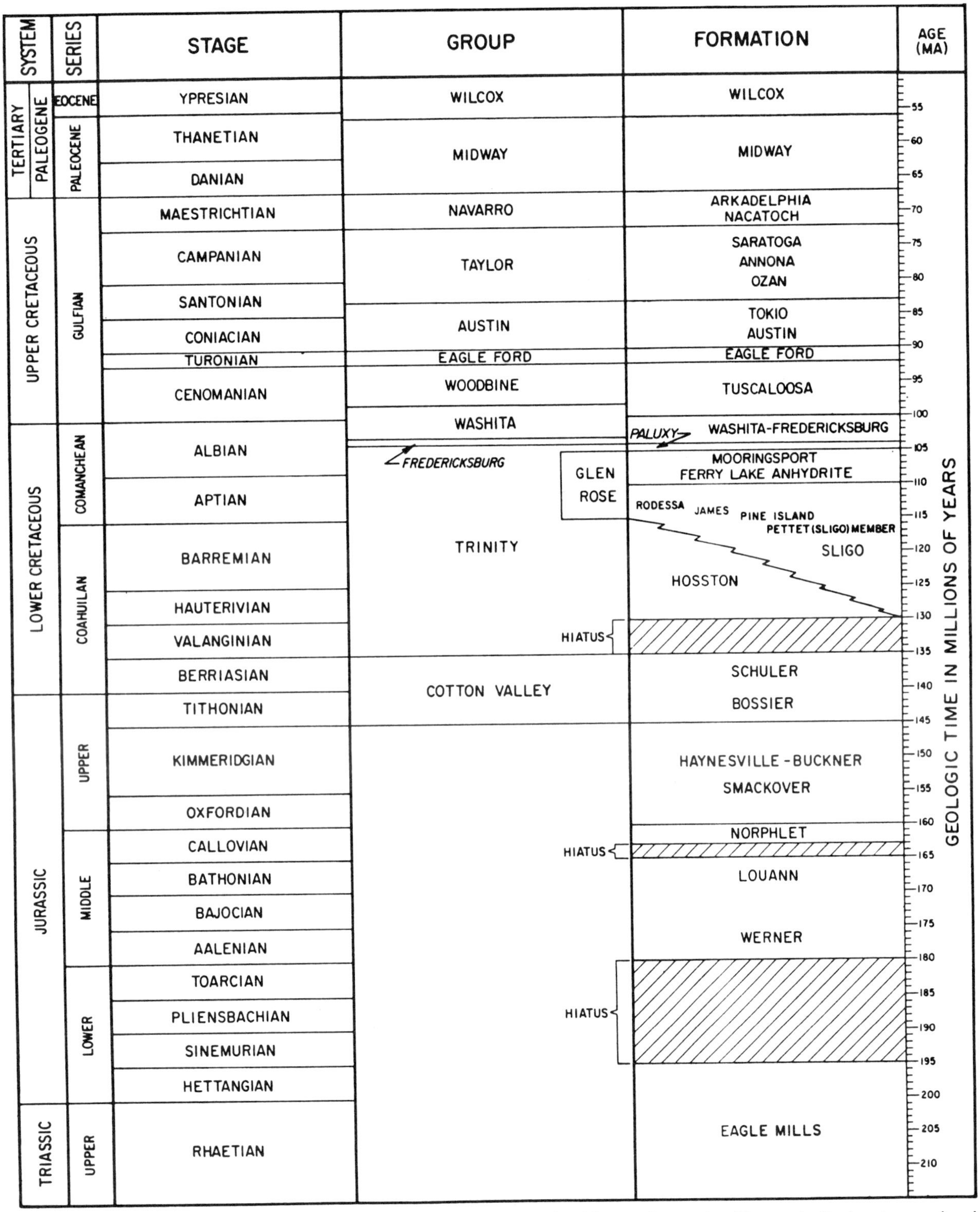

SYSTEM	SERIES	STAGE	GROUP	FORMATION	AGE (MA)
TERTIARY — PALEOGENE	EOCENE	YPRESIAN	WILCOX	WILCOX	55
	PALEOCENE	THANETIAN	MIDWAY	MIDWAY	60
		DANIAN			65
UPPER CRETACEOUS	GULFIAN	MAESTRICHTIAN	NAVARRO	ARKADELPHIA / NACATOCH	70
		CAMPANIAN	TAYLOR	SARATOGA / ANNONA / OZAN	75 / 80
		SANTONIAN	AUSTIN	TOKIO / AUSTIN	85
		CONIACIAN			90
		TURONIAN	EAGLE FORD	EAGLE FORD	
		CENOMANIAN	WOODBINE	TUSCALOOSA	95 / 100
LOWER CRETACEOUS	COMANCHEAN	ALBIAN	WASHITA	PALUXY → WASHITA-FREDERICKSBURG	105
			FREDERICKSBURG	MOORINGSPORT / FERRY LAKE ANHYDRITE	110
		APTIAN		GLEN ROSE / RODESSA JAMES PINE ISLAND / PETTET (SLIGO) MEMBER / SLIGO	115
	COAHUILAN	BARREMIAN	TRINITY	HOSSTON	120 / 125
		HAUTERIVIAN			130
		VALANGINIAN	HIATUS		135
		BERRIASIAN	COTTON VALLEY	SCHULER	140
JURASSIC	UPPER	TITHONIAN		BOSSIER	145
		KIMMERIDGIAN		HAYNESVILLE - BUCKNER / SMACKOVER	150 / 155
		OXFORDIAN			160
	MIDDLE	CALLOVIAN		NORPHLET / HIATUS	165
		BATHONIAN		LOUANN	170
		BAJOCIAN			175
		AALENIAN		WERNER	180
	LOWER	TOARCIAN		HIATUS	185
		PLIENSBACHIAN			190
		SINEMURIAN			195
		HETTANGIAN			200
TRIASSIC	UPPER	RHAETIAN		EAGLE MILLS	205 / 210

GEOLOGIC TIME IN MILLIONS OF YEARS

Figure 4. Chronostratigraphic section of North Louisiana. (From Shreveport Geological Society, 1987.) *Note:* The Schuler Formation of the Cotton Valley Group has only recently been assigned to the Lower Cretaceous; most authors prior to the 1980s assigned it to the Upper Jurassic. Also note that outcropping beds of the Eocene Claiborne Group (including the Cook Mountain Formation) that lie immediately above the Wilcox Group are not shown.

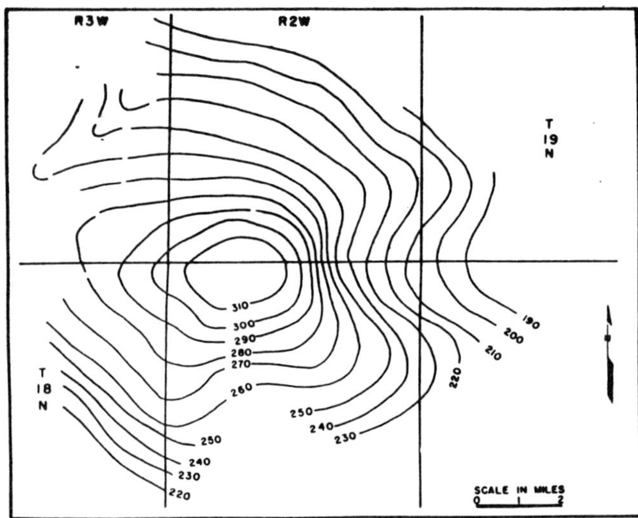

Figure 5. Surface structure map (1931) on top of Minden beds (Eocene). Contour interval, 10 ft. (From Walker, 1953.)

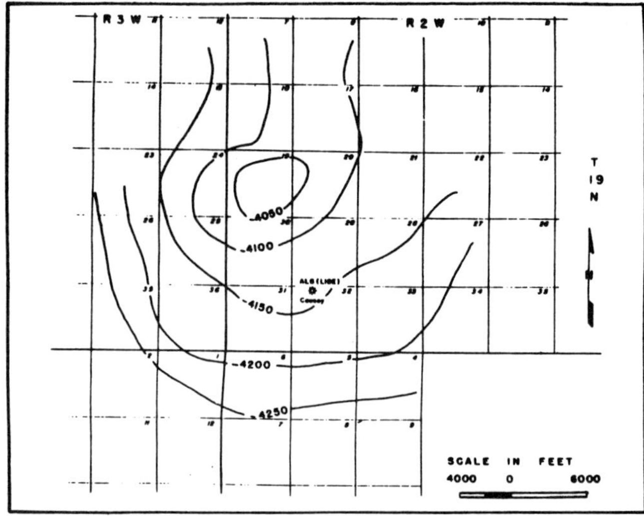

Figure 7. Reflection seismograph map (1934) showing contours on the base of the Ferry Lake Anhydrite. Contour interval, 50 ft. (From Walker, 1953.)

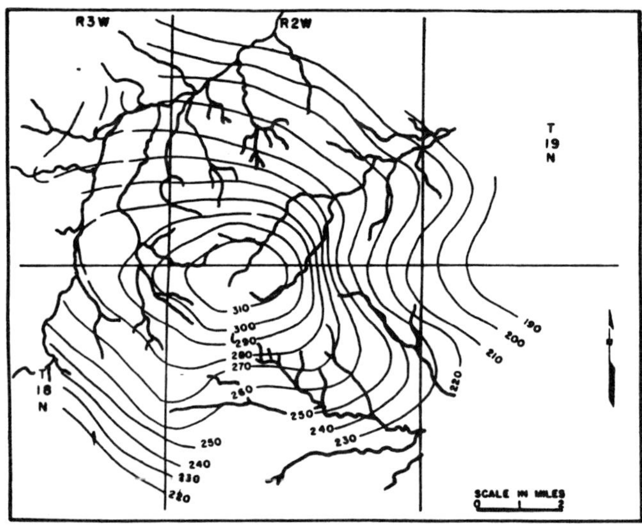

Figure 6. Same as Figure 5 with surface drainage pattern superimposed. Contour interval, 10 ft.

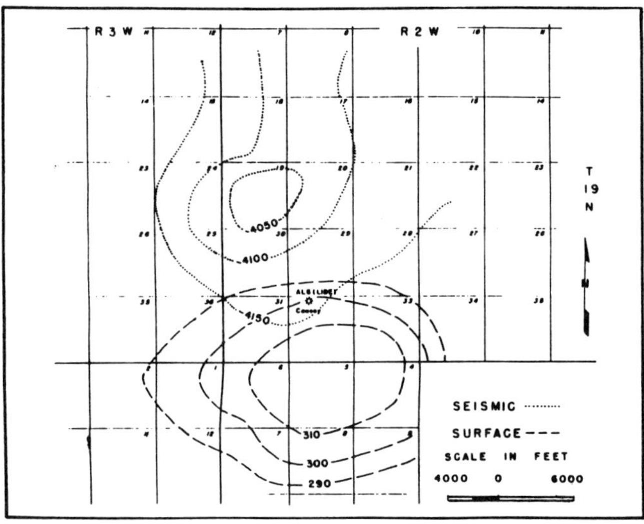

Figure 8. Surface structure on top of Minden beds compared with seismic structure on base of the Ferry Lake Anhydrite. Contour interval, 10 ft. (From Walker, 1953.)

structure (Figures 3 and 8). The well was drilled to a total depth of 4687 ft (1429 m) and abandoned on 18 April 1936. It was deepened to 5822 ft (1775 m) in 1937 by Lide and Greer and "was completed on September 5, 1937, through perforations from 5,316 to 5,322 ft (1620 to 1622 m) for 32,729,000 cubic feet (916,412 m³) of gas and a considerable amount of salt water per day" (Walker, 1953). The producing sand was the Causey sandstone member of the Pine Island Formation (Figure 4), about 50 ft (15 m) above the top of the Sligo Formation. Wisdom (1968) reported that "No connections were made to the Lide well and, after several unsuccessful attempts to shut off the salt water, it was plugged on November 6, 1942."

For this reason it is not considered the discovery well for the field.

The 1936 seismic survey (Figure 9), started prior to the drilling of the No. 1 Causey, was more sophisticated than the 1934 survey and showed a subsurface structure of the top of the James Limestone very similar to the subsurface map made of the James in 1953 (Figure 10). One more seismic survey was made in 1944 (Figure 11), and its results were little changed from those in the 1936 survey (Walker, 1953).

The final geophysical survey in the area by Arkansas Louisiana Gas Company was a gravity meter survey made in 1947 (Figure 12), the results

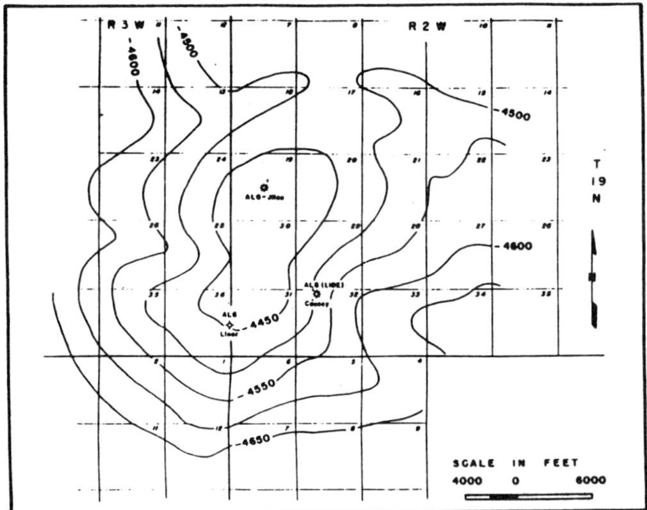

Figure 9. Reflection seismograph map (1936) showing contours on top of the James Limestone. Contour interval, 50 ft. (From Walker, 1953.)

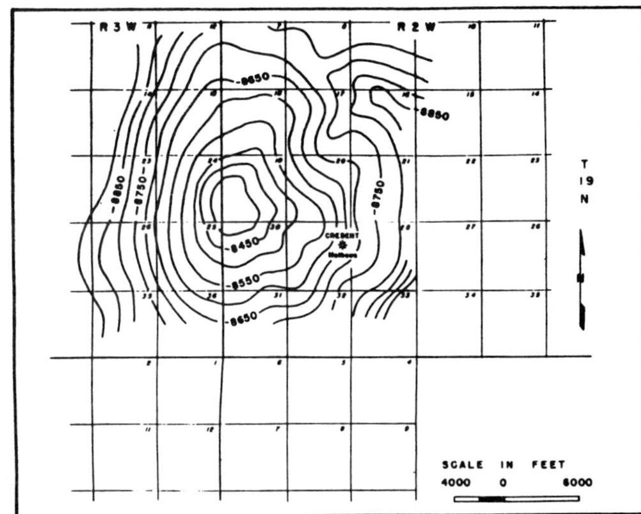

Figure 11. Reflection seismograph map (1944) showing contours on the Bodcaw sandstone in the Cotton Valley Group. Contour interval, 50 ft. (From Walker, 1953.)

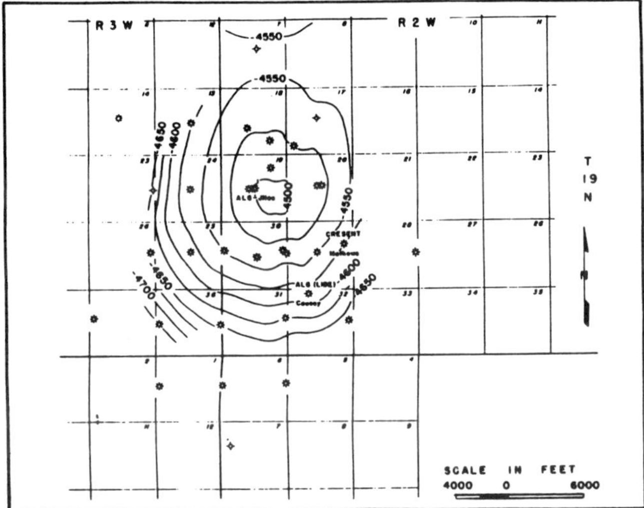

Figure 10. Subsurface map on top of the James Limestone defined by drilling as of 1953. Contour interval, 25 ft. (From Walker, 1953.)

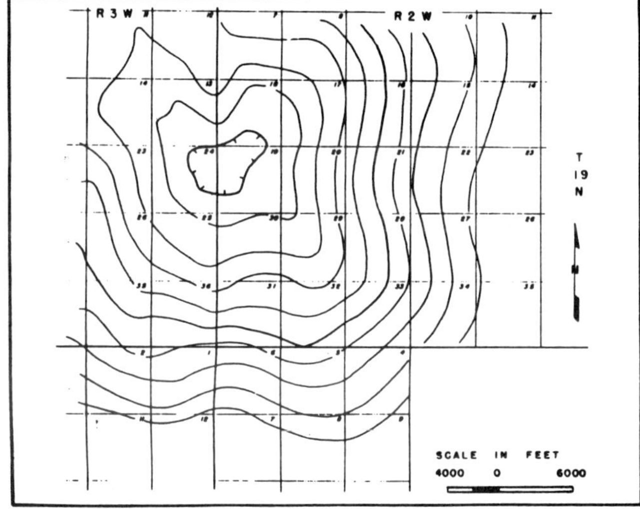

Figure 12. Observed gravity map (1947) of the Ruston field. Contour interval, 0.2 mgal. (From Walker, 1953.)

of which are remarkably similar to the seismic results (Walker, 1953). The negative anomaly indicates the presence of a salt uplift beneath the field.

Other important papers on the field include Breedlove et al. (1953), Kamb (1945), and Shreveport Geological Society (1946).

Discovery

The official discovery well for the Ruston field is the Arkansas Louisiana Gas Company No. 1 W. F. Jiles located 50 ft (15 m) south and 83 ft (25 m) east of the center of Sec. 19, T19N, R2W, in Lincoln Parish, Louisiana (Wisdom, 1968) (Figure 3). "The Jiles well was completed in October 1943 from the Jiles sandstone of the Lower Cretaceous Hosston Formation (approximately 350 ft below the top of the Hosston)" (Wisdom, 1968) from perforations 5878 to 5896 ft (1792–1797 m) using 108 shots (Figure 4). The total depth of the well was 5985 ft (1824 m) and the elevation of the well was 238 ft (72.5 m), derrick floor. Initial production from the well was 45 MMcf (1.26 million m^3) of gas per day, open flow, with shut-in pressure of 2400 psi (16,548 kPa) (Shreveport Geological Society, 1946).

By 1946, three other gas-producing zones had been discovered in the field: the Causey sandstone, the Pettet Limestone, and the James sandstone. Gas reservoirs in the Causey sandstone of the Pine Island

Formation and the Pettet Limestone Member of the Sligo Formation were both discovered in the Arkansas Louisiana Gas Company No. 1 J. C. Dowling Unit well in Sec. 30, T19N, R2W, completed on 25 May 1944. The Causey sandstone is about 50 ft (15 m) above the top of the Sligo, and the Pettet porosity zone is about 35 ft (11 m) below the top of the Sligo. The Causey was perforated from 5190 to 5200 ft (1582–1585 m) and the Pettet Limestone was perforated from 5262 to 5272 ft (1604–1607 m) with a dual initial potential of 15 MMcf (420,000 m^3) of gas per day. James sandstone production was discovered in the Arkansas Louisiana Gas Company No. 1 John E. Mitchell Unit well in Sec. 20, T19N, R2W on 28 April 1948. It was perforated in the interval 4750–4757 ft (1448–1450 m) and flowed 72 MMcf (2 million m^3) of gas with rock pressure of 1984 psi (13,680 kPa). This well was dually completed from the Mitchell sandstone of the upper Hosston.

Cotton Valley production was not found until 9 October 1948, when the Crescent Drilling Company drilled the No. 1 C. M. Mathews in the northeast quarter of Sec. 29, T19N, R2W based on subsurface control in the Trinity Group and the Hosston Formation. It was drilled to a total depth of 9390 ft (2862 m) and completed on 5 May 1948 in the "D" sandstone member of the Cotton Valley Group from perforations between 8796 and 8806 ft deep (2681–2684 m). This zone tested 6.7 MMcf (187,600 m^3) of gas per day, open flow, and some condensate (Walker, 1953). However, the scout ticket on this well records an initial potential of only 56 bbl of oil per day, flowing, on $^{24}/_{64}$-in. choke with tubing pressure of 2800 psi (19,306 kPa).

A total of 16 horizons have proved productive in the Ruston field since it was discovered in 1943. Twelve of these, which include the Rodessa, James, middle Hosston, lower Hosston, Cotton Valley "C," Bodcaw, Vaughn, Price, McCrary, McFearin (Davis), Feazel, and Smackover "Gray" sandstone, were discovered after 1948. Many of the horizons such as upper, middle, and lower Hosston have numerous subhorizons or zones that are grouped together for convenience in mapping and simplicity in nomenclature (see *Reservoir* section).

Table 1 lists names, locations, discovery dates, producing horizons and their geologic ages, perforated intervals, and initial potentials for all discovery wells in the Ruston field.

Post-Discovery

Fracturing techniques that were developed in the late 1950s are used in the field to enhance the production significantly. Petrophysical logging in the early development of the field consisted of an electric log and microlog, but logging programs during later development utilized induction, sonic, gamma ray, and other types of logs for better formation evaluation.

All Smackover, Cotton Valley, and Hosston production is natural gas with minor amounts of condensate, developed on 640 ac (259 ha) well spacing. Drilling programs commonly specify approximately 1100 ft (335 m) of 8⅝-in. (21.9 cm) surface casing to be used in an average 7000 ft (2134 m) Hosston test and approximately 2200 ft (671 m) of 8⅝-in. surface casing in a 9500 ft (2896 m) Cotton Valley test. Upon completion, the producing interval is perforated with 4 shots per foot, usually followed by stimulation and fracturing of most of the lower Hosston and Cotton Valley sandstones. CO_2 fracturing is performed on most thin sandstones and hydraulic fracturing on thicker sandstones. The logging suites generally include a dual induction-sonic log, a density-neutron-gamma ray log, and commonly a microlog.

DISCOVERY METHOD

The Ruston field serves as a good example of a systematic search for new hydrocarbon reserves in a developing region, utilizing a combination of sedimentary trend mapping, surface geologic mapping, and seismic and gravity surveys. These methods led to the discovery of more than 600 bcf (17 billion m^3) of natural gas and several million barrels of oil and condensate from 16 major horizons and numerous additional subhorizons. Based on these figures, and using the relationship of 6000 ft^3 of gas being equivalent to 1 bbl of oil, Ruston has ultimate recoverable reserves of approximately 100 million bbl oil equivalent.

It is not likely that current exploration methods would be any better able to discover the Ruston field than those used in the 1930s and 1940s. It is a classic domal feature produced by a deep-seated salt uplift, detectable by seismology and gravity methods. The only physiographic evidence of the existence of the surface structure mapped by Oldham 3 mi (4.8 km) south-southeast of the subsurface structure in 1931 consists of an inconspicuous knoll surrounded by a radial drainage pattern (Figure 6). There are no surface faults in the immediate vicinity; consequently, there have been no oil seeps reported in the literature.

STRUCTURE

Tectonic History

The Ruston field is in the North Louisiana portion of the Gulf Coast basin or geosyncline. The Gulf Coast region is classified by Bally and Snelson (1980) as basin type 1143, or an Atlantic-type passive margin (shelf, slope, and rise) that straddles continental and oceanic crust, overlying an earlier back-arc basin.

Table 1. Discovery wells for all producing horizons in the Ruston gas field, Lincoln Parish, Louisiana.

Discovery Well	Location	Date	Producing Horizon	Perforations ft (m)	Initial Potential
Arkansas Louisiana Gas Company No. 1 W. F. Jiles	Sec. 19, T19N, R2W	10/26/1943	Upper Hosston Jiles sandstone (L. Cretaceous)	5878–5896 (1792–1797)	45,000 MCFGPD, dry gas, RP 2400 psi
Arkansas Louisiana Gas Company No. 1 J. C. Dowling Unit	Sec. 30, T19N, R2W	5/25/1944	Causey sandstone, Pine Island Fm. (L. Cretaceous)	5190–5200 (1582–1585)	Est. 3–4 MMFGPD, dry gas, RP 2118 psi
Arkansas Louisiana Gas Company No. 1 J. C. Dowling Unit	Sec. 30, T19N, R2W	5/25/1944	Pettet limestone zone of Sligo Fm. (L. Cretaceous)	5262–5272 (1604–1607)	Est. 20 MMCFGPD, dry gas, RP 2118 psi
Arkansas Louisiana Gas Company No. 1 J. Mitchell	Sec. 20, T19N, R2W	5/6/1948	James sandstone (L. Cretaceous)	4750–4757 (1448–1450)	F/72 MMCFGPD, RP 1984 psi
Crescent No. 1 C. M. Mathews	Sec. 29, T19N, R2W	10/9/1948	Cotton Valley "D" sandstone (U. Jurassic)	8796–8806 (2681–2684)	F/56 BDPD, $^{10}/_{64}$-in. ck, TP 2800 psi
Arkansas Louisiana Gas Company No. 2 J. C. Dowling	Section 30, T19N, R2W	1/17/1949	Cotton Valley Bodcaw sandstone (U. Jurassic)	8760–8810 (2670–2685)	F/187 BCPD, $^{3}/_{8}$-in. ck, TP 2345 psi
Arkansas Louisiana Gas Company No. 1 Della Colvin	Sec. 18, T19N, R2W	8/3/1949	Cotton Valley Vaughn sandstone (U. Jurassic)	8809–8838 (2685–2694)	F/390 BDPD, $^{24}/_{64}$-in. ck, TP 2604 psi
Southwest Gas Producing Company No. 1 O. Herring	Sec. 14, T19N, R3W	10/26/1951	Cotton Valley Feazel/Davis (U. Jurassic)	9466–9476 (2885–2888)	F/51 BOPD, $^{24}/_{64}$-in. ck, TP 850 psi
Arkansas Louisiana Gas Company No. 2 J. P. Graham	Sec. 2, T18N, R3W	6/20/1955	Middle Hosston sandstone (L. Cretaceous)	7071–7093 (2155–2162)	COF (Est.) 8000 CFGPD, RP 2930 psi
Southwest Gas Producing Co. No. 1 Lewis Estate	Sec. 13, T19N, R3W	7/6/1959	Cotton Valley McFearin/Davis (U. Jurassic)	9131–9144 (2783–2787)	F/1800 MCFGPD plus 79 BDPMM, $^{1}/_{4}$-in. ck, TP 2025 psi
Atlantic Refining Co. No. 1 Hattie Price	Sec. 7, T18N, R3W	7/7/1959	Cotton Valley Price sandstone (U. Jurassic)	9936–9944 (3028–3031)	F/2200 MCFGPD plus 75 BDPMM, $^{14}/_{64}$-in. ck, TP 1975 psi
Pure Oil Company No. 1 John W. Colvin Unit	Sec. 33, T19N, R3W	6/25/1961	Cotton Valley McCreary/Davis (U. Jurassic)	9564–9574 (2915–2918)	F/3580 MCFGPD plus 432 BDPD, $^{12}/_{64}$-in. ck, FTP 4365 psi
Harvey Broyles (Vintage) No. 1 Mattie Sue Barham Unit	Sec. 12, T19N, R3W	6/28/1961	Lower Hosston sandstone (L. Cretaceous)	7915–7922 (2412–2415)	F/170 BOPD, $^{16}/_{64}$-in. ck, FTP 240 psi
J. C. Trahan No. 1 A. A. Pankey Unit	Sec. 31, T19N, R3W	12/14/1961	Cotton Valley "C" sandstone (U. Jurassic)	9253–9261 (2820–2823)	COF 2350 MCFGPD plus 46 BDPMM, $^{1}/_{4}$-in. ck, FTP 3125 psi
Arkansas Louisiana Gas Company No. 5 Giles	Sec. 19, T19N, R2W	2/5/1969	Gloyd limestone, Rodessa Formation (L. Cretaceous)	4494–4507 (1370–1374)	psi F/1668 MCFGPD, $^{12}/_{64}$-in. ck, psi TP 1700
Arkansas Louisiana Gas Company No. 6 Giles	Sec. 30, T19N, R2W	6/30/1979	Smackover "Gray" sandstone (U. Jurassic)	11,907–11,966 (3629–3647)	F/821 MCFGPD plus 10 BW, $^{10}/_{64}$-in. ck, TP 1571 psi

It is classified by Klemme (1971) as basin type IIC c/IV. The IIC portion defines it as a Continental Multicycle basin, crustal collision zone-plate margin closed. The c/IV defines it as a Delta basin, Tertiary to Recent.

The development of the Gulf Coast region is genetically related to the origin of the Ouachita orogenic belt that extends from central Arkansas through southeast Oklahoma to the Marathon region of South Texas. Many models have been proposed for the development of the Ouachita belt and the subsequent development of the Gulf Coast geosyncline. Some models, such as that of Burgess (1976), have envisioned subduction of the African–South American plate beneath the North American plate, whereas other, more recent models envision a "southward-dipping subduction of an Atlantic-type continental margin (the southern margin of North America) beneath either an island arc system, a microcontinent, or a larger continental mass" (Houseknecht, 1983).

The evolution of the Gulf Coast area according to Burgess (1976) is summarized in Figures 13 and 14 by means of a series of generalized cross sections extending from southeastern Oklahoma across the Sabine uplift to southwestern Louisiana.

Regional Structure

The Ruston field is situated in central Lincoln Parish, on the northeast flank of the North Louisiana Salt basin and southwest of the Monroe uplift. West of the North Louisiana Salt basin is the Sabine uplift, which straddles the Texas-Louisiana boundary. South Arkansas has several graben trends; one is the eastward extension of the Mexia-Talco trend of Texas, and the other is the State-Line trend along the Louisiana-Arkansas border. These and other major structures in the vicinity are illustrated on Figure 15. Salt movement was wholly or partially responsible for many of the petroleum-bearing structures in North Louisiana and South Arkansas, as it was for the Ruston field.

Local Structure

The Ruston field is a salt uplift as shown by a strong gravity minimum (Figure 12) and the nearly circular pattern exhibited by structural contour maps (Figure 10). There is no indication of salt piercement from drilling data or the seismic cross section (Figure 16), hence the structure was probably formed as a salt pillow. The beginning of growth of the uplift is uncertain but likely started in the Upper Jurassic when sediment thickness over the Louann Salt (Middle Jurassic) was sufficient to cause salt movement upward. The movement continued into the Lower Cretaceous based on thickness data, depositional trends, and petroleum maturation data (see burial-history plot, Figure 17).

An isopach map from the base of the Ferry Lake Anhydrite to the base of the Cotton Valley "B" limestone (Figure 18) shows about 200 ft of thinning. Growth apparently continued to the end of the Lower Cretaceous, whereupon erosion created a major angular unconformity prior to the deposition of Upper Cretaceous rocks.

Minor faults with very small displacements have been reported in the Ruston field, but they do not show up on the structure maps. However, a fault is interpreted on the seismic cross section (Figure 16) starting beneath the Cotton Valley "B" limestone and extending into the Louann Salt.

STRATIGRAPHY

The Ruston field is underlain by Tertiary and older sedimentary deposits (Figure 4). Not shown on Figure 4 are the outcropping beds of the Claiborne Group that lie immediately above the Wilcox. A number of wells in South Arkansas and North Louisiana have also penetrated the upper part of the Morehouse Formation of Paleozoic age.

The deepest formation reached by drilling at Ruston is the Louann Salt of Middle Jurassic age (Bathonian), a unit that was very important in forming many productive ridges and domes in the North Louisiana basin. This is followed by the Norphlet Formation, also Middle Jurassic (Callovian), which is composed mostly of thin, red, sandy shales. It also contains a red conglomerate; red, pink, gray, and white sandstone; and red and gray siltstone. At the Cotton Valley field in Webster Parish, the formation consists of dark gray dolomite and a white sandstone with frosted quartz grains (Berryhill et al., 1968).

The Smackover Formation, Late Jurassic (Oxfordian) in age, is about 1500 ft (457 m) thick and was deposited as a result of a major marine transgression throughout the North Louisiana–South Arkansas region. The lower part is dense, basinal limestone with interbedded argillaceous limestone and shale beds (Berryhill et al., 1968). The middle Smackover is dark-gray to tan argillaceous limestone, intercalated with calcareous sandstones in the northern part of the basin. The two most important of these are called the Smackover "C" sandstone and the "Gray" sandstone. These sandstones appear to be deep-water fan deposits (Judice and Mazzullo, 1982).

The upper Smackover contains many east–west-trending oolite bars that may be fairly extensive or very local. The bars range from very shaly units with low permeability to very well sorted oolitic limestone with excellent porosity and permeability. The oolite bars, with good porosity and permeability, formed across Middle to Late Jurassic "highs."

The Haynesville Formation of Late Jurassic (Kimmeridgian) age overlies the Smackover and consists of anhydrite at its base, grading upward into sandstone, limestone, red shale, and siltstone. In extreme northern Louisiana and southern Arkansas, the anhydrite becomes thicker and contains interbedded red shale, forming the Buckner Formation. Much of the Haynesville represents a regression or a return to marine, marginal marine, and fluvial deposition.

The Cotton Valley Group conformably overlies the Haynesville except on some local structures where the contact appears to be unconformable. It is subdivided into two formations, the Bossier and the Schuler, which, until recently, were both considered

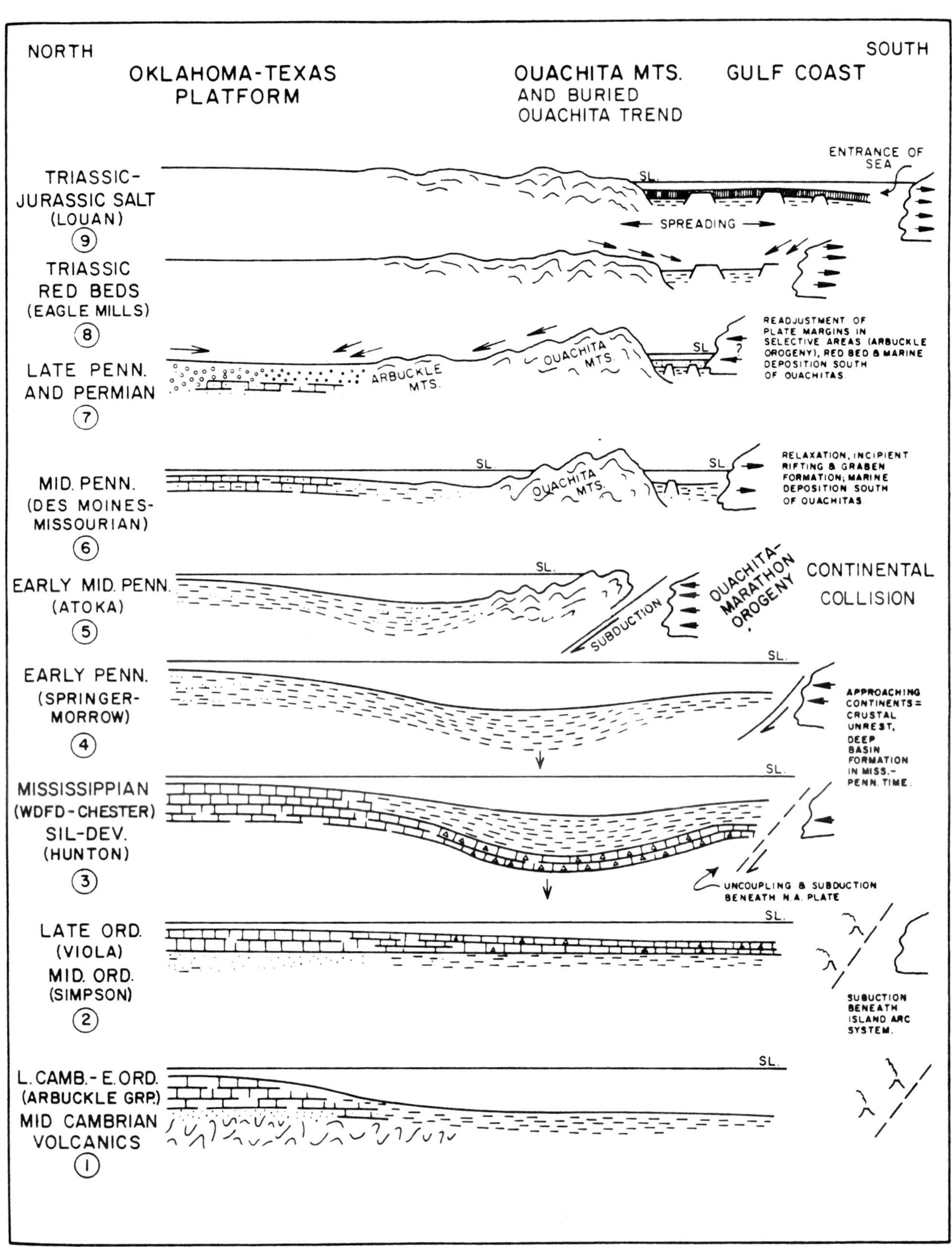

Figure 13. Early geologic evolution of Gulf Coast area, Cambrian through Triassic and Louann Salt deposition. Cross section from southwestern Louisiana through eastern Texas and into southeastern Oklahoma. (From Burgess, 1976.)

161

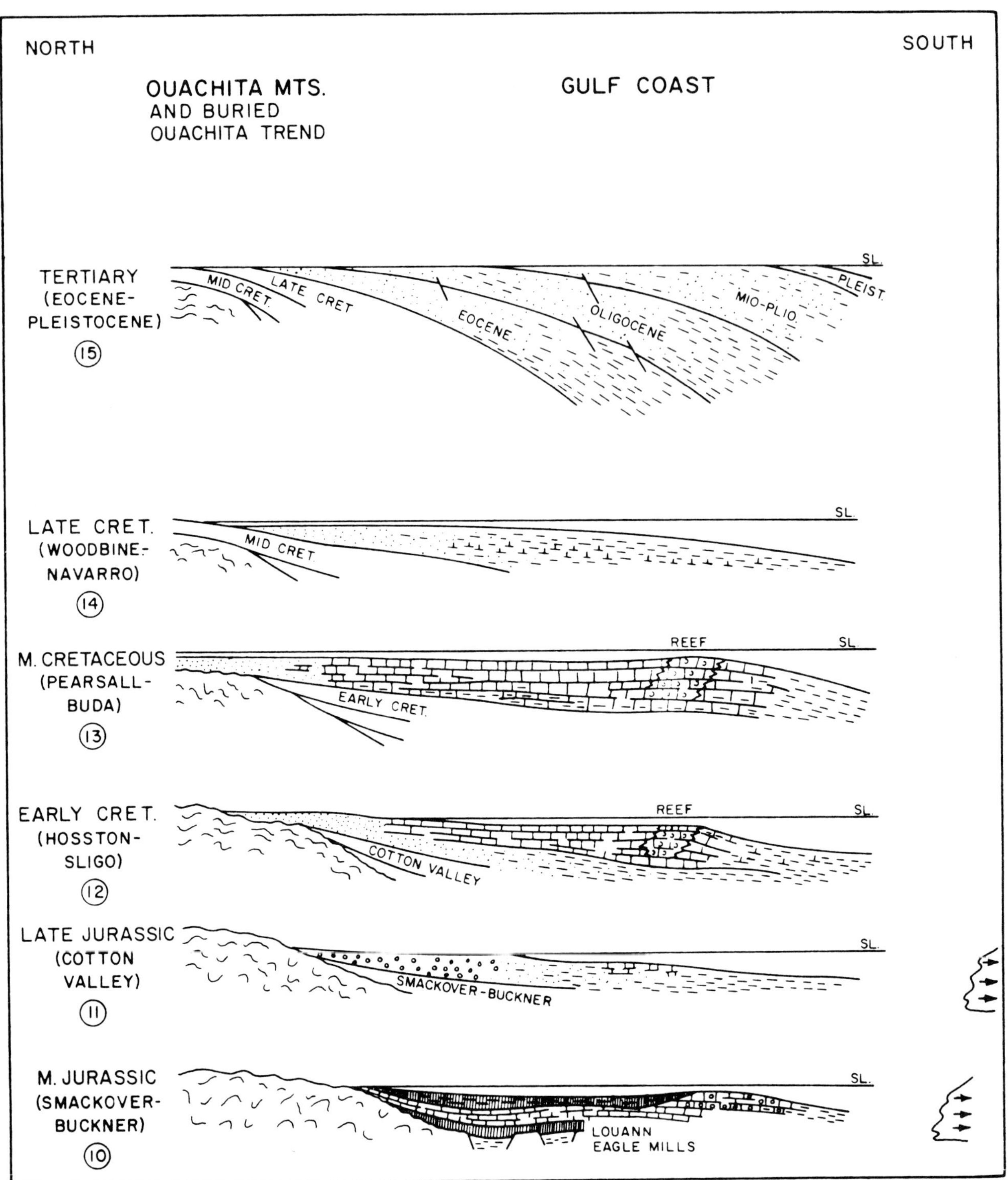

Figure 14. Later geologic evolution of Gulf Coast area, Jurassic through Tertiary (see Figure 13). (From Burgess, 1976.)

Late Jurassic in age. However, the Bossier is now placed in the Upper Jurassic (Tithonian) and the Schuler in the Lower Cretaceous (Berriasian) by many authors (Figure 4). The entire Cotton Valley interval represents a major regression that was interrupted by minor transgressions. Blanket sands were deposited across subsiding deltaic complexes with other thick deltaic and shelf sand complexes (Thomas and Mann, 1966; Eversull, 1985).

The Cotton Valley contains a high percentage of quartz sand delivered to the shelf by wave-dominated delta systems located in the vicinity of the Sabine uplift on the northwest and near the Monroe uplift on the northeast (Eversull, 1985). The sands were

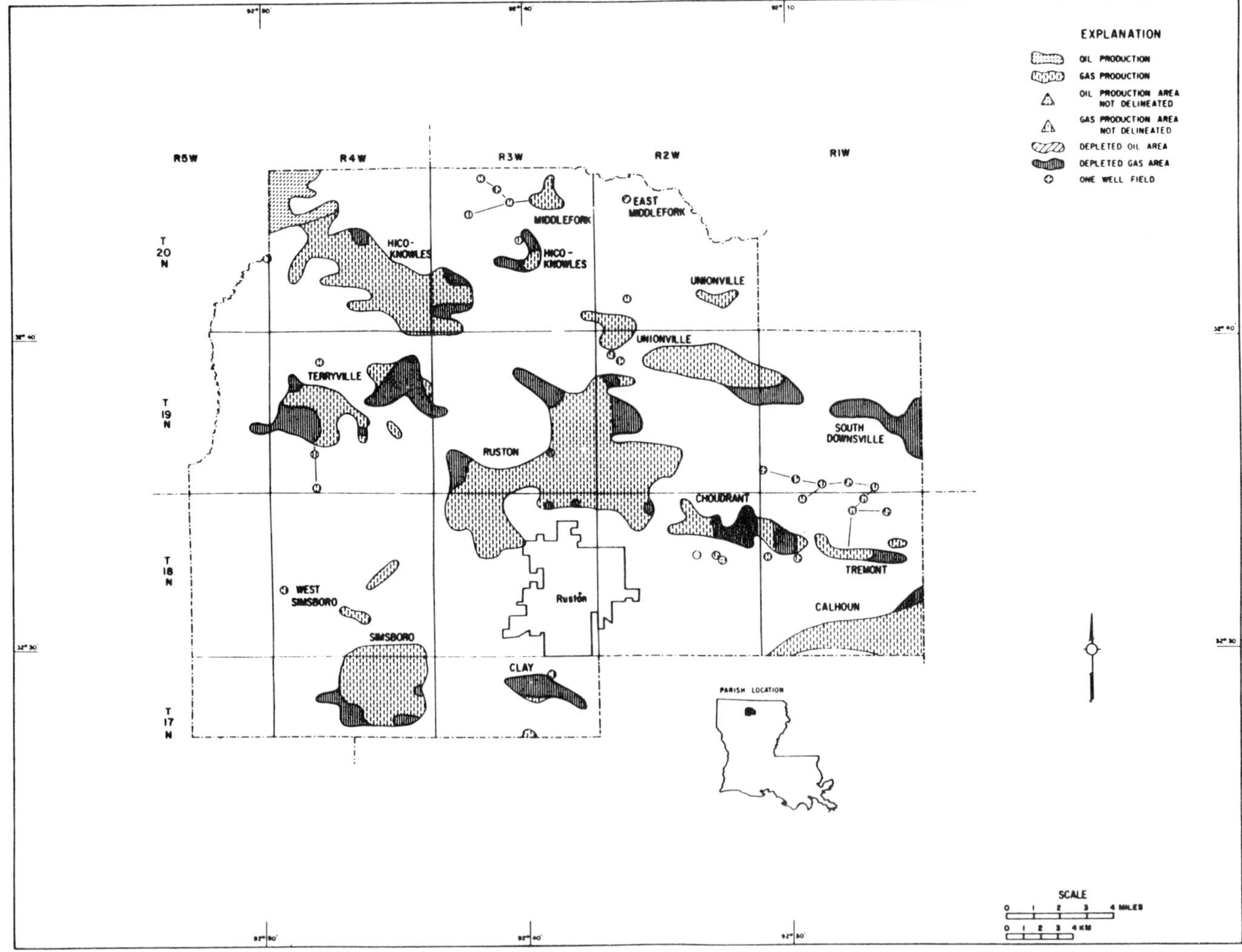

Figure 15. Diagrammatic map outlining major structural features of the southern Atlantic and northern Gulf Coastal province. 1, Ruston field. 2, Mexia-Talco trend. 3, State-Line trend. (From Murray, 1961.)

reworked by marine processes forming blanket sands updip from a massive depocenter named the Terryville (Mann and Thomas, 1964; Thomas and Mann, 1966). Numerous blanket sands are present in the upper Cotton Valley, the better known ones being the Vaughn, Bodcaw, and "D" sandstones. These sandstones merge northward into the Hico Shale, providing excellent updip stratigraphic pinch-outs across the basin (Thomas and Mann, 1966, figure 6).

The upper 300 to 400 ft (91 to 122 m) of the Cotton Valley, called the Knowles Limestone, consists of interbedded argillaceous limestones and gray shales.

The Hosston Formation is Early Cretaceous (Hauterivian–Barremian) in age and varies in thickness from 2000 to 4000 ft (610–1219 m) in north Louisiana. The Hosston is predominantly a redbed facies in the updip area of the basin and grades southward (basinward) into fluvial, deltaic, and shelf sandstones, dark marine shales, and shelf carbonates.

The Sligo Formation is also Early Cretaceous in age and reflects a transgression and a return to marine conditions in the basin. The Sligo consists of gray to brown argillaceous limestones and some interbedded sandstones. Oolitic bank and reef facies, known among oilmen as the Pettet "Lime," also occur within the formation.

The Pine Island shale and the James Limestone are Early Cretaceous (Aptian) in age. The Pine Island is a gray, marine shale with a local sandstone unit known as the Causey sandstone which, at one time, produced hydrocarbons in several wells in the Ruston field. The James Limestone is composed of an oolitic to sandy gray limestone with an interbedded fine-grained sandstone. Reef buildups occur in the James in some areas in the downdip part of the basin.

The Rodessa Formation is Early Cretaceous (Aptian) in age and consists of an oolitic to a coquinoid limestone, calcareous shale, anhydrite stringers, and sandstone (Berryhill et al., 1968). The Rodessa has been subdivided into five members, which are, from oldest to youngest, Young, Dees, Gloyd, Hill, and Kilpatrick. All of these consist of limestone with shale interbeds except the Hill, which is a quartz sandstone.

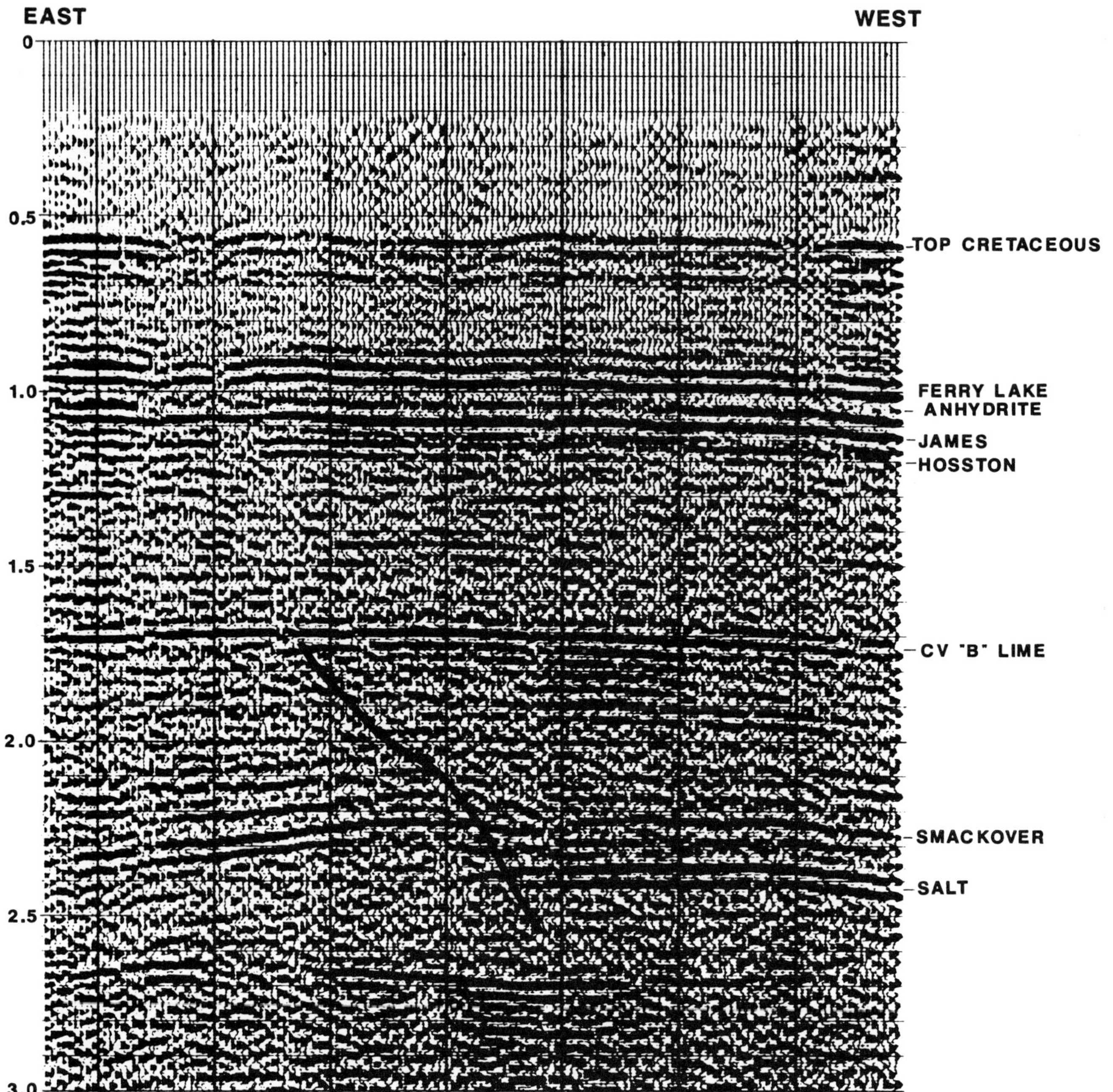

Figure 16. Seismic section across the Ruston field. Note the normal fault starting beneath the Cotton Valley "B" limestone and extending into the Louann Salt. (Compliments of Geosource.)

TRAPS

There are more than 40 sandstone or limestone zones producing natural gas and oil from seven formations in the Ruston field: the Smackover Formation; the Schuler Formation of the Cotton Valley Group; the Hosston Formation; the Sligo Formation; the Pine Island Formation; the James Limestone; and the Rodessa Formation. Trapping mechanisms for these zones are summarized below.

The Smackover "Gray" sandstone produced minor quantities of gas at Ruston from one well drilled just south of the crest of the Ruston dome. It is likely a domal trap, but it has been penetrated in only a few wells and structural conditions are uncertain.

The Cotton Valley contains numerous structural and stratigraphic traps at Ruston, making it the most prolific gas-bearing interval. Structural cross section A–A' (Figure 19) shows the relationship of the various Cotton Valley producing sandstones in the Ruston field.

The lowest of the Cotton Valley producing units is called the McFearin-Davis member, which has a total of nine different productive sandstones. The

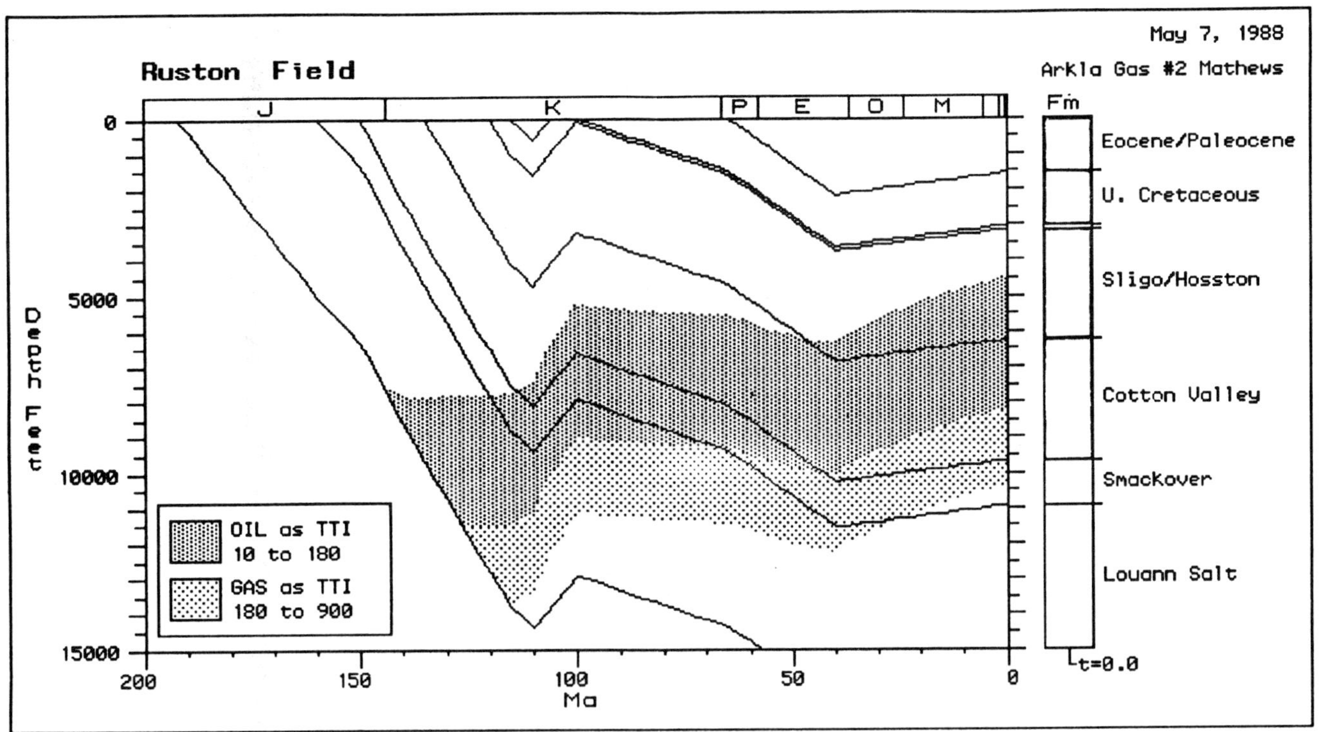

Figure 17. Burial-history plot of the Ruston field based on the Arkla Gas Company No. 2 Mathews, Sec. 29, T19N, R2W, Lincoln Parish, Louisiana. (Plot was prepared by Renae Wilkinson of Platte River Associates, Denver, Colorado.)

McFearin sandstone or McFearin "H" sandstone (Arkla Gas terminology) produces from a closed domal trap (Figure 20). All other traps in the McFearin-Davis interval are stratigraphic pinch-outs across the Ruston dome.

The Cotton Valley Vaughn consists of two sandstones that yield gas on the crest of the Ruston dome. The sandstones have two distinctly different gas-water levels owing to the porosity and permeability differences in the two units and the small area in which they are joined (Figure 21).

The Cotton Valley Bodcaw sandstone in the Ruston field forms a structural trap (Figure 22) at the crest of the dome with the original gas-water level at a subsea depth of 8642 ft (2634 m). The upper seal on the trap is a thick, impermeable marine shale.

The Cotton Valley "D" sandstones (Figure 23) are the most prolific gas-distillate producers at Ruston field, both in terms of productive acreage and ultimate recoveries. The "D" sandstones at the Ruston field form roughly east-west stratigraphic pinch-outs across the south flank of the Ruston field (Figures 19 and 23). Upper and lateral seals are interbedded marine limestones and shales.

The Cotton Valley "C" sandstone reservoir (Figure 23) is a stratigraphic trap located in the extreme southwest part of Ruston field. It is a discontinuous, isolated sandstone, usually less than 10 ft (3 m) thick, crossing a southwest-plunging nose extending off the crest of the field. The upper and lateral seal of the

"C" sandstone is the impermeable Cotton Valley "B" or Knowles Limestone.

The Hosston Formation is subdivided into upper, middle, and lower units (Figure 24). Hosston sandstones were deposited as fluvial-deltaic channel sands that are gas-productive where they cross the Ruston dome (Figures 25, 26, and 27). All of the sandstones have downdip gas-water levels, and almost all upper and lateral seals are shales or impermeable sandstones. The dome appears to have been growing throughout Hosston time as indicated by interval thinning on the crest of the structure.

The lower part of the Sligo Formation contains a small, isolated bar of porous oolitic limestone ("Pettet Lime") on the crest of the dome, forming a limited reservoir. Away from the crest it grades into a dense, nonporous limestone. Production from the Sligo has been approximately one-third oil and two-thirds natural gas. No water level is present in the reservoir.

The Causey sandstone of the Pine Island Formation is present in most wells in the field, but porosity is developed in only a few wells on the crest and slightly downdip from the crest of the dome. Production has been predominantly natural gas.

The James sandstone forms a structural trap at Ruston (Figure 28) with sandstone present over most of the structure and shale providing the upper seal on the reservoir. A gas-water level is present around the flank of the structure at approximately 4545 ft

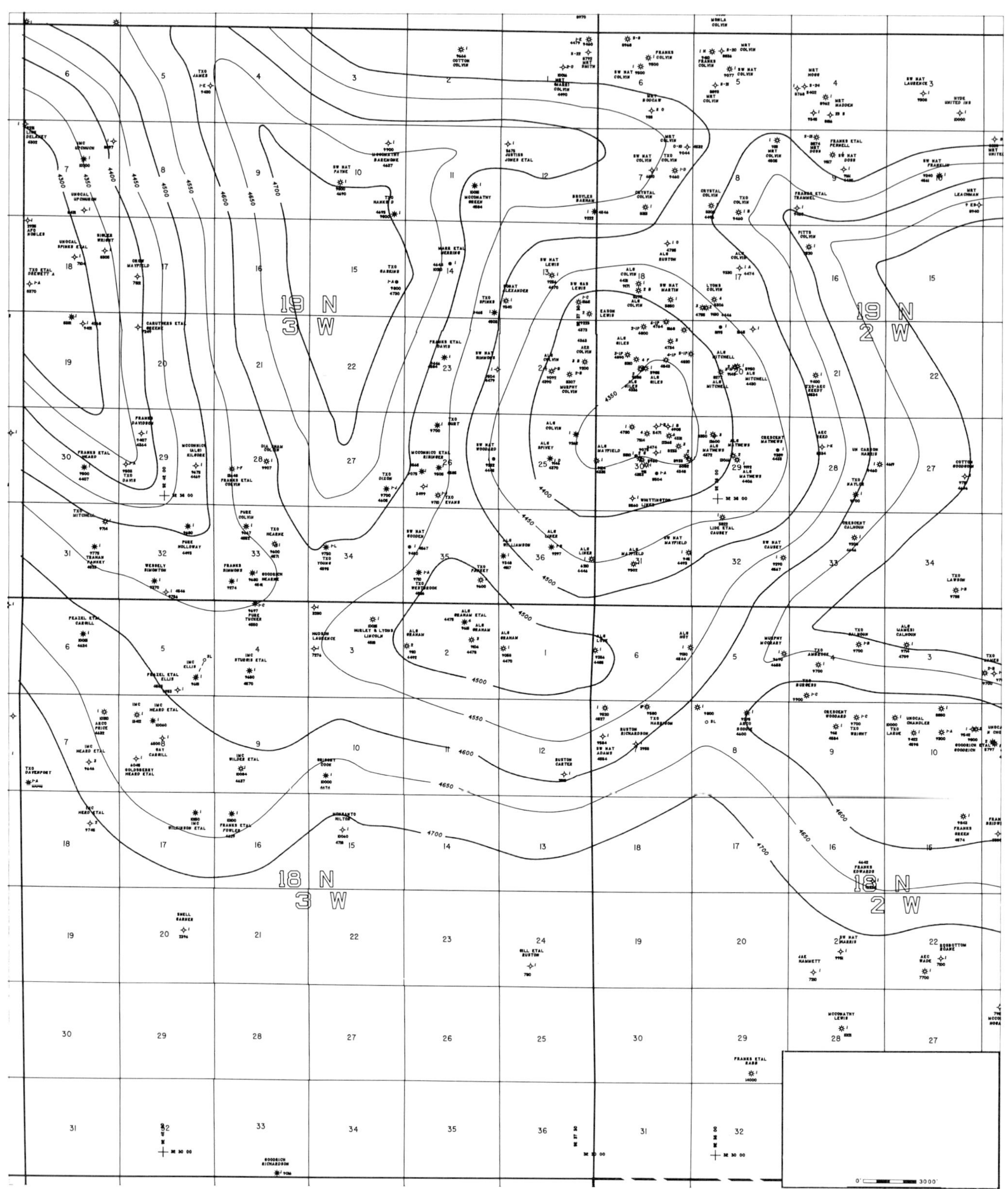

Figure 18. Isopach map from the base of the Ferry Lake Anhydrite to the base of the Cotton Valley "B" limestone. Contour interval, 50 ft.

166

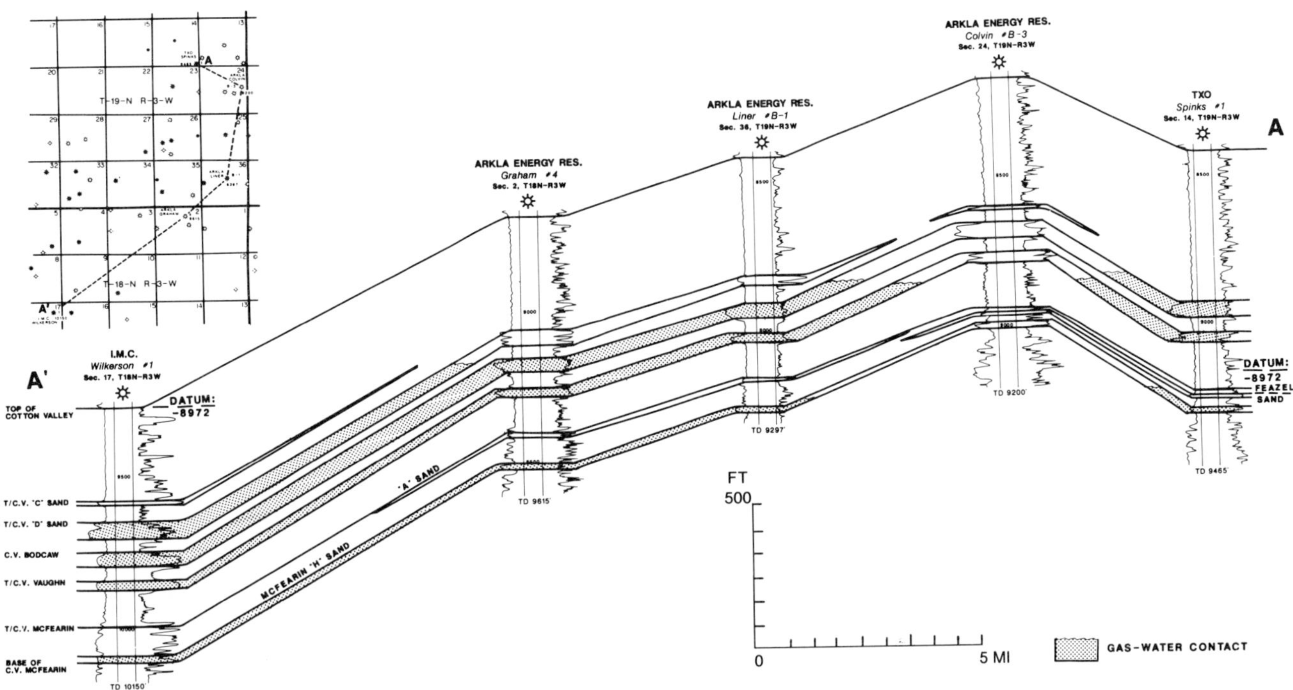

Figure 19. Structural cross section A–A' of the Cotton Valley producing intervals in the Ruston field. Datum, –8972 ft.

(1385 m) subsea. The James is now a gas-storage reservoir.

The Gloyd limestone of the Rodessa Formation has yielded a very minor amount of natural gas from a domal trap.

RESERVOIRS

Ruston field has produced or is presently producing hydrocarbons from approximately 44 sandstone or limestone zones in the Smackover, Cotton Valley, Hosston, Sligo, Pine Island, James, and Rodessa. Productive horizons as of 1989 are in the interval between 4500 and 12,100 ft (1371–3688 m) deep.

Lithologic descriptions of the McFearin-Davis sandstone, Bodcaw sandstone, Cotton Valley "D" sandstone, and James sandstone are based on scanning electron microscope (SEM) and petrographic microscopic analysis of core samples from two wells, the Murphy Oil Co. No. 1 McCrary, Sec. 5, T18N, R2W, and the Arkansas Louisiana Gas Co. No. 5 Giles, Sec. 19, T19N, R2W. Descriptions of other producing horizons are based on core reports or sample logs.

The Smackover Formation had produced 300 MMCFG from the "Gray" sandstone as of 1987. The "Gray" sandstone has porosities in the 7% to 9% range and permeabilities probably less than 1 md. It averages 30 to 45 ft (9–14 m) thick and is drilled in 640 ac (259 ha) units. The lower Smackover shales are rich enough in organics to make them prime candidates as source rocks for Smackover gas.

The Cotton Valley Group has 19 producing zones: Mcfearin-Davis, nine; Feazel, one; McCrary, one; Price, one; Vaughn, two; Bodcaw, one; "D" sandstone, three; and "C" sandstone, one.

The nine sandstone units in the McFearin-Davis produce gas and condensate from 41 wells (Figure 20) and have wide ranges of thicknesses, porosities, and permeabilities. Each unit is separated by shale breaks. The rocks are fine- to medium-grained, light brownish gray to brownish gray quartz wackes with a clay matrix of 6% to 10%. The framework averages 98% quartz and 1% fine-grained rock fragments, with the remainder chert, organically derived calcareous grains, feldspar, zircon, amphiboles, muscovite, and opaques. The primary intergranular porosity has been significantly reduced by a syntaxial quartz overgrowth cement. Other cement includes an intergranular and replacement dolomite averaging 7%, calcite averaging 1%, and a trace of pyrite (Figure 29). There are some oversized pores, but most are at least partially filled with vermicular clay and carbonate (Figure 30).

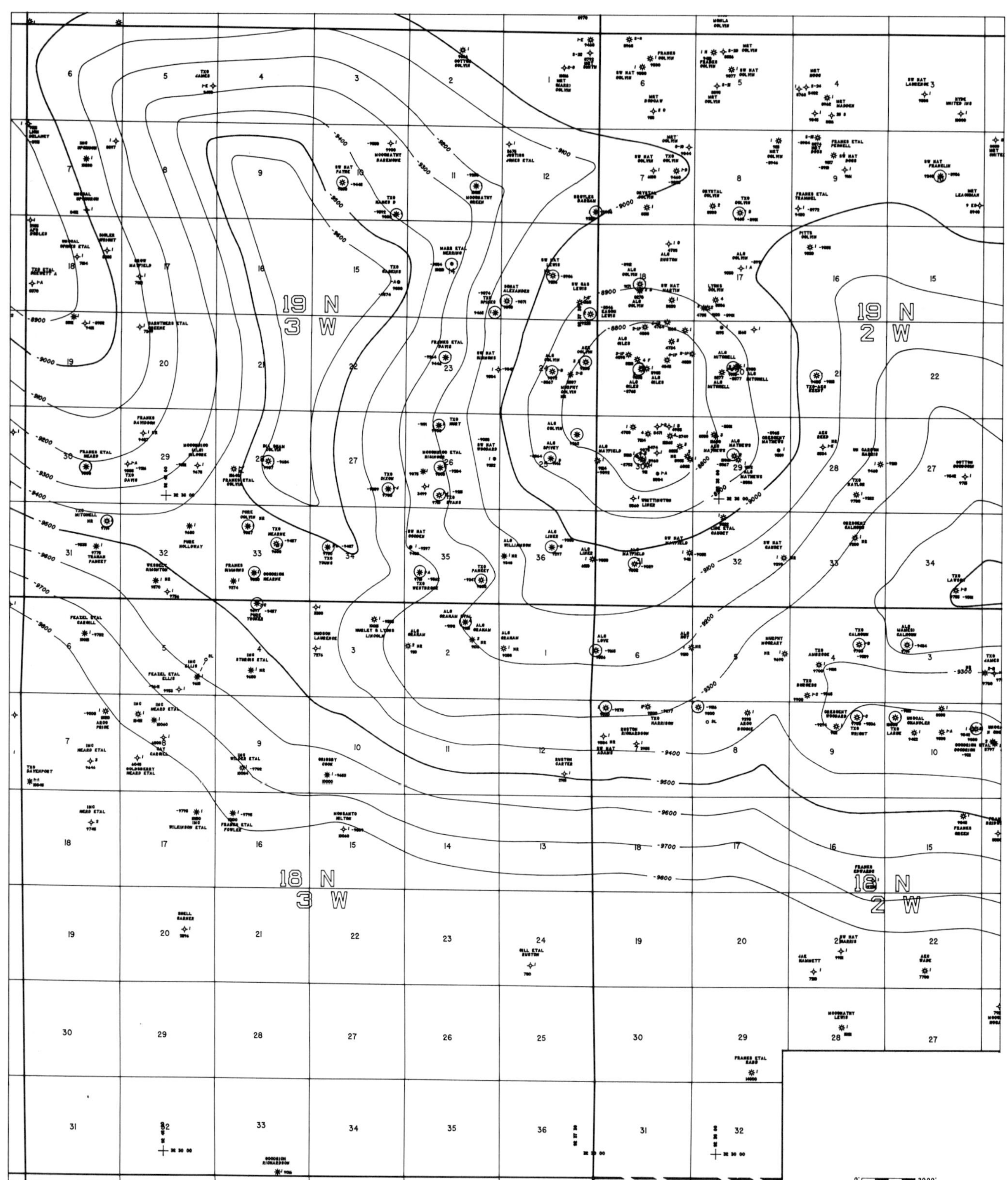

Figure 20. Structure map (1989) on top of the McFearin/ Davis "H" sandstone with McFearin/Davis producing wells circled. Contour interval, 100 ft.

168

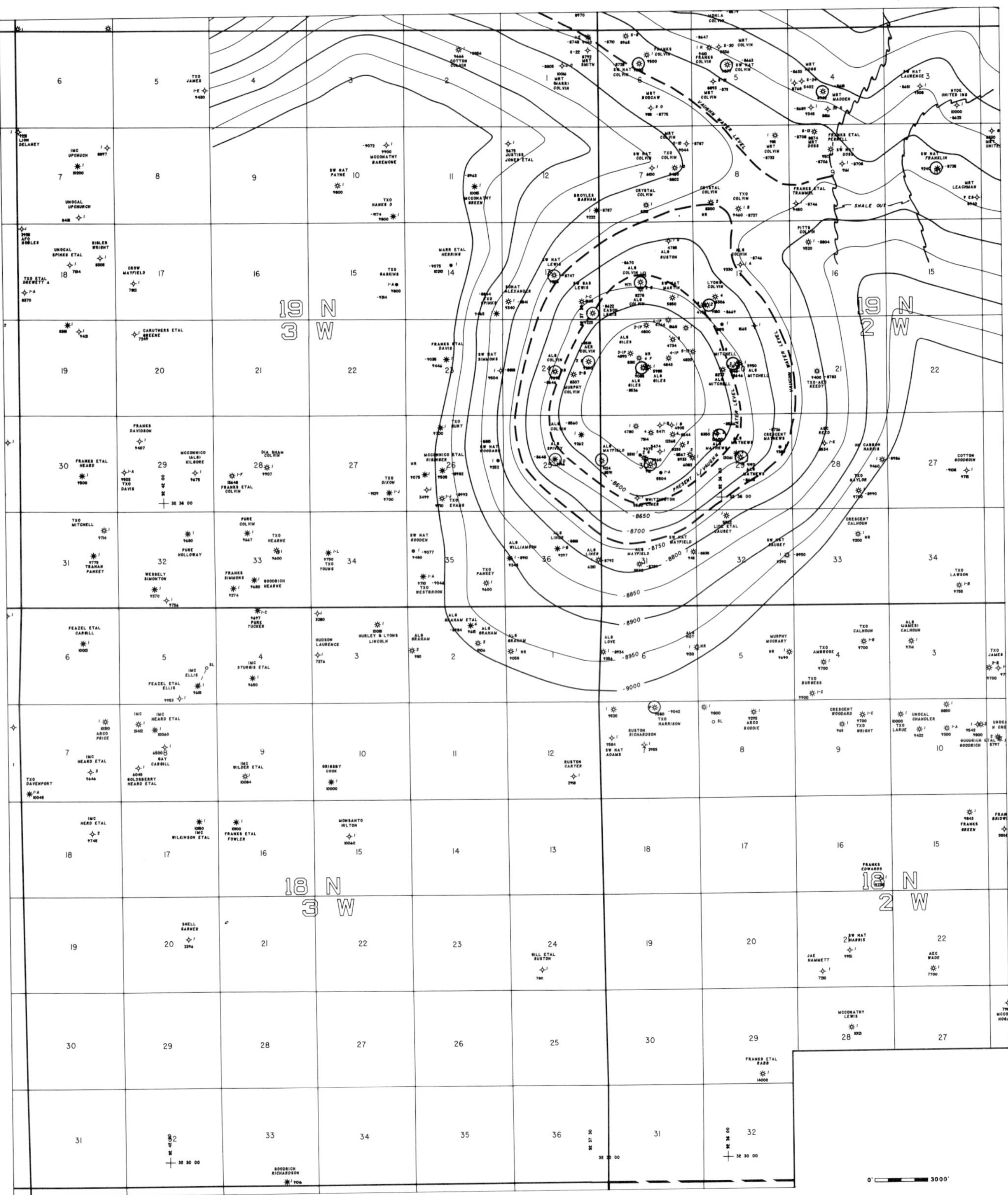

Figure 21. Structure map (1989) on top of the Cotton Valley Vaughn sandstone. Vaughn sandstone producing wells are circled. Because there are two separate Vaughn sandstones, two water levels are shown. See text for explanation. Contour interval, 50 ft.

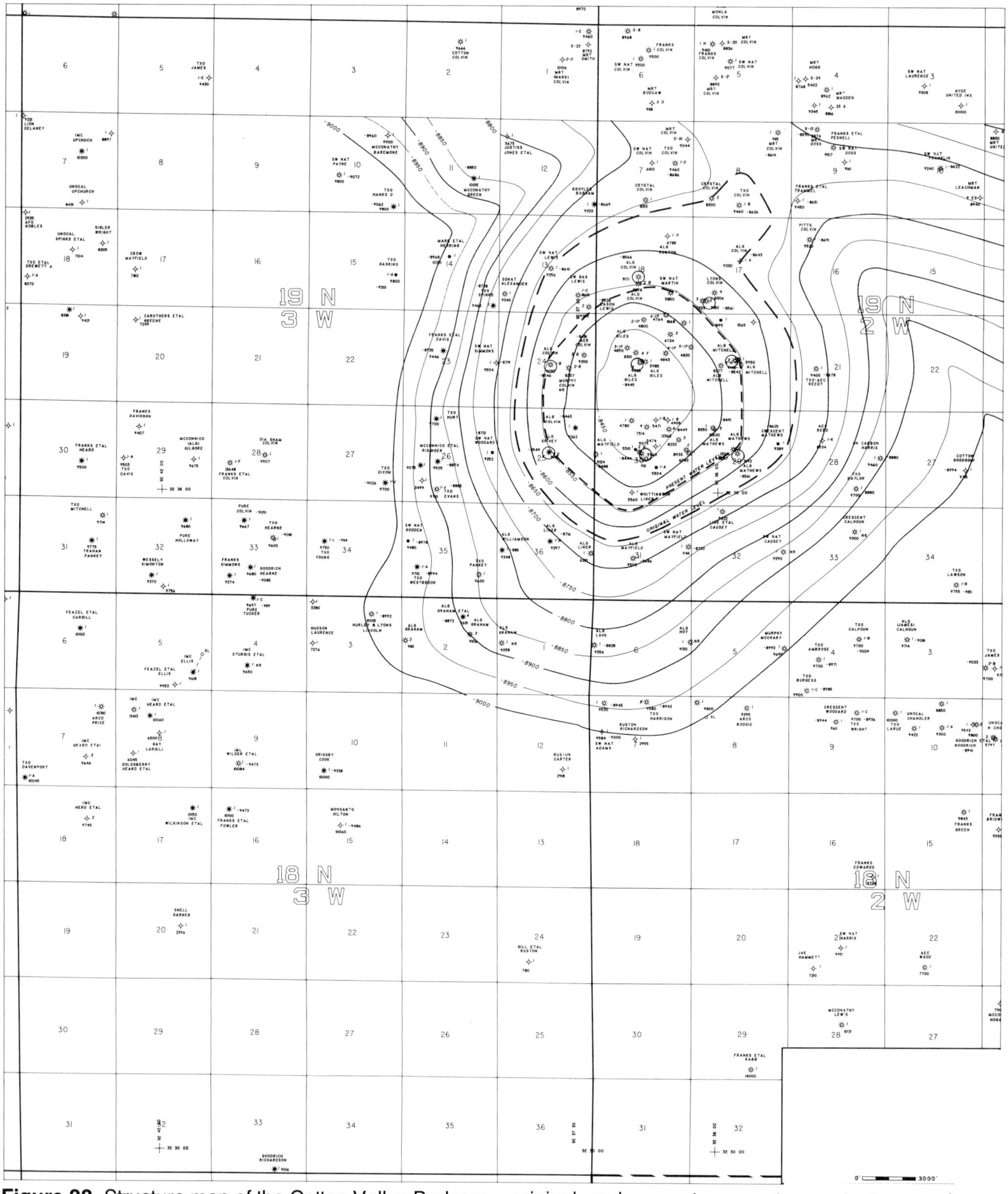

Figure 22. Structure map of the Cotton Valley Bodcaw sandstone with Bodcaw producing wells circled. The original and current gas-water contacts are shown. Contour interval, 50 ft.

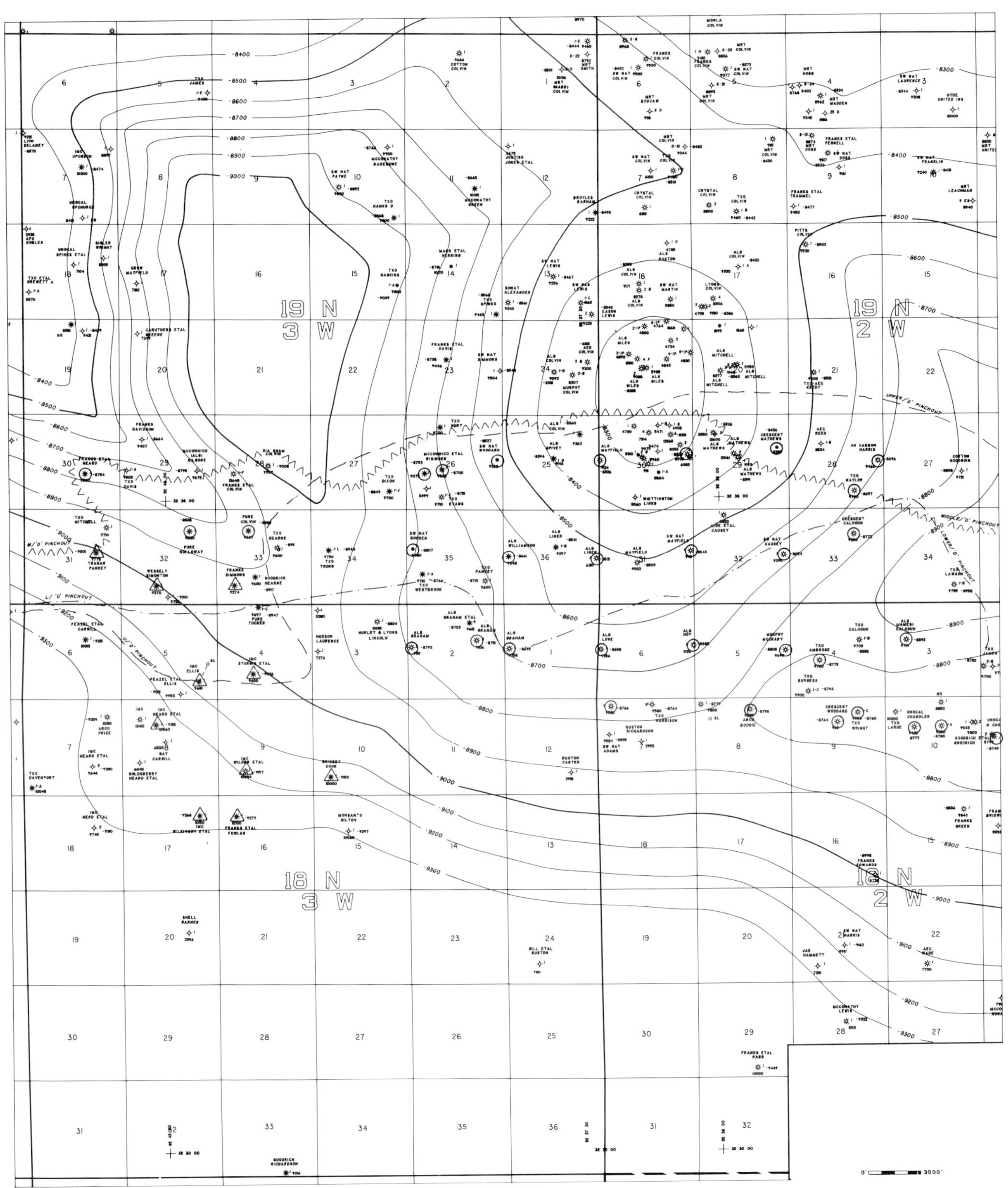

Figure 23. Structure map (1989) on the base of the Cotton Valley "B" limestone showing "C" sandstone producing wells (triangles) and "D" sandstone producing wells (circles). Updip pinch-outs of the upper, middle, and lower "D" sandstones are also shown. Contour interval, 100 ft.

171

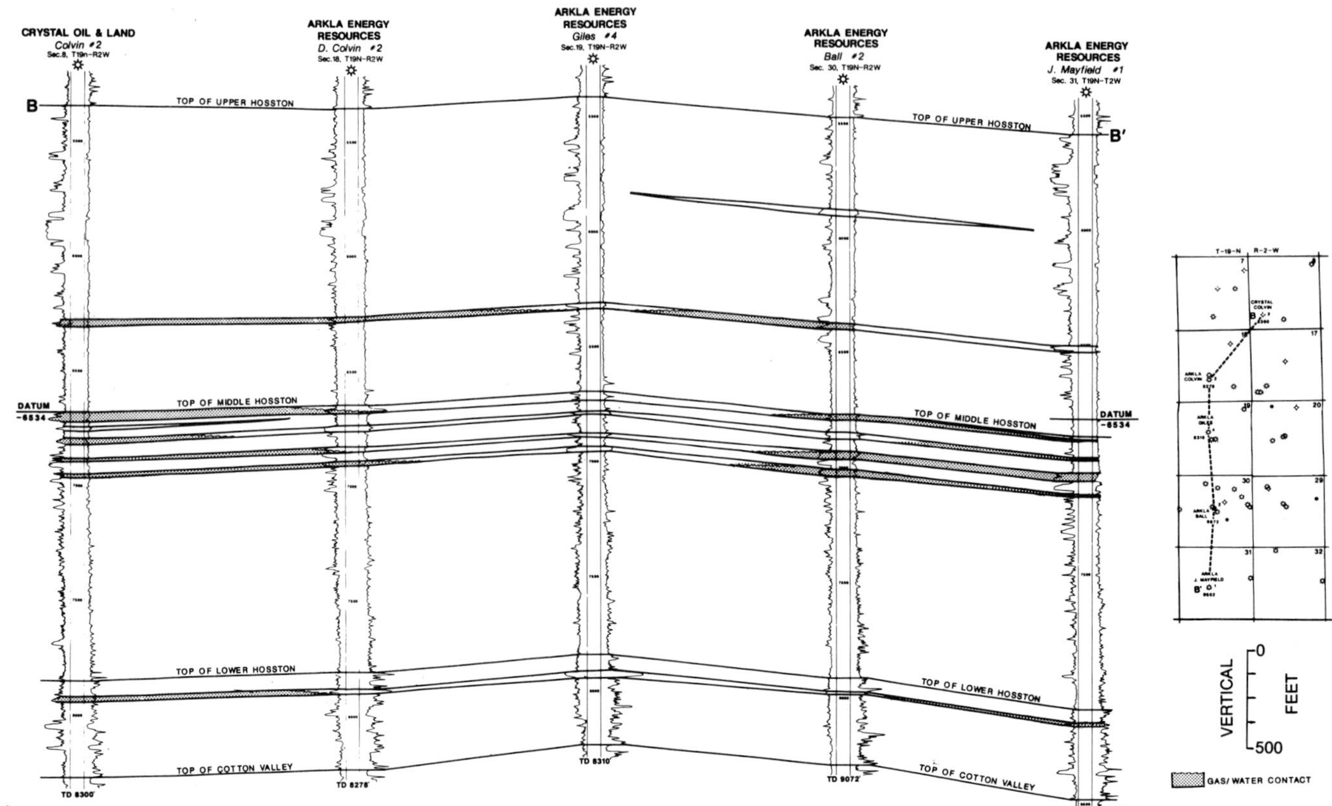

Figure 24. Structural cross section B–B′ showing the upper, middle, and lower Hosston producing intervals in the Ruston field. Datum, –6534 ft.

The McFearin-Davis "H" sandstone has produced gas and condensate from ten wells on the crest of the structure. Total production from 1959 to 1 January 1987 has been approximately 31 BCFG plus 181,000 bbl of condensate. Porosities measured from productive zones range from 9% to 13%. The rock has a fairly uniform thickness on-structure, averaging between 12 and 15 ft (3.65–4.6 m). The original reservoir pressure was 4200 psi (28,959 kPa), and pressure in 1987 was 1400 psi (9653 kPa). Associated condensate has a gravity of 67° to 68° API and a gas to condensate ratio of 46,785:1. The original size of this unit was 5425 ac (2197 ha), with an estimated ultimate recovery of 37.8 BCFG and 260,000 bbl of condensate.

The remaining McFearin-Davis sandstone producing units are similar, but on the average the porosities and permeabilities are lower. The sandstones are usually 6 to 10 ft (2–3 m) thick and require fracture stimulation to produce hydrocarbons. All McFearin-Davis wells are drilled in 640 ac (259 ha) units.

The Feazel and McCrary sandstones are defined separately from the McFearin-Davis by the Louisiana Conservation Commission. The porosities range from 12% to 16%. Permeabilities are unknown. The sandstones average 12 ft (3.65 m) thick, with reservoir pressures about 4400 psi (30,338 kPa). These sandstones have yielded 13.6 BCFG plus 43,000 bbl of condensate from the Feazel and 12 BCFG and more than 200,000 bbl of condensate from the McCrary between 1951 and 1987.

The Vaughn sandstone produces from an average depth of 8875 ft (2705 m) from 13 wells for a total production of 97.5 BCFG plus 86,000 bbl of condensate between 1949 and 1987. It is subdivided into the upper and lower Vaughn, with a combined average thickness of 35 ft (10.7 m). The original reservoir size of the Vaughn was 4935 ac (1999 ha), with an estimated ultimate recovery of 122 BCFG. Average porosities are 12% to 15%. The gas to condensate ratio is approximately 32,857:1 with a gravity of 62° API. The Vaughn is drilled on a 640 ac (259 ha) spacing.

The Bodcaw is a significant Cotton Valley natural gas reservoir with a cumulative production of 81 BCFG and 62,500 bbl of condensate from eight wells

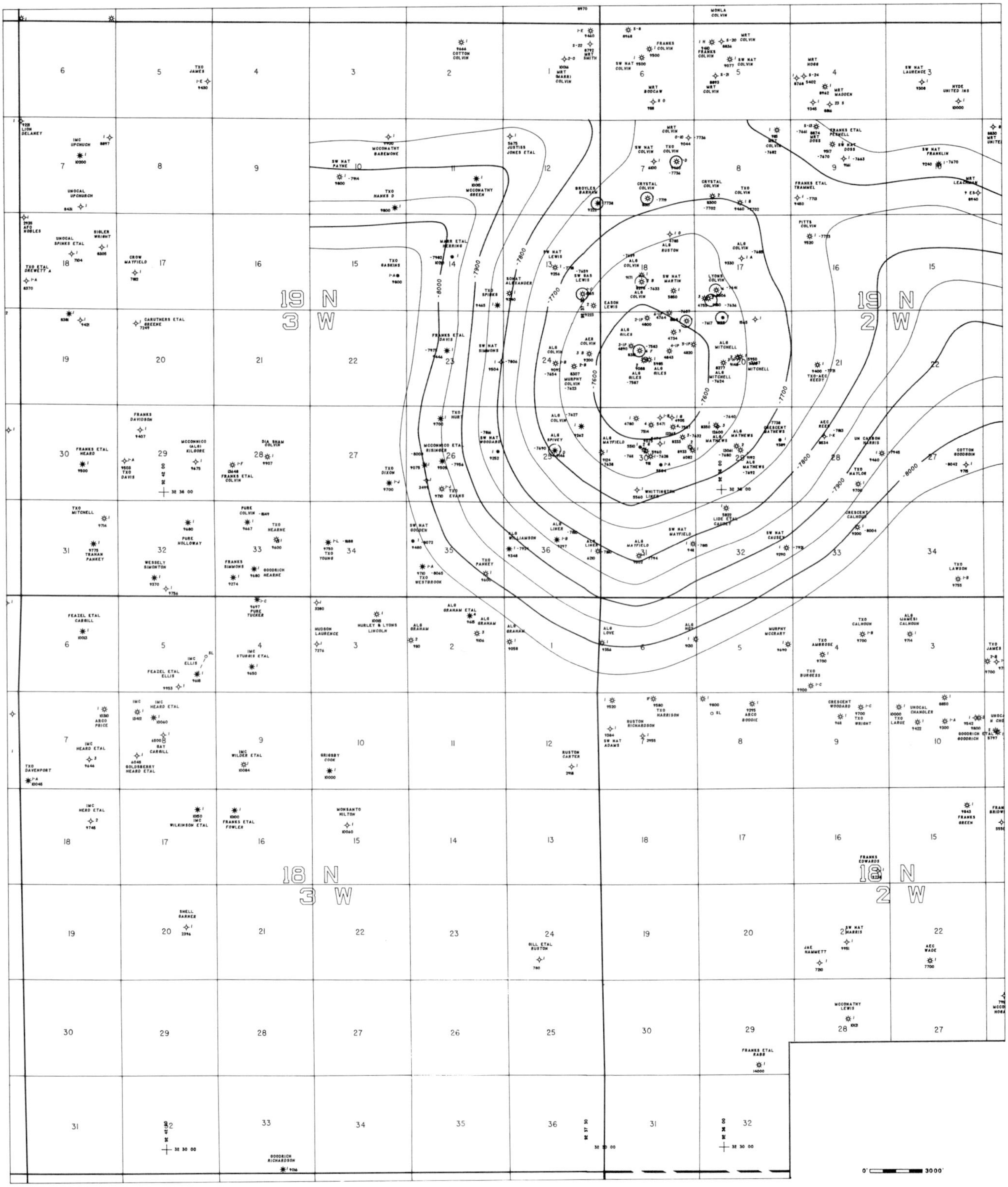

Figure 25. Structure map (1989) on top of the lower Hosston with lower Hosston producing wells circled. Contour interval, 50 ft.

173

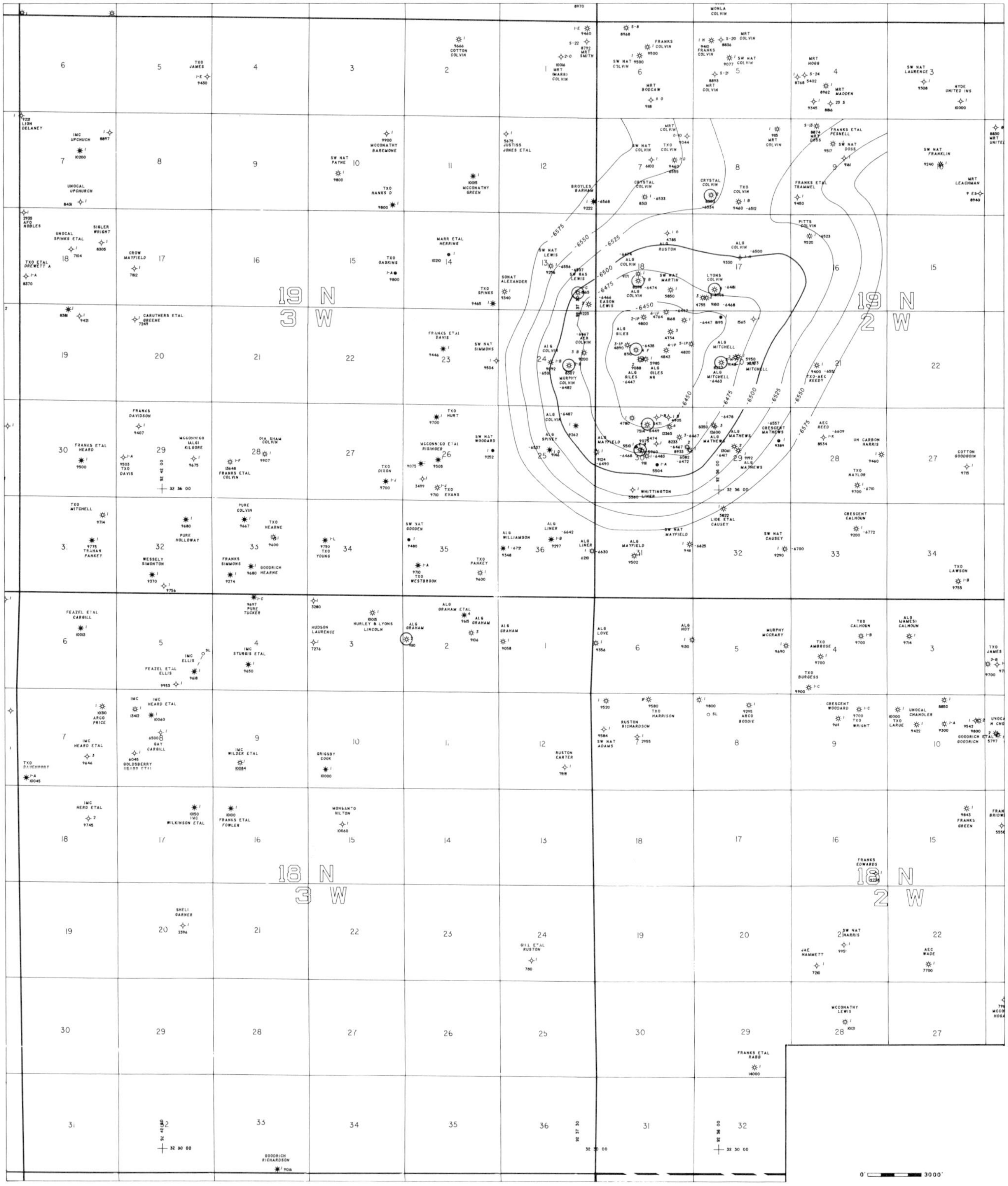

Figure 26. Structure map (1989) on top of the middle Hosston with middle Hosston producing wells circled. Contour interval, 25 ft.

174

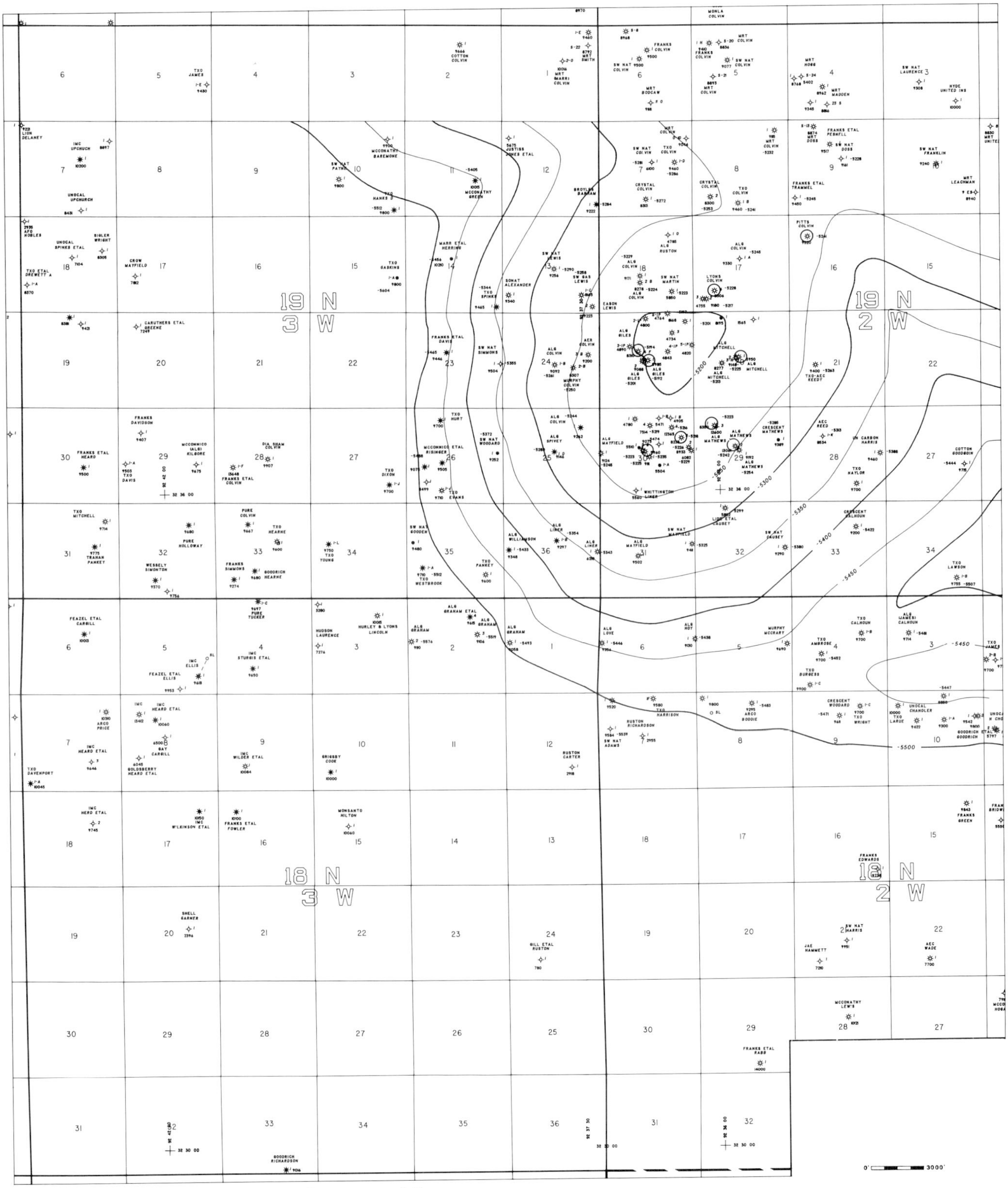

Figure 27. Structure map (1989) on top of the upper Hosston with upper Hosston producing wells circled. Contour interval, 50 ft.

175

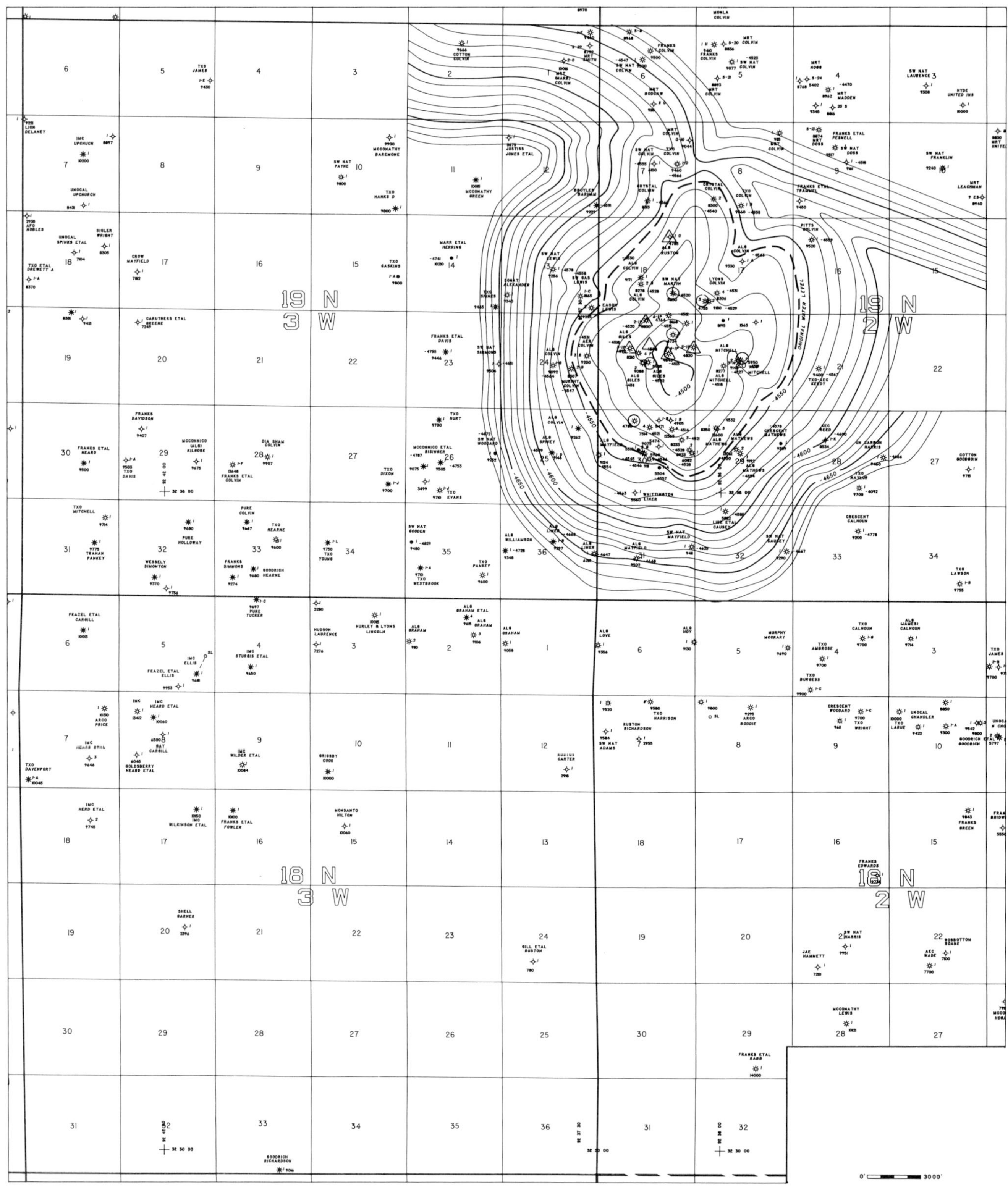

Figure 28. Structure map (1989) on top of the James sandstone showing James producing wells (circles) and James gas-storage wells (triangles). Contour interval, 10 ft.

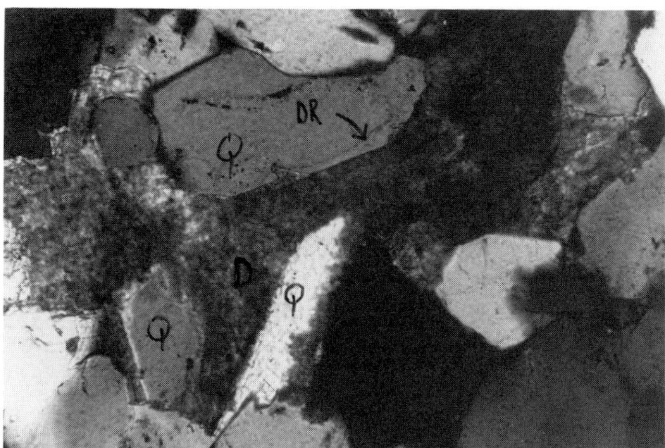

Figure 29. Photomicrograph of Cotton Valley McFearin/Davis sandstone. Intergranular dolomite (D) is replacing quartz (Q). Note dust ring (DR) on quartz grain with syntaxial overgrowth. Crossed nicols. Photomicrograph is 0.61mm × 0.40mm.

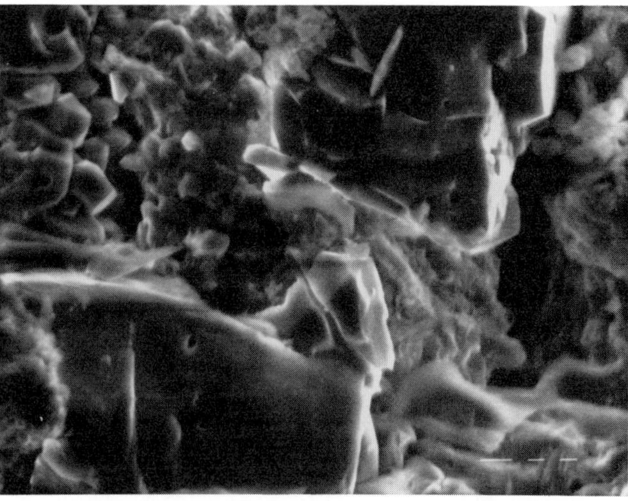

Figure 31. SEM photomicrograph of Bodcaw sandstone showing euhedral quartz overgrowths reducing intergranular porosity. Long bar is 10 microns.

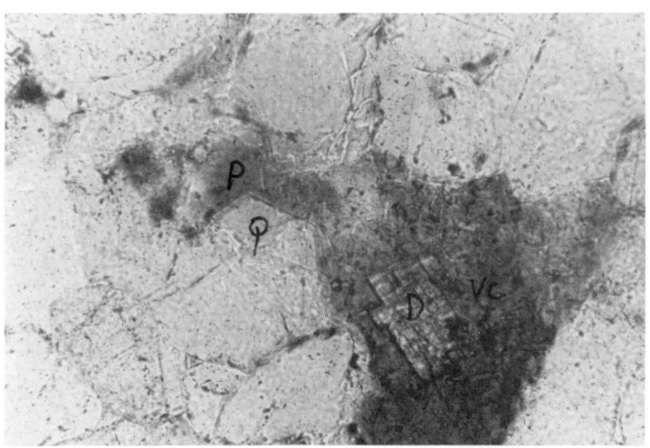

Figure 30. Photomicrograph of Cotton Valley McFearin/Davis sandstone showing oversize pore partially filled with dolomite (D), vermicular clay (VC), secondary quartz (Q), and porosity (P). Plane light. Photomicrograph is 0.61mm × 0.40mm.

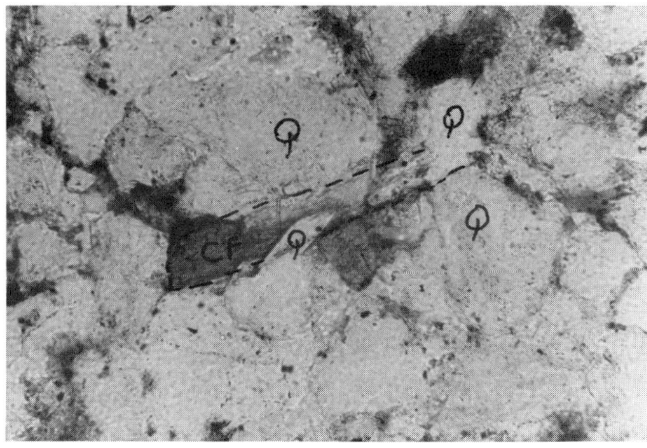

Figure 32. Photomicrograph of Bodcaw sandstone. Dashed outline of calcite fossil fragment (CF) showing quartz (Q) replacement. Plane light. Photomicrograph is 0.61mm × 0.40mm.

between 1949 and 1987. The Bodcaw sandstone averages 40 ft thick where it is productive. The typical sandstone is a fine-grained quartz arenite to quartz wacke with an average clay matrix of 5%. It is a light brownish gray color with occasional thin black streaks. The rock framework averages 93% quartz with a trace of chert, and the remainder consists of 4% fine-grained rock fragments, 3% feldspars (both plagioclase and potash), with traces of fossil fragments, amphiboles, muscovite, tourma-

line, zircon, rutile, glauconite, and opaques. The sands are cemented primarily with syntaxial quartz overgrowths, reducing the intergranular porosity to an average of 15% (Figure 31). Additional cement includes 4% calcite, 1% pyrite, and less than 1% dolomite, siderite, barite, and vermicular clay. Calcite bivalve shells are commonly corroded and sometimes partially replaced by quartz (Figure 32). Secondary porosity is minor with some honeycombed-feldspar grains and occasional clay shrinkage.

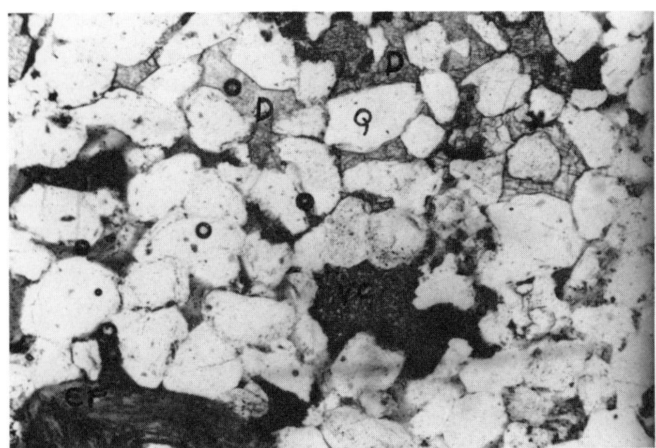

Figure 33. Photomicrograph of Cotton Valley "D" sandstone. Quartz grains (Q) and calcareous bivalve fragment (CF) cemented by quartz overgrowths and dolomite (D) with some interstitial vermicular clay (VC). Plane light. Photomicrograph is 1.56mm × 1.03mm.

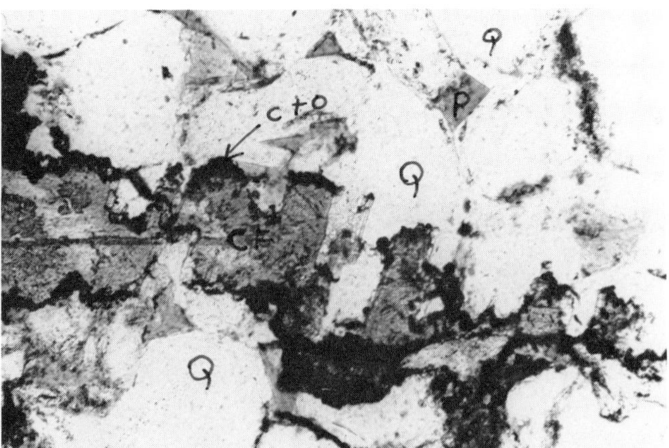

Figure 35. Photomicrograph of Cotton Valley "D" sandstone. Stylolite formed along calcareous fossil fragment (CF) with associated concentrations of clay and organics (?) (C+O). Note euhedral quartz (Q) and intergranular porosity (P). Plane light. Photomicrograph is 0.61mm × 0.40mm.

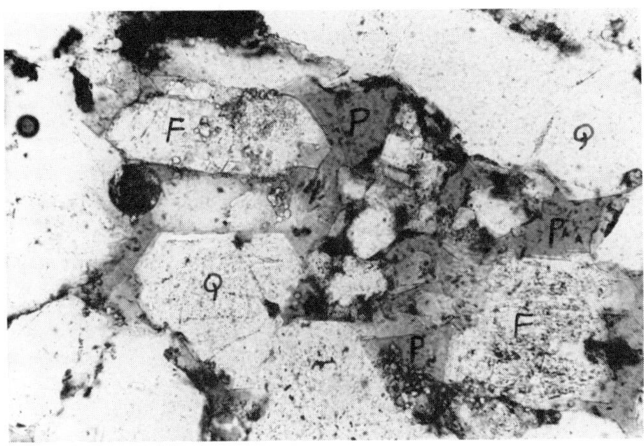

Figure 34. Photomicrograph of Cotton Valley "D" sandstone showing feldspars (F), intergranular porosity (P), and euhedral quartz (Q) produced by secondary overgrowths. Plane light. Photomicrograph is 0.61mm × 0.40mm.

Figure 36. SEM photomicrograph of Cotton Valley "D" sandstone. Dolomite (D) cement and euhedral quartz (Q) overgrowths reduce primary intergranular porosity. Long bar is 100 microns.

The Bodcaw sandstone reservoir is largely depleted with an ultimate recovery of only 82 BCFG expected. It is a pressure-depletion drive reservoir that has had a pressure drop from its original 4100 psi (28,270 kPa) in 1949 to 500 psi (3448 kPa) in 1987. The average porosity is 15%, and the average water saturation is 28%. Condensate is 57° API. The original reservoir covered 3562 ac (1442 ha), with a gas column of almost 200 ft (61 m) above an original water level at 8642 ft (2634 m) subsea.

The Cotton Valley "D" sandstones have produced from 22 wells for an average cumulative production of 12 BCFG per well between 1948 and 1987. The "D" sandstone is actually three distinct sandstone units that shale out updip across the south flank of the structure. Each interval averages 15 ft (4.6 m) thick with the middle and lower sandstones converging downdip into a single interval 20 to 30 ft (6–9 m) thick in the south part of the field. The rocks are fine- to medium-grained, medium gray to brownish gray or olive black quartz wackes and quartz arenites with a clay matrix comprising between 2% and 28% of the rock. The framework is 95% quartz with some chert, 3% feldspars, 2% fine-grained rock fragments, and trace amounts of calcareous fossil fragments, muscovite, amphiboles, tourmaline, zircon, and opaques (Figures 33 and 34). The porosity (Figures 35 and 36) varies from 8% to 20% in the producing wells. Authigenetic dolomite (3%), calcite (1%), and lesser amounts of vermicular clay, pyrite, siderite, and barite are also present (Figure 36). Primary intergranular porosity is the major type followed by secondary porosity shown by oversized pores, some corroded grains, and honeycombed feldspars.

The Cotton Valley "D" sandstones are the most prolific gas producers in the Ruston field, yielding 195 BCFG and more than 600,000 bbl of condensate from their discovery in 1948 to the beginning of 1987. The gas is high in methane, has a moderate amount of ethane, and small amounts of propane, CO_2, and other gases (Table 2). The upper and middle sandstones are the most productive, pinching out about 2 mi (3.2 km) updip from the lower sandstone. Permeabilities average 200 md and porosities average 16%.

The Cotton Valley "C" sandstone has yielded natural gas from ten wells in the southwest part of the field for a cumulative production of 13.5 BCFG and 229,000 bbl of condensate and a gas to condensate ratio of 43,000:1. Production spans the period from 1961 to the end of 1987. The average thickness of the sandstone is 12 ft (3.65 m), and the sandstone has an average of 16% to 17% porosity and 600 md permeability. The original reservoir pressure was 5250 psi (36,198 kPa) at 9550 ft (2911 m), slightly overpressured for this area.

The Hosston Formation has produced from 21 sandstone units in 24 wells in the Ruston field (Figures 25-27). Some of the wells, especially on or very near the crest of the structure, have produced from multiple pay zones in all three intervals. The lower Hosston produces only very near the crest of the structure.

Seven sandstones are productive in the lower Hosston, the majority yielding natural gas and several yielding oil with gravities in the 40° API range. Other than in one upper Hosston sandstone and in the lower Sligo, they provide the only oil production in Ruston field. The best of the lower Hosston sandstones, known as the "B" sandstone, yields natural gas with an estimated ultimate potential of 16.6 BCFG.

The middle Hosston has yielded natural gas from six zones for a cumulative production of 46.1 BCFG between 1955 and 1987. The sands are distributary channel deposits varying in thickness from 5 to 30 ft (1.5-9 m) at an average drilling depth of 6800 ft (2073 m). The sandstones have porosities averaging 18%. Permeabilities are unknown. Bottom hole pressures average 3100 psi (21,375 kPa).

The upper Hosston has yielded a cumulative production of 15 BCFG and 40,000 bbl of condensate from seven sandstones since its discovery in 1943 to the beginning of 1987. The sandstones are 5 to 40 ft (1.52-12.2 m) thick and have an average drilling depth of 5450 ft (1661 m). Porosities are usually in the 15-20% range and permeabilities are in the 10-200 md range. Bottom hole pressures range from 2500 to 2700 psi (17,238-18,617 kPa).

The lower Sligo, or Pettet Limestone Member, has produced from two wells on the crest of the Ruston structure for a total of 146,000 bbl of oil and 1.4 BCFG from 1944 to 1987. The productive zone is a gray, porous, oolitic limestone that contains vertical fractures in cores. The porosity averages 18% and the permeability averages 500 md. The productive limestone is a small oolite bar found at a drilling depth of 5450 ft (1661 m) on the crest of the structure. The base of the oolitic limestone marks the transition into the Hosston sand-shale sequence.

The Causey sandstone of the Pine Island Formation is a fine- to medium-grained sandstone that has produced a small amount of natural gas (no record of the actual amount could be found) from two wells near the crest of the Ruston structure. The porosity ranges from 15% to 18%. Permeability values are not available.

The James sandstone is a light gray to brownish gray, fine- to very fine grained quartz wacke with interstitial clay comprising between 9% and 34% of the rock (Figures 37 and 38). Cumulative production from this horizon totaled 19 BCFG from six wells from 1948 to the time it was converted to a gas-storage unit in 1969. The average drilling depth to the formerly productive zone is 4700 ft (1433 m), and it averages 15 ft (4.6 m) thick. The porosities range from 23% to 25% in the producing wells. The original reservoir pressure was 1984 psi (13,680 kPa), and the water level was originally at 4525 ft (1379 m) subsea. The James sandstone gas storage unit is operated by Arkansas Louisiana Gas Company with input and output wells shown on Figure 28.

The Gloyd limestone of the Rodessa Formation has yielded a minor amount of gas (no record of the amount was found) from the Ruston field from one well. Core data are not available, but Trowbridge sample logs indicate the Gloyd is pseudo-oolitic to oolitic, fossiliferous limestone.

SOURCE ROCKS

According to personnel of Arkla Exploration Company there is no geochemical information available for determining the identity of source rocks in the Ruston field. However, it is likely that hydrocarbons in each producing horizon were generated internally, i.e., Smackover gas probably came from Smackover shales, Cotton Valley gas from Cotton Valley shales, etc. It can be speculated that migration into the various traps started when the rocks became heated enough to generate hydrocarbons, which would be about 120 Ma for the Smackover, 110 Ma for the Cotton Valley, and about 50 Ma for the Hosston and Sligo, based on the burial history of the Ruston field (Figure 17).

No information was available on total organic carbon, kerogen type, maturation data (vitrinite reflectance or paleotemperature data from pollen coloration) from any of the Ruston producing horizons.

Table 2. Gas analysis, Cotton Valley "D" sandstone, Arkansas Louisiana Gas No. 1 V. V. Williamson, Sec. 36, T19N, R3W. (From Wisdom, 1968, Table II.)

Component	Separator Gas		Separator Liquid		Combined Stream		
	Mol. %	Gal/Mcf	Mol. %	Liq. Vol. %	Mol. %	Gal/Mcf	Sp. Gr.
N_2	0.21	—	—	—	0.20	—	0.0019
CO_2	2.25	—	—	—	2.22	—	0.0339
Methane	86.87	—	11.74	4.78	85.71	—	0.4748
Ethane	6.06	—	5.98	3.66	6.06	—	0.0629
Propane	2.49	0.685	5.78	3.82	2.54	0.699	0.0387
Isobutane	0.47	0.154	2.21	1.73	0.50	0.163	0.0100
N-butane	0.71	0.224	5.35	4.05	0.78	0.246	0.0156
Isopentane	0.31	0.113	4.77	4.19	0.38	0.139	0.0095
N-pentane	0.23	0.083	4.20	3.66	0.29	0.105	0.0072
Hexane	0.40	0.174	59.97	74.11	1.32	0.677	0.0551
Total	100.00	1.433	100.00	100.00	100.00	2.029	0.7096

Specific gravity—heptane plus	.7454	
Mol. weight—heptane plus	121	

Test Data Summary

Well-head flowing pressure	1,050 psig	Net condensate rate	34.55 OPO
Separator residue gas	2230 Mcfd	Net condensate rate	15.49 bbl/MMcf
Separator pressure	505 psig	Water rate	6.00 BPO
Separator temperature	94 °F	Water rate	2.69 bbl/MMcf
Flash factor (separator liquid)	82.09 %		

(Gallons Per Mcf)

	14.696 psi Corrected	15.025 psi Corrected
Propane	0.699	0.715
Butanes	0.409	0.418
Pentanes plus	0.921	0.942
Total	2.029	2.075

EXPLORATION CONCEPTS

Ruston field was discovered as a result of surface exploration in a fairway established by prior exploration starting in the Shreveport, Louisiana, area and proceeding eastward. Geologic mapping of a surface structure north of Ruston (but south of the field) stimulated interest in the locality that led to geophysical surveys and ultimately to drilling and discovery.

By the 1940s, geologists recognized the strandline nature of the Cotton Valley sandstones and began to use this information to explore for additional structural and stratigraphic traps. The Hosston sandstones were recognized as having a fluvial-deltaic origin and they, too, became prime targets for exploration, even though many geologists were wary of these sands in the early days of ArkLaTex (adjoining areas of Arkansas, Louisiana, Texas) exploration. It was not uncommon for geologists to refer to the Hosston as "Tragic Peak" in place of its East Texas name, Travis Peak.

Most of the important Cotton Valley and Hosston reserves in North Louisiana had been found by the early 1960s, and a paper by Pate (1963) presented the results of Cotton Valley exploration to that time. Maps and sections from his paper showed the updip limits of the Vaughn, Bodcaw, and "D" sandstones and their relationship to regional structures in North Louisiana.

Hosston and Cotton Valley exploration since the mid-1960s has consisted mostly of: (1) extensions of developed fields; (2) new stratigraphic plays such as the updip, less permeable extensions of the Davis sandstones in North Louisiana; (3) development of "tight" Cotton Valley sandstones in northwest Louisiana and East Texas; and (4) search for additional stratigraphic and structural traps in East Texas, North Louisiana, Mississippi, and Alabama from a variety of formations. Seismic exploration has played only a minor role in this exploration because most of the new reserves being found are in stratigraphic traps associated with known structures. Many of the newer discoveries in both the Hosston and Cotton valley have been in low porosity-low permeability reservoirs, making massive sand fracturing treatments necessary for commercial production.

In recent years more wells in this trend have been drilled into or through the Upper Jurassic Smackover Formation in search of hydrocarbon reserves in oolitic limestones (Smackover "B" limestone) and the "C" and "Gray" sandstones.

Only a small number of pre-Jurassic tests have been drilled in the ArkLaTex region, and only two of these were in North Louisiana. The CVOC No. 1 Hunt-

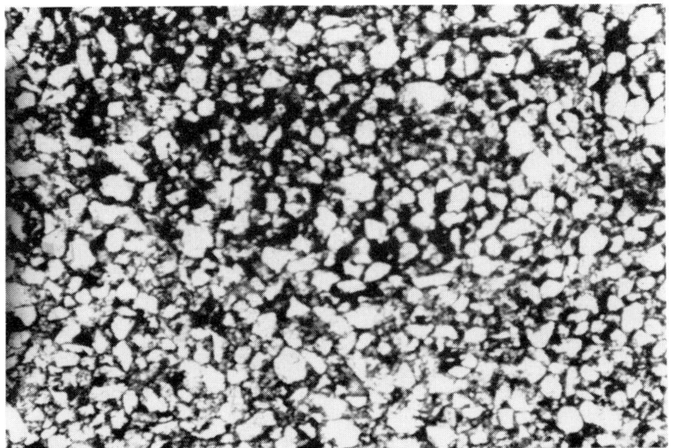

Figure 37. Photomicrograph of James sandstone with abundant clay matrix. Plane light. Photomicrograph is 1.56mm × 1.04mm.

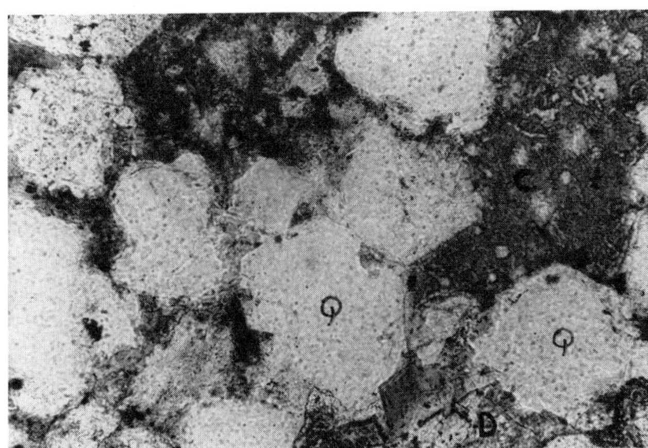

Figure 38. Photomicrograph of James sandstone. Subhedral quartz grains (Q) with calcite (C) and dolomite (D) cement. Some grain boundaries appear corroded. Plane light. Photomicrograph is 0.61mm × 0.40mm.

Hope in Sec. 24, T21N, R10W, Cotton Valley field, Webster Parish, Louisiana, drilled through 3589 ft (1095 m) of Louann Salt and 5086 ft (1551 m) of Eagle Mills (Triassic) red silts and shales before reaching a total depth of 20,395 ft (6220 m). The Union Producing Company No. A-1 Tensas Delta in Sec. 8, T22N, R4E, Morehouse Parish, Louisiana, reached a total depth of 10,475 ft (3195 m) in the Morehouse Formation of probable upper Paleozoic age (Berryhill et al., 1968). The lithology of the Morehouse consists principally of interbedded gray shales, siltstones, and carbonates.

The prospects for additional drilling below the Louann Salt are inhibited by the difficulty in obtaining seismic data below the salt, paucity of lithologic information, and high cost of drilling deep tests.

ACKNOWLEDGMENTS

Arkla Gas Company provided much of the data and field maps contained in this paper, and additional support was provided by these employees or divisions of Arkla: Gary Boland, Robert Mahoney, Greg Madden, and members of the computer mapping department. Arkla Gas Company and Louisiana Tech University paid for reproduction of illustrations, and Louisiana Tech University provided the funding for preparing thin sections and making SEM photographs. Dr. Gary Zumwalt of the Louisiana Tech Geosciences Department made the SEM photographs and helped interpret the results. Mr. Raymond Bruce of Core Laboratories, Inc., in Shreveport, Louisiana, provided the permeability plugs from cores of the Cotton Valley Davis, Bodcaw, and "D" sandstones and the James sandstone that were used in preparing thin sections and making SEM photographs. The burial-history plot was prepared by Renae Wilkinson of Platte River Associates, Denver, Colorado.

REFERENCES CITED

Bally, A. W., and S. Snelson, 1980, Realms of subsidence: Canadian Society of Petroleum Geologists Memoir 6, p. 9–94.

Berryhill, R. A., W. L. Champion, A. A. Meyerhoff, G. C. Sigler, and others, 1968, Stratigraphy and selected gas-field studies of North Louisiana—stratigraphy, *in* Natural gases of North America: American Association of Petroleum Geologists Memoir 9, v. 1, p. 1099–1137.

Breedlove, R. L., J. P. Jones, and A. M. Jackson, 1953, Ruston field, Lincoln Parish, Louisiana, *in* Shreveport Geological Society Reference Report v. 3, n. 2, p. 88–94.

Burgess, W. J., 1976, Geologic evolution of the Mid-Continent and Gulf Coast areas—a plate tectonics view: Gulf Coast Association of Geological Societies Transactions, v. 26, p. 132–143.

Eversull, L. G., 1985, Depositional systems and distribution of Cotton Valley blanket sandstones in northern Louisiana: Gulf Coast Association of Geological Societies Transactions, v. 35, p. 49–57.

Houseknecht, D. W., ed., 1983, Tectonic-sedimentary evolution of the Arkoma basin and guidebook to deltaic facies, Hartshorne Sandstone: SEPM Midcontinent Section, v. 1.

Judice, P. C., and S. J. Mazzullo, 1982, Gray sandstones (Jurassic) in Terryville field, Louisiana—basinal deposition and exploration model (abs.): American Association of Petroleum Geologists Bulletin, v. 66, p. 1432.

Kamb, H. R., 1945, Ruston gas field, Lincoln Parish, Louisiana: American Association of Petroleum Geologists Bulletin, v. 29, n. 2, p. 226–227.

Klemme, H. D., 1971, What giants and their basins have in common: Oil and Gas Journal v. 69, n. 9, 10, 11: pt. I, p. 85–90; pt. II, p. 103–110; pt. III, p. 96–100.

Louisiana Geological Survey, 1980, Parish atlas of Louisiana oil and gas fields: Louisiana Geological Survey, Folio series 4, 64 p.

Mann, C. J., and W. A. Thomas, 1964, Cotton Valley Group (Jurassic) nomenclature, Louisiana and Arkansas: Gulf Coast Association of Geological Societies Transactions, v. 14, p. 143–152.

Murray, G. E., 1961, Geology of the Atlantic and Gulf Coastal Province of North America: New York, Harper and Bros., 692 p.

Pate, B. F., 1963, Significant North Louisiana Cotton Valley stratigraphic traps: Gulf Coast Association of Geological Societies Transactions, v. 13, p. 177–183.

Shreveport Geological Society, l946, Ruston field, Lincoln Parish, Louisiana: Shreveport Geological Society Reference Report, v. 1, p. 183–186.

Shreveport Geological Society, 1987, Report on selected oil and gas fields, Ark-La-Tex and Mississippi: Shreveport Geological Society Reference Volume VII, 153 p.

Thomas, W. A., and C. J. Mann, 1966, Late Jurassic depositional environments, Louisiana and Arkansas: American Association of Petroleum Geologists Bulletin, v. 50, p. 178–182.

Walker, J. R., 1953, Exploration history (prior to the Cotton Valley discovery) of the Ruston field, Lincoln Parish, Louisiana: Geophysics, v. 19, n. 1, p. 124–138; reprinted in Geophysical Case Histories, v. 2, p. 341–355, 1954.

Wheeler, C. T., Jr., 1963, Producing horizons of North Louisiana oil and gas fields, *in* Shreveport Geological Society Reference Report, v. 5, p. 1–8.

Wisdom, C. F., 1968, Ruston field, Lincoln Parish, Louisiana, *in* Natural gases of North America, American Association of Petroleum Geologists Memoir 9, p. 1138–1142.

Appendix 1. Field Description

Field name .... *Ruston field*

Ultimate recoverable reserves .... *614 bcf + 1,732,000 BC + 1,600,000 BO*

Field location:

    Country .... *U.S.A.*
    State .... *Louisiana*
    Basin/Province .... *North Louisiana Salt Province/Gulf Coast*

Field discovery:

    Year first pay discovered .... *Lower Cretaceous Hosston Formation–Jiles sandstone*   *1943*
    Year second pay discovered ...... *Lower Cretaceous Pine Island Formation–Causey sandstone*   *1944*
    Third pay .... *Lower Cretaceous Sligo Formation–Pettet Limestone Member*   *1944*

Discovery well name and general location:

    First pay .... *Jiles No. 1, C Sec. 19, T19N, R2W*
    Second pay .... *J. C. Dowling, No. 1, SE NE Sec. 30, T19N, R2W*
    Third pay .... *J. C. Dowling, No. 1, SE NE Sec. 30, T19N, R2W*

Discovery well operator .... *Arkansas Louisiana Gas Co.*
    Second pay .... *Arkansas Louisiana Gas Co.*
    Third pay .... *Arkansas Louisiana Gas Co.*

IP in barrels per day and/or cubic feet or cubic meters per day:

    First pay .... *45,000,000 CFGPD (est.) (Hosston Sand)*
    Second pay .... *4,000,000 CFGPD (est.) (Causey Sand)*
    Third pay .... *20,000,000 CFGPD (est.) (Sligo Formation)*

All other zones with shows of oil and gas in the field:

Age	Avg. Depth to Top in ft (m)	Formation	Type of Show
Lower Cretaceous	*4500 (1373)*	*Rodessa (Gloyd)*	*Gas*
	4775 (1456)	*James*	*Gas*
	5100 (1556)	*Causey*	*Gas*
	5450 (1662)	*Sligo*	*Oil*
	5360 (1635)	*U. Hosston*	*Oil/gas*
	6650 (2028)	*M. Hosston*	*Gas*
	7810 (2382)	*L. Hosston*	*Oil/gas*
Jurassic	*8650 (2638)*	*Cotton Valley "C" ss*	*Gas*
	8730 (2663)	*Cotton Valley "D" ss*	*Gas/condensate*
	8815 (2689)	*Cotton Valley Bodcaw ss*	*Gas/condensate*
	8920 (2721)	*Cotton Valley Vaughn ss*	*Oil/gas*
	9030 (2754)	*Cotton Valley Price ss*	*Gas/condensate*
	9080 (2769)	*Cotton Valley McCrary ss*	*Gas/condensate*
	9100 (2776)	*Cotton Valley McFearin/Davis ss*	*Gas/condensate*
	9150 (2791)	*Cotton Valley Feazel ss*	*Gas/condensate*
	12,100 (3691)	*Smackover "Gray" ss*	*Gas*

Geologic concept leading to discovery and method or methods used to delineate prospect

A reconnaissance surface geologic survey, started in late 1930 by Mr. A. E. Oldham of the Arkansas-Louisiana Gas Company, located a surface structure centered in the west corner of T19N, R2W. Surface structure stimulated a seismograph survey in 1934 that showed a subsurface structure centered in Sec. 19, T19N, R2W, about 3 mi north-northwest of the surface structure.

Structure:

Province/basin type .... *Bally 1143, Klemme IIC c/IV*

Tectonic history

The Ruston field structure lies in the North Louisiana Salt basin, between the Monroe and Sabine uplifts, encompassing many salt supported structures generated by the Louann Salt, which began moving in the Upper Jurassic Period. The Ruston structure was formed by a salt swell or salt pillow. It is not a piercement structure.

Regional structure

Ruston field lies just south of the Unionville nose and east of the Terryville field anticline. A very slight dip reversal separates Ruston field from the Choudrant field structure to the southeast.

Local structure

Salt-generated dome with approximately 200 ft (60 m) of closure at Cotton Valley (Jurassic) level and 50 ft (15 m) of closure at Hosston-James (Lower Cretaceous) level. The structure plunges to the southwest, forming a significant nose.

Trap:

Trap type(s)

Majority of zones are structurally trapped but several zones (Cotton Valley Davis, Cotton Valley "C," Cotton Valley "D" sandstones and various Hosston sandstones) are combination structural-stratigraphic traps.

Basin stratigraphy (major stratigraphic intervals from surface to deepest penetration in field):

Chronostratigraphy	Formation	Depth to Top in ft (m)
Eocene	*Wilcox*	*1050 (320)*
	Midway	*1500 (458)*
Upper Cretaceous	*Saratoga*	*2420 (738)*
	Annona	*2500 (763)*
	Buckrange	*2800 (854)*
	Tuscaloosa	*3250 (991)*
Lower Cretaceous	*Ferry Lake*	*4000 (1220)*
	Rodessa	*4230 (1290)*
	James	*4650 (1418)*
	Causey	*5150 (1571)*
	Sligo	*5200 (1586)*
	Hosston	*5450 (1662)*
Upper Jurassic	*Upper Cotton Valley**	*8350 (2547)*
	*Lower Cotton Valley***	*9100 (2776)*
	Bossier (equiv.)	*9650 (2943)*
	Smackover	*11,520 (3514)*
	Louann	*13,030 (3974)*
**Upper Cotton Valley includes*	*"B" Limestone*	
	"C" Sandstone	
	"D" Sandstone	
	Bodcaw Sandstone	
	Vaughn Sandstone	
	Davis (McFearin) Sandstone	
***Lower Cotton Valley includes*	*Roseberry Sandstone*	
	Sexton Sandstone	
	Taylor Sandstone	

Reservoir characteristics:

Number of reservoirs .... *16 (approx.)*

Formations .... *Rodessa; James; Pine Island (Causey Sand); Sligo; upper, middle, and lower Hosston; Cotton Valley; Smackover*

Ages .... *Lower Cretaceous, Upper Jurassic*
Depths to tops of reservoirs .... *See zones with oil and gas in the field (above)*
Gross thickness (top to bottom of producing interval) .... *7500 ft (2290 m)*
Rodessa-Smackover gray sand

Net thickness—total thickness of producing zones
Thickening and thinning of sands plus pinch-outs in Hosston and Cotton Valley preclude making a valid estimate of net or total thickness of producing zones

Lithology
Rodessa: limestone, fossiliferous, gray
James: sandstone, light gray, fine grained, argillaceous, well sorted
Causey: sandstone, gray, fine to medium grained, fair sorting
Sligo: limestone, gray, oolitic, vertical fractures, hard
Hosston: sandstone, range from very fine, well sorted, well rounded to coarse, poorly sorted and angular; color may be white, brown, or red
Cotton Valley: sandstones, generally very fine to fine grained, well sorted, white to gray, commonly calcareous
Smackover: "Gray" sandstone is gray to white, fine to medium grained, slightly porous, slightly friable, calcareous

Porosity type
Lack of core data for determination of porosity type; probably mostly intergranular in sandstones; average: Rodessa, 26%; James, 20%, 75 md; Sligo, 16%; Hosston, ±15%, 50 md; Cotton Valley, ±15%, 25 md; Smackover, 7–9%
Average porosity *16%*
Average permeability .... *25–50 md*

Seals:

   Upper
      Formation, fault, or other feature .... *Formation*
      Lithology .... *Shales and nonporous limestones and sandstones*
   Lateral
      Formation, fault, or other feature .... *Formation*
      Lithology .... *Shales and nonporous limestones and sandstones*
Source .... *Unknown*

Appendix 2. Production Data

Field name .... *Ruston field*

Field size:

   Proved acres .... *32,320 ac (13,100 ha)*
   Number of wells all years .... *109 (125 diff. zones)*
   Current number of wells *75*
   Well spacing .... *Gas, 640 ac (259 ha); oil, 80 ac (32 ha)*
   Ultimate recoverable .... *614 BCFG + 1,732,000 bbl condensate + 1,600,000 bbl oil*
   Cumulative production .... *544 BCFG + 1,252,000 bbl condensate + 1,516,701 bbl oil*
   Annual production .... *9.6 BCFG + 29,700 bbl oil (condensate production unknown)*
   Present decline rate *13.5%*
      Initial decline rate .... *Cotton Valley, 19%; Hosston, 22%*
      Overall decline rate .... *Cotton Valley, 11%; Hosston, 16%*
   Annual water production .... *NA*
   In place, total reserves .... *767.5 BCFG + 3,900,000 bbl condensate + 5,300,000 bbl oil*
   In place, per acre-foot .... *Cotton Valley, 1014 MCFG/ac-ft; Hosston, 945 MCFG/ac-ft*

Primary recovery .... *614 BCFG + 1,732,000 bbl condensate + 1,600,000 bbl oil*
Secondary recovery .... *None*
Enhanced recovery .... *None*
Cumulative water production .... *NA*

Drilling and casing practices:

Amount of surface casing set .... *Cotton Valley, 2200 ft 9⅝-in. for 9500 ft test;*
Hosston, 1100 ft 9⅝-in. for 7000 ft test
Casing program .... *9⅝-in. surface; 5½-in. casing; 2⅞-in. tubing*
Drilling mud .... *Dispersed, water based*
Bit program .... *8¾-in. insert button bits*
High pressure zones .... *None*

Completion practices:

Interval(s) perforated .... *4 shots per foot*
Well treatment .... *Acidation if necessary; frac the tighter sands if necessary;*
CO_2 frac on Hosston and most Cotton Valley completions

Formation evaluation:

Logging suites .... *Dual induction-sonic; density-neutron-gamma; microlog, sidewall cores if needed*
Testing practices .... *DST where practical; formation test where pressure depletion is questionable*
Mud logging techniques .... *Catch 2 ft samples averaged over 10 ft; circulate important breaks*
in excess of 6 ft; watch chlorides closely in zones of importance

Oil characteristics*:

**Oil is currently being produced from only three wells in the Ruston field; most of the production is gas with variable amounts of condensate.*

Type .... *Paraffinic*
API gravity .... *42°*
Base .... *Paraffin (paraffin accumulation on rods is a problem)*
Initial GOR .... *ND*
Sulfur, wt% .... *ND*
Viscosity, SUS .... *ND*
Pour point .... *ND*
Gas-oil distillate .... *ND*

Field characteristics:

Average elevation .... *200 ft (60 m)*
Initial pressure .... *Cotton Valley, 4400 psi (27,580 kPa); Hosston, 3000 psi (20,685 kPa)*
Present pressure .... *Highly variable*
Pressure gradient .... *0.47 psi/ft (0.108 kg/cm²/m)*
Temperature .... *Smackover, 300°F (149°C); Cotton Valley, 220°F (104°C); Hosston, 160°F (71°C)*
Geothermal gradient .... *0.0178°F/ft (0.032°C/m)*
Drive .... *Pressure depletion, some partial water drive*
Oil column thickness .... *Variable with different reservoirs*
Oil-water contact .... *Different with each reservoir*
Connate water .... *Highly variable*
Water salinity, TDS .... *Approximately 200,000 ppm*
Resistivity of water .... *Cotton Valley, 0.018; Hosston, 0.021 (ohm-m at BHT)*
Bulk volume water (%) .... *NA*

Transportation method and market for oil and gas:

Oil: Enron Oil Trading and Transportation Company, Scurlock Oil Company
Gas: Arkansas Louisiana Gas Company, Lear Gas (formerly Producers Gas Company), Delhi Pipeline Company, Kerr-McGee Corporation, Louisiana Gas Purchasing Corporation, and Mississippi River Transmission Company

Raudhatain Field—Kuwait
Arabian Basin

PHILIP BRENNAN
Consulting Geologist
Longboat Key, Florida

FIELD CLASSIFICATION

BASIN: Arabian
BASIN TYPE: Foredeep
RESERVOIR ROCK TYPE: Sandstone and
  Limestone

RESERVOIR AGE: Cretaceous
PETROLEUM TYPE: Oil
TRAP TYPE: Anticline

RESERVOIR ENVIRONMENT OF DEPOSITION: Deltaic/Littoral to Shallow Marine
TRAP DESCRIPTION: Ovoid Dome

LOCATION

The Raudhatain field lies within the Arabian basin and in the State of Kuwait, an independent sheikdom situated in the northeastern part of the Arabian peninsula at the western side of the head of the Persian (Arabian) Gulf. Kuwait is bounded on the east by the waters of the Gulf; on the south by the Kingdom of Sa'udi Arabia; and on the west and north by the Republic of Iraq (Figure 1).

The small State of Kuwait (surface area 6880 mi²; 17,800 km²) contains a number of important oil fields, including the Greater Burgan, Minagish, Sabriya, and Raudhatain fields. Raudhatain field lies in the northern part of Kuwait at lat. 29°52′ N, long. 47°45′ E, 14 mi (22 km) south of the Kuwait/Iraq international boundary (Figure 2). The field is located in typical desert terrain at elevations ranging from 130 to 215 ft (40–65 m) and encompasses a surface area of 30 mi² (78 km²). The Raudhatain field produces from clastic and carbonate reservoirs of Early Cretaceous age. Original recoverable reserves are estimated at 8.8 billion bbl of oil (11 billion bbl of oil equivalent, including reservoir gas converted to STB) (Carmalt and St. John, 1986). Raudhatain field was ranked as 18th among the largest known oil fields of the world by Halbouty et al. (1970) and as 27th by Carmalt and St. John (1986).

HISTORY

Pre-Discovery

In 1934, the Ruler of Kuwait granted a 75-year petroleum concession to the Kuwait Oil Company, jointly owned by Gulf Oil Corporation and Anglo-Iranian Oil Company. Oil was discovered on the concession during 1938 at Burgan in southeastern Kuwait, 65 mi (105 km) south of the site of the future Raudhatain discovery (Figure 2). The Burgan discovery well tested 31.8° API crude oil from a sandstone reservoir of early Late Cretaceous age, and subsequent drilling confirmed the presence of a giant accumulation of oil in clastic reservoirs ranging in age from Albian to early Cenomanian. These were the "Burgan sands," later named the Burgan and Wara formations. Development of the Burgan field was delayed by World War II, and exports of crude oil did not commence until 1946 (Owen and Nasr, 1958).

During 1948, Basra Petroleum Company (an affiliate of the Anglo-American-Dutch-French Iraq Petroleum Company) drilled a discovery well on a geophysical prospect at Zubair in the southern desert of Iraq, 25 mi (40 km) north of the Raudhatain area (Figure 2). The Zubair-1 tested 26° API crude oil from a Cretaceous carbonate at 7400 ft (2255 m) and 35° API crude oil from a sandstone-shale section at total depth of 11,171 ft (3405 m). The deeper producing section at Zubair proved to be of Early Cretaceous age (Owen and Nasr, 1958; Jamil, 1975) and was given the name Zubair Formation. Appraisal drilling at Zubair quickly confirmed the presence there of a giant oil field.

The presence of commercial oil in upper Lower to lower Upper Cretaceous strata at Burgan, south of Kuwait Bay, and in Lower Cretaceous strata in southern Iraq focused attention upon the undrilled area of northern Kuwait between the northern shore of Kuwait Bay and the Kuwait-Iraq border, a distance of 40 mi (65 km) (Figure 2). This is a desert area in which Recent aeolian sands and gravels overlie slightly older sands, gravels, and clays of the Kuwait Group of Miocene to Pleistocene age. The land surface

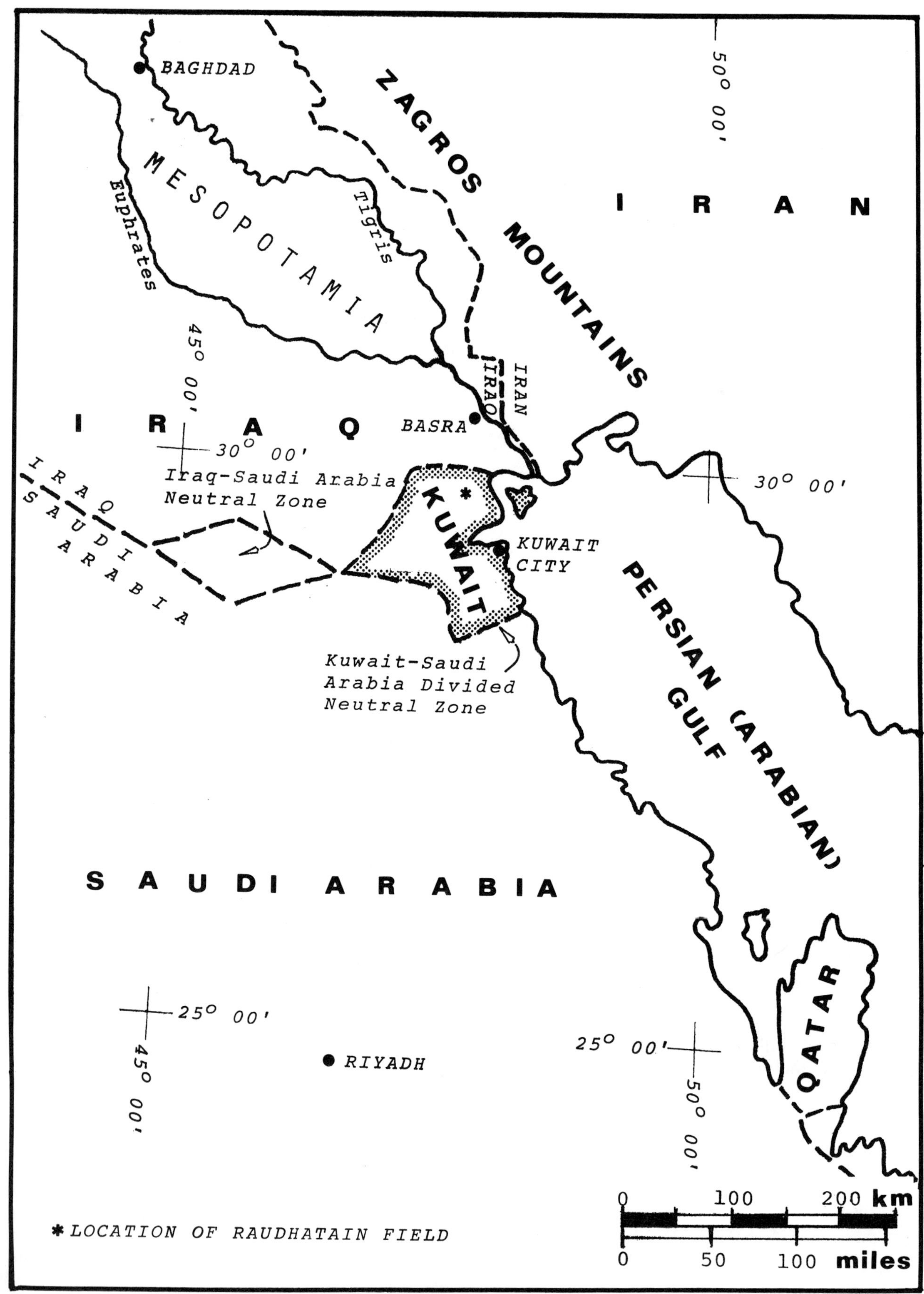

Figure 1. Regional location map of Kuwait. This small Persian (Arabian) Gulf state contains several giant oil fields.

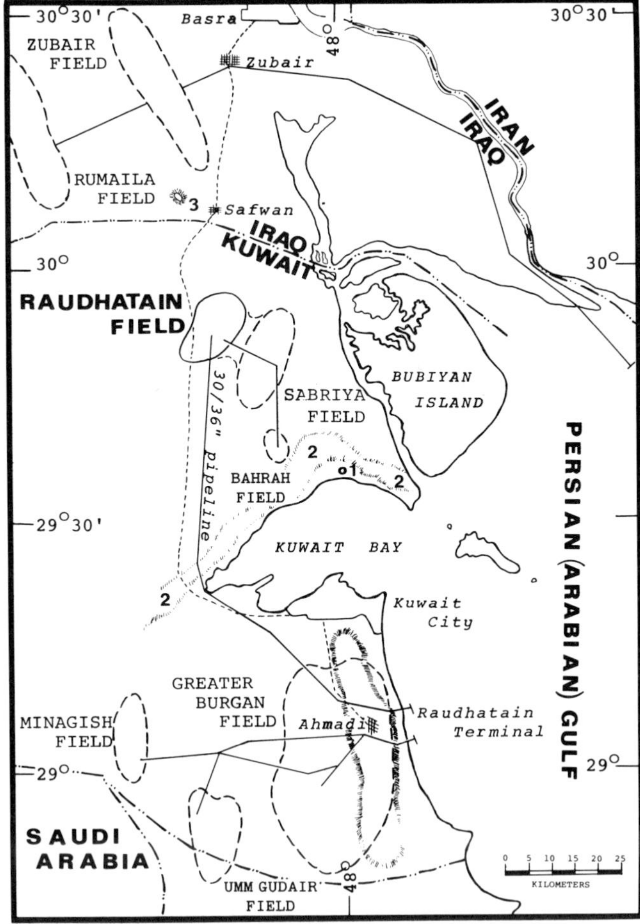

Figure 2. Raudhatain field lies in desert terrain in northern Kuwait, a few miles south of the Kuwait/Iraq frontier. The field is linked by pipeline to an export terminal on the Persian (Arabian) Gulf shore of southeastern Kuwait. 1, site of Bahrah-1 dry hole and seepages; 2, Jal az-Zor scarp; 3, Jabal Sana'am salt plug.

slopes gently away toward the north and northeast from a prominent topographic scarp, the Jal az-Zor, developed along the northern edge of Kuwait Bay (Figure 2). This scarp, which faces south and rises to a maximum elevation of 475 ft (145 m), possibly represents faulting at depth (Owen and Nasr, 1958; Holzer, 1968).

No obvious surface structures were present in the area north of the Jal az-Zor scarp, so that geophysical exploration would clearly be required. A limited gravityometer/surface magnetometer survey had been carried out prior to World War II but had revealed no anomalies of interest. During 1949, a reconnaissance grid of widely spaced reflection seismic traverses was shot north of the scarp in an attempt to pick up a postulated northerly extension of the "Burgan trend." Indications of subsurface structure approximately midway between the northern shore of Kuwait Bay and the Iraq frontier led to a more detailed reflection seismic survey during the cool-weather season of 1951–1952. This resulted in the outlining of a large subsurface high, showing in excess of 500 ft (150 m) of closure, which became the Raudhatain prospect (Figure 3). Velocity control in the area was rudimentary at this time, but the seismic mapping horizon was believed to approximate the top of the upper Lower Cretaceous Mauddud Formation (Figure 4), already established as an important structural mapping horizon within the producing section at Burgan field, 65 mi (105 km) to the south.

Regional dip toward the north was already clearly established. The top of the Mauddud Formation occurred at 3450 ft (1050 m) subsea at Burgan in southeastern Kuwait; at 6800 ft (2070 m) subsea at the Bahrah dry hole on the northern shore of Kuwait Bay (Figure 2); and at 8400 ft (2560 m) subsea at Zubair in southern Iraq. At Raudhatain, the seismic prediction for the top of the Mauddud Formation was approximately 6400 ft (1950 m) subsea. Although the seismic depth conversion was based on limited velocity control, the indicated reversal in regional dip to the north of the scarp suggested the presence of a large subsurface structure in the Raudhatain area (Figure 3).

Discovery

The development of the Raudhatain subsurface prospect in northern Kuwait confirmed the tentative geologic concept of a regional structural trend or axis extending from south to north through Kuwait into southern Iraq. The closed subsurface high outlined by seismic in the Raudhatain segment of this trend (Figure 3) offered a logical target for early wildcat drilling.

The Raudhatain-1 wildcat well (Figure 3) was spudded during September 1954, using a National 125 rig with conventional derrick. This drilling unit was "skidded" in the upright position, on Athey tracks, from the Burgan field to the Raudhatain location, a rolling distance of 99 mi (160 km) (Milton and Davies, 1965). The initial well at Raudhatain was designed to evaluate the Cretaceous section to at least the level of the "third pay" of the Lower Cretaceous Zubair Formation, as defined in the Zubair field to the north. This horizon was anticipated to occur at drilling depths of 9500 to 10,500 ft (2900–3200 m) at the Raudhatain site.

The sandstones of the Wara Formation, immediately overlying the Mauddud Formation and oil-productive at the Burgan field, were not present at Raudhatain. The Raudhatain-1 wildcat well encountered the top of the Mauddud Formation at a drilling depth of 7374 ft (2248 m) (7204 ft [2195 m] subsea) more than 800 ft (244 m) low to the seismic prediction. In spite of this lower-than-predicted structural elevation, a drill-stem test of the uppermost 47 ft (17 m) of the Mauddud flowed crude oil at a rate "in excess of 1000 b/d" (Milton and Davies, 1965), suggesting the presence of a commercial oil accum-

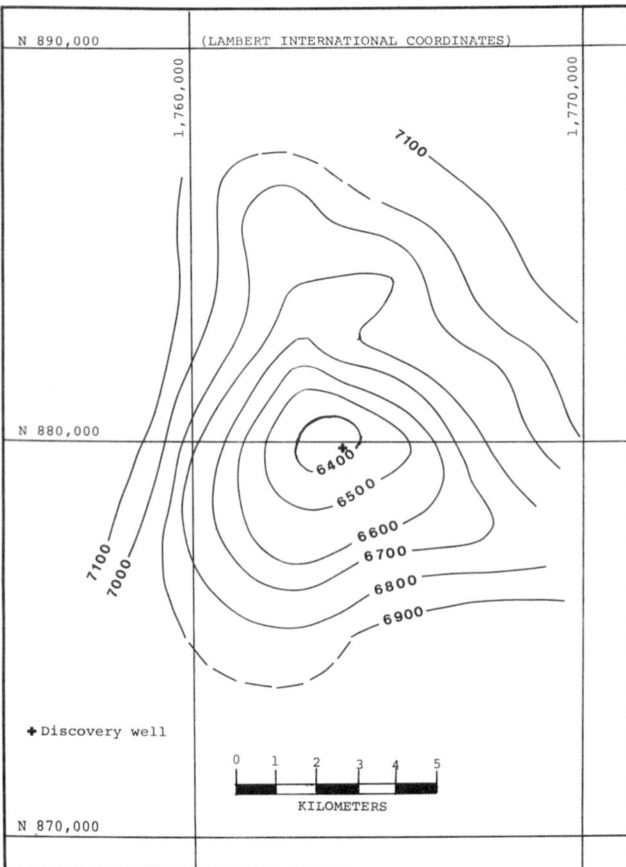

Figure 3. The original seismic prospect on which the Raudhatain discovery was drilled was believed to approximate a limestone horizon at upper Lower Cretaceous level. Contours shown are in feet below sea level; the location of the Raudhatain discovery well is marked by a cross. (After Milton and Davies, 1965.) Contour interval is 100 ft.

ulation. As had been anticipated from regional information, the carbonates of the Mauddud Formation had thickened toward the north, from 25–30 ft (7–9 m) at Burgan to 160 ft (50 m) at Bahrah to 330 ft (100 m) at Raudhatain. The upper three-quarters of this Mauddud section proved to be oil-bearing at Raudhatain-1.

Following the initial successful drill-stem test, the Raudhatain discovery well was cored continuously through the remainder of the Mauddud Formation and through the underlying clastics of the Burgan Formation, which also proved oil-bearing. The carbonates of the Shuaiba Formation, immediately underlying the Burgan Formation and comprising the uppermost unit of the Lower Cretaceous Thamama Group, were water-bearing. But the Shuaiba was in turn underlain by 1380 ft (420 m) of sandstone, siltstone, and shale of the Zubair Formation, and this proved oil-productive throughout its thickness. All told, the Raudhatain-1 well drilled and cored a gross 2370 ft (720 m) of oil-bearing section (Milton and

Davies, 1965) and was the first well in the area to establish commercial oil production in both the Zubair Formation, which produced at the Zubair field in southern Iraq to the north, and the Burgan Formation, which produced at the Greater Burgan field in southeastern Kuwait to the south (Figures 2 and 4).

The Raudhatain-1 discovery well was tested extensively through casing over a period of several months and was officially completed in November 1955 as a "potential outlet from the Zubair Formation" (Milton and Davies, 1965). The well was then shut in, pending the construction of production facilities at the field site and of a pipeline outlet to an export terminal. The Arabian Gulf shore directly east of Raudhatain was not suitable for terminal construction because of tidal inlets and the extensive mud flats of Bubiyan Island. The Raudhatain pipeline would therefore run south and southeast, skirting the head of the shallow Kuwait Bay, to the Gulf littoral south of Kuwait City, where the Kuwait Oil Company terminal of Mena al-Ahmadi was located (Figure 2). This terminal was already handling large volumes of crude oil from the Greater Burgan field.

Post-Discovery

The construction of new terminal facilities at Mena al-Ahmadi North, the laying of the 67 mi (108 km) 30 in. diameter pipeline from the field to the terminal, and the construction and commissioning of a 200,000 BOPD-capacity gathering center at Raudhatain field occupied the years 1956–1959. During the same period, an extensive program of exploration/development drilling was carried out in the field area, guided by the results of additional reflection seismic programs shot during 1954–1955. Step-out wells were drilled to the southeast, north, southwest, and northeast of the discovery to extend and define the limits of the productive area. Development wells were drilled within the proven area on a primary spacing pattern of 600 ac drainage units to develop productive capacity against the start-up of field-wide production. No crude oil for export was drawn from the field during this period, although limited amounts were produced for test and other purposes.

Sustained testing of the Raudhatain discovery well had confirmed the presence of a number of discrete oil reservoirs separated by varying thicknesses of impermeable section. Major oil accumulations were present in the upper part of the Mauddud Formation carbonates, in sandstones developed within the upper and lower sections of the Burgan Formation, and in sandstones developed within the upper and middle sections of the Zubair Formation (Figure 4). Oil accumulations were also present in more limited porosity developments within the upper and lower shale sections of the Zubair Formation (Milton and Davies, 1965; Al-Rawi, 1981). During subsequent drilling, minor amounts of crude oil were tested from thin porosity developments within the pre-Zubair

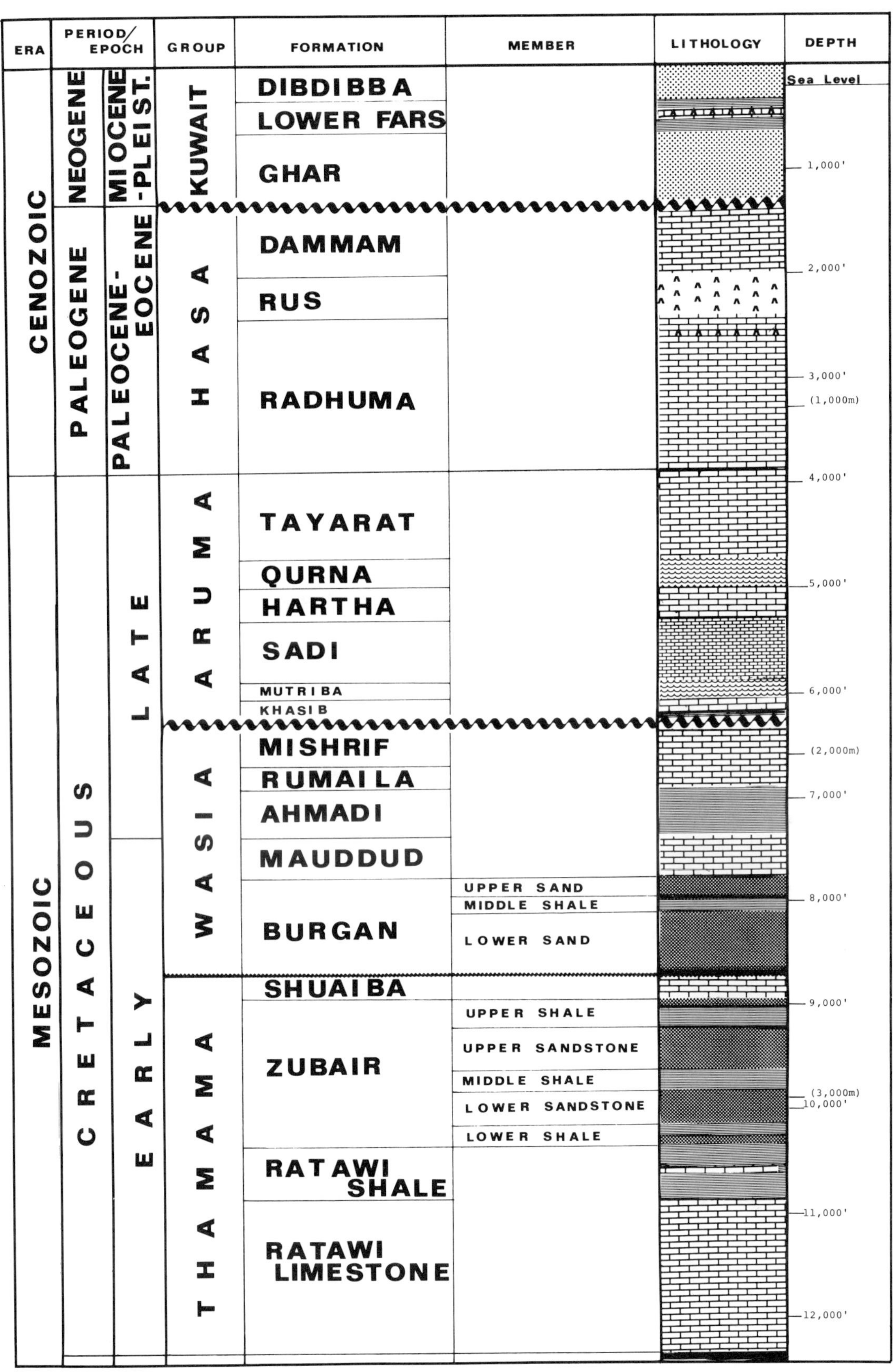

Figure 4. Generalized stratigraphic column for Raudhatain field area. The field produces from the carbonates of the Mauddud and the clastics of the Burgan and Zubair formations.

Ratawi Formation (Figure 4). The crude oil in the main reservoirs was originally undersaturated and showed a steady increase in API gravity (and a more general decrease in sulfur content) from reservoir to reservoir with depth.

The upper reservoirs of the field in the Mauddud and Burgan formations were separated from the lower reservoirs of the Zubair Formation by the porous to cavernous but nonproductive dolomitic limestones of the Shuaiba Formation, approximately 300 ft (90 m) thick at Raudhatain. The Shuaiba was a known "lost circulation" zone, and prior experience at the Greater Burgan field had pointed up the hazard of a well penetrating the Shuaiba section with oil reservoirs exposed up the hole. In wells drilled to the Zubair reservoirs at Raudhatain, early practice was to set a protective string of 9⅝-in. intermediate casing in the lower part of the Burgan Formation and then to drill ahead through the Shuaiba and Zubair formations to set a long string of 7-in. production casing in the upper part of the underlying Ratawi shale section.

The Raudhatain-1 discovery well had been drilled to a total depth of 10,301 ft (3140 m) in the Ratawi Formation (Figure 4); the oil-productive zones in the Mauddud and Burgan formations had been cased off and the well had been completed in the clastics of the Zubair Formation, which carried crude oil of approximately 35° API gravity. Because of an increasing market demand for heavier crude, however, the Kuwait Oil Company elected to concentrate the initial field development effort on the important reservoir in the lower member of the Burgan Formation, which contained 32.5–33.5° API crude with sulfur content approximately 2% by weight. This was comparable in quality to the crude oil already being produced at the Greater Burgan field in southeastern Kuwait (Milton and Davies, 1965).

The lower Burgan reservoir occurred at drilling depths of 8100–8650 ft (2470–2635 m) across the Raudhatain structure; the reservoir contained a defined oil-water contact at 8460 ft (2580 m) subsea and had a gross productive section of 560 ft (170 m) above water in the crestal area of the field. Development drilling to this reservoir did not involve penetration of the Shuaiba carbonate section. This permitted use of a modified casing program, comprising a surface string set into Eocene carbonates at 1200± ft (365 m); a 9⅝-in. intermediate string set into shales of the Ahmadi Formation below 7000 ft (2130 m) to seal off lost circulation and/or water-influx zones in Cretaceous–Paleogene carbonates; and a 7-in. string of production casing set through the oil-water contact in the lower Burgan reservoir at about 8700 ft (2650 m) (Figure 4).

Of the first 30 wells drilled in the Raudhatain field, 23 were completed in the reservoirs of the Burgan Formation; one as a dual producer from the Mauddud and upper Burgan, two as single completions in the Mauddud, and three in the reservoirs of the Zubair Formation. The remaining well, a long southwesterly step-out, was unproductive but provided information on structural closure in that direction. The three wells completed in the Mauddud were all located on the southeast flank of the Raudhatain structure; structural closure at this point appeared insufficient to contain the Mauddud oil column, which apparently continued across the structural saddle into an adjacent high. This situation is discussed in greater detail in the section on *Trap/Reservoir Stratigraphy*.

The Raudhatain field was ready to go on stream toward the end of 1959; during 1960, the field produced 33,623,376 bbl of oil, an average of 92,120 BOPD. Of this total, 92% was drawn from the lower Burgan reservoir. During 1961, average daily production passed the 100,000 BOPD mark; by 1962, the production level had reached 129,000 BOPD and by 1963 133,000 BOPD. In each year, the proportion of the production drawn from the lower Burgan reservoir remained essentially constant at 92%.

Cumulative oil production from Raudhatain field reached 1 billion bbl during 1975; 58 wells had been drilled within the field area at this time, of which 49 were producing. Cumulative production of gas during this same period was approximately 150 bcf. The level of oil production peaked at 310,000 BOPD during 1979 and declined sharply thereafter; however, this post-1979 decline was partially a matter of production restraints imposed by the Government of Kuwait working within OPEC production quotas. There is little doubt that producing capacity at Raudhatain substantially exceeds the annual production level of the field during more recent years. These current production rates are believed to lie in the range of 125,000–150,000 BOPD. At its current production rate, the Raudhatain field should pass the 2 billion bbl cumulative production mark during the early 1990s.

DISCOVERY METHOD

The Raudhatain oil discovery of 1955 resulted from wildcat drilling on a subsurface high outlined and defined by reflection seismic exploration. Bedrock throughout the area consisted of essentially flat-lying deposits of Neogene to Recent age; there were no obvious indications of surface structure, although a minor topographic high was suggested by the pattern of present-day consequent drainage (Milton and Davies, 1965). There were no surface indications of hydrocarbons in the immediate area of the Raudhatain prospect; the nearest known surface seepages lay 23 mi (37 km) to the south on the northern shore of Kuwait Bay (Figure 2).

The positioning of the seismic surveys that led to the outlining of the Raudhatain prospect was not, however, fortuitous. The discovery of oil at Burgan, in southeastern Kuwait, and at Zubair, in southern Iraq, had focused attention on the undrilled area of northern Kuwait that lay between these two important Cretaceous oil discoveries. The seismic exploration north of Kuwait Bay was designed to pick up a possible northerly extension of the "Burgan axis," in the hope that additional closed structures would prove to be located along this trend. A component of regional geologic thinking was therefore involved in the Raudhatain discovery.

The initial seismic survey shot north of the bay during 1949 was of a reconnaissance nature, with east-west profiles 6 mi (10 km) and north-south profiles up to 18 mi (30 km) apart, yielding very large loops. Deep weathering presented a problem and record quality was poor. Nevertheless, an anomalous subsurface trend was mapped, and this was developed into a closed subsurface high through more detailed seismic work carried out during 1951–1952 (Figure 3).

Even this detailed survey lacked adequate velocity control. The nearest well to Raudhatain, drilled during 1936–1937 at the site of the Bahrah oil seepages 23 mi (37 km) to the south, had reached a total depth of 7950 ft (2420 m) in strata of inferred Early Cretaceous age but had not been logged or shot for velocities prior to abandonment. The Raudhatain seismic prospect was mapped on a subsurface horizon believed to approximate the top of the Mauddud Formation carbonates, and the Raudhatain-1 wildcat test was drilled on the basis of this interpretation. The initial wildcat encountered the top of the Mauddud low to the seismic prediction, but a porosity development in the upper part of the formation tested crude oil at commercial rates. Deeper drilling confirmed the presence of a substantial reserve of crude oil in clastic reservoirs lower in the Cretaceous section, including the section originally designated as the principal objective of this exploratory well.

With velocity and formation-elevation data from the first two wells on the structure in hand, additional seismic profiles were shot and integrated into existing coverage. The structural interpretation resulting from this work was used successfully to guide the early step-out and development drilling programs. Raudhatain field offers a good example of a major oil discovery resulting from the intelligent use of reflection seismic programs in a regional geologic exploration play.

STRUCTURE

Geologic Setting

Outcrops of Precambrian igneous and metamorphic rocks extend eastwards from the shore of the Red Sea into the interior of the Arabian peninsula to form the Arabian shield. East and northeast of the shield, the broad Arabian shelf extends for 500 mi (800 km) to the foreland of the Zagros fold belt, described by Ala (1982) as the tectonized northeastern margin of the Middle East basin (Figure 5).

The Arabian shelf has been subdivided by Aramco (1975) into an interior homocline and an interior platform, corresponding to the stable shelf and unstable shelf sectors of Beydoun and Dunnington (1975). In the interior homocline or stable shelf, Paleozoic and Mesozoic strata dip away from the shield at rates reflecting the influence of the underlying basement rocks. Further to the east, the interior platform or unstable shelf is characterized by the presence of basically flat-lying Tertiary deposits that no longer obviously reflect basement configuration.

The interior platform or unstable shelf includes the present-day coastal zone of eastern Arabia, the subsurface of the Persian (Arabian) Gulf, and the greater part of the riverine lowlands of southeastern Iraq. Among the basins developed along this sector is the North Arabian Gulf basin, located at the head of the Gulf between the present Arabian shore and the Zagros foreland zone of Iran and extending northwestwards into Iraq. Northern Kuwait, including the area of Raudhatain field, lies within the interior platform or unstable shelf at the margin of the North Arabian Gulf basin (Figure 5).

The Raudhatain oil field structure is one among a number of individual culminations developed along a prominent anticlinal ridge that plunges gently northwards through Kuwait into southern Iraq (Figure 5). These individual highs, in which the principal oil accumulations of the area are localized, tend to be of large areal extent and substantial structural relief. Other oil fields along this 150 mi (240 km) Kuwait arch or axis include, from south to north, Fuwaris, Wafra, Umm Gudair, Minagish, Greater Burgan, Bahrah, Sabriya, Zubair, and Rumaila.

Postulated reasons for structural growth along the Kuwait axis include compressive stress generated by shortening of the bottom profile of the subsiding North Arabian Gulf basin acting against the stable western margin (Adasani, 1967); halokinesis (Fox, 1959); and halokinesis in combination with basement tectonics (Murris, 1980). The proposed halokinesis assumes the mobilization of the deep-seated "Infra-cambrian" Hormuz Salt Series, known from regional studies to underlie the Kuwait area at depth. There is evidence of salt piercement to the present-day land surface, with associated "Hormuz-type" metasedimentary rocks at Jabal Sana'am in southern Iraq, only 15 mi (24 km) north of Raudhatain field (Figures 2 and 5).

Raudhatain Structure

The Raudhatain structure is basically domal in plan, with gentle and regular flank dips (Figure 6). According to Milton and Davies (1965), the structure exhibits a quasi-radial pattern of small normal faults of limited throw. This geometry suggests a primary origin by uplift, even though the original structure may have been modified at some later date by

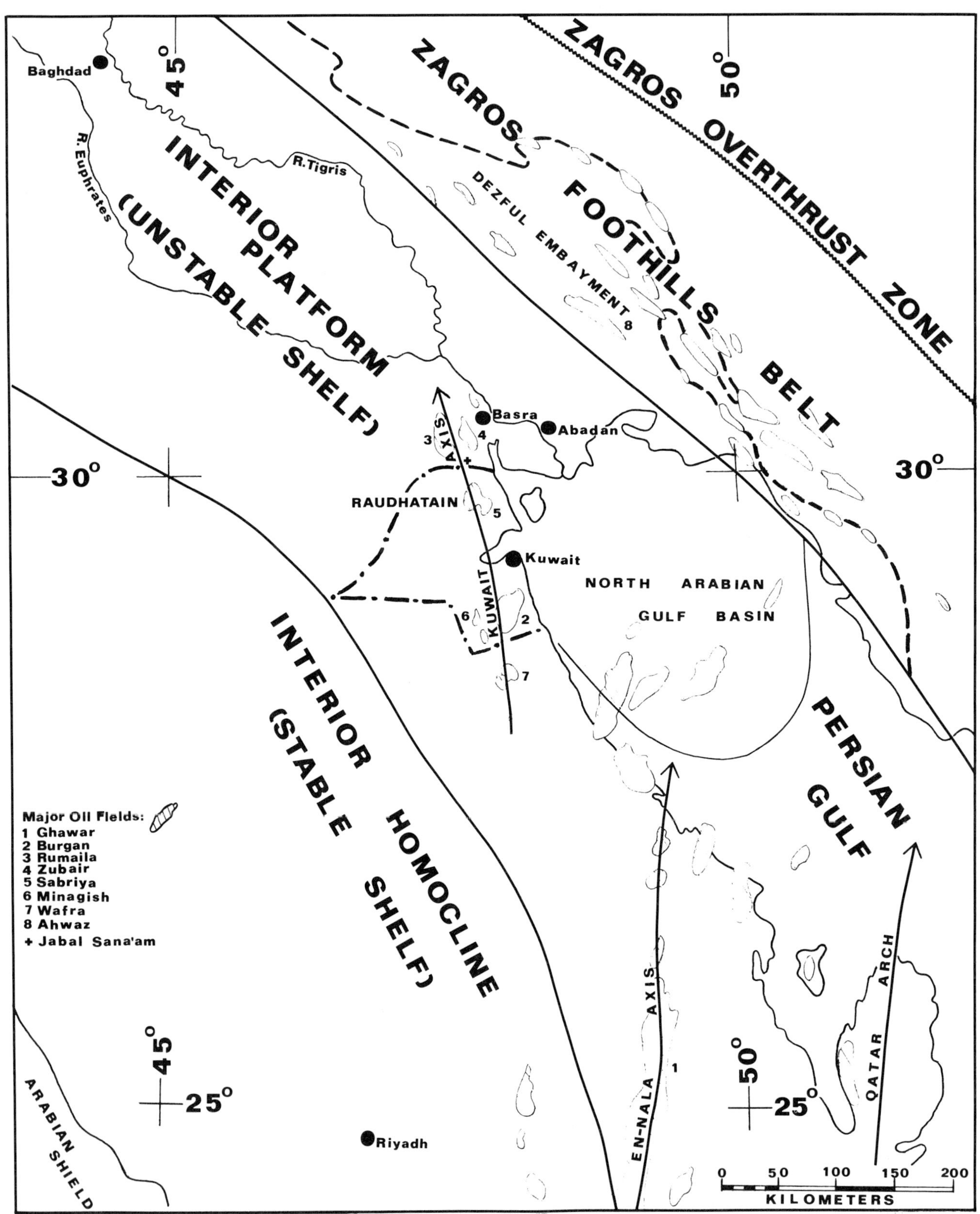

Figure 5. Raudhatain field lies toward the outer edge of the Arabian platform, on a structural axis that plunges north through Kuwait into southern Iraq. The oil field structure was formed during the Cretaceous but may have been slightly modified by the effects of the Tertiary Zagros orogeny.

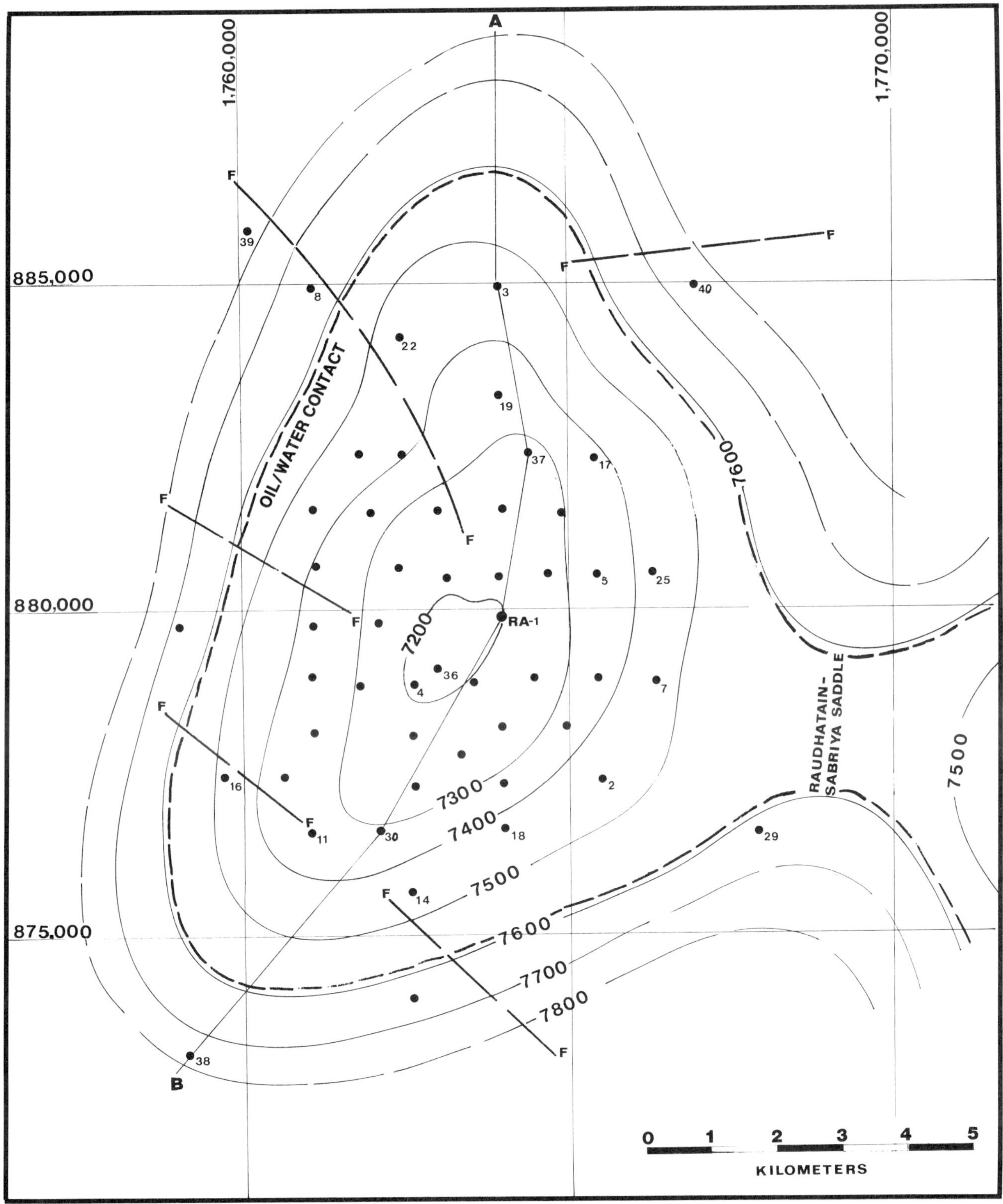

Figure 6. Raudhatain field structure, contoured in feet below sea level on top of the upper Lower Cretaceous Mauddud Formation. The oil column in the Mauddud continues across the Raudhatain-Sabriya saddle into the adjacent Sabriya oil field structure. FF, small normal faults of limited throw. Contour interval is 100 ft.

195

compressive forces. Such uplift could have been caused by the movement of basement blocks or by the injection of intrusive material; however, the overall aspect of the Raudhatain fold favors intrusion by deep-seated halite as the more likely mechanism of uplift.

Milton and Davies (1965) suggested that structural growth of the Raudhatain feature began during early Late Cretaceous time. Al-Rawi (1981) confirmed that structural growth had occurred during deposition of the clastics of the Lower Cretaceous Zubair Formation, and that Zubair deposition had itself been influenced by pre-existing structure. It therefore appears that structural growth at Raudhatain began not later than early Early Cretaceous time.

Once established, structural growth was evidently continuous; virtually all formations show some degree of thinning from flank to crestal areas. Fairly rapid growth during early Late Cretaceous time led to (probable) erosion of the upper part of the Mishrif Formation (Figure 4), while similar growth during late Late Cretaceous time led to depositional thinning of the Hartha-Tayarat section. The rate of growth appears to have been gentle; isopachous mapping of the Aruma Group yields a flank-to-crest variation in thickness of 1530 to 1420 ft (465 to 430 m), a depositional thinning of only 110 ft (33 m) or approximately 7% (Milton and Davies, 1965). Something of a balance may have existed between the rates of sedimentation and structural growth, permitting the deposition of a reduced sedimentary section across the high while creating few hiatuses within the overall sequence.

Structural growth persisted into Paleogene and Neogene time, rejuvenating the post-Eocene unconformity surface created by the regional Paleogene uplift that terminated marine deposition in the area. Isopachous mapping of the post-unconformity sections indicates a thinning of the continental clastics of the Kuwait Group, while examination of present-day drainage patterns led Milton and Davies (1965) to conclude that structural growth might still be active in the Raudhatain area.

Milton and Davies (1965) suggested that the axial trend of the Raudhatain fold may have been modified slightly by compressive stress during post-Cretaceous time, the effect being to displace the axis from an original northeasterly into a more northerly trend in the northern part of the structure (Figure 6). This compression was probably a distant effect of the intense Zagros tectonic episode that occupied much of Neogene time in Iran and Iraq. The effects of the Zagros orogeny along the Arabian margin were limited by distance, but renewed vertical uplift of the Raudhatain structure may well have been stimulated at this time.

STRATIGRAPHY

The stratigraphy of the Kuwait–southern Iraq area was first formalized by Owen and Nasr (1958) utilizing subsurface data from the Burgan field of southeastern Kuwait and the Zubair field of southern Iraq (Figure 2). The Raudhatain field lies between the Burgan and Zubair areas, and the sedimentary section penetrated at Raudhatain is in general similar to that present in the Zubair (and Rumaila) fields of southern Iraq; the section can be fitted into the stratigraphic framework developed by Owen and Nasr (1958). The stratigraphy of the Raudhatain area has been discussed by Milton and Davies (1965), Milton (1967), Adasani (1967), and Al-Rawi (1981).

The total thickness of sedimentary section underlying the Raudhatain area probably exceeds 30,000 ft (9250 m), including sediments ranging in age from "Infracambrian" to Recent. However, the section is known in detail only to the base of the Cretaceous at drilling depths in the order of 13,000 ft (3960 m) (Figure 4).

The Cretaceous section at Raudhatain has a drilled thickness of approximately 9000 ft (2740 m). In ascending order, the Cretaceous formations are organized into the Thamama, Wasia, and Aruma groups (Figure 4) (corresponding to deposits of Neocomian to Aptian; Albian to Turonian; and Coniacian through Maastrichtian age). In southeastern Kuwait, these three major groups of Cretaceous strata are separated by a minor unconformity between the Thamama and Wasia and a major unconformity between the Wasia and Aruma (Figure 4).

It was recognized early that the magnitude of the Aruma/Wasia unconformity decreased toward the north, being represented in southern Iraq "only by a stratigraphic condensation of sediments" (Owen and Nasr, 1958). Milton and Davies (1965) concluded that there was no evidence of angular unconformity at the top of the Wasia section at Raudhatain, suggesting that the Wasia-Aruma transition might be conformable. Later workers (Loutfi and Jaber, 1970; Al-Rawi, 1981) have confirmed the presence of an unconformity between the Wasia and Aruma, but one affecting only the uppermost formation (Mishrif Formation) of a basically complete Wasia section.

The contact between the Wasia and Thamama groups at Raudhatain has been considered conformable by Milton and Davies (1965) and other workers. The uppermost formation of the Thamama Group is the Shuaiba, a section of recrystallized dolomitic limestone with well-developed secondary porosity; according to Milton and Davies (1965) the Shuaiba section is "riddled with large caverns." The probability exists that the Shuaiba surface was subjected to subaerial weathering prior to the deposition of the clastics of the overlying Burgan Formation and that at least a disconformity exists at the top of the Thamama carbonate section.

The lower part of the Lower Cretaceous section in northern Kuwait–southern Iraq comprises up to 1500 ft (460 m) of neritic limestone overlain by approximately 500 ft (150 m) of greenish-gray to black shale, containing stringers of limestone in its lower part and of sandstone in its upper part. These sections have traditionally been referred to as the Ratawi Limestone and the Ratawi Shale, sometimes as separate formations, sometimes as members of a multimember Ratawi Formation. The Ratawi *sensu lato* is of early Early Cretaceous age and forms the lower part of the Thamama Group. The Ratawi limestone-shale section is overlain by the clastics of the Zubair Formation (Figure 4).

The Zubair Formation at Raudhatain averages 1380 ft (420 m) in thickness and consists of clean, poorly sorted, predominantly fine-grained, subangular to subrounded quartz sandstones, interbedded with dark shales and minor siltstones. The Zubair is an important oil-producing section at Raudhatain and will be treated in greater detail under *Reservoir Stratigraphy*.

The Zubair clastics are overlain with apparent conformity by the dolomitic limestones of the Shuaiba Formation, averaging 300 ft (90 m) in thickness across the structure. The Shuaiba contains much recrystallized rudist material and appears to reflect a shallow-marine transgression affecting the area during middle Early Cretaceous time.

The Wasia Group at Raudhatain averages 2450 ft (750 m) in thickness and includes, in ascending order, the Burgan, Mauddud, Ahmadi, Rumaila, and Mishrif formations. Of these, the clastic Burgan and carbonate Mauddud sections contain important oil reservoirs and will be discussed in greater detail under *Reservoir Stratigraphy*.

The widespread and characteristically thin Mauddud Formation carbonates reflect a brief shallow-marine transgression toward the end of Early Cretaceous time (comparable to the earlier transgression represented by the carbonates of the Shuaiba Formation). At Burgan in southeastern Kuwait, the Mauddud transgression was succeeded by a brief return to sand deposition, represented by the oil-bearing Wara Formation. At Raudhatain, the oil-bearing Mauddud carbonates are conformably overlain and capped by the dark marine shales of the Ahmadi Formation, apparently including the equivalents of both the Wara and Ahmadi sections of the Burgan area.

The shales of the Ahmadi Formation are 400 ft (120 m) thick at Raudhatain and are conformably overlain by argillaceous limestones, marls, and shales of the Rumaila Formation, assigned an early Late Cretaceous age by El-Naggar and Al-Rifaiy (1972). The lithologies of the Mauddud, Ahmadi, and Rumaila formations represent a progression toward deeper-marine conditions following deposition of the littoral clastics of the Burgan Formation (Figure 4).

During early Late Cretaceous time, a return to regressive conditions resulted in the deposition of limestones of increasingly shallow-water aspect. These constitute the Mishrif Formation, the uppermost formation of the Wasia Group. Because of shoaling to the point of emergence, the Mishrif section was subject to truncation along positive axes, resulting in a major regional unconformity (Figure 4). The effects of this regional unconformity are severe in southern Kuwait, less so in northern Kuwait and southern Iraq (Owen and Nasr, 1958). The difference probably reflects concomitant regional tilting toward the north that caused Cretaceous formations to thicken in that direction, so mitigating the effects of the erosion. At Raudhatain, more than 400 ft (120 m) of Mishrif section is preserved across the structure, as against the 526 ft (155 m) of thickness quoted for the type section in the Zubair field to the north (El-Naggar and Al-Rifaiy, 1972).

The Mishrif at Raudhatain is unconformably overlain by limestones and shales of the transgressive Khasib Formation, lowermost unit of the Upper Cretaceous Aruma Group. This group is approximately 2500 ft (760 m) thick at Raudhatain and includes, in ascending order, the Khasib, Mutriba, Sadi, Hartha, Qurna, and Tayarat formations (Figure 4).

The formations of the Aruma Group are the record of a shallow-marine transgression across the partially eroded surface of the underlying Wasia Group. The Khasib and Mutriba formations are thin (100–150 ft; 30–45 m) developments of marly limestones and calcareous shales; they are overlain by a much thicker (up to 700 ft; 215 m) section of light-colored, chalky to marly limestones of the Sadi Formation. The Sadi is overlain by organic to detrital limestones, with subordinate shales and marls, of the Hartha Formation, which is overlain in turn by marls and shales, with subordinate limestones, of the Qurna Formation. The uppermost formation of the Aruma Group is the Tayarat Formation, up to 850 ft (260 m) thick in the Raudhatain area and composed of dolomitic to anhydritic limestones with interbeds of black, pyritic shales. The carbonates of the Tayarat are porous and present a drilling problem in the Raudhatain field area in the form of saltwater flows.

Sedimentation in the Raudhatain area was continuous from Cretaceous into Paleogene time; the upper Upper Cretaceous Tayarat Formation is conformably overlain by shallow-water limestones of the Radhuma Formation, lowermost unit of the Paleocene–middle Eocene Hasa Group, 2600 ft (790 m) thick at Raudhatain. The Radhuma carbonates are cavernous and present a drilling hazard in the form of "lost circulation" zones. The section also contains beds of anhydrite and passes conformably upwards into the massive anhydrite of the Rus Formation, an evaporite development widespread in the shallow subsurface of the western shore of the Gulf. The Rus evaporites are overlain with probable disconformity by nummulitic limestones of the middle Eocene Dammam Formation, uppermost formation of the Hasa Group (Figure 4).

Marine deposition in the area was terminated by regional uplift at the close of middle Eocene time (Aramco, 1975), initiating a cycle of erosion that would persist through late Eocene and all of Oligocene time. The resulting unconformity, cut into the carbonates of the Dammam Formation, is of regional significance throughout the area of Kuwait (Figure 4).

The silicified post-Eocene unconformity surface is overlain by a section of continental sands and gravels separated into lower and upper sequences by a section of anhydritic and gypsiferous clays, marls, and shallow-water limestones. The lower and upper clastic sections, which are lithologically indistinguishable (Milton and Davies, 1965), have been named the Ghar and Dibdibba formations, while the intervening evaporitic interval has been correlated with the lower Fars (Gachsaran Formation) of Iranian terminology (Owen and Nasr, 1958; Milton and Davies, 1965; James and Wynd, 1965). The entire post-Eocene section is approximately 1200 ft (365 m) thick and is referred to as the Miocene–Pleistocene Kuwait Group (Figure 4). The upper clastic section, the Dibdibba Formation, is an important local aquifer in northern Kuwait, supplying brackish water for drilling and other purposes.

TRAP

The hydrocarbon trap at Raudhatain field is structural, a domal uplift elongated slightly along a north-northeast to south-southwest axis. At the oil-water contact in the upper reservoir of the Burgan Formation, the producing structure measures approximately 9 by 6 mi (14 by 10 km) and at the level of a deeper reservoir within the upper part of the Zubair Formation, 8 by 4.5 mi (13 by 7 km). The productive area covers some 30 mi^2 (78 km^2), just over 19,000 acres (Figure 6).

Regional dip in the area is toward the north, and it was understood early that critical closure on the Raudhatain structure would be along the southern, regionally updip, flank. Closure toward the southwest was defined at 625 ft (190 m) by the drilling of the RA-28 step-out well, but minimum closure on the structure proved to lie along the southeast flank, where a structural saddle connected the Raudhatain high to an adjacent structure given the name Sabriya (now the Sabriya field) (Figure 6). Limited drilling in the area of this saddle suggested that the closure at this point would approximate 500 ft (150 m) (Milton and Davies, 1965) and that this would be insufficient to isolate the oil column established for the uppermost of the Raudhatain field reservoirs, in the upper part of the Mauddud Formation (Figure 7); this situation is discussed more fully in the section on *Reservoir Stratigraphy*. Apart from the area of the Raudhatain-Sabriya structural saddle (Figure 6), dips on all flanks of the Raudhatain structure are regular and gentle, rarely exceeding 3°.

Continuous structural growth at Raudhatain apparently began not later than early Early Cretaceous time (Al-Rawi, 1981), with relatively rapid uplift occurring during early Late Cretaceous time (Milton and Davies, 1965). The uppermost cap rock of the field (the Ahmadi Formation) was deposited during the early Late Cretaceous, and the Raudhatain trap was probably completely in place by mid-Late Cretaceous time.

Reservoir Stratigraphy and Characteristics

The producing interval at Raudhatain is of Early Cretaceous age and consists of two thick sections of oil-productive clastics, separated by a nonproductive section of marine carbonate and overlain by a second, oil-productive, section of marine carbonate. The lower clastic section comprises the Zubair Formation; the Zubair is overlain by the nonproductive carbonates of the Shuaiba Formation. The upper clastic section, immediately overlying the Shuaiba, is the Burgan Formation; the Burgan is in turn overlain by the oil-productive Mauddud Formation carbonates (Figures 4 and 7).

The entire interval from the base of the Zubair to the top of the Mauddud averages 2900 ft (880 m) in thickness and contains nine separate oil reservoirs, of which four may be classed as containing major oil accumulations. Five of these nine reservoirs occur within the Zubair Formation and two within the Burgan Formation; these reservoirs are separated and capped by intraformational shale developments. The remaining two reservoirs occur within the Mauddud Formation; the upper of these is capped by the marine shales of the overlying Ahmadi Formation (Figure 7).

The Mauddud Formation at Raudhatain consists of up to 330 ft (100 m) of light-colored, organic to detrital, locally pseudo-oolitic limestone with minor shale breaks. The lower part of the formation is tight; the upper part contains two oil zones separated by a thin but apparently continuous impermeable zone.

The upper Mauddud reservoir occurs at a subsea depth of about 7200 ft (2195 m) at the Raudhatain crest; the producing interval has an average thickness of approximately 180 ft (55 m), of which three-quarters may be considered net pay. The zone has porosities in the range 16–22%, permeabilities averaging 20 md, and irreducible water saturation of approximately 20%. The reservoir contains undersaturated crude oil averaging 28.7° API with sulfur content 3.28 wt. % and original solution GOR 606 SCF/bbl (Table 1). The upper Mauddud reservoir is full to spill point; its oil column is 560 ft (170 m) thick and extends through the structural saddle on the southeast flank of the Raudhatain structure into the adjacent Sabriya high. In the trough of the saddle, and elsewhere along the flanks of the Raudhatain structure, the upper Mauddud oil column is limited by an oil-water contact at a subsea elevation of 7760 ft (2365 m) (Figures 6 and 7).

The lower Mauddud reservoir is approximately 80 ft (25 m) thick and has reservoir and oil properties generally similar to those of the upper Mauddud. The oil column in the lower reservoir is 275 ft (80 m) thick and extends across the crest of the structure

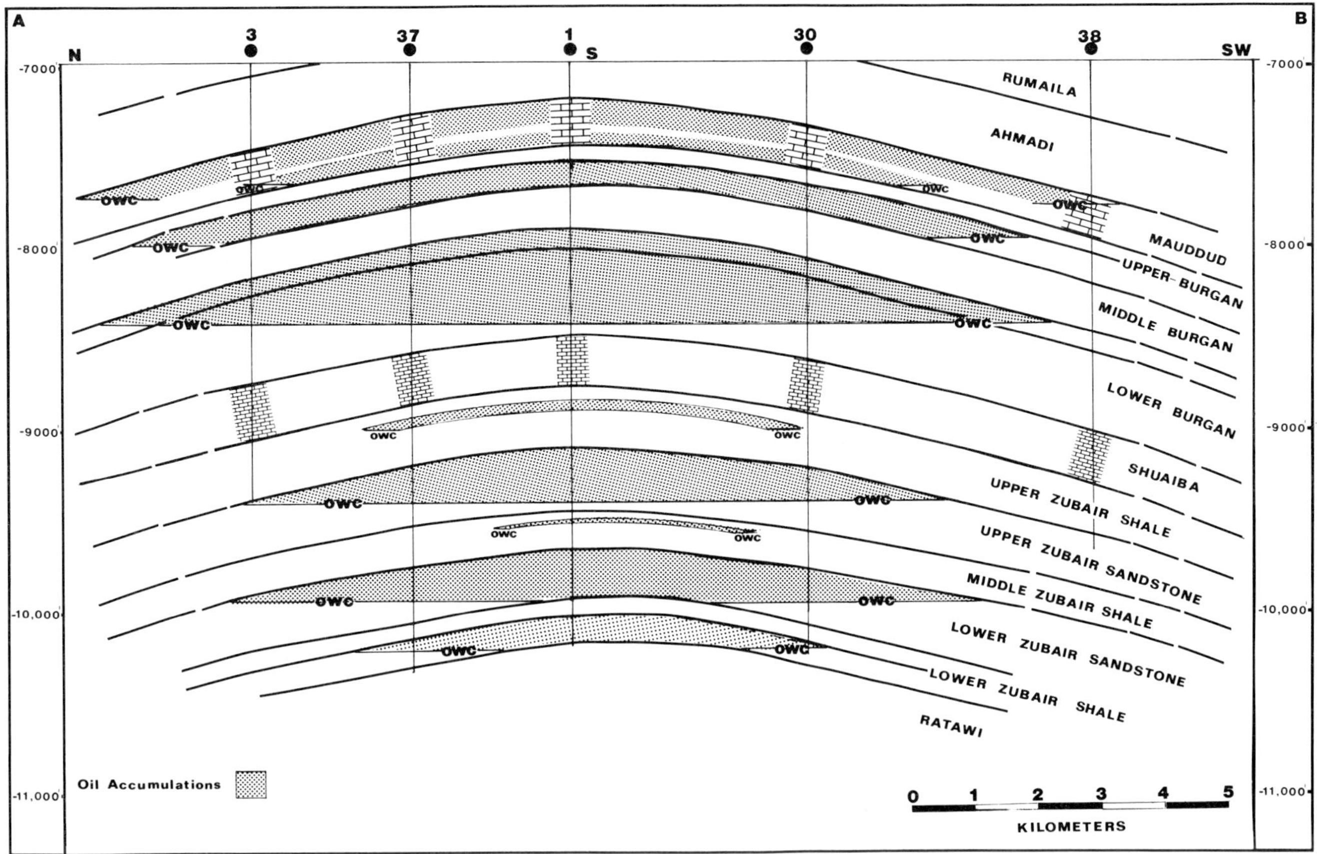

Figure 7. North to southwest structural section through Raudhatain field (refer to Figure 6 for line of section). The upper reservoirs of the field are full to spill point. (Some well data from Adasani, 1967.) Depths in feet subsea. OWC, oil-water contact.

to be limited by a rim of edge water along the flanks at a subsea elevation of 7660 ft (2335 m), which is within the vertical dimension of the oil column in the upper Mauddud (Figure 7).

The Burgan Formation underlies the Mauddud carbonates and consists of approximately 950 ft (290 m) of soft, clean, porous, well-sorted quartz sandstones of littoral to possibly deltaic aspect, interbedded with siltstones and dark shales. The sandstones show little in the way of secondary cementation within the oil zones but have been subjected to diagenetic alteration within the aquifer sections.

The Burgan Formation at Raudhatain has been informally divided into three members, termed the "Upper Sand," "Middle Shale," and "Lower Sand." The "Upper Sand/Middle Shale" section appears to correlate, at least in part, with the "Third Sand Member" of the Burgan Formation at the Greater Burgan field (where the type section of the formation occurs), while the "Lower Sand" of the Raudhatain section appears to be the equivalent of the Burgan "Fourth Sand Member" of the Greater Burgan field type section (Milton and Davies, 1965; Adasani, 1967).

The oil reservoir in the upper Burgan sandstone section at Raudhatain occurs at a subsea elevation of 7540 ft (2300 m) at the structural crest; the oil zone has an average thickness of 120 ft (40 m), of which approximately 70% is permeable. Average porosity is 24% and average permeability is 1035 md; irreducible water saturation is 26%. The reservoir contains undersaturated crude oil averaging 32.7° API with sulfur content 3.75 wt. % and original solution GOR 737 SCF/bbl (Table 1). The oil column in the reservoir is 460 ft (140 m) thick and is limited along the flanks of the structure by an oil-water contact at a subsea elevation of approximately 8000 ft (2440 m). Along the southeastern flank, the oil column in the upper Burgan reservoir extends out into the Raudhatain-Sabriya saddle but does not appear to pass into the adjacent Sabriya high. The upper Burgan sandstone reservoir is underlain by a thick shale development that forms the seal for the deeper—and larger—oil accumulation in the lower sandstone section of the Burgan Formation.

The lower Burgan reservoir occurs at a subsea elevation of 7900 ft (2410 m) at the crest of the Raudhatain structure and comprises some 600 ft (180 m) of quartz sandstone containing a thin but persistent shale unit in the upper part and with a few feet of shaly section at the base. The sandstone section contains an oil column 560 ft (170 m) thick

Table 1. Raudhatain field, summary of reservoir fluid characteristics. Data from Adasani (1967), Al-Rawi (1981), and other sources. Reservoir terminology follows Al-Rawi (1981); refer to Figure 9.

RESERVOIR	API GRAVITY	GOR SCF/BBL	SULPHUR WT.%	FVF	SATURATION PRESSURE/ DEPTH SUBSEA
MAUDDUD LS	29.0	608	3.28	1.362	2278 psig @ 7493 ft (2284m)
U. BURGAN SS	32.7	737	3.75	1.431	2471 psig @ 7559 ft (2304m)
L. BURGAN SS	33.0	796	2.08	1.464	2741 psig @ 8232 ft (2509m)
U. ZUBAIR SH	32.5	848	2.00	1.522	2856 psig @ 8844 ft (2696m)
U. ZUBAIR SS	32.0	765	2.00	1.478	2649 psig @ 9188 ft (2800m)
L. ZUBAIR SS	33.6	1365	2.00	1.743	4329 psig @ 9814 ft (2991m)
L. ZUBAIR SH	36.8	1109	2.00	1.640	3157 psig @ 10,046 ft (3061m)
"RATAWI SH"	42.0	2264	>2.00	2.387	4060 psig @ 10,337 ft (3150m)
"RATAWI LS"	44.0	2772	>2.00	2.740	4280 psig @ 10,619 ft (3266m)

at the crest, limited by a horizontal oil-water contact at 8460 ft (2580 m) subsea. The reservoir is full to spill point at the minimum closure on the southeast flank and the oil column extends out into the trough of the Raudhatain-Sabriya saddle. The thin shale unit in the upper part of the reservoir may form a lateral seal for the greater part of the oil column; there is, however, an oil accumulation in the equivalent reservoir in the adjoining Sabriya structure (Figure 2).

The lower Burgan reservoir has an average porosity of 24.5%, uniformly high permeabilities locally reaching thousands of millidarcys, and irreducible water saturations in the range 15–20%. Porosity and permeability values decrease below the oil-water contact because of diagenetic alteration in the zone of hypersaline bottom water. The producing interval contains undersaturated crude oil averaging 33° API with sulfur content 2.08 wt. % and original solution GOR 796 SCF/bbl (Table 1). The lower Burgan possesses an active water drive and is considered the most important of the Raudhatain field reservoirs.

The aquifer section of the lower Burgan reservoir is directly underlain by dolomitic limestones of the Shuaiba Formation, which has an average thickness of 300 ft (90 m) across the Raudhatain structure.

The Shuaiba carbonates are recrystallized but exhibit well-developed secondary porosity, including large caverns, probably as a result of subaerial weathering prior to the deposition of the Burgan Formation clastics. The Shuaiba does not produce oil at Raudhatain field, and the porous carbonate section, overlain and underlain by oil zones, presents a drilling hazard in the form of potential loss of circulation. The Shuaiba section at Raudhatain contains shows of tar and residual oil and possesses a rudimentary shale cap at the base of the overlying Burgan Formation (Milton and Davies, 1965). The lack of an oil accumulation in the Shuaiba at Raudhatain has not been satisfactorily explained, particularly since the carbonate zone is a prolific oil-producing section elsewhere in the Gulf region.

The Zubair Formation underlies the Shuaiba with apparent conformity. The Zubair is the lower of the two major clastic sequences of the Raudhatain field area; the formation has an average thickness of 1380 ft (420 m) and consists of quartz sandstones intercalated with siltstones, shales, and minor limestones (Al-Rawi, 1981). The sandstones are soft and clean, subangular to subrounded, predominantly fine grained and poorly sorted. In the oil-bearing sections, the sandstones show little in the way of secondary cementation, and their primary intergranular porosity is well preserved (Figure 8A). In the aquifers underlying the various Zubair Formation oil pools, the sandstones exhibit marked diagenetic alteration, with extensive plugging of the intergrain pores by secondary minerals (Al-Rawi, 1981) (Figure 8B). The shales interbedded with the sandstones are commonly dark gray to black.

Adasani (1967) divided the Zubair Formation at Raudhatain field into six informal members: "Upper Zubair Shales"; "Upper Zubair Sandstones"; Middle Zubair Shales"; "Middle Zubair Sandstones"; "Lower Zubair Shales"; and "Lower Zubair Sandstones" (Figure 9). Major oil accumulations are present in the "Upper and Middle Sandstone"

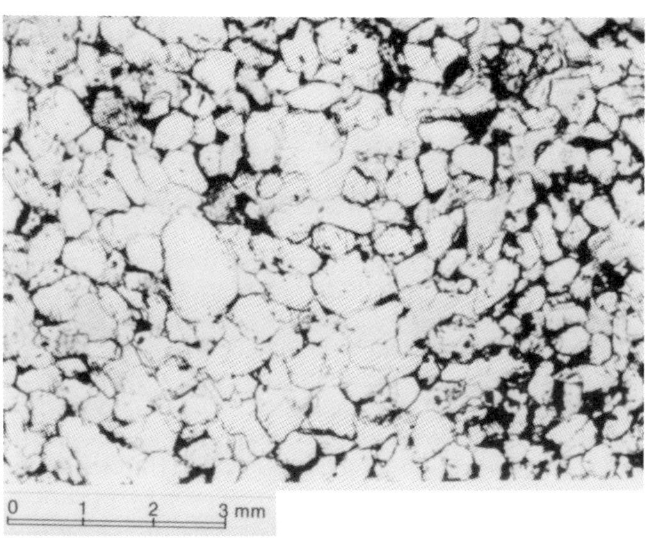

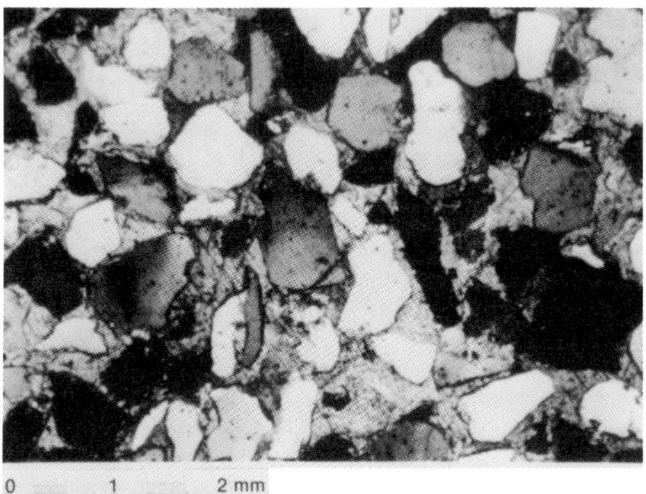

Figure 8. Primary porosity and permeability are well preserved in the oil sections of the clastic reservoirs, but have been severely affected by diagenesis in the aquifer sections. In the Zubair Formation, the reduction in primary porosity may be from in excess of 20% (A) to less than 10% (B). (A) shows upper sandstone member reservoir, ordinary light. (B) shows lower sandstone member aquifer with plugged porosity, crossed nicols. (From Al-Rawi, 1981.)

THICKNESS	LITHOLOGY	(ADASANI, 1967)	(AL-RAWI, 1981)
		SHUAIBA FORM.	*SHUAIBA FORM.*
0'		*UPPER ZUBAIR SHALE MEMBER*	*UPPER ZUBAIR SHALE MEMBER*
500'		*UPPER ZUBAIR SAND MEMBER*	*UPPER ZUBAIR SANDSTONE MEMBER*
		MIDDLE ZUBAIR SHALE MEMBER	*MIDDLE ZUBAIR SHALE MEMBER*
1000'		*MIDDLE ZUBAIR SAND MEMBER*	*LOWER ZUBAIR SANDSTONE MEMBER*
		LOWER ZUBAIR SHALE MEMBER	*LOWER ZUBAIR SHALE MEMBER*
1500'		*LOWER ZUBAIR SAND MEMBER*	
		RATAWI FORMATION	*RATAWI FORMATION*

Figure 9. Correlation of subdivisions of the multireservoir Zubair Formation as proposed by Adasani (1967) and Al-Rawi (1981). Al-Rawi's system is followed on Figure 7 and elsewhere in the text. Large and small solid circles indicate major and minor oil pools.

sections, a smaller accumulation in the "Lower Sandstone" section, and minor accumulations in sandstone bodies developed within the "Upper and Middle Shale" sections (Figure 7).

Al-Rawi (1981) following Gharib (1979) divided the Zubair into only five members, based largely upon sand-shale ratios. Al-Rawi's uppermost three members correspond to those of Adasani (1967); however, Adasani's "Middle Zubair Sandstones" are the equivalent of Al-Rawi's "Lower Zubair Sandstones," while Adasani's "Lower Zubair Shales" and "Lower Zubair Sandstones" together correspond to Al-Rawi's "Lower Zubair Shales" (Figure 9).

The uppermost of the Zubair Formation oil reservoirs occurs at an elevation of about 8800 ft (2680 m) subsea at the crest of the Raudhatain structure, in a sandstone body developed within the upper Zubair shales. The reservoir is overlain and underlain by shales and is isolated from the shallower, water-bearing carbonates of the Shuaiba and from the deeper, oil-bearing clastics of the upper Zubair sandstone section. The isolated upper reservoir is approximately 100 ft (30 m) thick with up to half this thickness comprising net pay; the oil column is 170 ft (50 m) thick and is limited by edge water at an elevation of 8970 ft (2735 m) subsea. The pay section has an average porosity of 23% and irreducible water saturation of about 30%; the section contains undersaturated crude oil averaging 32.5° API with sulfur content 2.0 wt. % and original solution GOR 848 SCF/bbl (Table 1).

The upper Zubair sandstone reservoir, largest and most productive of the Zubair Formation oil zones, lies slightly deeper in the section at 9055 ft (2760 m) subsea. The zone has an average thickness of 380 ft (115 m) of which up to three-quarters may be net pay; the oil column in the reservoir is 325 ft (100 m) thick and lies above a horizontal oil-water contact at 9380 ft (2860 m) subsea. The oil reservoir has an average porosity of 21%, permeabilities ranging from 400 to 4000 md, and irreducible water saturation of about 19%; the undersaturated crude oil present in the reservoir averages 32° API with sulfur content 2.0 wt. % and original solution GOR averaging 765 SCF/bbl (Table 1). This upper sandstone reservoir has a weak water drive, and pressure maintenance by gas injection has been in effect since shortly after the inception of production from the zone.

The upper Zubair sandstone reservoir is underlain by more than 200 ft (60 m) of shale. This shale section contains three minor oil accumulations in thin but continuous sandstone developments, isolated from one another and each limited by edge water. The lowest of the three is separated by shale from the larger oil accumulation lying slightly deeper in the section.

This is the "Middle Zubair Sandstone" member of Adasani (1967), equivalent to the "Lower Zubair Sandstone" member of Al-Rawi (1981), the second major reservoir of the Zubair Formation. The reservoir occurs at an elevation of 9630 ft (2935 m) subsea at the crest of the Raudhatain structure and contains an oil column 310 ft (95 m) thick, limited by a horizontal oil-water contact at 9940 ft (3030 m) subsea. Reservoir characteristics are essentially similar to those of the upper Zubair sandstone section, with the irreducible water saturation being slightly higher. The zone contains crude oil averaging 33.6° API with sulfur content 2.0 wt. % and original solution GOR 1365 SCF/bbl (Table 1). As with the other producing zones of the Zubair, this reservoir appears to possess a weak water drive and has been subjected to pressure maintenance by gas injection.

The lowermost reservoir of the Zubair Formation actually consists of two sandstone sections separated by a thin shale but sharing a common oil-water contact. The top of the productive section occurs at approximately 10,000 ft (3050 m) subsea, and an oil column of 200 ft (60 m) is present above water at 10,200 ft (3110 m) subsea. The combined reservoir has an effective thickness of approximately 80 ft (25 m), with an average porosity of 20% and variable but high permeability; irreducible water saturation approaches 28%. The reservoir contains undersaturated crude of 35.6–37.9° API with sulfur content 2.0 wt. % and original solution GOR 1100 SCF/bbl. The lower Zubair pay zone contains the lightest crude oil of the overall producing interval but has contributed little to the field's producing total (Table 1).

Oil and Field Characteristics

Nine separate oil reservoirs occur within the Mauddud-Burgan-Zubair section at Raudhatain field; they contain an aggregate gross pay thickness in excess of 1800 ft (550 m), of which two-thirds may reasonably be classed as net pay. The oil columns in these nine reservoirs have a combined thickness of almost 2900 ft (880 m).

Crude oils in the Raudhatain field are of asphaltic-paraffinic mixed base and show a steady increase in API gravity (decrease in specific gravity) with depth, from 28.5° API in the uppermost (Mauddud Formation) reservoir to 37.9° API in the lowermost pay zone of the Zubair Formation. This trend continues with depth below the base of the main producing section; crude oil tested in minor quantities from sandstone and carbonate stringers in the pre-Zubair Ratawi Formation below 10,200 ft (3110 m) subsea (the deepest of the field oil-water contacts) ranges in gravity from 42.0° to 44.0° API (Table 1).

Sulfur content by weight percent is greater in the heavier crudes of the upper part of the producing section and lower in the lighter crudes of the lower reservoirs (Table 1). Adasani (1967) has shown that sulfur content varies directly with increasing specific gravity within each major reservoir of the field.

The crude oil in all major reservoirs appears to have been undersaturated at the inception of production; saturation pressures were generally well below initial reservoir pressures and original solution GORs were moderate (Table 1). A possible exception to this pattern is the reservoir in the sandstone section of the middle Zubair in which saturation pressure, solution GOR, and formation volume factor (FVF) were substantially higher than in other Zubair reservoirs.

Pressure behavior during the early years of production confirmed the presence of an active water drive in the Burgan Formation reservoirs, especially in the lower sandstone member. This was extremely useful, since the Burgan reservoirs contained a large proportion of the total field reserve and were the source of most of the early production. The water drive mechanism was apparently much weaker in the case of the Zubair and Mauddud reservoirs, and these reservoirs have been subjected to pressure maintenance by gas injection since early in the life of the field. Pressure maintenance by water injection, at least on an experimental basis, has been attempted since 1985.

Sources: Origin and Migration of the Oil

The original recoverable reserve at Raudhatain field has been estimated at 8.8 billion STB of crude oil plus 13.19 tcf of natural gas, an oil-equivalent reserve of 11.0 billion bbl (Carmalt and St. John, 1984). A recovery factor of 39% has been suggested for the main reservoirs of the field; this translates into an original in-place reserve of 22.6 billion bbl of crude oil plus 33.8 tcf of natural gas, an oil-equivalent of 28.14 billion bbl.

The Raudhatain structure is one among several individual culminations developed along the structural ridge (Kuwait axis) that plunges gently north for approximately 150 mi (240 km) through Kuwait into southern Iraq (Figure 5). Many of these culminations contain major oil fields, including the Greater Burgan field, ranked as the world's second largest known oil field by Halbouty et al. (1970) and by Carmalt and St. John (1986); the position of Greater Burgan and other fields in relation to Raudhatain is shown on Figures 2 and 5. All the fields along the Kuwait axis carry oil in Cretaceous reservoirs, oil of sufficiently similar fundamental composition as to suggest a possible common origin (Strong, 1959; Kent and Warman, 1972). Total recoverable reserves along the Kuwait axis will exceed 120 billion bbl, suggesting an original oil in place of at least twice that quantity. The generation and accumulation of so large a volume of hydrocarbons must surely be of interest to all petroleum geologists.

The Cretaceous depocenter lay northeast of the Raudhatain area (Fox, 1959) and the oil-bearing clastic formations of Kuwait and southern Iraq give way to deeper-water deposits in that direction. The Ratawi and Zubair formations finger out into the shales and marls of the lower Lower Cretaceous Gadvan Formation; the Burgan Formation passes laterally into the bituminous shales and limestones of the upper Lower–lower Upper Cretaceous Kazhdumi Formation (James and Wynd, 1975). Both the Gadvan and Kazhdumi sections are exposed in outcrop along the Zagros mountain front of Iran (Figures 5 and 10).

Ayres et al. (1982) concluded that early Lower Cretaceous source rocks could have yielded the oil contained in lower Upper Cretaceous reservoirs in northern Sa'udi Arabia. These same lower Lower Cretaceous rocks (Gadvan Formation) would seem to offer a logical source for the oil contained in the Zubair Formation reservoirs in fields in southern Iraq and northern Kuwait, including Raudhatain field. Ala (1982) identified the upper Lower–lower Upper Cretaceous Kazhdumi Formation as containing mature organics and as providing the source for crude oil contained in Cretaceous and Tertiary reservoirs in southwestern Iran. Brennan (1990) concluded that lateral migration from a Kazhdumi source could have charged the upper Lower–lower Upper Cretaceous clastic reservoirs at Greater Burgan field in southeastern Kuwait. The Kazhdumi Formation offers a possible source for the crude oil in the Burgan and Mauddud reservoirs at Raudhatain field.

In the main depositional area of the Kazhdumi Formation, the minimum depth of burial required to initiate generation of hydrocarbons was not attained until Eocene time (Ala, 1982), and significant quantities of oil may not have been generated until late in Paleogene time. Assuming a Kazhdumi source for the oil in the Mauddud and Burgan reservoirs at Raudhatain, post-Cretaceous migration of oil into the structure is indicated. It may be noted that Kent and Warman (1972) dated the migration of oil into the Greater Burgan structural complex, located well to the south of Raudhatain but on the same regional axis, as post-Cretaceous, Eocene or even later.

Al-Rawi (1981) has argued for an Early Cretaceous migration of oil into the Zubair Formation reservoirs at Raudhatain field. Al-Rawi (1981) postulated the existence at Raudhatain of an Early Cretaceous paleotrap, in which an oil accumulation became "frozen in" through diagenetic reduction of porosity in the aquifer sections of the principal Zubair reservoirs (refer to Figure 8B). As structural growth proceeded, a portion of the oil trapped in the Zubair sandstones was able to migrate vertically upwards through fault channels into the higher Burgan and Mauddud reservoirs. As a result of this upward migration, the upper reservoirs of the structure have been charged to (or almost to) spill point while the lower reservoir sections are less completely filled (Figure 7). Entrapment of oil in the uppermost of the Raudhatain field reservoirs, the Mauddud, would not have been possible prior to the deposition of the overlying marine shales of the Ahmadi Formation, which is of early Late Cretaceous age (Owen and Nasr, 1958) (Figure 4).

If all or much of the oil in the Mauddud and Burgan reservoirs has been emplaced through vertical migration from an earlier and deeper accumulation in the Zubair Formation, then a Kazhdumi source for the oil must be considered unlikely. The logical source for the oil in such an earlier Cretaceous accumulation would appear to be the lower Lower Cretaceous Gadvan Formation, considered by James and Wynd (1965) to represent the basinal equivalent of the Zubair and Ratawi formations and comprising several hundreds of feet of dark shales and argillaceous limestones (Figure 10). The Gadvan is the partial equivalent of the Garau Formation, which Ala (1982) has identified as containing mature organics. Ala (1982) suggested the Garau as the source for the oil contained in early Early Cretaceous-aged reservoirs in the general area.

EXPLORATION AND DEVELOPMENT CONCEPTS

The first wildcat well in the territory of Kuwait, Kuwait Oil Company Bahrah-1, was drilled during 1936–1937 on the northern shore of Kuwait Bay (Figure 2). The well was located on a small surface anticline, associated with minor seepages of oil and gas, lying just south of the Jal az-Zor scarp, which runs along the northern side of the bay (Figure 2).

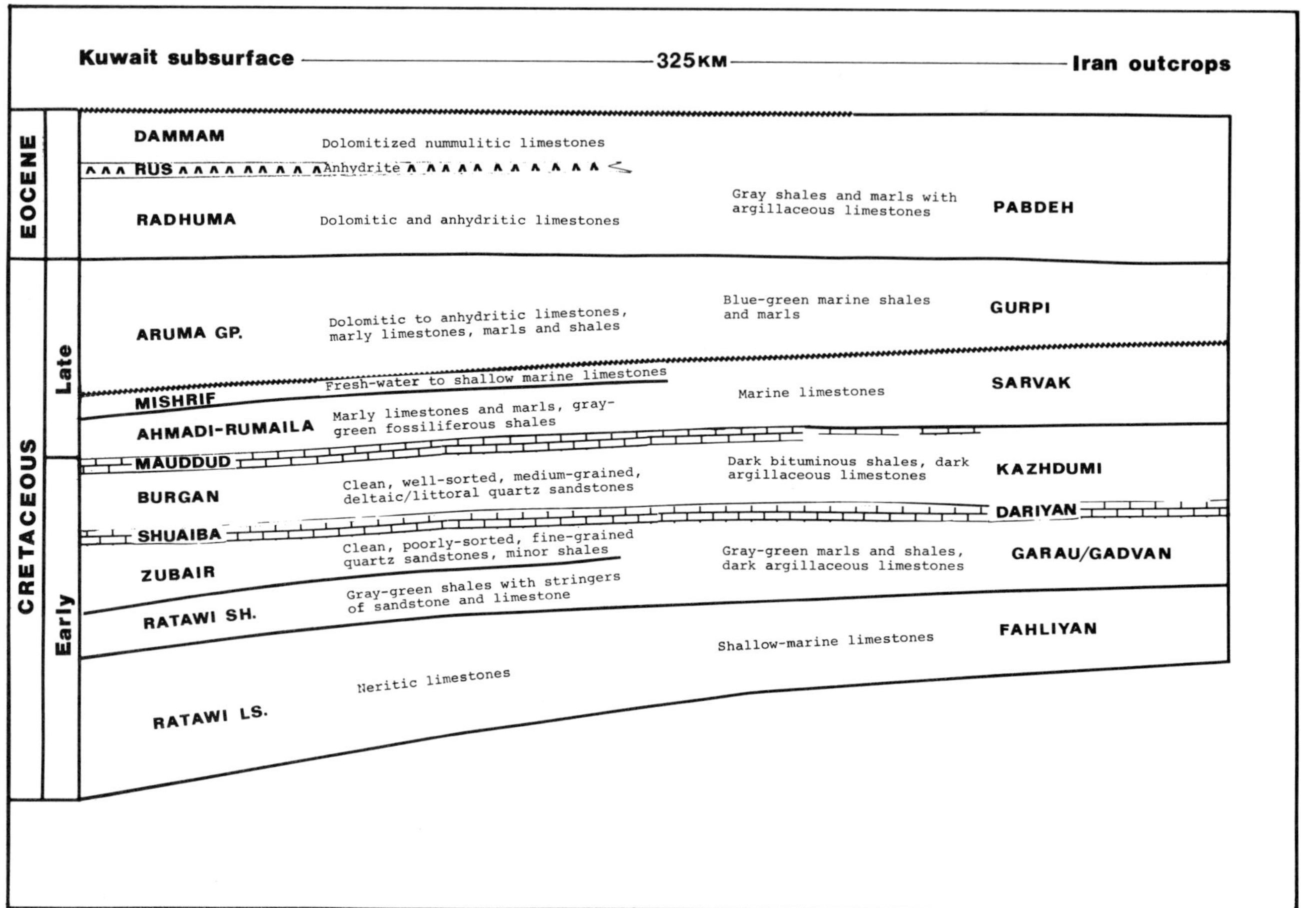

Figure 10. Stratigraphic relationship between reservoir sections present at Raudhatain field, Kuwait, and possible source-rock sections presently exposed in Iranian outcrop area (Zagros Mountains).

The Bahrah-1 well failed to discover oil or gas but did establish the presence of at least 7950 ft (2420 m) of sedimentary strata, ranging in age from Recent to late Early Cretaceous; some minor oil staining was reported in early Late Cretaceous carbonates (Milton, 1967; Brennan, 1990).

Following completion of Bahrah-1 as a dry hole, the focus of exploration shifted to southeastern Kuwait, where additional oil and gas seepages were known and where geophysical surveys were in progress. During 1938, oil was discovered in upper Lower to lower Upper Cretaceous sandstone reservoirs at Burgan. Development of the Burgan discovery was delayed by World War II but accelerated from mid-1945. Development drilling quickly confirmed the presence at Burgan of a giant oil accumulation.

During 1948, oil was discovered in lower Lower Cretaceous sandstone reservoirs at Zubair in southern Iraq (Figure 2). The presence of commercial oil accumulations in lower Lower Cretaceous reservoirs in southern Iraq and in upper Lower–lower Upper Cretaceous reservoirs in southeastern Kuwait, served to focus attention on the undrilled area of northern Kuwait, where the Raudhatain field is now located. An initial seismic reconnaissance survey was shot in the area north of the Jal az-Zor scarp (Figure 2) during 1949, designed to identify any northerly extension of the "Burgan trend." A subsurface structural lead was outlined, and was developed into a drillable anomaly by subsequent seismic work during 1951–1952 (Figure 3).

Before the subsurface structure was tested by the drill, it was understood that oil might feasibly be present in both the upper Lower–lower Upper Cretaceous (Wara-Mauddud-Burgan) and the lower Lower Cretaceous (Zubair) sections. The sandstones of the Zubair Formation were designated the official target of the initial test well, since prior experience at Zubair field suggested the likelihood of higher-quality crude oil being present in that zone. When the Raudhatain-1 wildcat was drilled during 1954–1955, it found the lower Upper Cretaceous (Wara Formation) sandstone section absent through facies change, but drilled on to discover oil in the underlying upper Lower Cretaceous carbonates and sandstones (Mauddud and Burgan formations) and in the deeper lower Lower Cretaceous (Zubair Formation) sandstones, as predicted.

There was nothing uniquely difficult in the outlining of the Raudhatain prospect once the area was accepted as prospective. The geophysical techniques of the time were adequate to define the area of interest, even in the face of extreme near-surface weathering conditions. Modern geophysical techniques, especially more accurate velocity control, would undoubtedly have produced a more reliable predrilling structural picture. But the Raudhatain structure proved large enough to be oil-bearing, even though the discovery well ran low to seismic prediction on key horizons.

An interesting aspect of Raudhatain field development was that the productive potential was largely established before the field was placed on stream. The reason for this lay in the need for increased export terminal capacity at Mena al-Ahmadi in southeastern Kuwait, together with the laying of a connecting pipeline (Figure 2). This occupied the period from the completion of the discovery well in November 1955 through April 1959. During this period, 23 delineation/development wells were drilled in the field area, an oil-gathering center was constructed, and field-wide pressure surveys and reservoir-fluid analyses were carried out (Milton and Davies, 1965). By June 1960, the theoretical combined potential of all completed oil wells was 269,500 BOPD and a maximum production rate into the gathering center of 200,000 BOPD had been achieved. Raudhatain field went on production during 1960 at an initial rate of approximately 100,000 BOPD and was on continuous production thereafter. Cumulative production passed the 1 billion bbl mark during 1975 (Table 2).

ACKNOWLEDGMENTS

The author wishes to thank Munim M. Al-Rawi for helpful comments and suggestions, and AAPG reviewers Howard H. Lester, Edward A. Beaumont, and J. Glenn Cole for constructive criticism of the

Table 2. Raudhatain field production statistics.

YEAR	ANNUAL PROD./BBL	AVG. BOPD	CUMULATIVE BBL
1960	33,623,796	92,119	33,623,676
1961	41,900,000	114,795	75,523,676
1962	47,200,000	129,315	122,723,676
1963	48,700,000	133,425	171,423,787
1964	53,700,000	147,123	225,123,787
1965	56,900,000	155,890	282,023,797
1966	61,400,000	168,219	343,423,400
1967	59,900,000	164,110	403,323,800
1968	71,200,000	195,069	474,523,800
1969	77,200,000	211,507	551,724,000
1970	77,900,000	213,425	629,624,000
1971	81,600,000	223,562	711,224,130
1972	83,900,000	229,863	795,124,125
1973	76,400,000	209,315	871,524,100
1974	73,000,000	200,000	944,524,100
1975	70,700,000	193,700	1,015,224,600
1976	73,000,000	200,000	1,088,224,600
1977	70,000,000	191,781	1,158,224,665
1978	90,200,000	247,123	1,248,424,560
1979	113,369,000	310,600	1,361,793,560
1980	70,671,300	193,620	1,432,464,860
1981	47,901,140	131,236	1,480,366,000
1982	33,821,200	92,661	1,514,187,200
1983	44,796,179	122,729	1,558,983,379
1984	46,632,822	127,761	1,605,616,201
1985	51,100,000*	140,000*	1,656,716,201*
1986			
1987			
1988			
1989			

Notes: * indicates estimated production. No separate production figures have been released for Raudhatain field since 1985. Production figures for the years 1980–1985 probably do not reflect actual producing capacity; field production has been subject to government restraint under OPEC quota guidelines.

text. Thanks are extended to the Society of Petroleum Engineers (Dallas, Texas) and to the Institute of Petroleum (London, England) for permission to utilize figures published in earlier papers.

REFERENCES CITED

Adasani, M., 1967, The North Kuwait oil fields: 6th. Arab Petroleum Congress, Kuwait, 39 p.

Ala, M. A., 1982, Chronology of trap formation and migration of hydrocarbons in Zagros Sector of Southwest Iran: American Association of Petroleum Geologists Bulletin, v. 66, n. 10, p. 1535–1541.

Al-Rawi, M. M., 1981, Geological interpretation of oil entrapment in the Zubair Formation, Raudhatain field: SPE Middle East Oil Technical Conference, Bahrain, SPE 9591, p. 149–159.

Arabian American Oil Company, 1975, Eastern Arabia and adjacent areas: Schlumberger Well Evaluation Conference, Arabia, p. 9–25.

Ayres, M. G., M. Bilal, R. W. Jones, L. W. Slentz, M. Tartir, and A. O. Wilson, 1982, Hydrocarbon habitat in main producing areas, Saudi Arabia: American Association of Petroleum Geologists Bulletin, v. 66, p. 1–9.

Beydoun, Z. R., and H. V. Dunnington, 1975, The petroleum geology and resources of the Middle East: Beaconsfield, England, Scientific Press Ltd., 99 p.

Brennan, P., 1990, Greater Burgan field, Kuwait, Arabia, *in* Treatise of Petroleum Geology Atlas of Oil and Gas Fields, Structural Traps I, Tectonic fold traps: Tulsa, American Association of Petroleum Geologists, p. 103–128.

Carmalt, S. W., and B. St. John, 1986, Giant oil and gas fields, *in* M. T. Halbouty, ed., Future petroleum provinces of the world, American Association of Petroleum Geologists Memoir 40, p. 11–54.

El-Naggar, Z. R., and I. A. Al-Rifaiy, 1972, Stratigraphy and microfacies of type Magwa Formation of Kuwait, Arabia; Part 1, Rumaila Limestone Member: American Association of Petroleum Geologists Bulletin, v. 56, p. 1464–1493.

Fox, A. F., 1959, Some problems of petroleum geology in Kuwait: Institute of Petroleum Journal, London, v. 45, p. 95–110.

Gharib, I. M., 1979, Geometry and depositional environment of the Zubair Formation, Raudhatain and Sabriya oil fields, northern Kuwait: Unpublished Thesis, Kuwait University, 59 p.

Halbouty, M. T., A. A. Meyerhoff, R. E. King, R. H. Dott, H. D. Klemme, and T. Shabad, 1970, World's giant oil and gas fields; geologic factors affecting their formation, and basin classification, *in* M. T. Halbouty, ed., Geology of giant petroleum fields: American Association of Petroleum Geologists Memoir 40, p. 502–528.

Holzer, H. F., 1968, Geology, general review, *in* Explanatory text to the synoptic geologic map of Kuwait: Vienna, Geological Survey of Austria, 87 p.

James, G. A., and J. G. Wynd, 1965, Stratigraphic nomenclature of Iranian Oil Consortium agreement area: American Association of Petroleum Geologists Bulletin, v. 49, p. 2182–2245.

Jamil, A. K., 1975, Hydrogeochemistry of brine waters of south oil fields, Iraq: 9th. Arab Petroleum Congress, Dubai, no. 123-B3, 21 p.

Kent, P. E., and H. R. Warman, 1972, An environmental review of the world's richest oil-bearing region—the Middle East: 24th International Geological Congress, sec. 5, p. 142–152.

Loutfi, G., and A. S. Jaber, 1970, Geology of the Upper Albian-Campanian succession in the Kuwait-Saudi Arabia Neutral Zone offshore area: 7th. Arab Petroleum Congress, Kuwait, no. 62-B3, 14 p.

Milton, D. I., 1967, Geology of the Arabian Peninsula: Kuwait: U.S. Geological Survey Professional Paper 560F, p. F1–F-7.

Milton, D. I., and C. C. S. Davies, 1965, Exploration and development of the Raudhatain Field: Institute of Petroleum Journal, London, v. 51, n. 493, p. 17–28.

Murris, R. J., 1980, Middle East: Stratigraphic evolution and oil habitat: American Association of Petroleum Geologists Bulletin, v. 64, p. 597–618.

Owen, R. M. S., and S. N. Nasr, 1958, Stratigraphy of the Kuwait-Basra area, *in* L. G. Weeks, ed., Habitat of oil: Tulsa, American Association of Petroleum Geologists, p. 1252–1278.

Strong, T. M. W., 1959, Some problems of petroleum geology in Kuwait: discussion: Institute of Petroleum Journal, London, v. 45, p. 95–110.

Appendix 1. Field Description

Field name .... *Raudhatain field*

Ultimate recoverable reserves .... *8.8 billion bbl oil; 11.0 billion bbl oil equivalent*

Field location:

    Country .... *Kuwait*

    Basin/Province .... *Arabian basin*

Field discovery:

    Year first pay discovered .... *Lower Cretaceous Mauddud Formation carbonates   1955*

    Year second pay discovered .... *Lower Cretaceous Burgan Formation clastics   1955*

    Third pay .... *Lower Cretaceous Zubair Formation clastics   1955*

Discovery well name and general location:

    First pay .... *Raudhatain-1, 29°52′N; 47°45′E*

    Second pay *Raudhatain-1*

    Third pay *Raudhatain-1*

Discovery well operator .... *Kuwait Oil Company (Gulf Oil-British Petroleum)*

    Second pay .... *Kuwait Oil Company*

    Third pay .... *Kuwait Oil Company*

IP in barrels per day and/or cubic feet or cubic meters per day:

    First pay .... *In excess of 1000 bbl (160 m³)/day*

    Second pay .... *No formal IP released*

    Third pay .... *No formal IP released*

All other zones with shows of oil and gas in the field:

Age	Formation	Type of Show
Early–Middle Miocene	*Lower Fars*	*Heavy oil*
Early Cretaceous	*Ratawi*	*Oil and gas*

Geologic concept leading to discovery and method or methods used to delineate prospect:
Postulated northward extension of productive "Burgan trend" confirmed by reconnaissance reflection seismic survey. Indicated subsurface anomaly developed into drilling prospect by subsequent detailed reflection seismic.

Structure:

    Province/basin type .... *Bally 221, Klemme 2Ca*

    Tectonic history
    Located in gently subsiding shelf environment from Paleozoic through most of Mesozoic time. Vertical uplift from early Early Cretaceous through Tertiary time, probably due to deep-seated halokinesis. Minor modification by compressive stress during post-Cretaceous time.

    Regional structure
    On outer margin of Arabian platform between Arabian shield/stable shelf to southwest and Mesopotamian (Zagros) geosynclinal belt to northeast.

    Local structure
    One of several major culminations developed along regional anticlinal ridge plunging gently north through Kuwait into southern Iraq.

Trap:

    Trap type(s)
    Structural trap: ovoid domal uplift with longer axis oriented NNE-SSW. Flank dips are gentle, rarely exceeding 3°. Minimum closure toward southeast where Raudhatain oil column is partially continuous across structural saddle into adjacent (Sabriya) structure.

Basin stratigraphy (major stratigraphic intervals from surface to deepest penetration in field):

Chronostratigraphy	Formation		Depth to Top in ft (m)
Pleistocene–Miocene	Kuwait Group		0
Eocene–Paleocene	Hasa Group		1200 (365)
Upper Cretaceous	Aruma Group		3800 (1160)
	Wasia Group:	Mishrif	6250 (1905)
		Rumaila	6700 (2040)
		Ahmadi	7000 (2130)
Lower Cretaceous		Mauddud	7400 (2255)
		Burgan	7750 (2360)
	Thamama Group:	Shuaiba	8700 (2650)
		Zubair	8950 (2730)
		Ratawi	10,300 (3140)

Location of well in field .... *Crestal area of dome*

Reservoir characteristics:

    Number of reservoirs .... *9 (4 major, 5 minor)*

    Formations .... *Mauddud (2); Burgan (2); Zubair (5)*

    Ages .... *Early Cretaceous (Albian–Barremian)*

    Depths to tops of reservoirs
    Mauddud, –7200 ft (–2195 m); Burgan, –7540 ft (–2300 m); L. Burgan, –7900 ft (–2410 m); Zubair, –8800 ft (–2680 m); Zubair, –9055 ft (–2760 m); Zubair, –9630 ft (–2925 m); Zubair, –10,000 ft (–3050 m)

    Gross thickness (top to bottom of producing interval) .... *2900 ft (880 m)*

    Net thickness—total thickness of producing zones
        Average .... *1200 ft (365 m)*
        Maximum .... *1750 ft (533 m)*

    Lithology
    Mauddud: organic to detrital, locally pseudo-oolitic limestone
    Burgan: clean, well-sorted, quartz sands, littoral/deltaic, interbedded with shales and siltstone
    Zubair: fine-grained, poorly sorted, subangular quartz sands intercalated with shales, siltstones, and minor limestones

    Porosity type .... *Mauddud, predominantly intercrystalline; Burgan and Zubair, intergranular*
    Average porosity .... *Mauddud, 18.5%; Burgan, 25.0%; Zubair, 21.0%*
    Average permeability .... *Mauddud, 20 md avg.; Lower Burgan and Zubair, 400–4000 md*

Seals:

    Upper
        Formation, fault, or other feature .... *Ahmadi Formation*
                (for reservoirs below the Mauddud, seals are intraformational impermeable developments)
        Lithology .... *Shale*
    Lateral
        Formation, fault, or other feature .... *Ahmadi Formation*
        Lithology .... *Shale*

Source:

    Formation and age .... *Kazhdumi Formation (Albian)*
    Lithology .... *Dark, polybituminous shale; subordinate argillaceous limestone*
    Average total organic carbon (TOC) .... *NA*
    Maximum TOC .... *NA, believed in excess of 5%*
    Kerogen type (I, II, or III) .... *II*
    Vitrinite reflectance (maturation) .... *NA*
    Time of hydrocarbon expulsion .... *Paleogene*

Present depth to top of source .... *Exposed at present-day surface*
Thickness .... *700 ft (210 m)*
Potential yield .... *NA*

Appendix 2. Production Data

Field name .... *Raudhatain field*

Field size .... *30 mi² (78 km²)*

   Proved acres .... *19,000 (7770 ha)*
   Number of wells all years *61*
   Current number of wells *41*
   Well spacing .... *Primary 600, secondary 200 to 66.66 ac*
   Ultimate recoverable .... *8800 million bbl (1400 million m³)*
   Cumulative production .... *2000 million bbl (318 million m³)*
   Annual production .... *Approx. 30 million bbl (4.8 million m³)*
   Present decline rate .... *NA*
     Initial decline rate .... *NA*
     Overall decline rate .... *NA*
   Annual water production .... *NA*
   In place, total reserves .... *22,600 million bbl (3600 million m³);*
oil-equivalent in place: 28,140,000,000 bbl

   In place, per acre-foot .... *NA*
   Primary recovery .... *NA*
   Secondary recovery .... *NA*
   Enhanced recovery .... *NA*
   Cumulative water production .... *NA*

Drilling and casing practices:
   Amount of surface casing set .... *±1250 ft (380 m)*

   Casing program
   Early wells: 13⅜-in. to 1250 ft (Dammam); 9⅝-in. to 7200 ft (Ahmadi); 7-in. to 8700 ft through Burgan; 5-in. to 10,300 ft through Zubair
   Later wells: 18⅝-in. to 1250 ft; 13⅜-in. to 5200 ft (Sadi); 9⅝-in. through Burgan; 7-in. through Zubair

   Drilling mud .... *Based on locally produced brackish water (6000 ppm)(Dibdibba Formation)*
   Bit program .... *NA*
   High pressure zones .... *Formation-water flows from Late Cretaceous Tayarat section*

Completion practices .... *Wells commonly drilled and cased, completed using workover rig*
   Interval(s) perforated .... *Substantial intervals; mainly single completions in Burgan or Zubair*
   Well treatment .... *Not required in principal field reservoirs*

Formation evaluation:
   Logging suites .... *ES/IES plus GR-N and sonic; CCL used for completions*
   Testing practices .... *DST on early wells; later wells drilled and cased through target reservoir(s) and production tested through casing/tubing*

   Mud logging techniques .... *NA*

Oil characteristics:
   Type .... *NA*
   API gravity .... *28.5–37.9° (increases with depth, reservoir by reservoir)*
   Base .... *Mixed asphaltic-paraffinic*

Initial GOR .... *602–1400 (highest in lower reservoirs of Zubair Formation)*
Sulfur, wt% .... *3.28–2.0 (general decrease with depth through reservoir section)*
Pour point .... *0°F (–18°C)*
Gas-oil distillate .... *Vol. % crude 24.1; aromatic vol. % 24; naphthene, vol. % 17*

Field characteristics:

Average elevation .... *130–215 ft (40–65 m)*
Present pressure .... *NA*
Pressure gradient .... *NA*
Temperature .... *170–205°F (75–95°C)*
Geothermal gradient .... *1.125°F/100 ft (0.0205°C/m)*
Drive .... *Gas and water (Mauddud, Zubair); water (Burgan)*
Oil column thickness .... *2900 ft (combined thickness)(880 m combined thickness)*
Oil-water contact .... *7760 ft (2365 m) highest; 10,200 (3110 m) lowest*
Connate water .... *Mauddud, 20%; Burgan, 15–24%; Zubair, 19–30%*
Water salinity, TDS .... *150,000–185,000 ppm*
Resistivity of water .... *NA*
Bulk volume water (%) .... *NA*

Transportation method and market for oil and gas:

30–36-in. pipeline (67 mi; 108 km) southeast to Mena al-Ahmadi North export terminal on Arabian Gulf littoral south of Kuwait City (Figure 2); thence by tanker to world market.

Hassi Messaoud Field—Algeria
Trias Basin, Eastern Sahara Desert

W. D. BACHELLER
International Consulting Geologist
Dallas, Texas

R. MICHAEL PETERSON
Terra Vac
Princeton, New Jersey

FIELD CLASSIFICATION

BASIN: Trias
BASIN TYPE: Foredeep
RESERVOIR ROCK TYPE: Sandstone
  (Quartzitic)
RESERVOIR ENVIRONMENT OF DEPOSITION: Alluvial Fan

RESERVOIR AGE: Cambrian
PETROLEUM TYPE: Oil and Gas
TRAP TYPE: Faulted Dome

LOCATION

Hassi Messaoud field, as shown in Figure 1, is located in Algeria within the eastern Sahara Desert. It is about 625 km (390 mi) south-southeast of the Mediterranean coastal city of Algiers and about 450 km (280 mi) west-southwest of the town of Gabes on the eastern coast of Tunisia.

HISTORY

Pre-Discovery

Prior to the drilling of the discovery wells at Hassi Messaoud, there was little enticement to explore the vast Sahara Desert except that it was a very large, unexplored area that appeared to be tectonically undisturbed. The absence of oil shows at the surface or in shallow wells also discouraged exploration. In 1953, however, the Algerian government granted hydrocarbon prospecting licenses for a 250,000 km² (96,500 mi²) area in the northern Sahara to two companies: Societe Nationale de Researche et d'Exploitation des Petroles en Algerie (SN REPAL) and Compagnie Francaise des Petroles (Algerie) (CFP[A]). These companies entered into a partnership agreement and began exploration efforts within the concession.

A regional gravimeter survey was conducted as the initial phase of the exploration. Due to the surface conditions of sand, gravels, and mountainous sand dunes, and perhaps to the unrefined state of the gravimetry tool, the results were poor at best. Seismic reflection surveys conducted at that time were subject to the same conditions and the results were virtually useless. Ultimately, a program of drilling deep wells in conjunction with an extensive reconnaissance refraction seismic survey was initiated. This resulted in the identification of both a significant thickness of sedimentary rocks and the major structural trends that exist in the area.

Discovery

The location of the discovery well at Hassi Messaoud, Md-1, was chosen by SN REPAL to test the large, gently dipping anticline that was interpreted from geophysical surveys. This well, identified on the structure map (Figure 2), was completed in May 1956 for 1125 BOPD of 43° API oil from Cambrian quartzitic sandstones at a depth of 3329 to 3468 m (10,922–11,377 ft). A few months later, CFP(A)'s well, Omj-1, drilled 8 km (5 mi) north of Md-1, was the "discovery well" on their concession. CFP(A)'s well was completed about 50 m (165 ft) structurally higher than well Md-1 for 2500 BOPD. Ultimately, it was evident that the east-west boundary between the two concessions split Hassi Messaoud field almost evenly into northern and southern regions.

Post-Discovery

It is unknown exactly how many wells have been drilled within the confines of the Hassi Messaoud field to date. It is known, however, that as many as ten rigs were used to develop the field in the early development phase. By the end of 1967, a total of

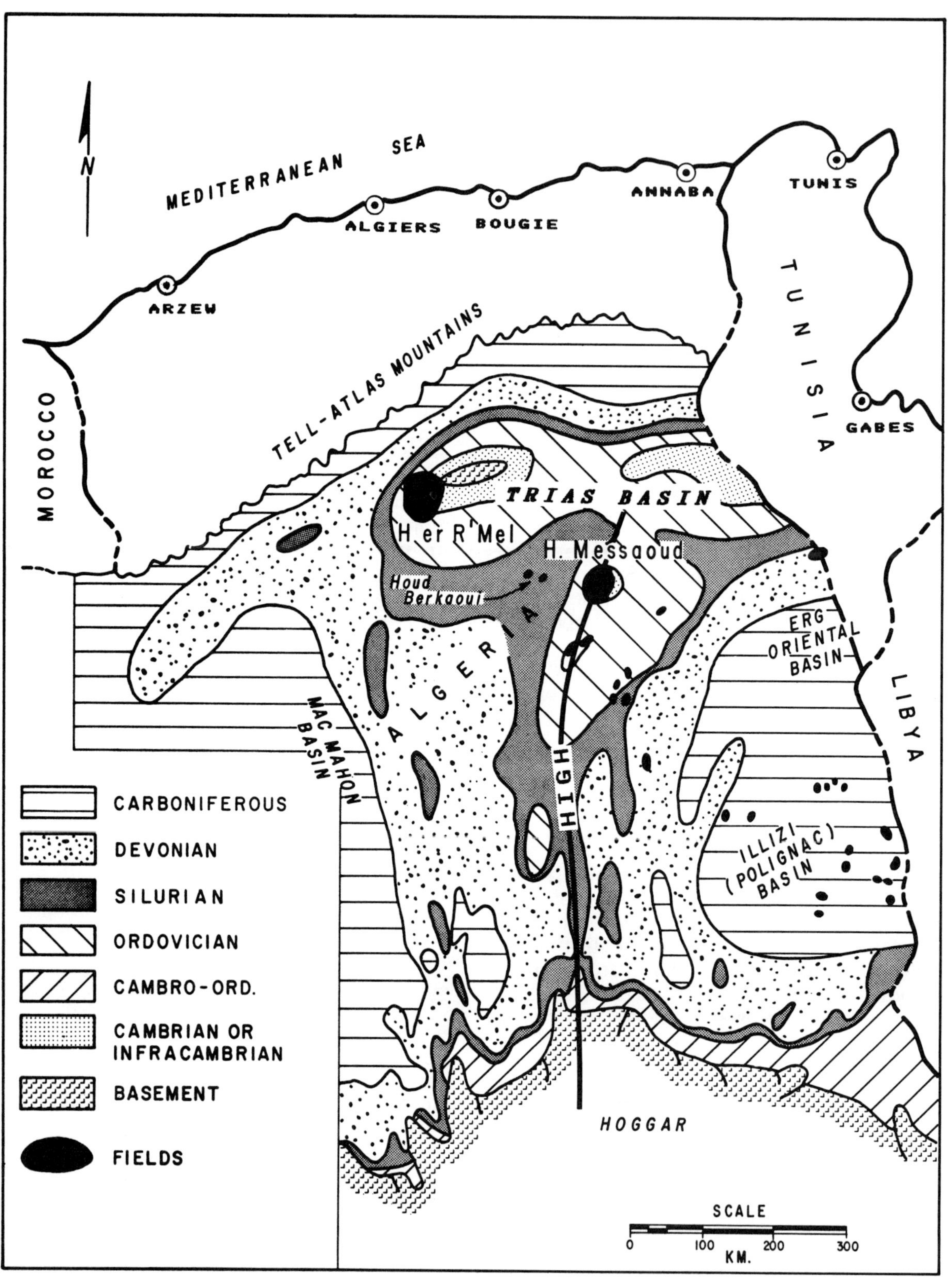

Figure 1. Subcrop map of Hercynian unconformity, Algerian Sahara desert. The Hassi Messaoud field is highlighted as well as other fields in the region. (Modified from Bishop, 1975.)

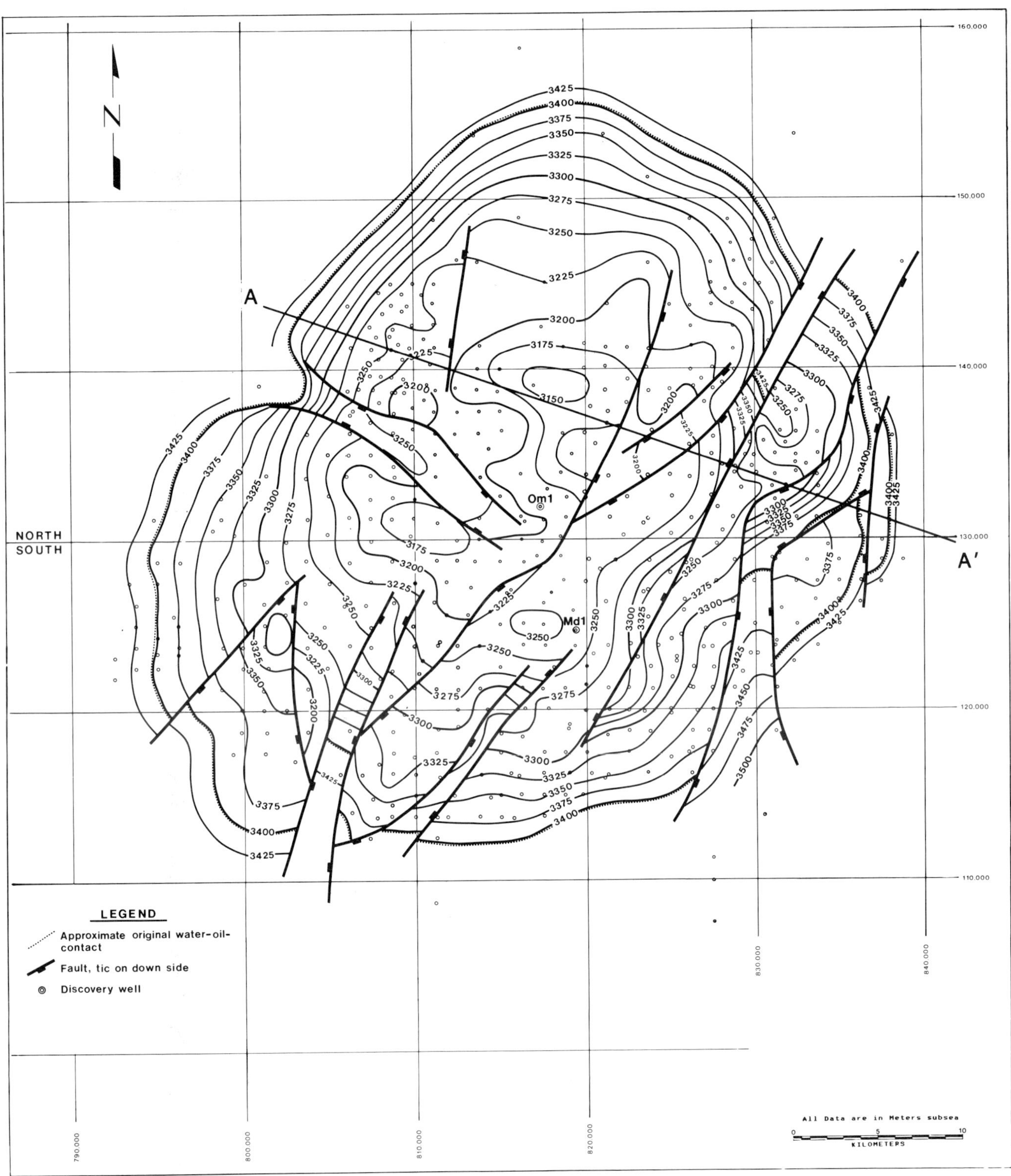

Figure 2. Structure map of Hassi Messaoud field. Structure is contoured on top of R-2 in meters subsea. Contour interval, 25 m. Discovery wells Md-1 and Om-1 are located on the south and north portions of the field, respectively. The cross section line (Figure 4) is oriented west-northwest to east-southeast across the northern part of the field.

178 wells had been drilled in the field. Since that time, it is estimated that at least 500 additional wells have been drilled in order to properly develop the reservoirs and provide the necessary wells for the various secondary recovery projects.

DISCOVERY METHOD

The delineation of some of the large subsurface structural features in the eastern Sahara desert area, including the Hassi Messaoud anticline, was accomplished through the use of regional seismic refraction surveys in conjunction with the drilling of judiciously located stratigraphic boreholes. Subsequent to the delineation of the Hassi Messaoud structure by these means, it was shown that the Hassi Messaoud structure could also be defined by using a comprehensive magnetometer survey.

STRUCTURE

Tectonic History

There is evidence of intense deformation of the pre-Late Cambrian metamorphic and granitic basement rocks at selected locations across the Sahara platform. The bulk of the Cambrian sandstones, however, was deposited on a broad, relatively stable, gently subsiding area. The general paleocurrent direction for most of the Cambrian rocks outcropping in the Hoggar Massif (south of the field) was to the north, implying the same to be true for the whole region. The incipient Hassi Messaoud structure was briefly active at the end of deposition of the R-a formation (Figure 3) where up to 50 m (165 ft) of that unit has been eroded from an area just south of the present-day structural crest.

The effects of the Hercynian orogeny are very pronounced at Hassi Messaoud field. The Cambrian and Ordovician rocks are locally characterized by extensive faulting and deep erosion. The subcrop map of the Hercynian unconformity (Figure 1) shows the widespread effect of the continuing tectonic events and erosion that have occurred in the Sahara Platform area during the Late Devonian through the end of the Permian Period. This activity resulted in the formation of the Messaoud–El Agreb anticlinorium that extends about 700 km (435 mi) in a northerly direction from the Hoggar Massif.

The final phase of the Hercynian orogeny was closed by the local deposition of Permian(?)–Triassic clastic sediments in the deeply eroded topographic low areas and the local extrusion of Triassic igneous rocks onto the landscape. De Lapparent (1961) considered these to be andesite flows. It is possible that the source of these flows and the source of the intruded igneous rocks found in several wells in the southwest part of the field may be the same.

The Sahara platform gently subsided from pre-Late Triassic to almost the end of the Cretaceous, with the entire area receiving a relatively thick sequence of evaporite and clastic sediments. Positive orogenic movement along the Messaoud–El Agreb high during the early Tertiary is postulated from the absence of rocks of this age in the area. The presence of flat-lying continental Miocene–Pliocene rocks suggests the tectonic activity had ceased by that time. The Hassi Messaoud area has remained tectonically inactive since.

Regional Structure

As shown on the index map (Figure 1), Hassi Messaoud field is situated in the Trias basin on the northern end of a 700 km (435 mi) long positive element extending north from the Hoggar Mountains of southern Algeria. This positive element, called the Messaoud–El Agreb high, separates the Mac Mahon basin on the west from both the Erg Oriental and Polignac basins on the east. This "Basin and High" complex is the result of both mild tectonic activity during the deposition of the early Paleozoic sediments on the Saharan Platform, which covered a large part of the north African continent, and the more intense tectonism of the Hercynian orogeny that began in Late Devonian (Balducchi and Pommier, 1970; Bishop, 1975; Chenevart, 1963; Fothergill, 1961; Malenfer and Tillous, 1963).

Local Structure

The Hassi Messaoud structure is a broad, flat-topped, oval anticline trending north-northeast to south-southwest, parallel to the major fault zones evident on the structure map (Figure 2). The extent of the field, based on the intersection of the oil-water contact (OWC) with the top of the R-a formation, is unknown due to the lack of well control around the fringes of the field. Based on data from the structure map on the top of the underlying R-2 formation, the field is about 46 km (29 mi) long by 35 km (22 mi) wide. By extrapolating the surface of the R-a formation to the original OWC, the maximum extent can be estimated for the R-a formation to be around 50 km (31 mi) long by 40 km (25 mi) wide.

It is believed that all of the significant faults in the field are vertical. Faulting and stratigraphic omissions are very rarely seen in the wells. The lack of fault cuts in the wells and the fact that the early reflection seismic surveys provided such poor quality data were the reasons why the faulting was not interpreted on the early seismic sections.

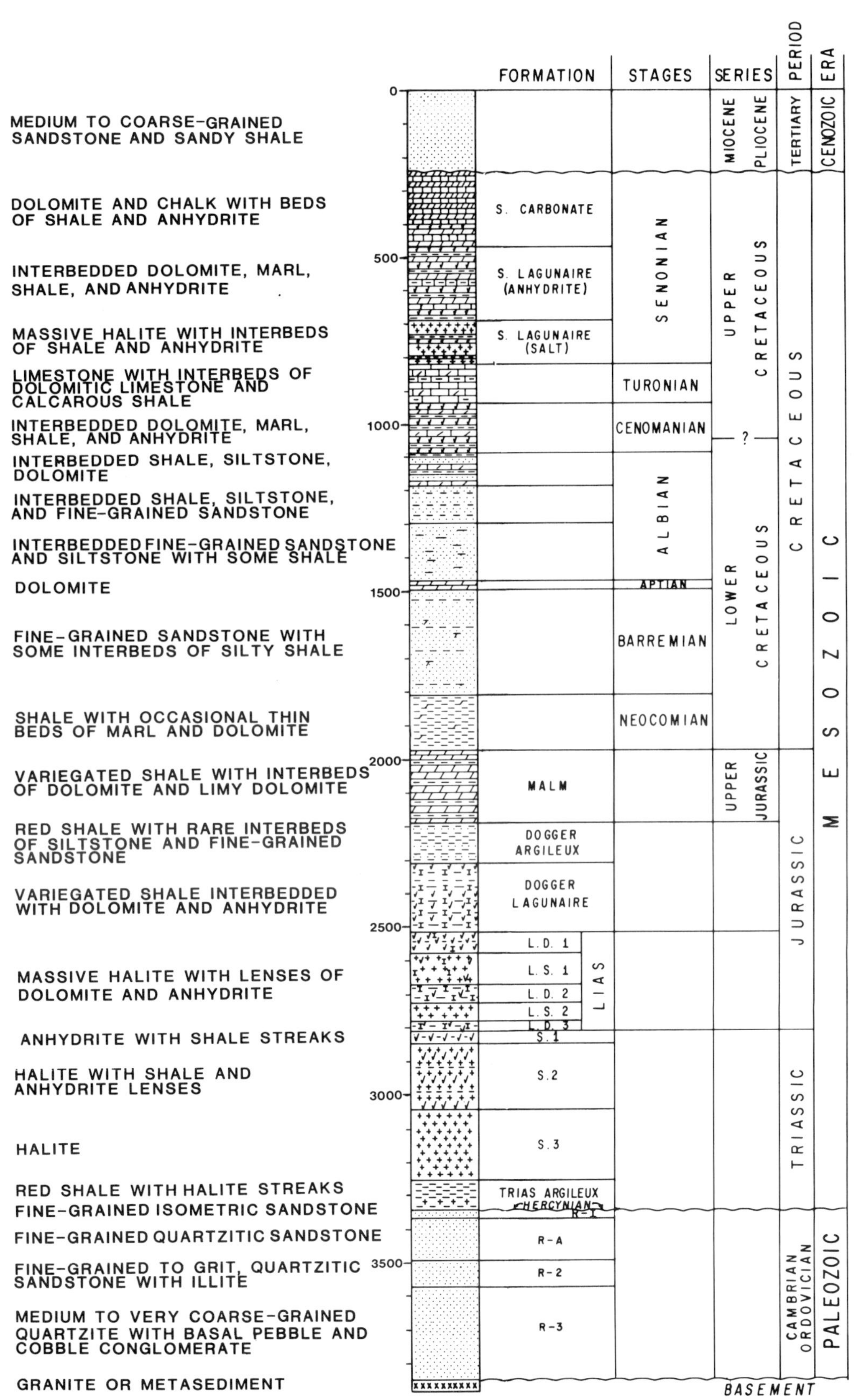

Figure 3. Generalized stratigraphic section, Hassi Messaoud field. The Cambrian-Ordovician section below the Hercynian unconformity is the producing interval in the field.

STRATIGRAPHY

Stratigraphy, Lithology, and Depths

A generalized stratigraphic section (Figure 3) is adapted from Balducchi and Pommier (1970), De Lapparent (1961), and Malenfer and Tillous (1963). This section identifies the various rock units that are normally encountered in the area.

Miocene-Pliocene

About 250 m (820 ft) of sand, sandstone, and sandy shale comprise the sediments of the Miocene-Pliocene.

Cretaceous

The upper 800 m (2625 ft) of the Cretaceous section include a very large amount of anhydrite, halite, limestone, and dolomite with minor interbedded shale. The Lower Cretaceous is a massive clastic unit with siltstone, shale, and fine-grained sandstone throughout its 850 m (2790 ft) thickness. A very widespread dolomite bed about 25 m (82 ft) thick occurs in the middle of this lower unit.

Jurassic

The upper third of the 850 m (2790 ft) thick Jurassic interval consists of massive dolomite. A thick shale section, identified on Figure 3 as the Dogger Lagunaire, separates the overlying dolomite from the massive salt and anhydrite unit below.

Triassic

The bulk of this 550 m (1800 ft) thick section is composed of massive anhydrite and halite overlying a basal clastic section. The thickness of this basal unit is highly variable due to the very irregular erosion surface upon which it was deposited. Extrusive(?) igneous layers are locally interbedded with the clastic rocks. The massive salt in the section has caused considerable problems both in well completions and in later collapsed casings.

Devonian

As shown on the Hercynian subcrop map (Figure 1), these sediments are present in the Algerian Sahara considerably beyond the limits of Hassi Messaoud field.

Silurian

The Hercynian erosion removed virtually all of the Silurian sediments to well beyond the limits of the Hassi Messaoud field. It is unknown where the erosion edge is located; however, some of the exploratory boreholes at least 25 km (16 mi) or more away have encountered Upper Silurian black, organic-rich shales and other sedimentary rocks in the section.

Ordovician

Although the various Ordovician rock units are eroded over virtually the entire broad crestal area of the Hassi Messaoud field, they are present in the area immediately around the field. Some of these formations are also present within the field limit because they were preserved in the narrow grabens seen on the structure map in the southern part of the field and in the northeast sector (Figure 2). Malenfer and Tillous (1963) note that well Md-55 has such a section.

According to Malenfer and Tillous (1963), the Ordovician section includes from Upper to Lower:

A. Argiles d'Azel.
B. Gres d'Ouargla.
C. Quartzites de Hamra, a fine-grained quartzite with low porosity and permeability.
D. Gres d'El Atchane, a 15 m (49 ft) thick fine-grained quartzitic sandstone having numerous shale partings and poor reservoir quality.
E. Argiles d'El Gassi, a black, glauconitic, graptolitic marine shale that is as much as 70 m (230 ft) thick at Hassi Messaoud.

Cambrian

The Cambrian stratigraphic sequence in the Hassi Messaoud field has been subdivided by Malenfer and Tillous (1963) into five major intervals. These include, from Upper to Lower, the (1) Zone des Alternances, (2) Zone Isometrique (R-i), (3) Zone Anisometrique (R-a), (4) Zone R-2, and (5) Zone R-3. The R-a is the primary reservoir unit with the R-i and R-2 included by these authors as secondary reservoirs.

1. Zone des Alternances—As its name suggests, this unit is composed of alternating layers of quartzites, siltstones, and black shales, totaling from 25 to 40 m (82–131 ft). This unit is considered the base of the Ordovician section by other investigators.

2. Zone Isometrique (R-i)—The R-i zone is a 50 m (164 ft) thick quartzitic sandstone unit characterized as a uniformly thick, well-sorted, medium-grained sandstone with interbeds of shale and siltstone. It contains abundant tigillites (a local name for scolite) throughout the section, except for the basal meter or so that includes winnowed debris from the underlying formation. The R-i unconformably overlies the R-a zone, and this contact equates to L'Homer's (1967) "Gamma Zero" marker in his fieldwide correlations. Where present above the original OWC, the R-i contains hydrocarbons but it is not considered a significant producing zone because of its poor reservoir characteristics. Despite the fact that this unit rarely produces, the top of the R-i is considered the top of the reservoir section.

Balducchi and Pommier (1970) included both the Zone des Alternances and the R-i zone as the Ordovician El Gassi sandstone in their correlation chart.

3. Zone Anisometrique (R-a)—Where not affected by the Hercynian and the pre-R-i zone erosion cycles, the R-a zone has a maximum total thickness of 150 m (492 ft) in the western portion of the field. The

R-a zone has been subdivided into five subzones. These subzones, from upper to lower, include Drain 4, Drain 3, Drain 2, Interdrain, and Drain 1. All subzones are essentially the same lithologically, with the exception of Drain 3, in that they all consist of highly cross-bedded, very poorly sorted, medium-grained to microconglomeratic quartzose sandstones with irregularly interbedded lenses of shaly siltstone.

Malenfer and Tillous (1963) have defined a "Drain" as that portion of the R-i zone that has a much better than average capacity for production. Expressed as Kh (millidarcy meters), as much as 70 to 80% of the total Kh of a well is localized in these "drains." Malenfer and Tillous have also found a strong similarity between continuous flowmeter curves and plots of cumulative Kh for the respective wells. Further, Balducchi and Pommier (1970) consider the variations in the producibility to be directly related to the lithology, with the best zones (drains) having a high median grain size, low clay content, and less than 15% silica as overgrowth cement.

Drain 3 differs in that it is the only subzone in the R-a that contains abundant tigillites and is generally finer grained and is not cross-bedded. Also, the primary clay mineral in Drain 3 is kaolinite. Drain 3 is L'Homer's (1967) "Median Zone."

L'Homer's (1967) study of the R-a interval at Hassi Messaoud and the outcrop study by Beuf et al. (1971) of equivalent sediments in the Hoggar Mountains indicate that these sediments were deposited in a series of alluvial fan and associated depositional environments. Considering this, the thickness and the continuity of the subzones are remarkably uniform on a fieldwide basis.

4. Zone R-2—This zone is a uniformly thick sequence of highly cross-bedded, medium- to coarse-grained quartzose sandstones very similar to the R-a zone sediments except that these have a much higher clay content, occurring as interstitial clay and irregular thin interbeds of shale. The clay minerals are predominantly illite with minor amounts of kaolinite. Because of the clays, this zone has poor reservoir qualities and is generally nonproductive. In the northeastern part of the field, however, the upper 20 m (65 ft) or so is included as part of the producing interval. Where it is not eroded, this zone is about 80 m (262 ft) thick.

5. Zone R-3—At least three wells (Md-2, Om-81, and Omg-57) have been drilled through the R-3 zone into the basement. Based on these boreholes, this zone is a highly cross-bedded, arkosic, very argillaceous quartzose sandstone and microconglomerate with poorly sorted and angular grains. The cement, predominantly illite, averages 30% of the rock content. The section also includes a pebble to cobble basal conglomerate. The R-3 section thickens from 275 m (902 ft) in well Md-2 in the south-central part of the field northward to 368 m (1207 ft) in well Omg-57 north of the field.

L'Homer (1967) describes pink granite in wells Md-2 and Om-81 and thin sandstone of the Infracambrian in well Omg-57 below the R-3 zone. The Mobil Field Research Laboratory dated the pink granite porphyry as Early Cambrian by establishing its age as 560± 25 million years by the Rb/Sr method (Balducchi and Pommier, 1970).

TRAP

Trap Type

Hassi Messaoud field is a structural trap, composed of a number of discrete producing areas separated by vertical sealing faults. Over the crest of the structure the vertical seal is a combination of the overlying Triassic shales and the effects of erosion at the Hercynian unconformity. The effectiveness of this eroded surface is evidenced by the virtual lack of hydrocarbons in the basal Triassic clastic section, which includes an appreciable amount of sandstone. On the flanks of the structure, where the Ordovician section has not been removed by Hercynian erosion, the vertical seal appears to be the combination of the Ordovician shales and the minor erosion surface between the Cambrian R-a formation and the Ordovician shales and tight sandstones.

Reservoirs

Three different but stratigraphically continuous Cambrian quartzitic sandstones comprise the reservoirs at Hassi Messaoud. These rocks occur below the Hercynian unconformity at the base of a largely Mesozoic section ranging roughly from 3250 to 3500 m (10,700–11,500 ft) thick.

Depositional Setting

Beuf et al. (1971) observed "tigillites" in the Cambrian outcrops in the Hoggar mountain area and described them in considerable detail. Tigillite is a local name for scolite or *Skolithos*, vertical borings or tube-like evidence of either plant or animal life, containing clastic material that is either finer or coarser grained than the surrounding rock. The tigillites observed in the outcrop ranged from 3 to 10 mm in diameter and when abundant, such as in the R-i zone and in Drain 3 of the R-a zone at Hassi Messaoud field, they can easily be identified in cores. Since tigillites are observed in shalier, finer-grained, planar-bedded sandstones, they probably indicate a shallow-marine or brackish water environment of deposition.

From the studies done by L'Homer (1967) and Beuf et al. (1971), the remaining part of the Cambrian section was deposited as a series of alluvial fans and associated fluvial units. Although typical individual channel-fill deposits are evident in the outcrop, they have not been traced in the subsurface.

Porosity Types and Diagenesis

The reservoir characteristics of the three productive zones are highly variable due to the different amounts of detrital clay, diagenesis of the clay minerals, and the amount of secondary silicification. Kulbicki and Millot (1963) identified two successive periods of diagenesis that altered the original constituents of the Cambrian sandstones. The earliest diagenesis altered the feldspars, micas, and illite clays to kaolinite. The second period of diagenesis changed the "fan" and "accordion" (or vermicular) kaolinite facies back into illite where the section was not protected by the oil saturation (Kulbicki and Millot, 1963).

Porosity, Permeability, Pressure Relationships

Balducchi and Pommier (1970) have summarized the averages and ranges of the porosity and permeability and other aspects of the three reservoirs, and an adaptation of their data is presented as Table 1. Slightly different petrophysical data presented by Malenfer and Tillous (1963) are also included in Table 1. This difference may be due to the seven years of additional data available to Balducchi and Pommier (1970). Although the actual number of cored wells is unknown, it is estimated that about 75% of all wells drilled within the confines of the field have been conventionally cored through the reservoir section.

Reservoir pressure was originally 483 kg/cm^2 (6870 psi) at 2926 m (9600 ft) and temperature of the oil at this depth is 119°C (246.2°F) (IP Review May 1962). This overpressured reservoir required heavy weighted drilling mud during drilling and completion operations. (The even more overpressured zones of Lower Jurassic anhydrites between depths of 2500 and 2600 m [7600–7900 ft] required still heavier mud weights, in excess of 2.2 kg/L [18.4 lb/gal].)

Liquid expansion was the original recovery mechanism. Bonfils (1963), however, studied the compressibility of the Hassi Messaoud reservoir rocks and considered that another complimentary recovery mechanism.

De Lavilleon (1965) described the objective of the gas injection secondary recovery scheme initiated as trying to maintain a reservoir pressure of at least 380 kg/cm^2 (5400 psi). DeClaude and Leduc (1965) identify the minimum reservoir pressure as 360 kg/cm^2 (5120 psi)

Fracture Importance

Although the field is separated into different producing areas through the sealing capabilities of the major vertical faults, fracturing does appear to have some influence on the producing characteristics of certain areas within the field. Specifically, the wells near the intersections of two fault/fracture zones have higher productivity than the average well in the field.

Pay Zone Thickness

The thickness of the uneroded R-i, R-a, and R-2 zones totals about 280 m (920 ft), but this figure cannot be considered the pay zone thickness. Due to the various combinations of the Hercynian and intraformational erosion cycles, the location of the OWC relative to the eroded areas, the variation in original thickness of the R-a zone across the field, and the lack of reservoir quality rock in much of the R-2 zone, the actual hydrocarbon column is much thinner. Originally, it probably was no more than 150 m (492 ft) thick and this thickness was confined to the west-central part of the field.

Hydrocarbon Characteristics

Balducchi and Pommier (1970) state that, "There is no gas cap in Hassi Messaoud and the oil is highly undersaturated. Some geographic differentiations of the crude are evident. Overall, it is an intermediate base crude with a mean gravity of 43° API, low viscosity, and low sulphur content. It is rich in gasoline."

The IP Review (1962) describes the Hassi Messaoud crude oil as "a light crude having an average density of about 0.80 and containing 0.6% sulphur."

Fluid Flow Rates

The two discovery wells for SN REPAL and CFP(A) had initial flow rates of 1125 and 2500 BOPD, respectively.

Hydrocarbon Sources

Source Beds

Balducchi and Pommier (1970) identified two zones in the Late Silurian black shale section as containing large quantities of organic carbon and being rich in kerogen. Thus, they concluded that these zones were the most likely source of the Hassi Messaoud hydrocarbons. The overlying Triassic shales were eliminated as source beds due to a paucity of contained organic materials.

Later geochemical analyses have identified both the Ordovician and the Late Silurian black shales as the source rock for the Hassi Messaoud oil (Poulet and Roucache, 1969). The Ordovician source rocks include the thick, black shales of the Argile d'el Gassi formation found in the few wells located around the periphery of the field. It is theorized that these Ordovician shales, and not the Silurian shales, were the main contributor of the Hassi Messaoud oil due to their relative proximity to the structure.

Migration Paths and Timing

Through their analysis of the paleotectonic events involving the Hassi Messaoud structure, Poulet and Roucache (1969) have concluded that there was "no structural closure on the southern flank of the Hassi Messaoud structure until Cretaceous Albian time. Thus, migration of hydrocarbons into the trap could not have taken place before the Cenomanian." This concept does not allow for the possibility of oil

Table 1. Generalized petrophysical properties for the Cambrian and Ordovician section. Data in [brackets] from Malenfer and Tillous (1963). All other from Balducchi and Pommier (1970). Conversions from metric to English units added.

Zone	Thickness m (ft)	Quartz (%)	Secondary Silica (%)	Clay Amount %	Type	Grain Size (microns)	Mean Porosity (%)	Mean Permeability (md)
El Gassi Shale	91 (298)	50	15	20	Illite	300–400	—	—
El Gassi Sandstone (R-i)	61 (200)	70	20	6	[Illite]	300	— [Mediocre]	0.0
R-a	122 (400)	70	14 [10]	8 [8/15]	Kaolin	300	8.0 [8/15]	0–100.0
U. R-2	34 (112)	65	9	15	Illite–Kaolin	400	11.0	0–100.0
L. R-2 (All)	52 (171)	60	5 [7]	30 [20]	Illite	450	— [12.5]	0.0 [>1.0]
R-3	274–366 (899–1201)	55–60	2–5	35–25	Illite	500–800	-	0.0

generation and migration at a slightly earlier time with the oil being trapped by the numerous vertical sealing faults that are present in the area or the possible effect of hydrodynamic gradients.

During their study, Poulet and Roucache (1969) also determined that 1500 m (5000 ft) was the critical thickness of overburden for the generation of hydrocarbons from the Ordovician and Silurian shales of the northern Sahara area. This figure is compatible with the world-wide depth of hydrocarbon generation and migration charts presented by Ala (1982). It is also compatible with the generalized structure cross section, Figure 4, that shows this depth of burial was attained during the early part of the Lower Cretaceous.

The combination of the subcrop map of Hercynian unconformity (Figure 1) and the generalized structural cross section shows the Hercynian unconformity truncating successively older rocks toward the center of the regional high trending northerly from the Hoggar Massif through the Hassi Messaoud structure. The Silurian subcrop is closer to the field area in the northwest direction, being only 25 km (16 mi) away from the edge of the field. Clearly, then, the shortest distance for the migration of the oil generated from the Silurian shales would be in the southwest direction.

The west-east structural cross section presented as figure 6 in the study by Malenfer and Tillous (1963) shows the Argile d'el Gassi present in wells Md-27, Md-41, and Md-201. Since these wells are from 1 to 7 km (0.6–4.2 mi) away from the original OWC at the top of the R-2 formation, the Ordovician source rocks are at the very periphery of the field. The

migration paths for these hydrocarbons would be toward the structural high from all sides of the field.

It has been suggested that since the Hassi Messaoud hydrocarbons were probably generated from both the Ordovician shales and from the Late Silurian black shales, it is likely that the small variations in the hydrocarbon composition in the many unique producing areas of the field are the result of differences in the relative amounts contributed by the different rock units.

Balducchi and Pommier (1970) suggest the actual migration path of the hydrocarbons was updip in the source rock until the unconformable surface was reached. The fluids migrated farther updip along the underside of the unconformity until they encountered porous and permeable zones in the Cambrian sediments, whereupon they then migrated downward into the reservoirs.

Maturation
Chemical analyses of the Hassi Messaoud crude have shown a high degree of maturation, which is characteristic of crude oil from Paleozoic source rocks (Balducchi and Pommier, 1970).

EXPLORATION AND DEVELOPMENT CONCEPTS

Despite the fact that the geophysical tools have been much improved since Hassi Messaoud was discovered, it is apparent that any regional exploration efforts should always include both gravimetric

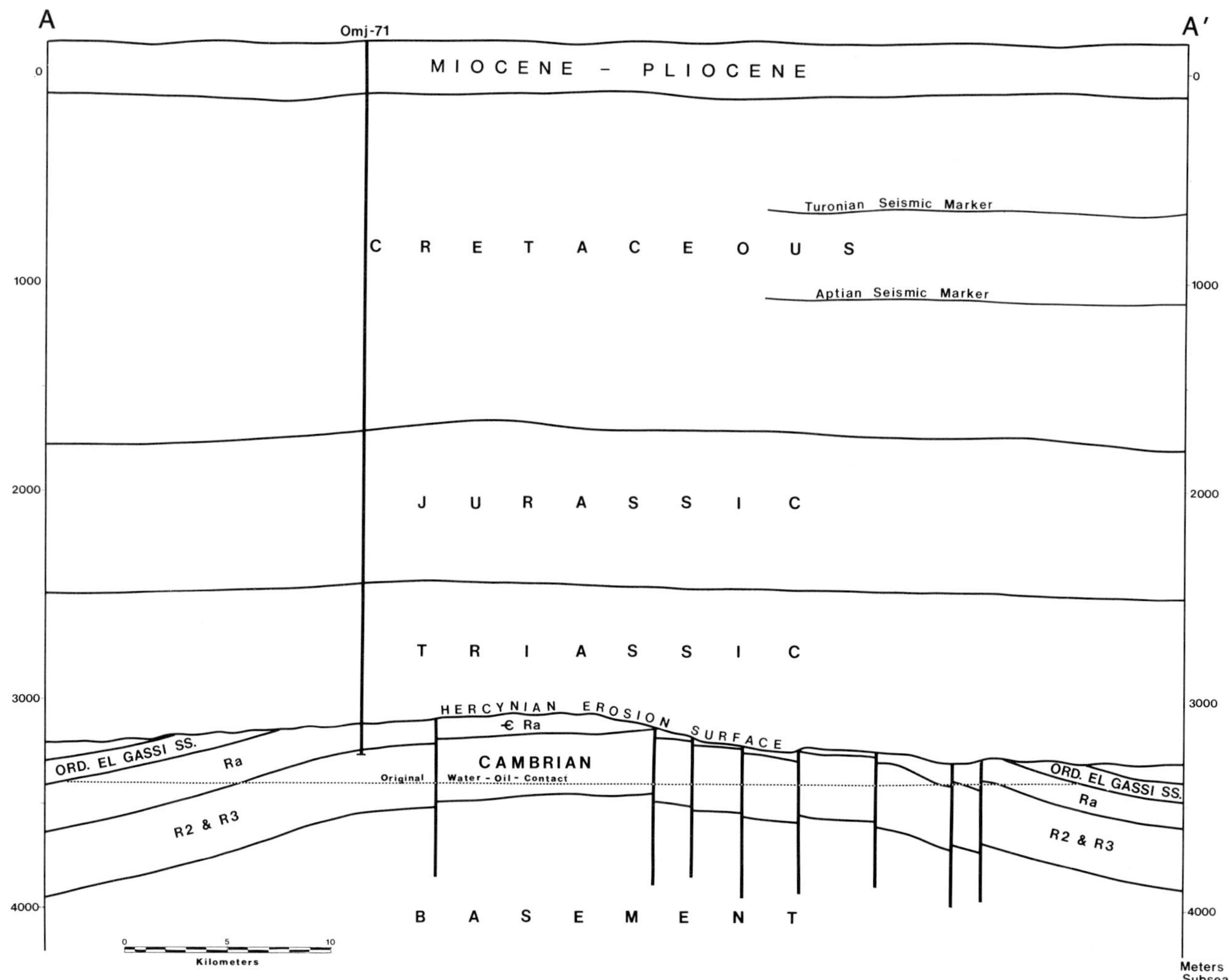

Figure 4. Generalized west-northwest to east-southeast structural section, Hassi Messaoud field. See Figure 2 for location of cross section. Producing intervals below the Hercynian unconformity as well as the original OWC are shown. Note that the Mesozoic and younger strata show little evidence of the underlying structure.

and magnetometer surveys along with the regional seismic survey. As was eventually learned, the magnetometer tool could have defined the broad structure of this field as well as the early refraction seismic surveys did. Clearly, then, when the three geophysical methods are used in combination, the resulting interpretations can be much more definitive.

Since it is necessary to have some knowledge of the stratigraphy in an unexplored area, the technique of drilling strategically located stratigraphic boreholes to supplement the geophysical surveys and integrate the results with data from outcrop investigations is a sound exploration strategy. In the Hassi Messaoud exploration, this technique confirmed the presence of a thick series of sedimentary rocks that eventually proved to contain both source and reservoir rocks. It is also necessary to have an adequate understanding of the regional geology in order to help decipher and understand the local geology.

The understanding of the local Hassi Messaoud geology and the correlation of the local stratigraphy with regional stratigraphy was enhanced by the availability of the many cores taken through the reservoir section. Thus, credit must be given to those who understood the need for such expenditures.

Finally, much credit must be given to those who were willing to take the great risks of drilling the first exploration stratigraphic tests in such a remote and forbidding area as the Sahara Desert.

REFERENCES CITED

Ala, M. A., 1982, Chronology of trap formation and migration of hydrocarbons in Zagros sector of southwest Iran: American

Association of Petroleum Geologists Bulletin, v. 66, n. 10, p. 1535–1541. (English)

Balducchi, A., and G. Pommier, 1970, Cambrian oil field of Hassi Messaoud, Algeria, *in* M. T. Halbouty, ed., Geology of giant petroleum fields, American Association of Petroleum Geologists Memoir 14, p. 477–488. (English)

Beuf, S., B. Biju-Duval, O. de Charpal, P. Rognon, O. Gariel, and A. Bennacef, 1971, Les Gres du Paleozoique Inferieur au Sahara: L'Institut Francais du Petrole Collection, Science et Technique du Petrole, n. 18, 464 p. (French)

Bishop, W. F., 1975, Geology of Tunisian and adjacent parts of Algeria and Libya: American Association of Petroleum Geologists Bulletin, v. 59, no. 3, p. 413–450. (English)

Bonfils, J., 1963, La compressibilite de la roche dans le mechanisme de la recouperation a Hassi-Messaoud: Revue de l'Institut Francais du Petrole, v. 18 Supplementary Issue, December, p. 145–161. (French)

Chenevart, C., 1963, Peiges structuraux et synclinaux perches de la subsurface paleozoique du Sahara: Ecologae Geol. Helv., v. 56, n. 2, p. 933–939. (French)

De Claud, C., and J. Leduc, 1965, Premiers resultats de l'injection de gaz a Hassi-Messaoud: 2nd ARTPF (Association Rech. Tech. Forage et Production) Colloquium, Rueil, France, 5/31/-6/4/65, Transaction paper no. 24, p. 427–434. (French)

De Lapparent, C., 1961, Hassi Messaoud, giant of the Sahara is tribute to scientific oil search: Oil and Gas International, v. 1, n. 1, p. 43–46. (English)

De Lavilleon, P., 1965, Gas injection favorable at Hassi Messaoud: World Petroleum, v. 36, n. 9, p. 30–32. (English)

Fothergill, C. A., 1961, Oil in Africa, Part 2, oil developments and aspects of petroleum geology in North and West Africa: Petroleum (London), v. 24, n. 6, p. 221–224. (English)

IP Review, 1962, The oilfield of Hassi Messaoud, the fruits of French endeavour: Institute Petroleum Review, v. 16, n. 185, p. 150–157. (English)

Kulbicki, G., and G. Millot, 1963, Diagenesis of clays in sedimentary and petroliferous series: 10th National Research Council Clays and Clay Minerals Conference, Austin, Texas, 1961, Proceedings 1963, p. 329–330 plus 6 pages of photomicrographs. (English)

L'Homer, A., 1967, Precisions sur la lithologie et la sedimentologie des gres du Cambrien (Zones Ri et Ra) a Hassi Messaoud Sud: Publication du Service Geologique de l'Algerie, Bulletin no. 35, p. 105–125. (French)

Malenfer, J., and A. Tillous, 1963, Etude du champ de Hassi-Messaoud, stratigraphie, aspect structural, etude de detail du reservoir: Revue de L'Institut Francais du Petrole, v. 18, n. 6, p. 851–867. (French)

Poulet, M., and J. Roucache, 1969, Etude Geochimique des Gisements du Nord Sahara (Algerie): Institut Francais du Petrole, v. 24, p. 615–644. (French)

SUGGESTED READING

Baude, M., and M. Tartera, 1963, Les problemes de completion et d'exploitation recontres par la S. N. Repal a Hassi-Messaoud: Revue de l'Institut Francais du Petrole, v. 18, Supplementary Issue, December, p. 589–604. (French)

Boultabi, A., 1985, Twenty years of hydrocarbon miscible gas injection in the Hassi-Messaoud field—Algeria: 3rd AGIP SPA et al Improved Oil Recovery Europe Meeting (Rome, 85.04.16–18) Proc V2, p. 97–107. (English)

Boultabi, A., J. B. Dorsey, J. G. Litherland, and A. A. Saadi, 1975, Past and predicted future performance of Hassi Messaoud field, Algeria under gas and water injection: 50th Annual SPE of AIME Fall Meeting Preprint no. SPE-5537, 16 p. (English)

Cassan, J. P., and J. Lucas, 1966, Diagenesis of the argillaceous sands of Hassi Messaoud (Sahara) silicification and dickitization: Alsace-Lorraine Serv. Carte Geol. Bulletin, v. 19, nos. 3-4, p. 241–253. (French)

Chiarelli, A., 1978, Hydrodynamic framework of eastern Algerian Sahara-influence on hydrocarbon occurrence: American Association of Petroleum Geologists Bulletin, v. 62, n. 4, p. 667–685. (English)

Dupal, P., 1970, Progress achieved in the drilling program at Hassi Messaoud North during the last ten years: Rev Ass Franc Tech Petrol no. 204, p. 45–50. (French)

Graf, P., and J. M. Neu, 1971, Problems connected with salt deposit formation at Hassi Messaoud: Rev Institut Franc Petrol Ann Combust Liquides, v. 26, n. 11, Nov., p. 985–1007. (French)

Hoffman-Rothe, J., 1966, Zur Stratigraphie und Tektonik des Paleozoikums der Algerischen Ostsahara: Geologische Rundschau, v. 55, n. 3, p. 736–774. (German)

Hossin, A., P. Delvaux, M. A. Quint, and M. Gondouin, 1970, Application to the Hassi-Messaoud Cambrian reservoir of a new quantitative interpretation method for shaly sands: 11th Annual SPWLA Logging Symposium Transaction Paper no. H, 28 p. (English)

IP Review, 1964, Field gas injection at Hassi Messaoud: v. 18, n. 210, p. 196–202. (English)

IP Review, 1964, S.N. Repal's plant at Hassi Messaoud South: v. 18, n. 212, p. 306–308. (English)

Kerbouc, P., 1968, Hydraulic fracturing in Well ONM 15, Hassi Messaoud field: 3rd ARTFP (Assn Rech Tech Forage + Prod) Symposium 68.09.23–26, Paper No. 38, 19 p. (French)

Khelil, C., and E. P. Miesch, 1973, Applications of material balance techniques in matching and predicting performance of the heterogeneous multilayer Hassi Messaoud field: 48th Annual SPE of AIME Fall Meeting Preprint no. SPE-4431, 16 p. (English)

Labussiere and Boyeldieu, 1964, Le log de densité de formation dans le champ d'Hassi-Messaoud: Association Francaise des Techniciens du Petrole Bulletin, no. 165, 5/31/64, p. 257–268. (French)

Lamau, H., and J. Bosio, 1965, Gas injection at Hassi Messaoud: Petroleum Engineer, v. 37, n. 6, p. 84–91. (English)

Lucas, J., and J. P. Cassan, 1967, Comportment diagenetique du quartz et de la dickite dans les gres d'Hassi-Messaoud (Sahara): 7th International Sedimentologic Congress Paper, 8/14–16/67, 3 p. (French)

Magloire, Philippi R., 1970, Triassic gas field of Hassi R'Mel, Algeria, *in* M. T. Halbouty, ed., Geology of giant petroleum fields: American Association of Petroleum Geologists Memoir 14, p. 489–501. (English)

Massa, D., M. Ruhland, and J. Thouvenin, 1972, Structure and natural fractures in the Hassi-Messaoud oil field (Algeria), Part 1, Observations of tectonic phenomena in the Tassili Najjers: Rev Institut Franc Petrol Ann Combust Liquides, v. 27, n. 4, p. 489–534. (French)

Massa, D., M. Ruhland, and J. Thouvenin, 1972, Structure and fracturing of the Hassi Messaoud field, Algeria, Part 2: Rev Institut Franc Petrol Ann Combust Liquides, v. 27, n. 5, p. 665–713. (French)

Megatli, A., A. Said, and D. G. Sarber, 1969, Exploration in Algeria; past, present, and future: The Exploration for Petroleum in Europe and North Africa, Institute of Petroleum p. 271. (English)

Odeh, A., 1975, Al Agreb-El Gassi oil fields, Central Algerian Sahara: American Association of Petroleum Geologists Bulletin, v. 59, n. 9, p. 1676–1684. (English)

Roberts, W. H., 1965, Hydrodynamics applied to petroleum exploration: Enciclopedia del Petrolio e del Gas Naturale, v. 5, p. 617–636. (Italian)

Soviche, G., G. Baron, and M. Tartera, 1965, Rock behavior during developments for the recovery of oil in producing the Hassi Messaoud reservoir: 2nd ARTFP Colloquium Transaction Paper no. 13, p. 229–246. (French)

S. N. Repal Direction Production, 1962, Lucky well is Hassi Messaoud: World Petroleum, v. 33, n. 2, , p. 26–28. (English)

Tartera, M., G. Baron, and G. Soviche, 1965, Tenu des terrains pendant l'exploitation en decouvert des puits de production d'Hassi-Messaoud: 2nd ARTFP (Assoc. Rech. Tech. Forage et Production) Colloquium (Rueil, France, 5/31–6/4/65) Transactions Paper no. 13, p. 229–246. (French)

Wells, M. J., 1970, Hassi Messaoud production capacity to expand 50 percent: World Petroleum, v. 41, n. 2, p. 18–21. (English)

Whiteman, A. J., 1971, Cambro-Ordovician rocks of Al Jazair (Algeria), a review: American Association of Petroleum Geologists Bulletin, v. 55, p. 1295–1335. (English)

Appendix 1. Field Description

Field name .... *Hassi Messaoud*

Ultimate recoverable reserves .... *9.056 billion bbl (1.440 billion m³) of oil (1980 est.);*
6.964 tcf (197.2 billion m³) of gas (1978 est.)

Field location:

    Country .... *Algeria*

    Basin/Province .... *Trias basin (also referred to as Oued Mya basin)*

Field discovery:

    Year first pay discovered .... *Cambrian sandstones   1956*

Discovery well name and general location:

    First pay .... *Md 1, approx. Lambert coordinates (m) 819,400 E, 124,700 N*

Discovery well operator .... *SN REPAL*

IP in barrels per day and/or cubic feet or cubic meters per day:

    First pay .... *1125 bbl (180 m³) of oil per day*

All other zones with shows of oil and gas in the field:

Age	Formation	Type of Show
Triassic	*Trias Argileux*	*Bitumen*

Geologic concept leading to discovery and method or methods used to delineate prospect:
Regional exploration of concession area using refraction seismic data in conjunction with expendable stratigraphic boreholes widely spaced in the basin to define major structural features and stratigraphy within basin.

Structure:

    Province/basin type .... *Bally 222, Klemme IIA*

    Tectonic history
    Stable platform during Cambrian–Silurian time; regional uplifts in middle and late Paleozoic (Hercynian) time characterized by faulting and igneous activity. Tectonic activity ceased by middle Tertiary.

    Regional structure
    Field located in east-central Trias basin on northern end of 800 km long Messaoud–El Agreb high.

    Local structure
    Broad, elliptically shaped, flat-topped, crest-eroded, vertically faulted anticline.

Trap:

    Trap type(s)
    Anticlinal structure with different producing areas separated by vertical faults and permeability barriers.

Basin stratigraphy (major stratigraphic intervals from surface to deepest penetration in field):

Chronostratigraphy	Formation	Depth to Top in ft (m)
Miocene–Pliocene	*Gravels and sand dunes*	*0*
Cretaceous	*Limestone and evaporite at top, clastics at base*	*600 (183)*
Jurassic	*Evaporites with minor limestone*	*6000 (1830)*
Triassic	*Evaporites overlying sandstone*	*8800 (2680)*
Ordovician	*Sandstone*	*10,700 (3260)*
Cambrian	*Fractured sandstone*	*11,000 (3353)*
Basement	*Granite porphyry and metasediments*	*12,600 (3840)*

Reservoir characteristics:

    Number of reservoirs .... *3 in 22 producing areas*

Formations .... *R-i, R-a, R-2*

Ages .... *Cambrian*

Depths to tops of reservoirs .... *R-i, 10,968 ft (3343 m); R-a, 11,050 ft (3368 m); R-2, 11,398 ft (3474 m)*

Gross thickness (top to bottom of producing interval)

R-i, 164 ft (50 m); R-a, 492 ft (150 m); R-2, 262 ft (80 m), maximum thickness of R-2 having reservoir quality rock in 65 ft (20 m)

Net thickness—total thickness of producing zones

    Average .... *Estimated at 200 ft (61 m)*

    Maximum .... *492 ft (150 m) (for western part of field)*

Lithology

R-i: thin-bedded, shallow-marine, fine-grained sandstone
R-a: fine- to coarse-grained fluvial sandstone
R-2: fine- to coarse-grained quartzarenite with conglomeratic lenses

Porosity type .... *Intergranular and fracture*

Average porosity .... *R-i, 5.4%; R-a, 7.4%; R-2, 12.5%*

Average permeability .... *R-i, 0.1 md; R-a, 2.2 md; R-2, 7.8 md*

Seals:

  Upper

    Formation, fault, or other feature ...... *Triassic in crestal area and Ordovician on uneroded flanks*

    Lithology .... *Alternating salt and shale above major unconformity*

  Lateral

    Formation, fault, or other feature .... *Vertical faults and poor reservoir qualities segment field into producing areas*

    Lithology .... *NA*

Source:

  Formation and age .... *Argile D'el Gassi, Ordovician and possibly rocks of late Silurian*

  Lithology .... *Black shale*

  Average total organic carbon (TOC) .... *NA*

  Maximum TOC .... *NA*

  Kerogen type (I, II, or III) .... *NA*

  Vitrinite reflectance (maturation) .... *NA*

  Time of hydrocarbon expulsion .... *No earlier than Late Cretaceous*

  Present depth to top of source .... *Source(s) probably exist around the periphery of the field about 10–40 km away; depths NA*

  Thickness .... *NA*

  Potential yield .... *NA*

Appendix 2. Production Data

Field name .... *Hassi Messaoud*

Field size:

  Proved acres .... *321,000 (43 km × 32 km)*

  Number of wells all years .... *>600 estimated*

  Current number of wells .... *NA*

  Well spacing .... *1961, 8 wells/100 km²; 1971, 16 wells/100 km²; 1978, 1.25 km per well with closer spacing in selected areas*

  Ultimate recoverable (1980 est.) .... *9.056 billion bbl (1.440 billion m³) of oil; 6.964 tcf (197.2 billion m³) of gas*

Cumulative production (through 1982) .... *3.4 billion bbl (545 million m³) of oil; cum. gas NA*

Annual production .... *1979 max., 206.8 million bbl (32.88 million m³) of oil; 1982, 162.3 million bbl (25.80 million m³) of oil; annual gas NA*

Present decline rate .... *NA*

Annual water production .... *NA*

In place, total reserves .... *25 billion bbl (4.0 billion m³) of oil*

Primary recovery .... *9000 million bbl (1439.8 million m³)*

Secondary recovery .... *NA*

Drilling and casing practices:

Amount of surface casing set .... *13⅜-in. at 760 ft (230 m)*

Casing program

9⅝-in. intermediate casing is set at 7600 ft to 7900 ft (2500–2600 m) in a 12¼-in. hole and is cemented in two stages to shut off water zones; intermediate 8½-in. hole is drilled through high pressure zone at 7900 ft (2600 m) to 10,000 ft (3350 m) where 7-in. casing is cemented; 5⁹/₁₆-in. hole cored and drilled to 10,200 ft (3340 m)

Drilling mud

Saturated saltwater mud to 7-in. casing point; bentonitic emulsion mud used to core; mud weight of 12.5 lb/gal (1.5 kg/L) used to drill the reservoir (overpressured reservoir)

Bit program

13⅜-in. casing set at top Upper Cretaceous about 760 ft (250 m); 12¼-in. conventional bits and diamond bits into top 180 ft (60 m) of Late Jurassic anhydrites, 7600 to 7900 ft (2500–2600 m); 8½-in., primarily diamond to about 10,000 ft (3050 m); 5⁹/₁₆-in. diamond core in reservoir to TD

High pressure zones .... *7600–7900 ft (2500–2600 m), continuing to total depth*

Completion practices:

Interval(s) perforated .... *Permanent open-hole completions with 2⅞-in. or 4½-in. tubing set*

Well treatment .... *Hydraulic fracturing of several wells on experimental basis only (through 1968)*

Formation evaluation:

Coring .... *Majority of wells have been conventionally cored through the reservoir section*

Logging suites .... *NA*

Testing practices .... *NA*

Mud logging techniques .... *NA*

Oil characteristics:

Type .... *Intermediate base*

API gravity .... *43.5–49°*

Base .... *Intermediate*

Initial GOR .... *790 ft³/bbl (250 m³/m³) at std. cond.*

Sulfur, wt% *0.13*

Viscosity, SUS .... *3 cs at 68°F (20°C)*

Pour point .... *–62°F (–52°C)*

Gas-oil distillate .... *NA*

Field characteristics:

Average elevation .... *525 ft (160 m)*

Initial pressure .... *6870 psi (483 kg/cm²) at 9700 ft (3200 m)*

Present pressure .... *NA*

Pressure gradient .... *NA*

Temperature .... *248–270°F (120–132°C) at 9700 ft (3200 m)*

Geothermal gradient .... *2.1°F/100 ft (3.1°C/100 m)*

Drive .... *Solution gas with some rock compression*

Oil column thickness .... *1017 ft (310 m)*

Oil-water contact .... *–11,155 to 11,350 ft (3400–3459 m)*

Connate water *19%*
Water salinity, TDS .... *400,000 ppm*

Transportation method and market for oil and gas:

Pipelines from field to Skikda on Mediterranean Coast (north); pipelines (one oil and three gas) to northwest through Hassi R'Mel to port of Arzew. Refinery at Hassi Messaoud supplies Sahara market with gasoline, kerosene, and refined oil.

Snapper Field—Australia
Offshore Gippsland Basin, Southeast Australia

P. N. GLENTON
Esso Australia Ltd.
Sydney, Australia

FIELD CLASSIFICATION

BASIN: Gippsland
BASIN TYPE: Failed Rift
RESERVOIR ROCK TYPE: Sandstone
TRAP TYPE: Faulted Anticline
RESERVOIR ENVIRONMENT OF DEPOSITION: Fluvial to Estuarine
TRAP DESCRIPTION: Elongate anticline with faulting (degree of faulting increases with depth)

RESERVOIR AGE: Eocene and Paleocene
PETROLEUM TYPE: Oil and Gas

LOCATION

Snapper field is located in the Gippsland basin in eastern Bass Strait, southeastern Australia (Figure 1). The field is 30 km offshore in 55 m of water, on the continental shelf between mainland Australia and Tasmania.

The Gippsland basin is the most prolific oil and gas province in Australia, with initial reserves of 3.7 billion barrels of oil and condensate, 0.6 billion barrels of liquified petroleum gas, and 8900 bcf of gas (i.e., 8.9 tcf).

Snapper is the second largest gas field in the Gippsland basin and is located along a major anticlinal trend that also hosts the slightly smaller Barracouta gas field. The major Gippsland basin oil fields—Kingfish, Mackerel, Fortescue, Cobia, and Halibut—are located about 35 km further offshore, to the southeast of the Snapper field.

The Snapper field covers an area of 50 km². Estimated ultimate recovery of gas from the main, top-of-Latrobe Group Snapper N-1 gas reservoir is 2140 bcf, and recovery of oil from the associated thin oil leg is expected to be about 11 million stock tank barrels. Additional reserves of about 6 million STB are expected to be recovered from various small reservoirs below the main accumulation.

HISTORY

Pre-Discovery

The first indication of oil in the Gippsland basin was in 1924, in a water well drilled in the onshore part of the basin near Lakes Entrance on the Gippsland coast of Victoria (Figure 1). This small oil pool was developed and produced 8000 barrels of 15.7° API oil between 1924 and 1956.

Regional aeromagnetic surveys and shipborne seismic surveys conducted during the 1950s by the Australian Federal Government's Bureau of Mineral Resources indicated a deep sediment-filled trough in the offshore part of the Gippsland basin (Leslie et al., 1976; Weeks and Hopkins, 1967).

In 1960, Haematite Exploration Pty. Ltd., now known as BHP Petroleum Pty. Ltd., acquired exploration permits over the whole Bass Strait area, the eastern part of which includes the Gippsland basin, on the advice of the American consulting geologist, Lewis Weeks. As his fee Weeks received a 2.5% royalty.

Haematite acquired additional aeromagnetic and marine seismic data, which permitted structure mapping and the recognition of some of the major structural features of the basin. In 1964, Esso Exploration and Production Australia Inc. farmed into the Haematite acreage and assumed the role of operator.

Offshore exploration drilling commenced in 1964. The first well resulted in the discovery of the major Barracouta gas field, followed soon afterward by the discovery of Marlin, the largest gas field in the basin. Kingfish, the largest oil field in the basin, was discovered in 1967 (Figure 1).

Discovery

Snapper field was discovered in 1968 by Snapper-1, which was drilled in two stages by the drillship "Discoverer II" and the semi-submersible "Ocean

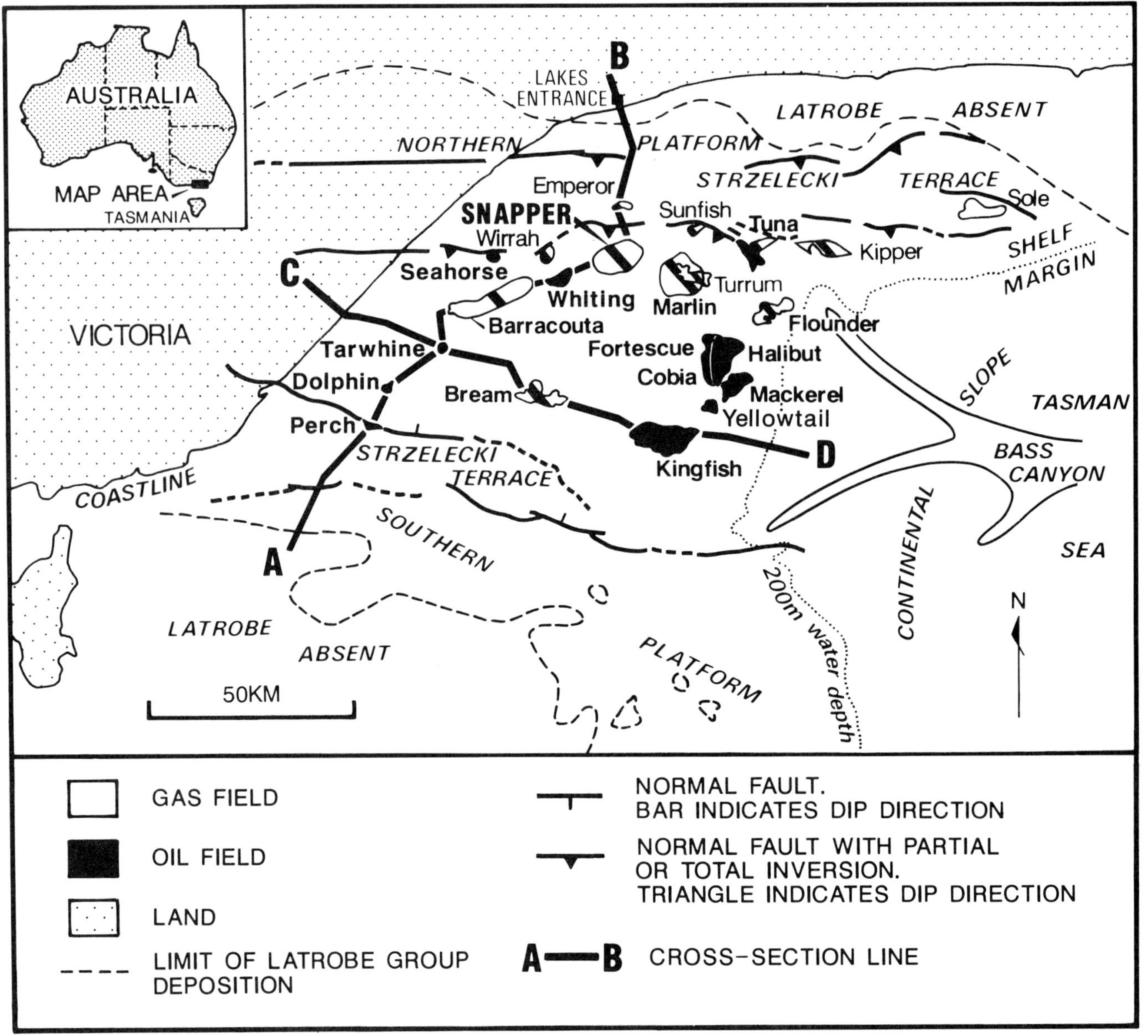

Figure 1. Producing fields (indicated by bold type) and other significant discoveries in the Gippsland basin. All the major fields are located in the central graben of the basin, between the northern and southern platforms. Cross section line AB refers to Figure 14, and CD to Figure 16.

Digger." The Snapper field is near the northern margin of the central graben of the basin (Figure 1), along an anticlinal trend to the northeast of the previously discovered Barracouta field. Snapper-1 was drilled to test a separate, major culmination at the top of the Latrobe Group on this trend (Figures 1 and 2) and to explore deeper within the Latrobe Group.

Snapper-1 discovered a major gas pool (97 m net gas sandstone) in Eocene sandstones at the top of the Latrobe Group, together with the thin associated oil leg (Figure 3). This zone is now known as the N-1 reservoir. A number of minor deeper Latrobe Group oil and gas zones were also penetrated before the well reached total depth of 3725 m subsea in Late Cretaceous (*N. senectus* spore/pollen zone) Latrobe Group sediments (Figure 4). Two production tests were carried out. The first, of a deep Latrobe Group zone, flowed 1.1 mmcf of gas with 5 BOPD, which was not economically attractive considering the depth and small estimated reserves of this reservoir. The second production test was in the top-of-Latrobe Group accumulation and flowed 4.8 mmcf/day with 61 barrels of condensate. The well was plugged and abandoned after testing.

The field was delineated by Snapper-2 (total depth 3142 m subsea) in 1969, and Snapper-3 (TD 3202 m subsea) in 1970. These wells confirmed the extent of the N-1 reservoir. This reservoir contains a maximum gas column of 205 m, and the associated

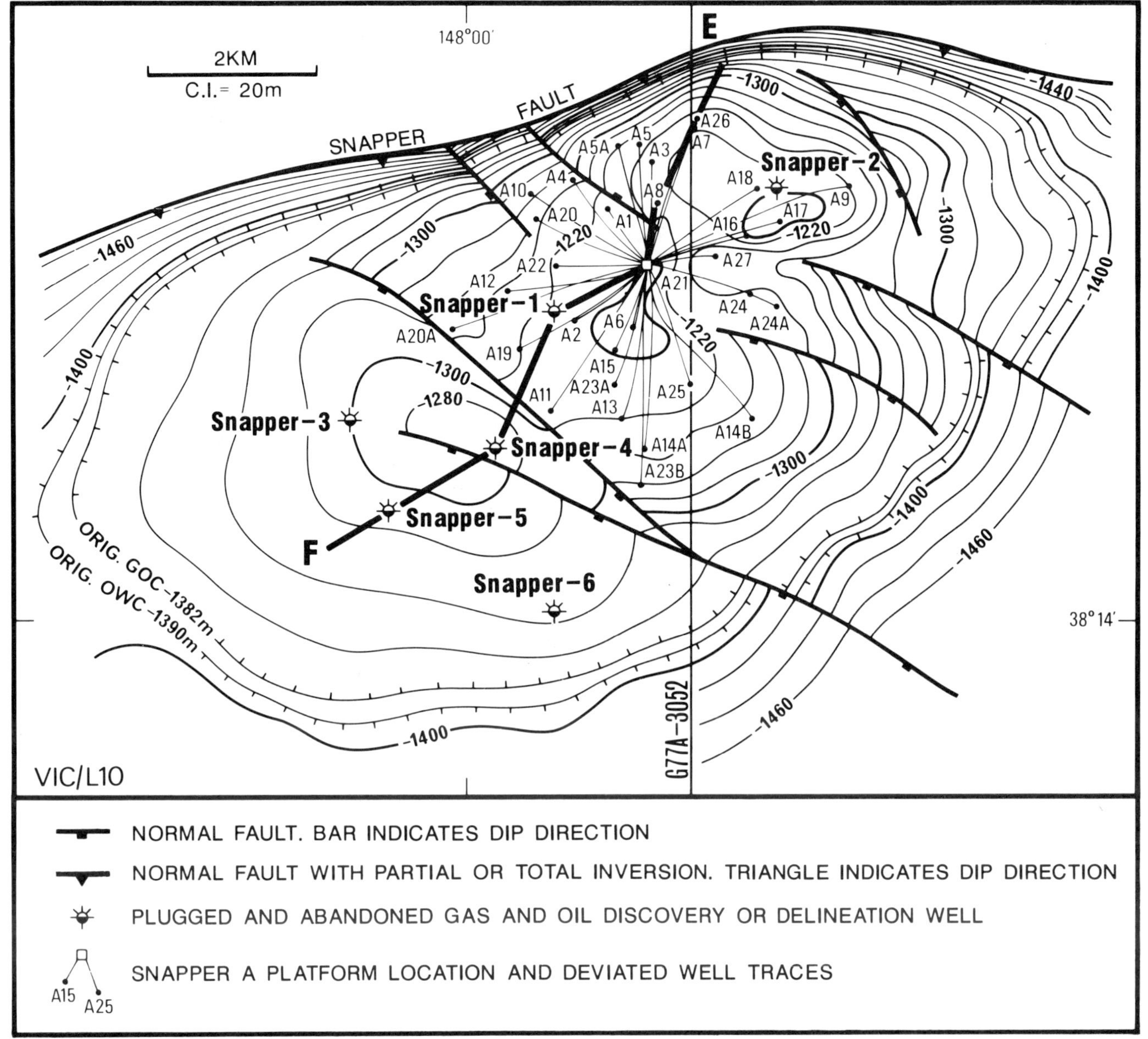

Figure 2. Structure map on the top of the Snapper N-1 reservoir (at the top of the Latrobe Group "coarse clastics"). The discovery well, Snapper-1, is located 1.2 km southwest of the Snapper A platform. Snapper A-21 is a vertical development/exploration well drilled beneath the platform. Cross section line EF refers to Figure 3. Seismic line G77A-3052 is shown on Figures 6 and 7.

oil leg ranges in thickness from 4 to 8 m. The original field gas-oil contact was at 1382 m subsea. No major deeper hydrocarbon zones had been intersected, but the results of Snapper-1 suggested that further exploration of the Latrobe Group would be justified during development of the N-1 reservoir.

Post-Discovery

The major N-1 gas accumulation at the top of the Latrobe Group provided justification for development of the Snapper field. The nearby Marlin and Barracouta gas fields had already been developed and were capable of providing the bulk of immediately required gas production. The initial Snapper development plans concentrated on maximizing production from the thin N-1 oil leg. The upper Latrobe Group reservoirs have a strong natural water drive, and future gas production would lead to rising contacts and a "smearing" out of the oil leg, with resultant loss of oil reserves, so early oil development was essential. Exploration for deeper Latrobe Group reservoirs was also planned.

The Snapper A jacket was launched in 1979, and development drilling commenced in 1981. Development drilling from the 27 conductor platform took

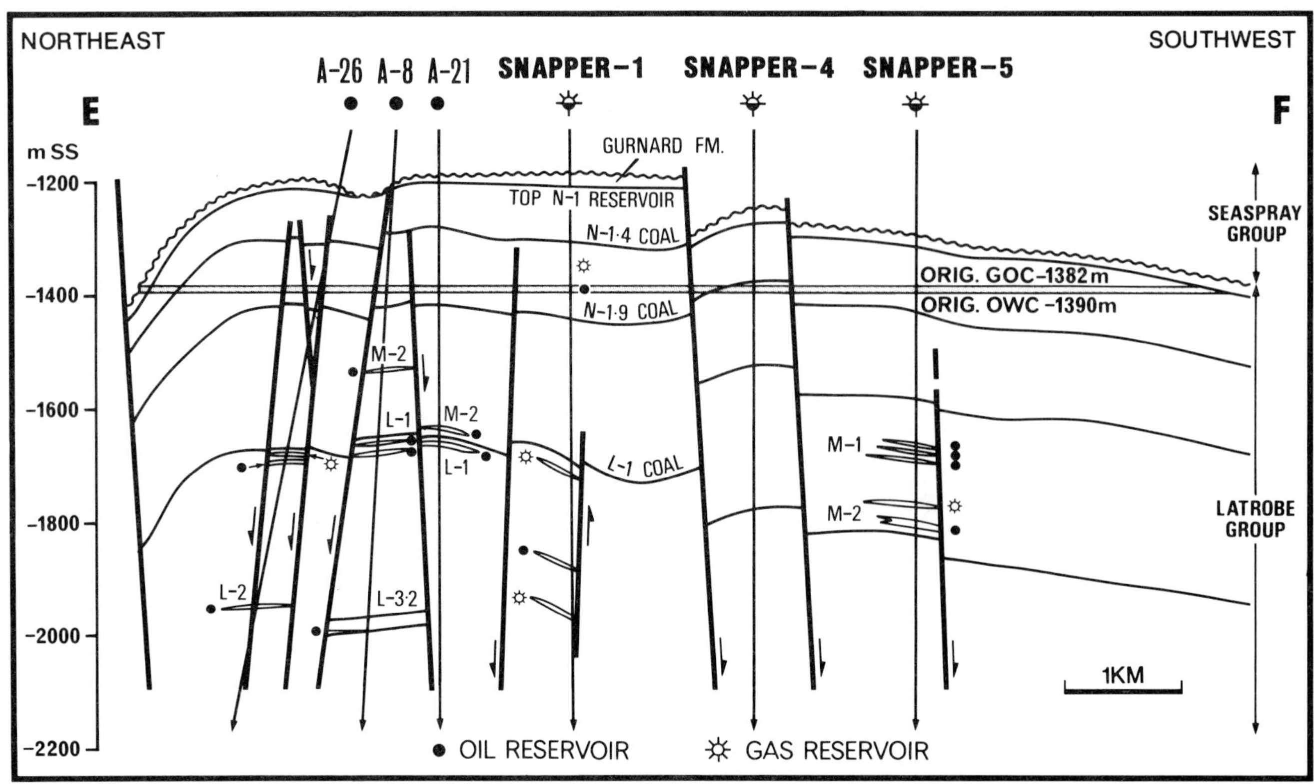

Figure 3. Schematic cross section of Snapper field, showing the N-1 reservoir and some of the more important deeper Latrobe Group reservoirs. Reservoir prefixes N, M, and L are taken (loosely) from the spore/pollen biostratigraphic zone of the reservoir (i.e., *N. asperus, M. diversus,* and *L. balmei*). The Seaspray Group includes the Lakes Entrance Formation and the Gippsland Limestone. Snapper-1, -4, and -5 are exploration and delineation wells; A-8, A-21, and A-26 are oil producers from reservoirs below the N-1 gas/oil zone. The line of the cross section is indicated on Figure 2.

place in two phases. From March 1981 to March 1983, 21 wells were drilled, resulting in 14 N-1 oil and five N-1 gas completions and four deeper Latrobe Group completions. (Two wells were dual completions; one was completed with a sliding sleeve enabling either gas or oil to be produced from different zones; and one well was not completed at that time.)

During this phase of drilling, the maximum horizontal reach from the platform to the N-1 gas-oil contact was 1860 m, with an average angle below the surface casing shoe of 62°.

Snapper A-21 was the first and deepest of the Latrobe Group exploration wells drilled from the platform. The well intersected a 9 m oil column, designated the L-1a oil reservoir, at –1660 m, which provided encouragement for future exploration. The L-1 reservoir had cumulative production to the end of 1989 of 2.0 million STB of oil. A-21 was drilled to a total depth of 3253 m subsea but intersected only minor, uneconomic hydrocarbon shows below the L-1a accumulation.

Economic deeper Latrobe Group oil zones were also penetrated in Snapper A-6, A-8, A-16, and A-18 during this phase of drilling, although the A-16 and A-18 zones were not completed until later.

Development drilling was resumed in 1986 and completed in mid-1987. The six wells drilled from the remaining conductors resulted in four N-1 oil completions and one N-1 gas producer, and three deeper Latrobe Group completions. (There were two dual completions.) Maximum horizontal reach from the platform during this phase of development was achieved by Snapper A-9, which intersected the field gas-oil contact 2480 m from the platform, with an average angle below the surface casing shoe of 71°. Unfortunately, this well was not a successful oil producer. A shale bed was present at the depth of the oil leg, and the well was completed as a gas producer.

At this stage, although all the conductors had been used, several areas of the N-1 oil leg within the platform-conformable area remained undeveloped. Five wells (N-1 gas or depleted N-1 oil producers from the earlier phase of drilling) were abandoned and redrilled. In four of these it was possible to plug the well, cut the production casing below the surface casing and retrieve it, and then redrill the well from the surface casing shoe. In the fifth, the surface casing could not be recovered and the well was redrilled from the conductor shoe.

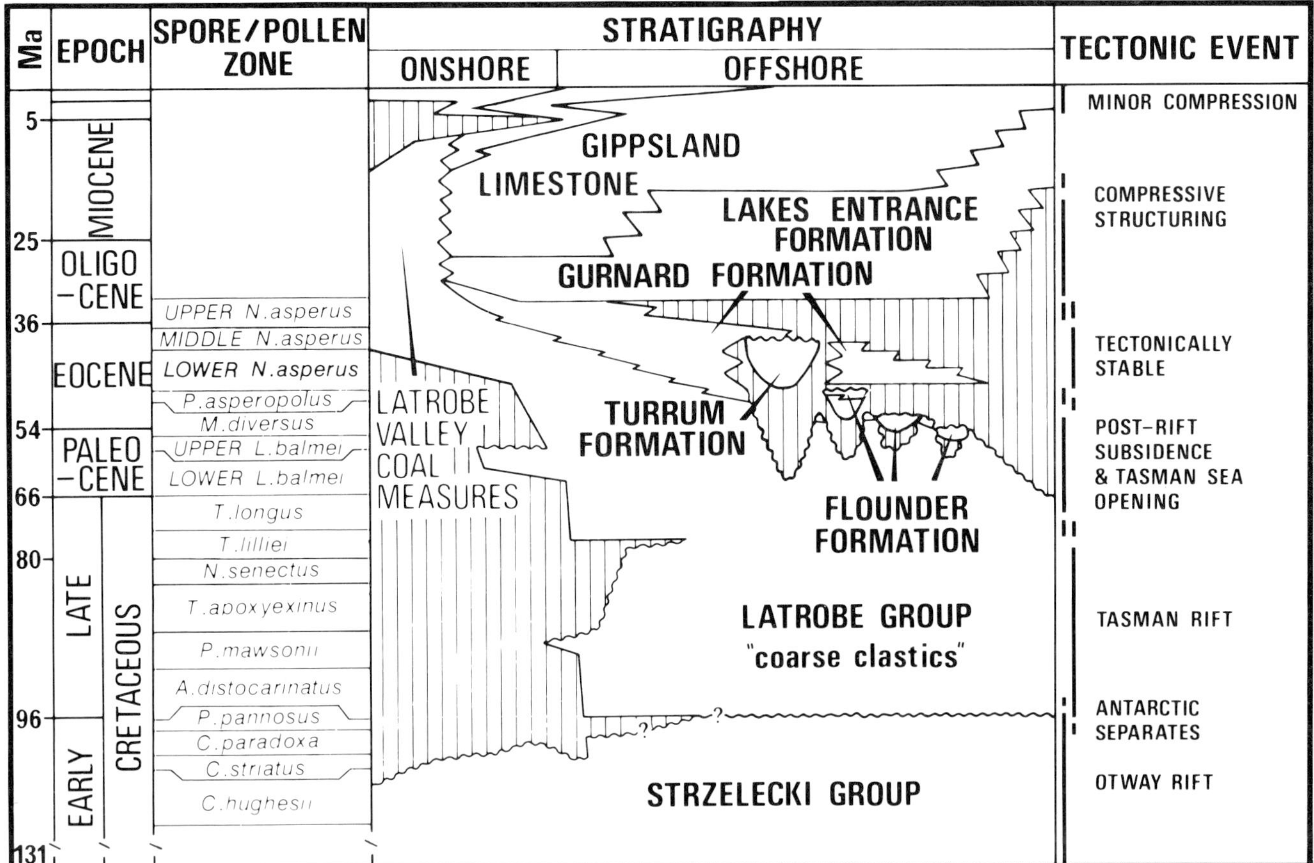

Figure 4. Generalized stratigraphy and tectonic history of the Gippsland basin, including spore-pollen biostratigraphic zones. The Otway Rift refers to the rifting of Australia from the Antarctic, and the Tasman Rift to the rifting of Australia from the Lord Howe Rise.

The redrills were technically difficult, requiring substantial change in direction and increase in angle. For example, Snapper A-14b required a 40° turn and an average angle of 69° from the surface casing shoe to the target.

Open-hole logging was possible in 10 of the last 11 wells, owing to the introduction of "push-down" logging systems. This permitted conventional wireline logging tools to be attached to the bottom of the drillstring and pushed to bottom, allowing a full suite of open-hole logs to be obtained. This was essential in order to define the thin N-1 oil leg and choose an appropriate perforation interval.

Three additional exploration wells were drilled from a floating rig after platform drilling had commenced. These were designed to test the Latrobe Group in the southwestern fault block, beyond the reach of platform wells (Figure 2). Snapper-5 was the most successful of these wells, discovering several oil zones in the *M. diversus* section and a number of deeper gas zones. However, these reservoirs have limited lateral continuity and have not yet been capable of supporting a commercial development.

Mapping during the development of the field was based on a 1 km grid of 48-fold data shot in 1977. Additional 48-fold data were acquired in 1984, reducing the grid to 0.5 km over most of the field. Strong reflections are obtained from the top of Latrobe Group unconformity and two thick coal seams in the upper Latrobe Group (the N-1.4 and N-1.9 Coals; Figures 3, 5, 6, and 7). Most of the seismic energy is absorbed by these coals and mapping of deeper horizons is difficult.

DISCOVERY METHOD

The stratigraphy of the onshore part of the Gippsland basin was known from surface mapping and from a number of wells. These had been drilled for oil exploration after the discovery in 1924 of the small Lakes Entrance oil field, and to investigate the thick brown coal deposits of the basin. (These are now extensively mined.)

Early aeromagnetic data indicated that a deep, sediment-filled graben exists in the offshore part of

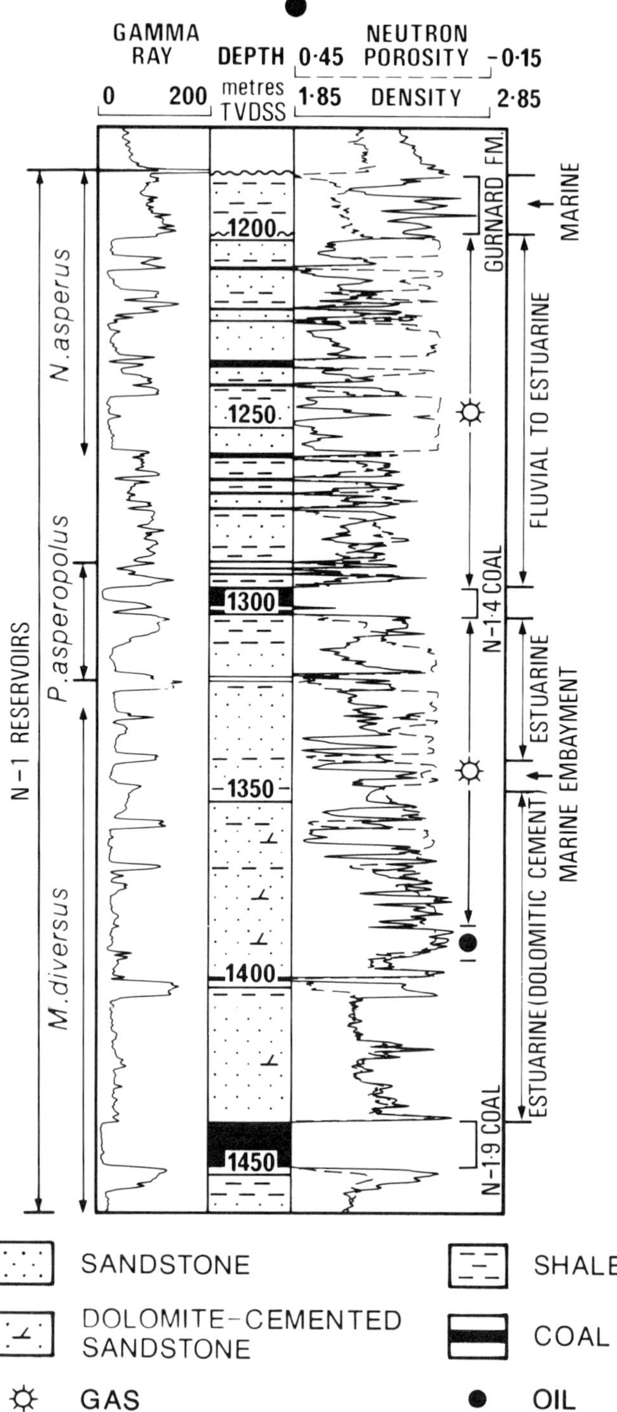

Figure 5. Stratigraphy of the N-1 reservoir sequence, Snapper A-13 well. Dolomite cementation (indicated by high density and low neutron porosity values) of the reservoir sandstones between the N-1.4 Coal and N-1.9 Coal was particularly severe in this well. The presence of this cement in the oil zone resulted in the well's producing only 18,000 STB of oil before depletion.

the Gippsland basin. During the early 1960s, seismic data permitted the mapping of large anticlinal or topographic features at the top of the Latrobe Group within this graben.

The presence of the graben in the offshore part of the basin was encouraging, as it suggested that a thick Tertiary to Cretaceous section would be present (by correlation with the much thinner equivalent section in the onshore part of the basin). It was considered likely that reservoir and source would be present in the graben, and that it would be deep enough to have generated hydrocarbons.

One recognized risk was the presence of a freshwater aquifer within the Latrobe Group in the onshore Gippsland basin. The onshore Lakes Entrance field contains heavy, biodegraded oil, but it was hoped that the offshore parts of the basin would not be affected. Snapper and other fields along the northern margin of the graben are now known to be underlain by this aquifer. The influx of fresh water is interpreted to have taken place after hydrocarbon emplacement (Kuttan et al., 1986) and did not affect hydrocarbon accumulation. The oil from several fields, including Snapper (Figure 8), has undergone some biodegradation (Burns et al., 1987), but not as severely as in the onshore Lakes Entrance accumulation.

During early offshore drilling, the major gas fields Marlin and Barracouta were discovered along the northern margin of the central graben. Seismic mapping allowed Snapper to be recognized as a separate culmination at the eastern end of the large Barracouta-Snapper anticlinal trend, and indicated that the structure persisted with depth. This was confirmed by the results of Snapper-1, -2, and -3, which were drilled prior to development.

STRUCTURE

Tectonic History

The development of the Gippsland basin is related to the separation of Australia from the Antarctic (Otway Rift), and the later separation of eastern Australia from the Lord Howe rise and opening of the Tasman Sea (Figure 9) (Rahmanian et al., 1990).

The inception of the basin occurred during the earliest Cretaceous (about 130 Ma), with the development of a pre-breakup depression along the southern margin of Australia. This depression developed into a complex graben system several hundred kilometers wide (Veevers, 1984), with continuous deposition between the Gippsland basin and the neighboring Bass and Otway basins to the west. In the Gippsland basin, these deposits are known as the Strzelecki Group (Figure 4), a thick sequence of lithic and volcanogenic clastic fluviatile sediments.

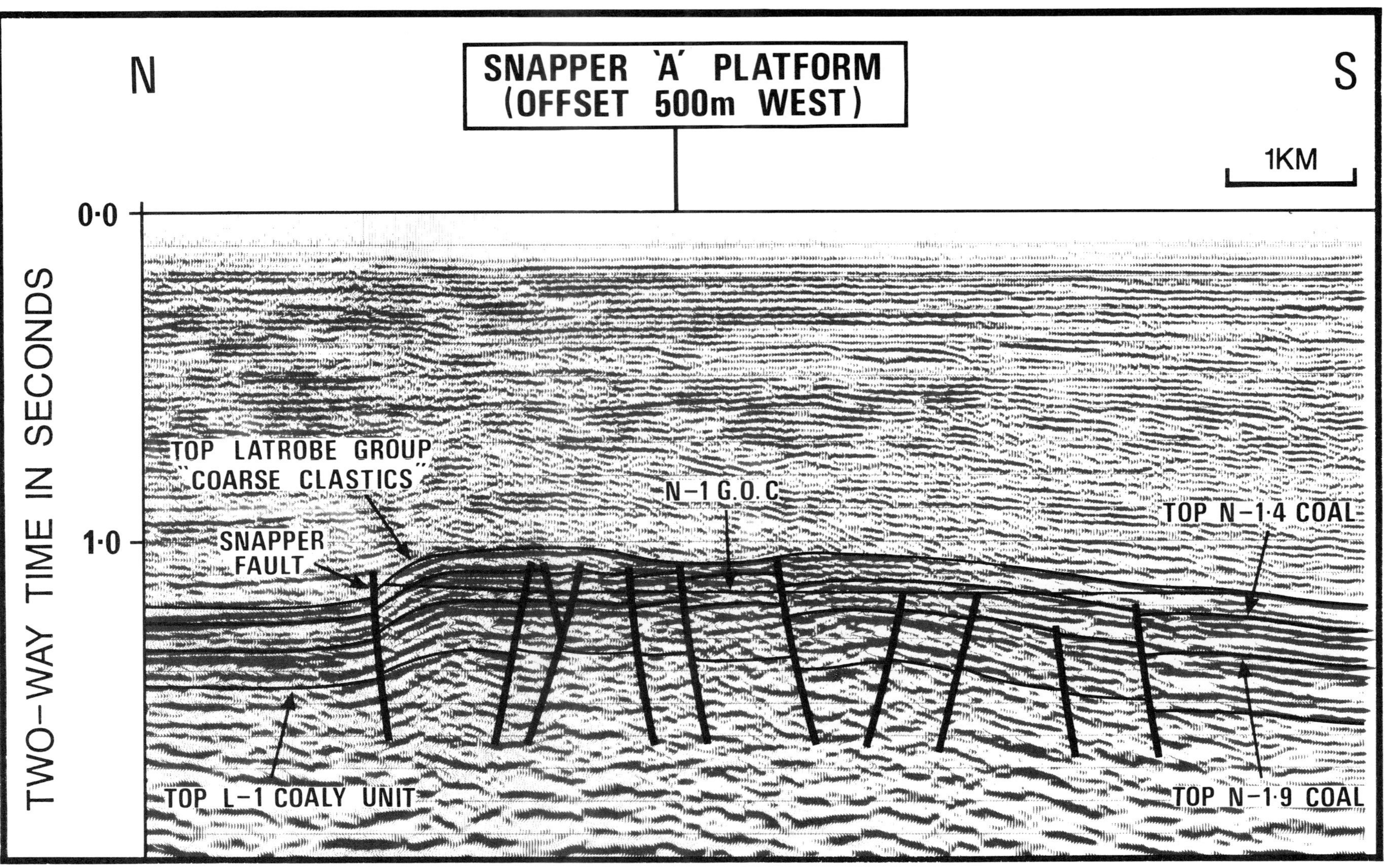

Figure 6. Interpreted seismic line G77A-3052. Location of the line is shown on Figure 2.

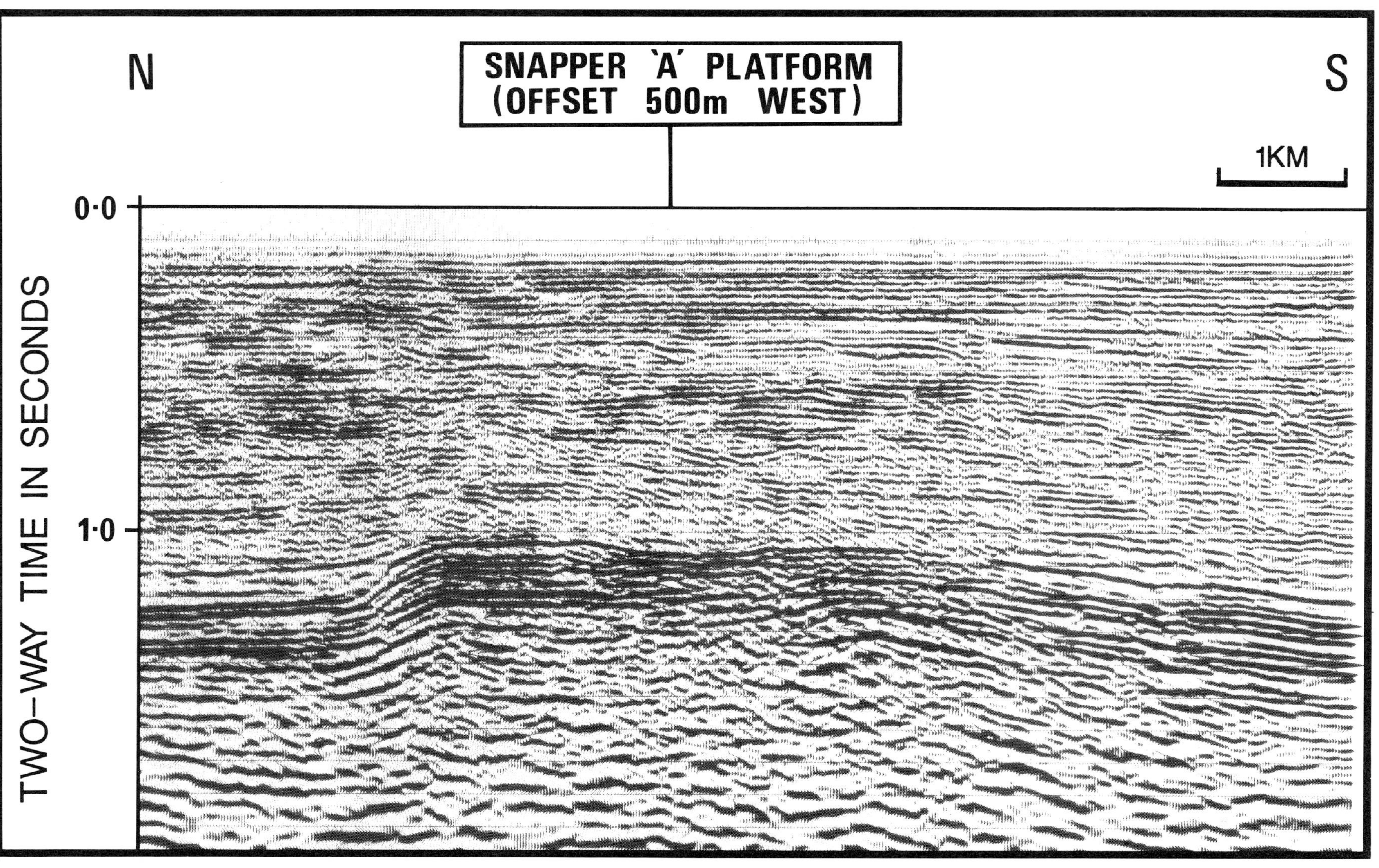

Figure 7. Uninterpreted seismic line G77A-3052.
Location of the line is shown on Figure 2.

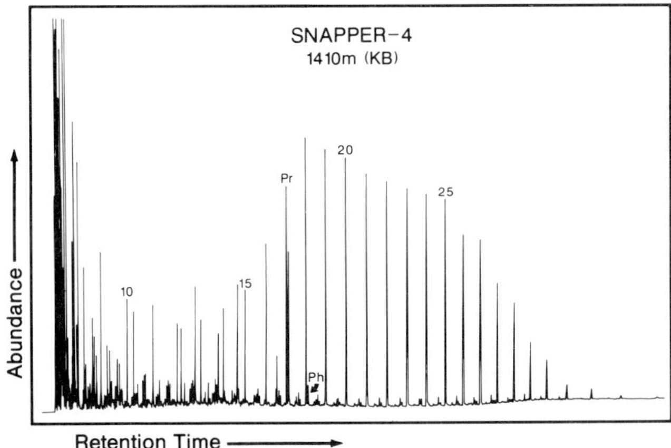

Figure 8. Chromatogram for N-1 oil leg in Snapper-4; sample from 1389 m subsea. A moderate degree of biodegradation is indicated, with loss of normal alkanes up to C_{17}. There has not been much effect on the isoprenoids pristane and phytane at this stage.

The Australian Plate separated from the Antarctic Plate at the end of the Early Cretaceous, with breakup occurring down the western side of Tasmania. The southern Victorian graben system in which the Otway, Bass, and Gippsland basins evolved represents a failed arm for this rift system, radiating from a triple junction west of Cape Otway. The breakup created a major unconformity at the top of the Strzelecki Group and led to the development of major structural ridges that separate the Otway, Bass, and Gippsland basins.

Latrobe Group deposition commenced during the Late Cretaceous in this failed arm. The system was 40 to 80 km wide, with northeast-southwest extension resulting in fault-controlled subsidence and northwest-southeast-oriented faults. Due to provenance change, the Latrobe Group deposits are quartzose, in contrast to the volcanogenic Strzelecki sandstones. Latrobe sedimentation was initially nonmarine, with the oldest marine deposits dated as Santonian.

Further development of the Gippsland basin was associated with rifting along the eastern margin of Australia (Tasman Rift). Breakup along the eastern margin of Australia (i.e., the opening of the Tasman Sea) occurred during the Campanian (about 80 Ma).

Sedimentation continued in the Gippsland basin as it evolved during the latest Cretaceous to Eocene into a marginal sag. Greater marine influence occurred during this time as the Tasman Sea encroached, leading to the deposition of shoreline and offshore marine sediments in the southeastern part of the basin, with coastal plain and alluvial plain sedimentation to the northwest.

Major compression commenced in the Late Eocene, toward the end of Latrobe Group deposition, and continued in some areas until mid-Miocene. Uplift associated with this structuring resulted in wide-spread erosion at the top of the Latrobe Group. The Snapper and Barracouta features are large anticlinal structures at the top of the Latrobe Group, and the Marlin, Kingfish, Fortescue, Cobia, and Halibut structures resulted from a combination of folding and erosion. Minor reactivation of compression occurred in the Pleistocene to recent.

After the commencement of the compressional phase, subsidence continued slowly, and marine marls and shales of the Oligocene Lakes Entrance Formation were deposited, forming a seal at the top of the Latrobe Group. This was followed by marine channeling and rapid deposition of the thick bioclastic wedge of the late Oligocene to Recent Gippsland Limestone (Figure 4).

Regional Structure

The Snapper anticline is located at the eastern end of the major Barracouta-Snapper anticlinal trend, along the northern margin of the central graben of the Gippsland basin (Figure 1). The anticlinal trend was formed during the late Eocene to mid-Miocene compressional phase.

Local Structure

At the top of the Latrobe Group, the Snapper structure is an east–northeast-trending anticline 13 km long by 6 km wide. The field is bounded to the north by the Snapper fault, a major, rift-margin normal fault, which underwent partial inversion during the late Eocene to mid-Miocene compressional phase. As a result of compression against this fault (Figures 2, 3, 6, and 7), the field is asymmetrical, with steeper dip on the northern flank than on the southern flank. Northwest-trending normal faults divide the field into a number of fault blocks.

Deeper in the Latrobe Group, the anticlinal shape of the structure is retained, but the degree of faulting increases with depth. Several of the fault blocks contain independent anticlinal closure, and minor hydrocarbon accumulations have been discovered in these crests.

STRATIGRAPHY

Regional stratigraphy of the Gippsland basin is summarized in Figure 4. The lithic and volcanogenic clastic sediments of the Early Cretaceous Strzelecki Group (not penetrated at Snapper) are unconformably overlain by the Late Cretaceous to Eocene quartzose sediments of the Latrobe Group "coarse clastics."

In the eastern part of the basin, Latrobe Group sedimentation was initially nonmarine, but later deposition occurred in an offshore marine to marginal marine wave-dominated environment. In the western area (including Snapper), sedimentation was mainly nonmarine. The Turrum and Flounder formations

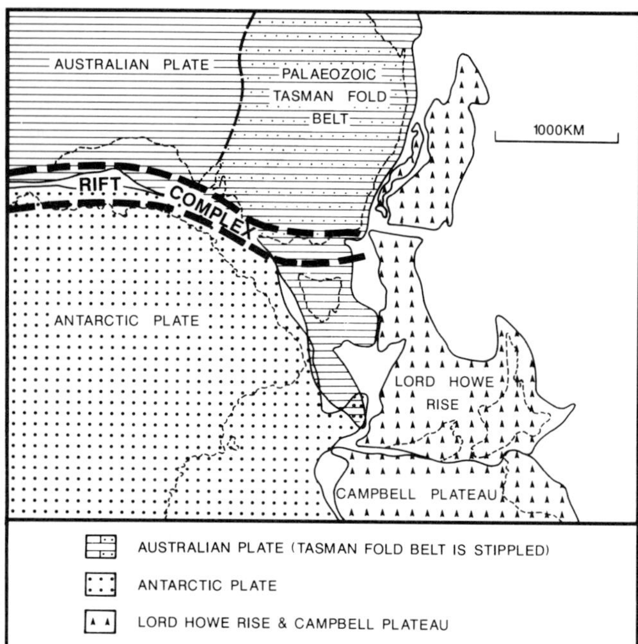
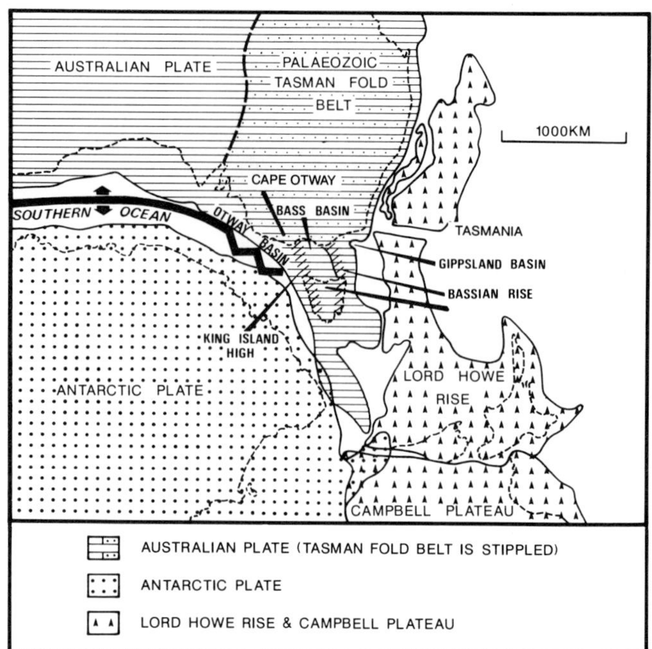

Figure 9. Plate tectonic evolution of the Gippsland basin. (A) Early Cretaceous. A broad rift complex exists along the southern margin of mainland Australia. (B) Mid-Cretaceous. The Southern Ocean has formed (at 95 Ma) and the Gippsland, Bass, and Otway basins are separate entities. (C) Latest Cretaceous. Breakup has occurred along the eastern margin of Australia (at 80 Ma) and the Gippsland, Bass, and Otway basins are left as failed rift basins. Used by permission of the Geological Society from "Sequence stratigraphy and the habitat of hydrocarbons, Gippsland Basin, Australia" by Rahmanian, V. D., Moore, P. S., Mudge, W. J., and Spring, D. E. in Special Publication No. 50—Classic Petroleum Provinces, 1990.

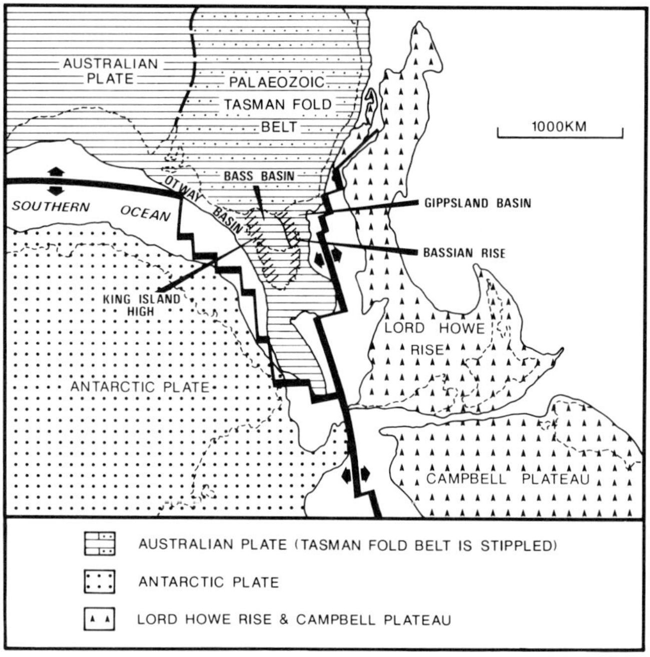

indicated in Figure 4 are incised-valley fill deposits formed late in Latrobe Group history. The Gurnard Formation is a diachronous, offshore marine, glauconitic unit that unconformably overlies the Latrobe Group "coarse clastics" in parts of the basin.

At Snapper, sedimentation from Late Cretaceous to late Paleocene took place in a coastal plain environment. The section consists of interbedded siltstone, sandstone, and thin coal seams. Sandstone beds are generally of limited lateral extent and typically were deposited as fining-upward point bars. During the Eocene, sedimentation was mainly fluvial to estuarine, with occasional marine incursion. The sandstone units deposited during this interval generally have good lateral continuity.

The oldest sediments penetrated at Snapper are Late Cretaceous Latrobe Group units (*N. senectus* spore/pollen zone; see Figure 4). No economic reservoirs were intersected in these deeper tight sandstones of the *N. senectus* and *T. lilliei* intervals.

Sandstone beds 10 m thick are present in the *T. longus* (Late Cretaceous) section, but reservoir quality is poor. The overlying lower *L. balmei* (Paleocene) section (Figure 10) contains the deepest economic reservoirs at Snapper, with good quality reservoir sandstones up to 25 m thick, though of limited areal extent. In the upper *L. balmei* and lower part of the *M. diversus* (Figures 10 and 11), sandstone beds are typically less than 10 m thick.

The N-1 reservoir interval (N-1.1 to N-1.9 units) (Figure 5) spans the Eocene *M. diversus, P. asperopolus,* and *N. asperus* zones. The style of sedimentation in this interval is different from that deeper in the Latrobe Group. By this time, the basin had become tectonically stable, with only limited intermittent movement on the deep-seated

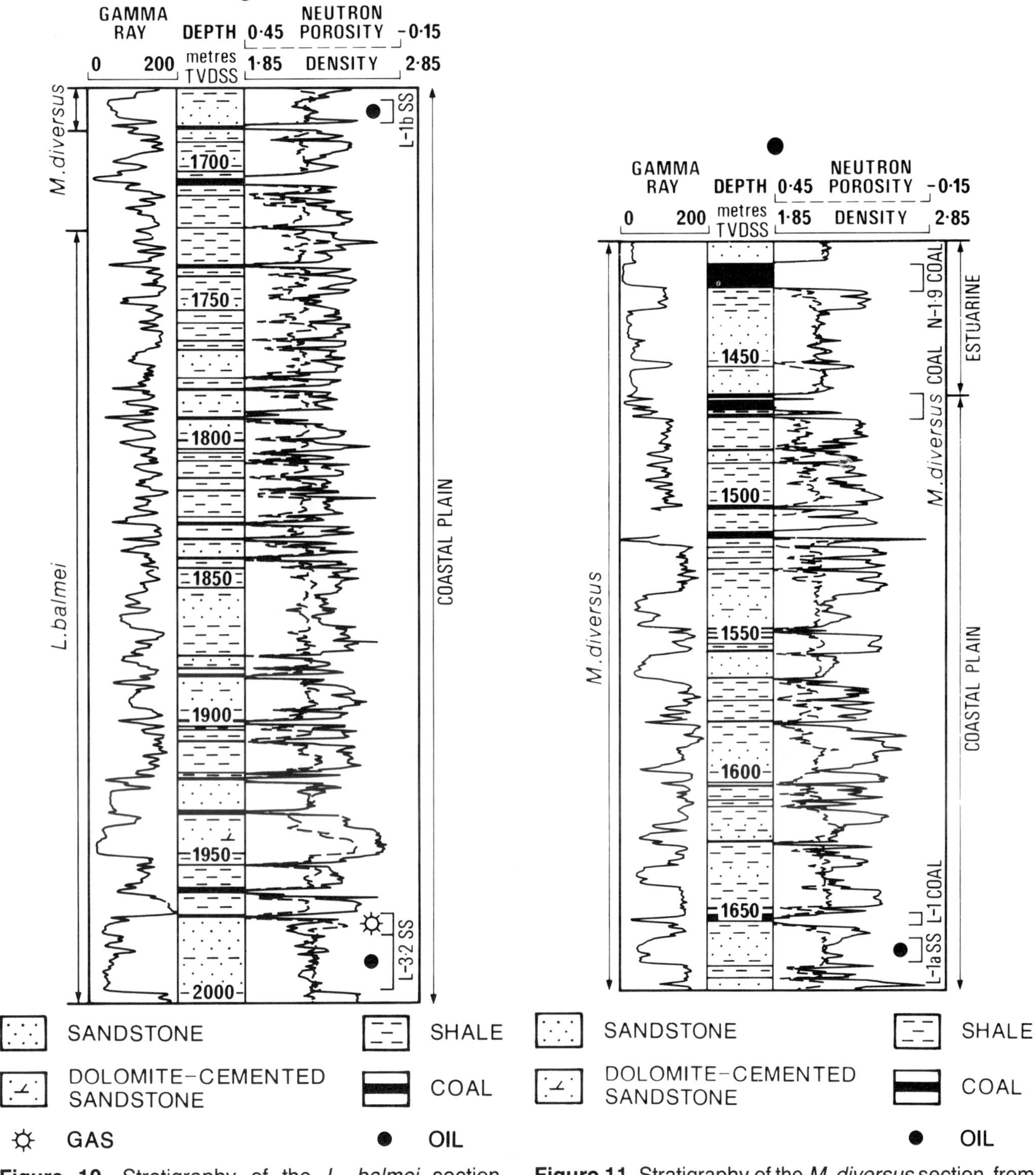

Figure 10. Stratigraphy of the *L. balmei* section, Snapper A-8 well. This interval immediately underlies that shown in Figure 11, with no overlap. The L-3.2 Sandstone in this well is unusually thick compared to most of the deeper Latrobe Group sandstones but is nevertheless of limited lateral extent. Its 17 m oil leg is the thickest of the deeper Snapper Latrobe Group oil zones. Cumulative production from this zone to the end of 1989 was 0.6 million STB of oil.

Figure 11. Stratigraphy of the *M. diversus* section, from the L-1 Coal to the N-1.9 Coal, Snapper A-21 well. This section immediately underlies the N-1 reservoir section. The L-1a Sandstone, an oil reservoir in this well, is typical of the deeper Latrobe Group reservoirs—relatively thin and of limited lateral extent. This reservoir has been the most productive of the deeper oil zones, with cumulative production to the end of 1989 of 2.0 million STB of oil.

northwest-trending faults that had earlier been active. As uniform subsidence continued, sedimentation was strongly influenced by eustatic sea-level changes. The N-1 reservoir sequence consists of stacked, thick fluvial and estuarine sandstones, minor siltstones, and several thick, continuous coal seams. The reservoir sandstones generally have good lateral continuity across the field. At least one marine incursion is recorded in the lower part of the interval (Figure 5). (Cores through this interval contain *ophiomorpha* burrows, and logs indicate a coarsening-up trend.)

Latrobe Group sedimentation ended after the deposition of the Gurnard Formation (Figures 3 and 5) (the N-1.0 reservoir unit at Snapper). This formation unconformably overlies the Latrobe Group "coarse clastics" and is a "sediment-starved" glauconitic unit deposited in an offshore marine environment.

The marine shales and marls of the overlying Lakes Entrance Formation provide the seal for the top of Latrobe Group N-1 hydrocarbon accumulation. These are overlain by calcareous sediments of the Gippsland Limestone. The Lakes Entrance Formation and Gippsland Limestone are collectively referred to as the Seaspray Group (Figures 3 and 4).

TRAP

The Snapper N-1 hydrocarbon accumulation at the top of the Latrobe Group is trapped within an anticlinal structure and sealed by the marine shale and marl of the Lakes Entrance Formation. Anticlinal folding occurred during the late Eocene to mid-Miocene compressional phase.

The original gas-oil contact was at –1382 m. The associated oil leg ranges in thickness from 4 to 8 m. The interpreted field-wide gas-oil contact resulted from communication across faults owing to the high sand to shale ratio of the N-1 reservoir sequence. However, the variation in thickness of the oil leg suggests that at this level there may be barriers, probably of very limited extent, between different fault blocks.

At the top of the Latrobe Group, the structure is mapped as being full of hydrocarbons to a synclinal spill point at the southwestern end of the field (Figure 2).

Deeper Latrobe Group hydrocarbon accumulations have been intersected in separate anticlinal culminations in each of the major fault blocks of the field. The discontinuous nature of these sandstones suggests that there could be a stratigraphic component to some of these traps. Deposition was influenced by contemporaneous movement on the northwest-trending faults, and this influenced sand continuity. The top seal for these reservoirs is provided by siltstone and coal units. It is possible that some faults in this section act as lateral seals.

RESERVOIRS

The N-1 reservoir is of generally excellent quality, with average porosity of 24% and average permeability of around 2 darcies. Sand deposition was predominantly in a high energy fluvial environment. The sandstones are typically quartzose, with minor amounts of feldspathic and lithic grains, and are classified as quartzarenite to subarkose using the Folk classification.

Porosity enhancement has occurred because of the dissolution of feldspathic grains, although dolomitic cement that was subsequently precipitated in some areas of the reservoir has caused local porosity reduction. The most dolomite-prone area is to the south and southeast of the Snapper platform, in the fluvial and estuarine units (between the N-1.4 Coal and N-1.9 Coal indicated on Figure 5). The presence of dolomitic cement has a profound effect on oil recovery. Production of 0.5 to 1.3 million barrels of oil has been possible from oil completions in uncemented sandstones. Completions in dolomitic sandstone have typically produced less than 100,000 barrels. Diagenetic processes in the Gippsland basin and Snapper field, especially dolomite cementation, have been discussed in previous publications (Bodard et al., 1984; Bodard and Wall, 1986; Glenton, 1988).

Dolomitic cementation at Snapper may have been related to the migration of early-generated gas from the thick coal seams at the base of the N-1 sequence. Carbon isotope values suggest a gas source for the carbon in the dolomite cement ($\delta^{13}C$ is about –19 PDB). The extensive dolomitization immediately above the N-1.9 Coal, decreasing in shallower units, suggests that gas generated from this coal may have been involved in the dolomite cementation.

Because the oil leg is only 4 to 8 m in thickness, targeting of oil development wells was critical. In several wells coal or shale was intersected at the depth of the oil leg and an oil completion was not possible (Figure 12).

Drainage radius for a typical N-1 oil producer is estimated at 400 to 500 m along strike. Coal and shale beds overlying the reservoir sandstones generally limit the drainage in the dip direction to less than this. Major faults may also act as barriers to oil drainage in some cases.

Coning of water was minimized by perforating a short interval as close as possible to the gas-oil contact. Satisfactory production was achieved from as little as 0.5 m vertical thickness of perforation zone. Although some wells produced with a very high gas-oil ratio, satisfactory oil production was achieved. An outstanding example was Snapper A-16, which produced 310,000 barrels of oil in addition to natural gas liquids from perforations immediately below the gas-oil contact, with an average gas-oil ratio of 62,500 standard cubic feet per barrel.

Oil recovery has been quite variable. The poorest well produced only 18,000 barrels before depletion, and the best 1.3 million barrels. The best seven wells

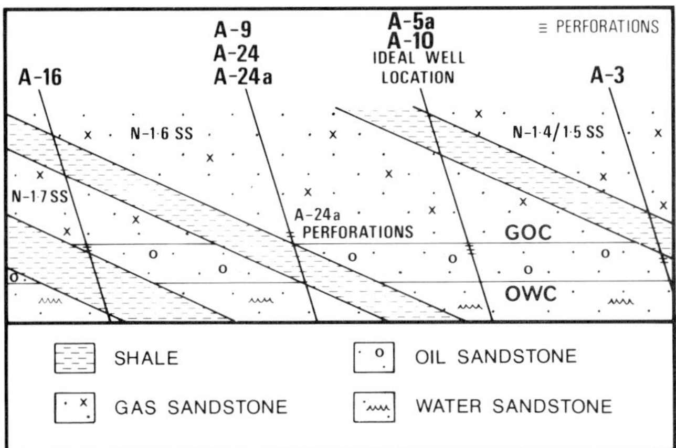

Figure 12. Cross section illustrating different types of intersection of the N-1 oil leg. It was expected that oil would come to the perforations in A-24a as the oil leg rises. However, no oil had been produced 2.5 years after perforation, suggesting that the oil leg could be absent in this part of the field. Note that this section is diagrammatic only and does not show an actual line of section through the field.

have produced more than 400,000 barrels each, with average production rate ranging from 730 to 1430 STB of oil per day, and average gas-oil ratio from 810 to 7340 scf/STB. Average water cut ranged from 3 to 31%.

The Snapper N-1 oil has an API gravity of 47°, pour point of 21°C, and wax content of 11% by weight. Solution gas-oil ratio is 400 scf/STB. The oil shows evidence of biodegradation (Burns et al., 1987), with significant loss of normal alkanes up to C17 (Figure 8). The bacteria were introduced by the freshwater aquifer that underlies Snapper and other fields in the western part of the offshore Gippsland basin (Kuttan et al., 1986).

Although the large N-1 gas reserve provided the initial economic justification for platform installation, gas production to date has been kept to a minimum. Barracouta and Marlin fields have been able to meet most gas commitments, with Snapper required mainly to meet peak winter demands. A significant part of the gas production from Snapper has actually been from oil wells that have been unavoidably produced at high gas-oil ratios. The few dedicated gas wells have been perforated in the upper part of the N-1 reservoir and have typical production rates of 35 mmcf of gas per day. As gas demand increases and oil wells water-out, more wells will be converted to gas production. It is expected that the gas reservoir will be thoroughly drained from the Snapper A platform.

The Snapper N-1 reservoir, like other upper Latrobe Group reservoirs, has a very strong bottom and flank water drive. Total pressure decline over the life of the field is expected to be only about 300

psi (Prasser, 1988). At the commencement of development drilling in 1981, the aquifer pressure was 48 psi less than it had been when the discovery well, Snapper-1, was drilled. This was due to production from other large fields in the basin (Seage et al., 1984) and resulted in expansion of the Snapper gas cap. The oil leg is calculated to have been pushed down as a result by 4 m, and has subsequently risen by about 6 m (by the end of 1989), owing to Snapper gas production. The observed rise is close to that predicted by material balance calculations, indicating good fluid communication throughout the gas reservoir.

The movement of the oil leg has provided strong incentive to maximize early production of oil, before the oil leg is "smeared out" and oil reserves lost. It has also meant that in some cases, upward reperforation is necessary to chase the rising oil column.

The deeper Latrobe Group reservoirs were deposited in a fluvial environment and are of limited lateral extent. Reservoirs in the Upper Cretaceous section have been tight and unable to sustain production, and abnormal pressure develops in the *N. senectus* section, below –2800 to –3000 m.

The L-3.2 reservoir, in the lower *L. balmei* (Paleocene) interval in Snapper A-8 (Figures 3 and 10), is the deepest commercially successful oil reservoir at Snapper. This zone contains a 17 m oil leg below a 4 m gas cap and had produced 0.6 MSTB of oil by the end of 1989, with no water cut. This sandstone is about 25 m thick, with porosity of 21%. Several nearby wells have penetrated correlatives of this sandstone above the L-3.2 oil-water contact, but these sandstones are structurally isolated and are water-wet. Another successful oil zone is the L-1a reservoir in Snapper A-21 (Figures 3 and 11). Although it is only 9 m thick, cumulative production to the end of 1989 was 2.2 MSTB.

Performance of other reservoirs has been varied, and none has been as good as the above two zones.

FAULTS

Mapping of major faults has been possible using seismic data, with a few additional minor faults recognized in well intersections.

The northeast-trending "Snapper fault" that bounds the field to the north is related to the major fault system that forms the margin of the deep central graben of the basin. Compressional roll and partial reversal of movement occurred on this fault during the Late Eocene to mid-Miocene compressional phase that caused the development of the Snapper anticline.

Northwest-trending faults within the field developed during the Late Cretaceous extension associated with rifting and continued to move during the latest Cretaceous to Eocene phase of marginal sag. Movement on many of these faults was diminishing by the time of deposition of the N-1 reservoir sequence.

The variation in thickness of the N-1 oil leg suggests that faults may have acted locally as barriers and that in places they act as drainage barriers for N-1 oil completions. Within the N-1 gas zone, good communication between fault blocks is interpreted to exist as a result of sand-to-sand juxtaposition.

SOURCE

A number of studies have indicated that the source of Gippsland basin gas and oil is organic matter of terrestrial, higher-plant origin and that maturation levels are the main control on hydrocarbon type in the major, top-of-Latrobe Group fields (e.g., Brooks and Smith, 1969; Alexander et al., 1987; Smith and Cook, 1984; Burns et al., 1984, 1987).

The source is interpreted to have been deep Latrobe Group shales and coaly units of Late Cretaceous age, typically Campanian or older. These were deposited in a coastal plain to lacustrine environment (Rahmanian et al., 1990). The best oil-source environments are interpreted to have been coastal plain swamps, lagoons, and possibly lakes. There is good correlation between gas chromatography—mass spectrometry traces of Upper Cretaceous source rocks and oil in Latrobe Group reservoirs (Alexander et al., 1987).

The source rocks contain significant oil-prone material, with typically 50% of the organic matter being oil-prone kerogen, dominated by biodegraded terrestrial material with minor spores and pollens. Coals and dispersed organic matter are dominated by conifer remains and have variable but often high (up to 30%) exinite contents (Smith and Cook, 1984).

The total organic content of the shales averages 2.4%, with a slightly shalier and richer sequence in the deep central part of the basin (Rahmanian et al., 1990). Source rocks are mostly type II/III and type III, although Campanian intervals are characterized by type II organic matter.

Change in source rock maturation level controls oil type and gas versus oil occurrence in the large upper Latrobe Group reservoirs (Figure 13). The small deeper Latrobe Group reservoirs do not show such a clear geographical correlation, and in many cases are probably influenced by the ability of cap rocks and fault zones to seal either gas or oil (Rahmanian et al., 1990). Oil types range from waxy and paraffinic to light and condensate-like. Early mature oils as found in Tuna and Flounder fields, and peak maturity oils, as in the major oil fields Kingfish, Mackerel, Fortescue, and Halibut, are generally undersaturated with respect to gas. Post-peak maturity oils are very light, trending toward condensates in character (Burns et al., 1987).

Oil maturation levels have been investigated using isotopic compositions of gas flashed from oil samples (Burns et al., 1984) and methylphenanthrene index (MPI) (Burns et al., 1987). These studies indicated that most of the Gippsland basin oils have been generated at maturity levels corresponding to vitrinite reflectance of $R_v = 0.9$–1.1%, higher than would be expected for classical marine source rocks but appropriate for oil generation from terrestrially derived organic matter. Light oils from the basin are interpreted to have been generated up to values of $R_v = 1.5\%$. Condensate and gas maturity levels correspond to the supermature state of hydrocarbon generation, with R_v greater than 1.8% (Burns et al., 1987). This leads to the interpretation of a similar source for both the gas and oil, with maturity being the main control over the types of hydrocarbon generated.

Top-of-Latrobe Group reservoirs are typically immature ($R_v = 0.4$–0.6), and very few wells in the basin have been drilled deep enough to intersect mature source rocks. This indicates considerable secondary migration. Along the Barracouta-Snapper anticlinal trend, vertical migration of around 2 km and lateral migration of several kilometers at least, are indicated. Deep faults and the sandy upper Latrobe Group section would have assisted migration. Most of the large top-of-Latrobe Group structures are interpreted to be full to spill, indicating that very large quantities of hydrocarbons have been generated.

In the Barracouta-Snapper area, the bulk of the source rock is supermature (Figure 13) and the oil window is relatively thin (Figures 14 and 15), resulting in generation mainly of wet gas. The large top-of-Latrobe Group structures are thus full or nearly full of gas. In contrast, to the southeast the oil window is relatively thick and the large top-of-Latrobe Group structures (such as the Kingfish field) are full of oil (Figure 16).

The Snapper N-1 oil is a waxy, paraffinic crude. Burns et al. (1987) proposed a mechanism to explain this type of association. Both the gas and oil could have started out as wet gas phase, but as the gas migrated up the section, the fall in temperature caused condensate to drop out of the gas phase and form a separate liquid or oil phase. This liquid extracted hydrocarbons from the section as it continued to migrate along the same pathways as the gas, thus acquiring a more oily character.

An alternative would be migration of gas and oil into the reservoir from source rocks at different maturation levels. This leads to the speculation that either the oil leg is a remnant, with much of the oil having been displaced out of the reservoir by gas; or that the control on spill point is different for gas and oil (e.g., a synclinal spill for the oil and a fault-leak spill point for the gas).

The Snapper N-1 oil has been modified by biodegradation in the reservoir (Figure 8; Burns et al., 1987). Biodegradation of oil by aerobic bacteria has occurred to differing degrees in a number of fields in the basin. Most of the affected reservoirs are in contact with meteoric water introduced by a

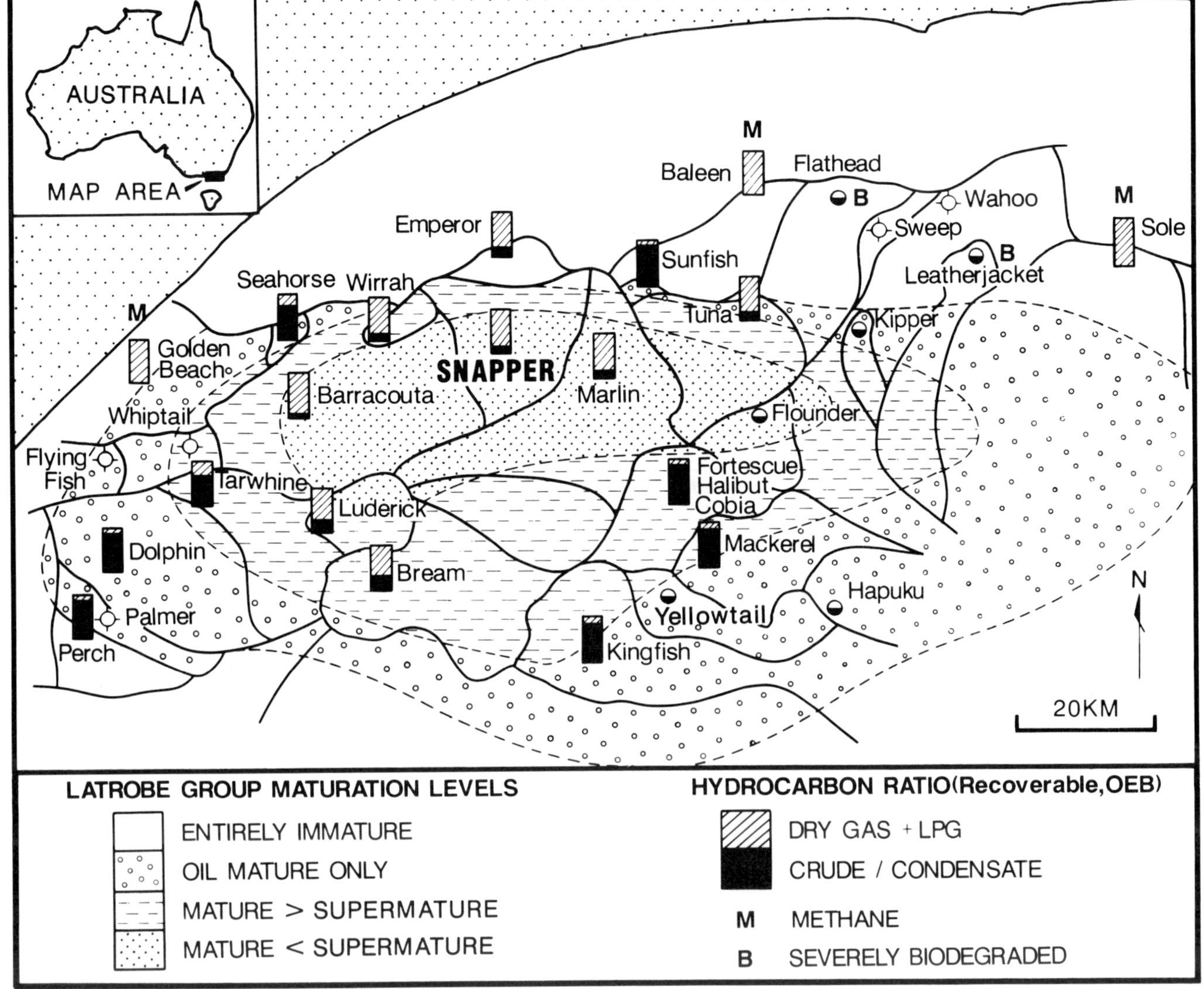

Figure 13. Relationship between top of Latrobe Group hydrocarbons, top of Latrobe Group drainage areas, and maturity levels within the Latrobe Group. The scale bar beside each field indicates the ratio of crude oil to dry gas plus liquified petroleum gas for that field, but does not indicate the relative sizes of the fields.

Used by permission of the Geological Society from "Sequence stratigraphy and the habitat of hydrocarbons, Gippsland Basin, Australia" by Rahmanian, V. D., Moore, P. S., Mudge, W. J., and Spring, D. E. in Special Publication No. 50—Classic Petroleum Provinces, 1990.

freshwater aquifer in the upper Latrobe Group in the western part of the offshore Gippsland basin (Burns et al., 1987; Kuttan et al., 1986). The few that are not in direct contact with meteoric water are very close to it, and it is possible that the extent of the aquifer may have been slightly greater in the past (for example, a sea-level fall would result in a greater hydraulic head and an expanded freshwater zone). Bacteria would have been introduced into the basin by this meteoric water. Burns et al. (1987) also noted that biodegradation did not occur within the freshwater zone in oil reservoirs when reservoir temperatures exceed 75 to 80°C.

EXPLORATION CONCEPTS

Snapper field is a large, seismically identified, anticlinal culmination at the top of the Latrobe Group, on the Barracouta-Snapper anticlinal trend. Anticlinal folding persists with depth, and numerous small deeper gas and oil reservoirs have been penetrated.

A number of similar features exist at the top of the Latrobe Group in the Gippsland basin, such as Barracouta gas field, and the much smaller Perch, Dolphin, and Tarwhine oil fields (Figure 1). The Marlin gas field is also an anticlinal structure,

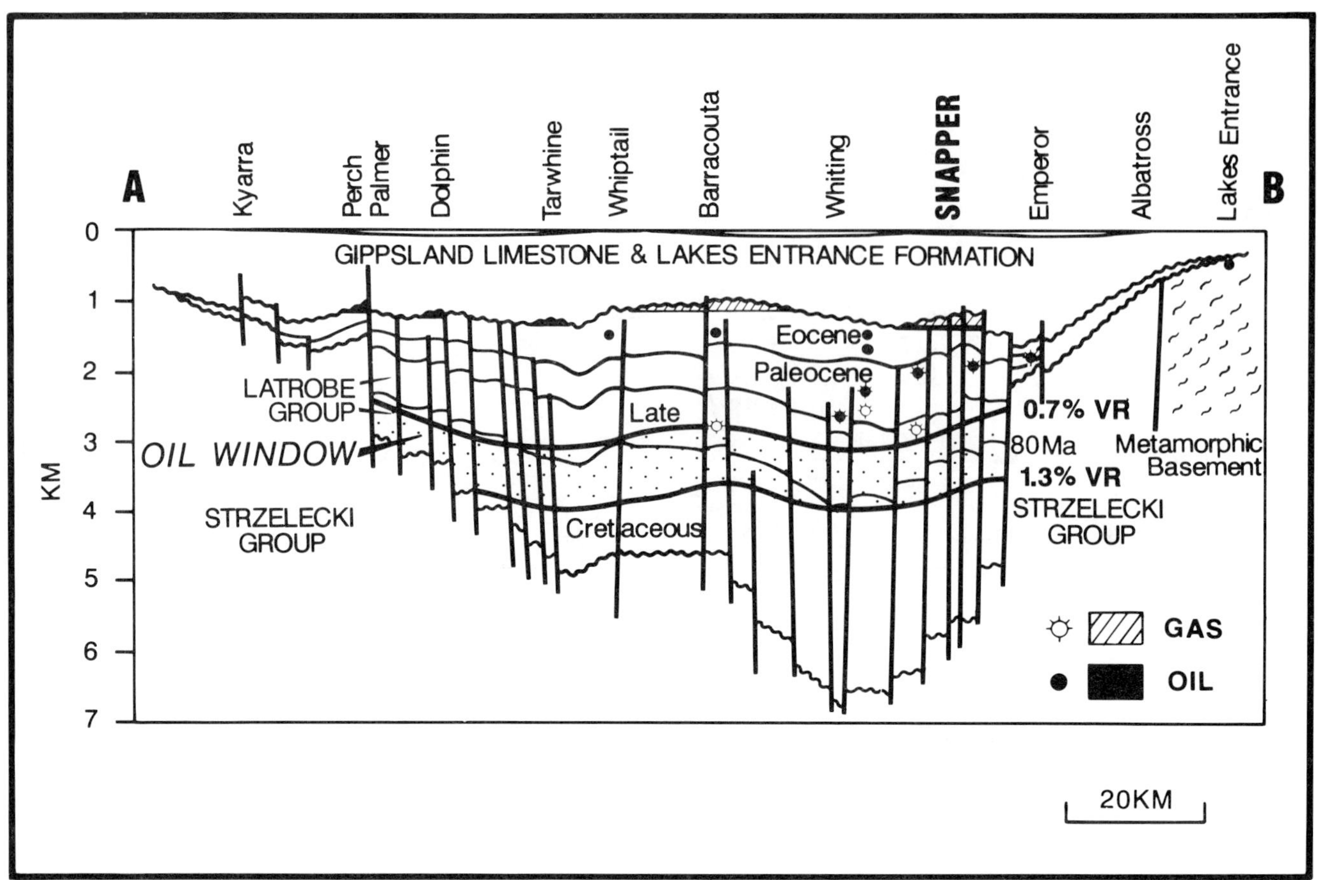

Figure 14. Regional cross section AB, showing estimated position of oil window (defined as 0.7% to 1.3% vitrinite reflectance). Location of the section is shown on Figure 1. Tarwhine contains a very light, high GOR oil. Snapper is along strike from Tarwhine, but at Snapper more of the Latrobe Group is supermature (Figure 13). Used by permission of the Geological Society from "Sequence stratigraphy and the habitat of hydrocarbons, Gippsland Basin, Australia" by Rahmanian, V. D., Moore, P. S., Mudge, W. J., and Spring, D. E. in Special Publication No. 50—Classic Petroleum Provinces, 1990.

although its northeastern flank was truncated by the mid-Eocene Marlin Channel.

All these structures have Eocene sandstone reservoirs and are sealed at the top of the Latrobe Group by the fine-grained marine sediments of the Lakes Entrance Formation.

The hydrocarbon source throughout the basin is interpreted to have been terrestrial plant material, with variations in hydrocarbon type being determined by maturation level, not source rock variations. This terrestrial source yielded great volumes of both gas and oil.

At Snapper, the entire Latrobe Group was deposited under predominantly nonmarine conditions. Sedimentation in the eastern part of the basin was under progressively more marine conditions. Sandstones of excellent reservoir quality are present in the upper Latrobe Group throughout most of the basin.

The type of play represented by Snapper has been thoroughly tested at this stage, and only small top-of-Latrobe Group closures remain to be drilled. Future exploration is likely to concentrate on deeper Latrobe Group plays (such as fault-dependent closures) and stratigraphic traps.

Although identification of the Snapper structure was not difficult and occurred very early in the exploration in the Gippsland basin, development of the thin oil leg presented geological and engineering challenges. Nevertheless, economic development of this oil leg within the platform conformable area was possible, once the installation of platform facilities had been justified by the large N-1 gas reserves. The N-1 oil reserves were sufficient to justify the additional oil facilities that were required.

Horizontal drilling technology was not available at the time of the Snapper development but could be appropriate in future developments of thin oil legs. However, an oil zone as thin as that in the Snapper N-1 reservoir would push drilling and survey accuracy to the limit and would present difficulties with gas and water coning. Contact movement owing to gas production and aquifer pressure changes would also have to be considered.

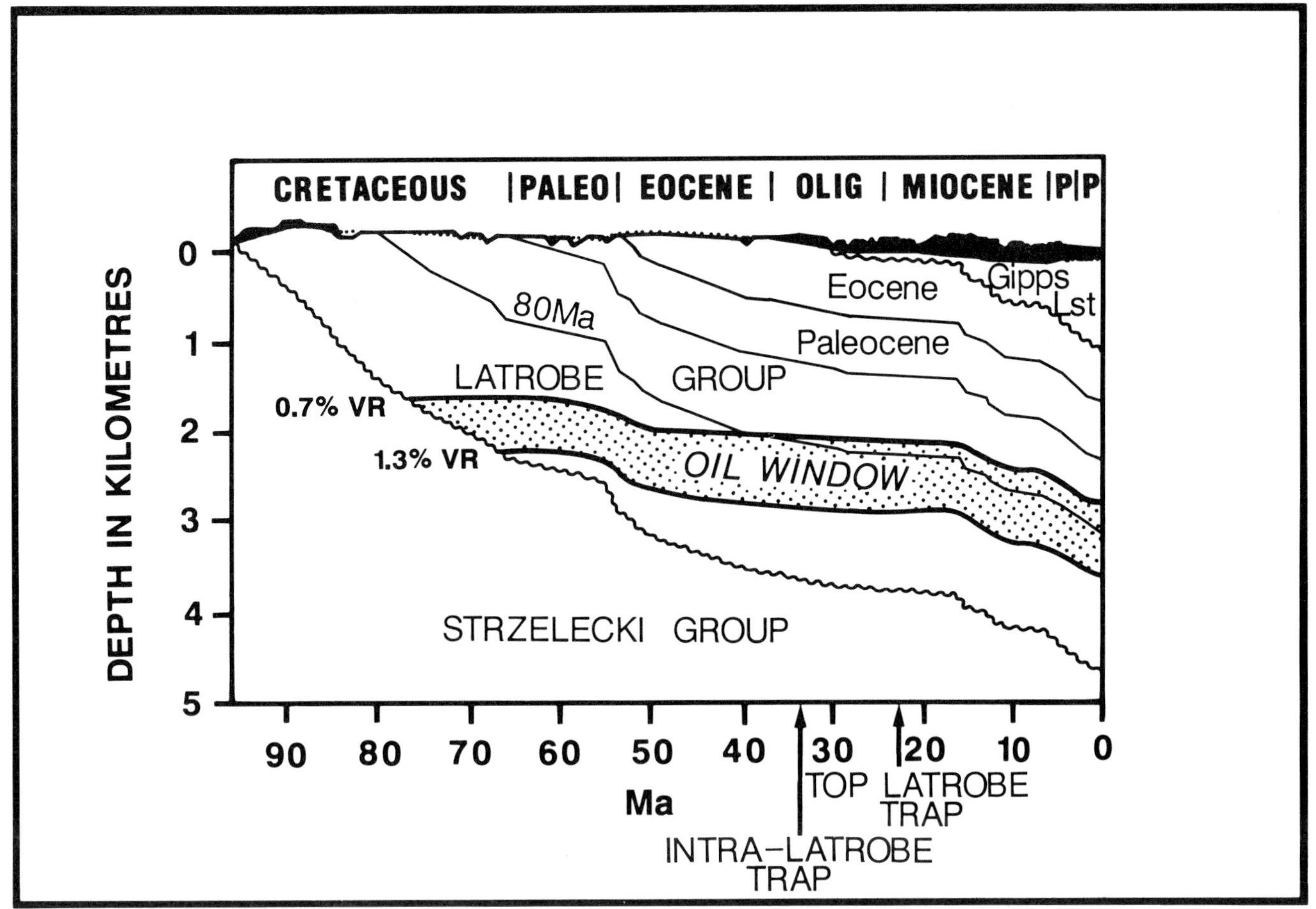

Figure 15. Timing chart for hydrocarbon generation, for Barracouta. At Barracouta, as at Snapper, there was high heat flow, resulting in a thin oil window and a large volume of supermature source rock, and the top-of-Latrobe Group structure is mostly gas filled. Used by permission of the Geological Society from "Sequence stratigraphy and the habitat of hydrocarbons, Gippsland Basin, Australia" by Rahmanian, V. D., Moore, P. S., Mudge, W. J., and Spring, D. E. in Special Publication No. 50—Classic Petroleum Provinces, 1990.

Despite the excellent reservoir quality, water coning in the conventional deviated Snapper development wells was restricted by appropriate perforation and production practices.

The accuracy required in well targeting was at the limit of that possible using seismic data. In the offshore environment, where well costs are high and platform facilities expensive, 3D seismic may have assisted in achieving optimum well locations and in modeling the distribution of dolomite cement (a major factor in reduction of oil reserves in some areas of the reservoir).

ACKNOWLEDGMENTS

This paper is the result of work by a number of employees of Esso Australia Ltd. and Exxon Production Research Company. Thanks are due to Ken Grieves, who was responsible for seismic mapping during the final development phase, Victor Rahmanian, whose regional stratigraphic studies were of great assistance, and Brodie Thomson for reservoir engineering data. Esso Australia Ltd. and BHP Petroleum Pty. Ltd. permitted publication of the paper. Thanks are also due to the London Geological Society for permission to publish figures from their Special Publication No. 50, *Classic Petroleum Provinces.*

REFERENCES CITED

Alexander, R., R. A. Noble, and R. I. Kagi, 1987, Fossil resin biomarkers and their application in oil to source-rock correlation, Gippsland basin, Australia: Australian Petroleum Exploration Association Journal, v. 23, n. 1, p. 53–63.

Bodard, J. M., and V. J. Wall, 1986, Sandstone porosity patterns in the Latrobe Group, offshore Gippsland basin, *in* R. C. Glenie, ed., Second South-Eastern Australia Exploration Symposium, 14–15 November 1985, Melbourne, 469 p.

Bodard, J. M., V. J. Wall, and R. A. F. Cas, 1984, Diagenesis and evolution of Gippsland basin reservoirs: Australian Petroleum Exploration Association Journal, v. 24, n. 1, p. 314–334.

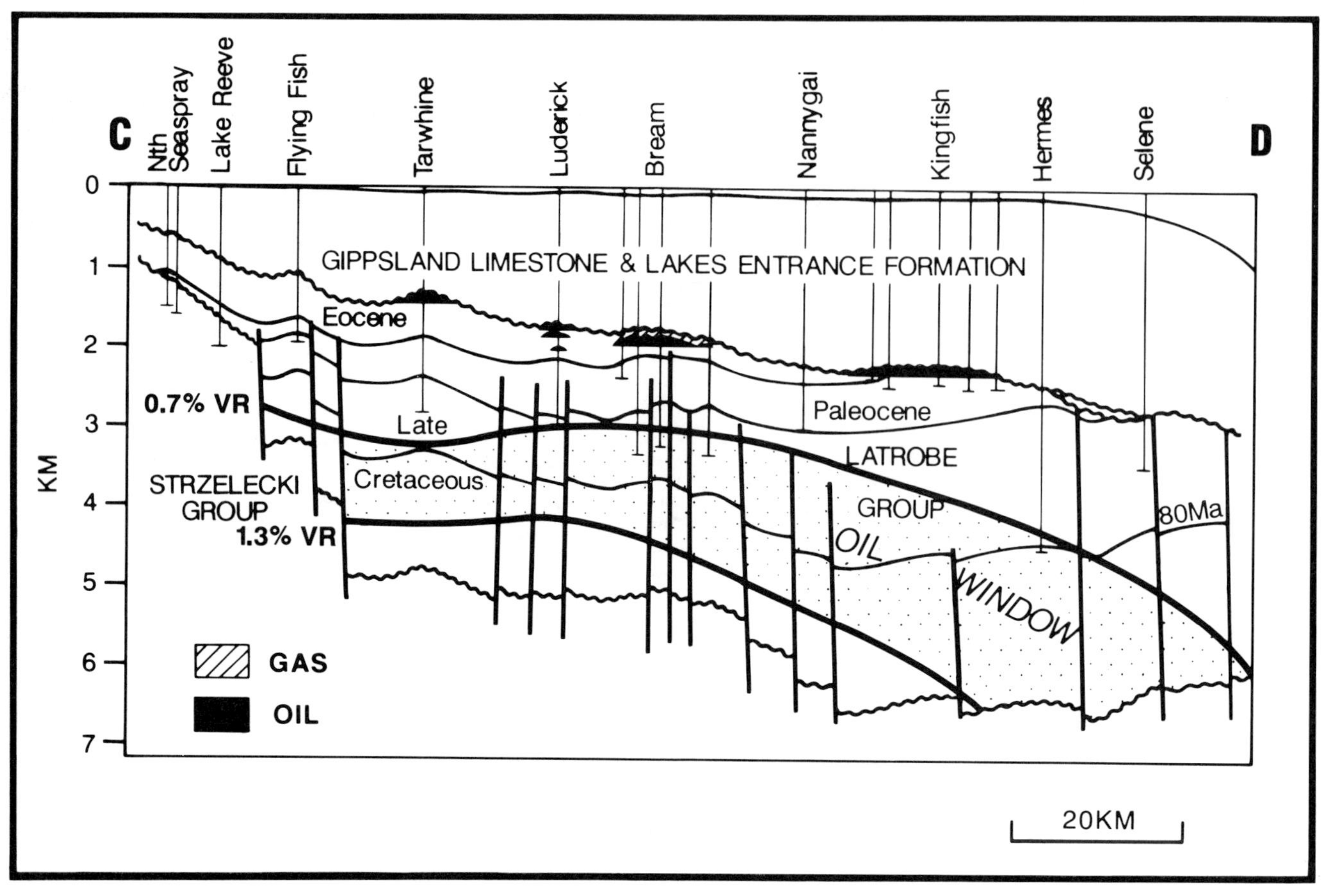

Figure 16. Regional cross section CD, showing estimated position of the oil window (defined as 0.7% to 1.3% vitrinite reflectance). Location of the section is shown on Figure 1. At Kingfish the oil maturation window is relatively thick, and the large top-of-Latrobe Group Kingfish structure contains only oil with no gas cap. Used by permission of the Geological Society from "Sequence stratigraphy and the habitat of hydrocarbons, Gippsland Basin, Australia" by Rahmanian, V. D., Moore, P. S., Mudge, W. J., and Spring, D. E. in Special Publication No. 50—Classic Petroleum Provinces, 1990.

Brooks, J. D., and J. W. Smith, 1969, Coalification and the formation of oil and gas in the Gippsland basin: Geochimica et Cosmochimica Acta, v. 33, p. 1183–1194.

Burns, B. J., A. T. James, and J. K. Emmett, 1984, The use of gas isotopes in determining the source of some Gippsland basin oils: Australian Petroleum Exploration Association Journal, v. 24, n. 1, p. 217–221.

Burns, B. J., T. R. Bostwick, and J. K. Emmett, 1987, Gippsland terrestrial oils—recognition of compositional variations due to maturity and biodegradation effects: Australian Petroleum Exploration Association Journal, v. 27, n. 1, p. 73–84.

Glenton, P. N., 1988, The Snapper development, Gippsland basin: Australian Petroleum Exploration Association Journal, v. 28, n. 1, p. 29–40.

Kuttan, K., J. B. Kulla, and R. G. Neumann, 1986, Freshwater influx in the Gippsland Basin: impact on formation evaluation: Australian Exploration Association Journal, v. 26, n. 1, p. 242–249.

Leslie, R. B., H. J. Evans, and C. L. Knight, eds., 1976, Economic geology of Australia and Papua New Guinea, Volume 3: petroleum: The Australasian Institute of Mining and Metallurgy Monograph Series, No. 9, 541 p.

Prasser, R. S., 1988, The challenge of thin oil development: a case study: 8th Offshore South East Asian Conference.

Rahmanian, V. D., P. S. Moore, W. J. Mudge, and D. E. Spring, 1990, Sequence stratigraphy and the habitat of hydrocarbons, Gippsland basin, Australia, *in* J. Brooks, ed., Classic petroleum provinces: Geological Society of London Special Publication No. 50.

Seage, N. W., J. Lau, and S. T. Koh, 1984, Development of the thin oil zone underlying the Snapper gas reservoir: 5th Offshore South East Asia Conference, 1–15 August.

Smith, G. C., and A. C. Cook, 1984, Petroleum occurrence in the Gippsland basin and its relationship to rank and organic matter type: Australian Petroleum Exploration Association Journal, v. 24, n. 1, p. 196–216.

Veevers, J. J., 1984, Phanerozoic earth history of Australia: Oxford, Clarendon Press.

Weeks, L. G., and B. M. Hopkins, 1967, Geology and exploration of the Bass Strait basins, Australia: American Association of Petroleum Geologists Bulletin, v. 51, n. 5, p. 742–760.

SUGGESTED READING

Beddoes, L. J. R., Jr., 1973, Oil and gas fields of Australia, Papua New Guinea and New Zealand: Tracer Petroleum and Mining Publications, Pty Ltd, Sydney, Australia, 392 p. Includes data on Australian oil and gas fields, including early data for Snapper.

Douglas, J. G., and J. A. Ferguson, eds., 1976, Geology of Victoria: Geological Society of Australia Special Publication No. 5, 528 p. Contains a summary of the regional and economic geology of Victoria, including the Gippsland basin. Revised edition published in 1988 by Victorian Division of Geological Society of Australia Inc. 663 p.

Fagg, K. G., 1985, Gas lift in Bass Strait: Australian Petroleum Exploration Association Journal, v. 25, n. 1, p. 107-113. Gas lift to improve recovery.

Shibaoka, M., J. D. Saxby, and G. H. Taylor, 1978, Hydrocarbon generation in Gippsland basin, Australia—comparison with Cooper basin, Australia: American Association of Petroleum Geologists Bulletin, v. 62, n. 7, p. 1151-1158.

Tissot, B. P., and D. H. Welte, 1984, Petroleum formation and occurrence, 2nd ed.: New York, Springer-Verlag, 699 p.

Weeks, L. G., 1978, ". . . a lifelong love affair"—The Memoirs of Lewis G. Weeks. This contains L. G. Weeks's account of his consultancy to Haematite Exploration Pty. Ltd. and his advice to acquire the Bass Strait acreage including the Gippsland basin.

Wilkinson, R., 1983, A thirst for burning: Sydney, Australia, David Ell Press; Includes an account of early exploration for hydrocarbons in the Gippsland basin and of L. G. Weeks's role.

Appendix 1. Field Description

Field name .... *Snapper field*

Ultimate recoverable reserves .... *2140 bcf of gas; 15–20 million STB of oil*

Field location:

    Country .... *Australia*

    State .... *Victoria (offshore)*

    Basin/Province .... *Gippsland basin/Bass Strait*

Field discovery:

    Year first pay discovered .... *Eocene Latrobe Group sandstones   1968*

    Year second pay discovered *1981*

    Third pay *1983*

Discovery well name and general location:

    First pay .... *N-1 gas and oil accumulation: Snapper-1; lat. 38°12′03″S, long. 148°00′50″E, approximately 30 km offshore and 80 km east of Sale*

    Second pay .... *L-1a oil zone: Snapper A-21; a vertical well drilled from Snapper platform (platform coordinates lat. 38°11′27″S; long. 148°01′27″E)*

    Third pay .... *L-2d oil zone: Snapper A-6; a deviated well (average 20° from vertical) drilled from the Snapper platform*

Discovery well operator .... *Esso Exploration and Production Australia Inc.*

    Second pay .... *Esso Exploration and Production Australia Inc.*

    Third pay .... *Esso Exploration and Production Australia Inc.*

IP in barrels per day and/or cubic feet or cubic meters per day:

    First pay .... *N-1 gas zone: 4.8 mmcf/day (production test, Snapper-1; 9.2 mmcf/day (Snapper A-5) N-1 oil zone: 690 STB/day (Snapper A-1)*

    Second pay .... *Snapper A-21, L-1a, 820 STB/day*

    Third pay .... *Snapper A-6, L-2d, 300 STB/day*

Note: Reservoir properties are highly variable.

All other zones with shows of oil and gas in the field:

Well	Age	Reservoir	Type of Show
Snapper A-6	*Eocene*	*M-2c*	*Gas cap, oil leg (production)*
	Paleocene	*L-2d*	*Oil (production)*
Snapper A-8	*Eoc/Paleoc*	*L-1b*	*Oil (production)*
	Paleocene	*L-3.2*	*Gas cap, oil leg (production)*
	Paleoc/L. Cret.	*T-1b*	*Oil (production)*
Snapper A-16	*Eocene*	*M-1a*	*Oil (production)*
Snapper A-18	*Eocene*	*N-1.9*	*Oil (production)*
Snapper A-21	*Eoc/Paleoc*	*L-1a*	*Oil (production)*
Snapper A-26	*Eoc/Paleoc*	*L-1g*	*Oil (production)*
	Paleocene	*L-2h*	*Gas cap, oil leg (production)*
Snapper A-27	*Eocene*	*M-1c*	*Interpreted gas cap, oil leg (logs)*
	Eocene	*M-1d*	*Interpreted gas cap, oil leg (logs)*
	Eocene	*M-2e*	*Oil (production)*
	Paleocene	*L-2g*	*Interpreted gas cap, oil leg (logs)*

Snapper-5	Eocene	M-1.10	Oil (wireline samples)
	Eocene	M-1.20	Interpreted oil (logs)
	Eocene	M-1.30	Oil (wireline samples)
	Eocene	M-2.20	Gas cap, oil leg (wireline samples)
	Eocene	M-2.30	Gas cap, oil leg (wireline samples)
	Eocene	M-2.70	Oil (wireline samples)

Note: The zones listed are deeper Latrobe Group oil or gas/oil zones. Only the more significant reservoirs have been listed. Reservoir prefix (N, M, L, and T) is taken (loosely) from its age, using spore/pollen zone (N. asperus, M. diversus, L. balmei, T. longus) (Figure 4).

Geologic concept leading to discovery and method or methods used to delineate prospect, e.g., surface geology, subsurface geology, seeps, magnetic data, gravity data, seismic data, seismic refraction, nontechnical:

Anticlinal trend in northern part of basin. Seismic mapping identified the large Snapper culmination on this trend. Three exploration/delineation wells were drilled prior to development.

Structure:

Province/basin type .... Bally III, Klemme III A (failed rift)

Tectonic history

Early Cretaceous: rifting between Australia and the Antarctic; Strzelecki Group sedimentation. Separation of Australia and Antarctic occurred at end of Early Cretaceous, leaving failed arm of the rift to the east, in which the Gippsland, Bass, and Otway basins evolved. Late Cretaceous: rifting between Australia and the Lord Howe Rise, leading to opening of the Tasman Sea. Latrobe Group sedimentation. Latest Cretaceous to Eocene: post-rift subsidence. In the offshore part of the Gippsland basin, Latrobe Group deposition ceased late in the Eocene. Late Eocene to mid-Miocene: compressive structuring. Pleistocene to Recent: Minor reactivation of compression.

Regional structure

The Snapper anticline is located along the northern margin of the central graben of the Gippsland basin. A major normal fault bounds the field to the north. Movement along this fault was reversed during the late Eocene–mid-Miocene compressive structuring.

Local structure

Snapper is an east–northeast-trending anticline, 13 km by 6 km, at top of Latrobe Group. It has steeper dips on the northern flank than on the southern flank. The amount of faulting increases with depth.

Trap:

Trap type(s)

(1) Anticlinal structure at top of Latrobe Group; (2) several separate smaller anticlinal culminations in separate fault blocks within the Latrobe Group.

Basin stratigraphy (major stratigraphic intervals from surface to deepest penetration in field):

Chronostratigraphy	Formation	Depth to Top in m
Oligocene to Recent	Gippsland Limestone	55 (seafloor)
	Lakes Entrance Formation	
Upper Eocene	Latrobe Group Gurnard Formation	1177
Upper Cretaceous	Latrobe Group to upper Eocene	
	"coarse clastics"	1195
Upper Cretaceous to	Strzelecki Group (not penetrated)	3725+
Lower Cretaceous		

Note: Depths are subsea and are given at crest of structure. Age of Gurnard Formation varies across the basin, becoming older to the southeast.

Reservoir characteristics:

N-1 gas and oil reservoir

Number of reservoirs .... 1 gas accumulation with associated thin oil leg

Formations .... *Latrobe Group, Gurnard Formation (N-1.0 reservoir unit)
and "coarse clastics" (N-1.1 to N-1.8 reservoir units)*

Ages .... *Eocene*

Depths to tops of reservoirs .... *1177 m*

Gross thickness (top to bottom of producing interval) .... *205 m of gas, 4–8 m of oil*

Net thickness—total thickness of producing zones

    Average .... *120 m*

    Maximum .... *135 m*

Lithology .... *Sandstone, fine to coarse quartzarenite to subarkose, mainly medium to coarse;
contains dolomitic cement in places*

Porosity type .... *Mainly intergranular primary porosity; some secondary porosity from feldspar
dissolution, and in some areas from partial dissolution of dolomite cement*

Average porosity *24%*

Average permeability .... *2000 md*

Deeper Latrobe Group oil reservoirs

    Number of reservoirs .... *20 significant, separate oil zones; numerous additional minor oil and gas zones*

    Formations .... *Latrobe Group "coarse clastics" (N-1.9, M, L, and T reservoirs)*

    Ages .... *Late Cretaceous to Eocene*

    Depths to tops of reservoirs .... *1420–2140 m*

    Gross thickness (top to bottom of producing interval) .... *730 m*

    Net thickness of producing zones

        Average .... *7 m*

        Maximum .... *17 m*

    Lithology .... *Sandstone, fine to medium quartzarenite to subarkose*

    Porosity type ... *Mainly intergranular primary porosity; some secondary porosity from feldspar dissolution*

    Average porosity *20%*

    Average permeability .... *Variable; tens to hundreds of md*

Seals:

N-1 reservoir

    Upper

        Formation, fault, or other feature .... *Lakes Entrance Formation*

        Lithology .... *Marl*

Deeper Latrobe Group reservoirs

    Upper

        Formation, fault, or other feature .... *Latrobe Group*

        Lithology .... *Shales and coals*

    Lateral

        Formation, fault, or other feature .... *In some cases, faults may provide lateral seal;
sand pinchouts also occur*

Source:

    Formation and age .... *Latrobe Group, Upper Cretaceous section*

    Lithology .... *Shale and coaly shales*

    Average total organic carbon (TOC) .... *2.3% (Snapper-5)*

    Maximum TOC .... *4.0% (Snapper-5)*

    Kerogen type (I, II, or III) .... *II/III*

    Vitrinite reflectance (maturation) .... $R_o = 0.9–1.1$ *(oil); 1.5+ (gas)*

    Time of hydrocarbon expulsion .... *Paleocene to present*

    Present depth to top of source .... *3000 m*

    Thickness .... *1000+ m*

    Potential yield .... *NA*

Appendix 2. Production Data

Field name .... *Snapper field*

Field size:

    Proved acres .... *12,850 ha at N-1 OWC*

    Number of wells all years .... *32 development, 6 exploration/delineation*

    Current number of wells .... *27 productive development wells*

    Well spacing .... *Avg. 65 ha within platform conformable area*

    Ultimate recoverable .... *11 MSTB (N-1 oil zone); 2140 bcf (N-1 gas zone);*
    6 MSTB (deeper Latrobe Group zones)

    Cumulative production (to end of 1989) .... *8.4 MSTB (N-1 thin oil leg); 230 bcf (N-1 gas zone);*
    3.2 MSTB (deeper Latrobe zones)

    Annual production

    Present decline rate *30%*

        Initial decline rate *30%*

        Overall decline rate *30%*

    Annual water production (in 1989) .... *535,000 bbl*

    In place, total hydrocarbons .... *3400 bcf (N-1 gas); 230 MSTB (N-1 oil leg, entire field);*
    55 MSTB (N-1, platform conformable)

    In place .... *7300 bbl/ha-m*

    Primary recovery .... *2140 bcf (N-1); 10 MSTB (N-1) (includes gas lifted oil)*

    Secondary recovery .... *NA*

    Enhanced recovery .... *NA*

    Cumulative water production .... *1.5 MSTB*

Drilling and casing practices:

    Amount of surface casing set .... *620–1200 mMD (610–850 m TVD)*

    Casing program
    Typical production well: 20-in. conductor; 10¾-in. surface casing; 7⅝-in. production casing
    High angle production well: 20-in. conductor; 13⅜-in. surface casing; 9⅝-in. intermediate casing; 7-in. production liner
    Exploration wells below the N-1 reservoir: 20-in. conductor; 13⅜-in. surface casing; 9⅝-in. intermediate/ N-1 production casing; 7-in. production liner for deeper zones

    Drilling mud .... *First development phase (first 21 development wells), freshwater polymer gel; second development phase (6 development wells and 5 redrills), seawater polymer gel*

    Bit program .... *Soft formation milled tooth bit (IADC Code 1-1-1, 1-1-4, 1-1-6) or PDC bit to top of Latrobe Group; compact insert bit through Latrobe Group (IADC Code 1-3-4, 1-3-7, 4-3-7)*

    High pressure zones
    Mild elevation in pressure at top of Latrobe Group due to the gas gradient effect of the N-1 gas accumulation (maximum column 205 m); abnormal pressure develops in Latrobe Group at 2800 to 3000 m (Late Cretaceous, deeper than commercial reservoirs); maximum pressure encountered was 13.5–14.0 ppg equivalent mud weight in Snapper-1 at total depth of 3725 m

Completion practices:

    Interval(s) perforated .... *N-1 gas, 6–12 m TVT (2⅛-in. Enerjet); N-1 oil, 0.5–2 m TVT; deeper Latrobe oil, 5–10 m TVT*

    Well treatment .... *None*

Formation evaluation:

    Logging suites
    First development phase (first 21 development wells): ISF-BHC-MSFL-GR-SP; FDC-CNL-GR; slim-hole tools or cased hole neutron logs were run in a few wells when conventional open-hole logs could not be obtained

SNAPPER

249

Second development phase (6 development wells and 5 redrills): DLL-MSFL-LDT-CNL-GR-SP (multi-combination), run on drillpipe-conveyed logging system when necessary in highest angle wells; cased hole GST run in one well instead of open-hole logs (due to hole conditions)

Testing practices .... *Wireline formation fluid sampling and pressure tests; Snapper-1 was production tested (N-1 gas reservoirs and one deep Latrobe Group zone)*

Mud logging techniques .... *Samples at 5 m intervals in the Latrobe Group; total gas and gas chromatography*

Oil characteristics:

Type .... *Paraffinic*
API gravity .... *47°*
Base .... *Paraffinic*
Initial GOR .... *400 SCF/STB*
Sulfur, wt% .... *<0.20% (Snapper-4, 1389 m, N-1 oil leg)*
Viscosity, SUS .... *NA*
Pour point .... *21°C*
Gas-oil distillate .... *NA*

Field characteristics (N-1 reservoir):

Average elevation .... *1280 m*
Initial pressure .... *2012 psig at 1326 m*
Present pressure .... *NA*
Pressure gradient .... *1.52 psi/m*
Temperature .... *73°C at 1326 m*
Geothermal gradient .... *0.055°C/m*
Drive .... *Water*
Oil column thickness .... *4–8 m (N-1 reservoir)*

Note: Maximum Latrobe Group oil column is 17 m in the L-3.2 reservoir in Snapper A-8.

Oil-water contact .... *1388 m subsea (N-1 original OWC)*
Connate water .... *N-1 gas zone, 15%; N-1 oil zone, 15%; deeper Latrobe Group zones, 25–50%*
Water salinity, TDS ...... *N-1 gas reservoir, 20,000 ppm NaCl equiv.; N-1 aquifer, 5000 ppm NaCl equiv.*
Resistivity of water .... *N-1 gas reservoir, 0.14 ohm-m at 74°C; N-1 aquifer, 0.70 ohm-m at 74°C*
Bulk volume water (%) .... *NA*

Transportation method and market for oil and gas:

Gas and oil separated on platform. Gas transported to onshore gas plant at Longford by 24/30-in. pipeline. LPG processed at Longford, then sent to Long Island Point by pipeline, then to refineries. Gas supplied to Victorian Gas and Fuel Corporation for Victorian gas market. Oil transported to Marlin to 10-in. pipeline, then to 24-in. Halibut pipeline and thence to Longford gas plant. Stabilized, then sent to Long Island Point for distribution via pipeline and tanker to refineries. The bulk of Gippsland basin production is used domestically.

Tirrawarra Field—Australia
Cooper Basin, Central Australia

C. G. SKILBECK
University of Technology, Sydney
Broadway, New South Wales

D. P. ANTHONY
M. R. FABIAN
Santos Limited
Adelaide, South Australia

J. W. HUNT
Esso (Australia) Limited
Sydney, New South Wales

FIELD CLASSIFICATION

BASIN: Cooper
BASIN TYPE: Rift
RESERVOIR ROCK TYPE: Sandstone
RESERVOIR ENVIRONMENT
   OF DEPOSITION: Fluvial (Braided Stream)

RESERVOIR AGE: Permian
PETROLEUM TYPE: Oil, Gas, and
   Condensate
TRAP TYPE: Faulted Anticline

LOCATION

Tirrawarra oil and gas field is located in the southwestern portion of the Permian–Triassic Cooper basin of central Australia (Figure 1). The Cooper basin is situated beneath the "channel country" of Sturt's Stony Desert, about 500 mi (800 km) north of Adelaide. The field lies approximately 30 mi (50 km) north of the Moomba processing facility and is the largest oil field so far discovered in the basin.

Tirrawarra field lies within a Lower Permian hydrocarbon-rich structural trend called the Patchawarra trough that occupies the western portion of the southern Cooper basin (Figure 1). The Patchawarra trough is an arcuate depression approximately 87 mi (140 km) long and 25 mi (40 km) wide. Other oil and gas fields in the vicinity include the Fly Lake–Brolga, Moorari, and Woolkina fields, which contribute subordinately to the reserves of the Patchawarra trough, containing only about 20% of the oil and 35% of the gas reserves trapped by the Tirrawarra structure.

Tirrawarra field is situated in the "Patchawarra Central Block" of Petroleum Exploration Licenses 5 and 6 and is operated by Santos Limited for a consortium that also includes Delhi Petroleum Pty Ltd (now owned by Esso Australia Ltd), Bridge Oil Limited, Barcoo Petroleum NL, Vamgas NL, and Sagasco Resources Limited.

The areal extent of the principal reservoir in Tirrawarra field is approximately 20 mi^2 (50 km^2) (Figure 2). Estimates of initial ultimate recovery of hydrocarbons from Tirrawarra field are 177 bcf (billion standard cubic feet) (5.0 billion Sm^3) of sales gas, 32 bcf (0.9 million Sm^3) of ethane, 26 million stock tank barrels (MMSTB) (4.1 million Sm^3) of liquified petroleum gas, and 42 MMSTB (6.7 million Sm^3) of crude oil and condensate.

HISTORY

Strata of the Cooper basin are not exposed at the surface; the rocks are situated between the craton of early to middle Paleozoic basement and overlying Jurassic to Cretaceous Eromanga basin strata. Early hydrocarbon exploration drilling in central Australia was aimed at the latter basin and it was this activity that eventually led to the discovery of the Cooper basin. The history of this exploration effort has been thoroughly documented by Sprigg (1986).

Pre-Discovery

The Great Artesian Basin of eastern central Australia, of which the Eromanga basin is a significant feature (Figure 1), has long been a source of artesian water for agricultural and domestic use.

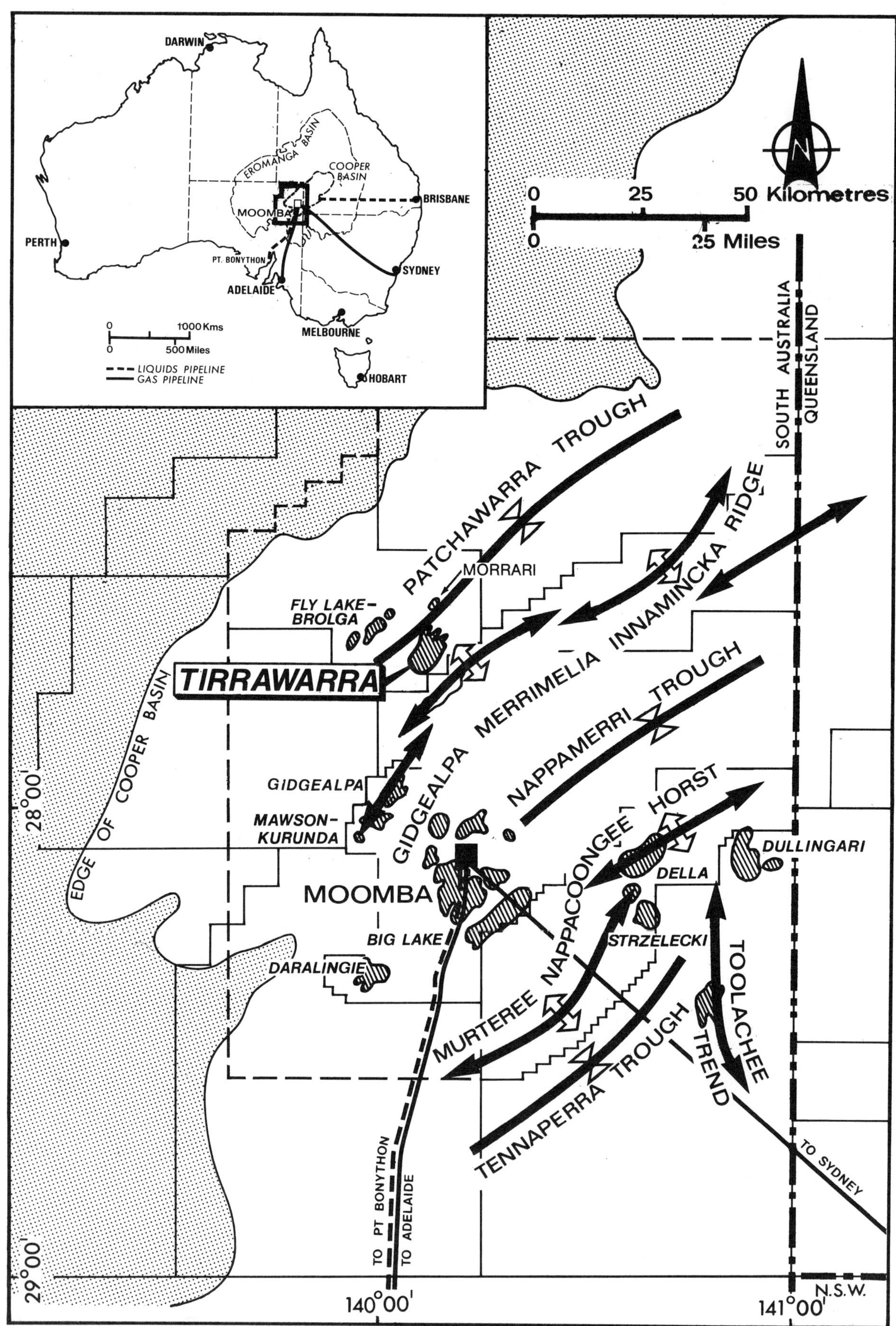

Figure 1. Map of generalized structural elements of the southern Cooper basin showing the location of Tirrawarra field and nearby fields. The locations of Gidgealpa field (initial Cooper basin discovery) and Moomba field are also indicated.

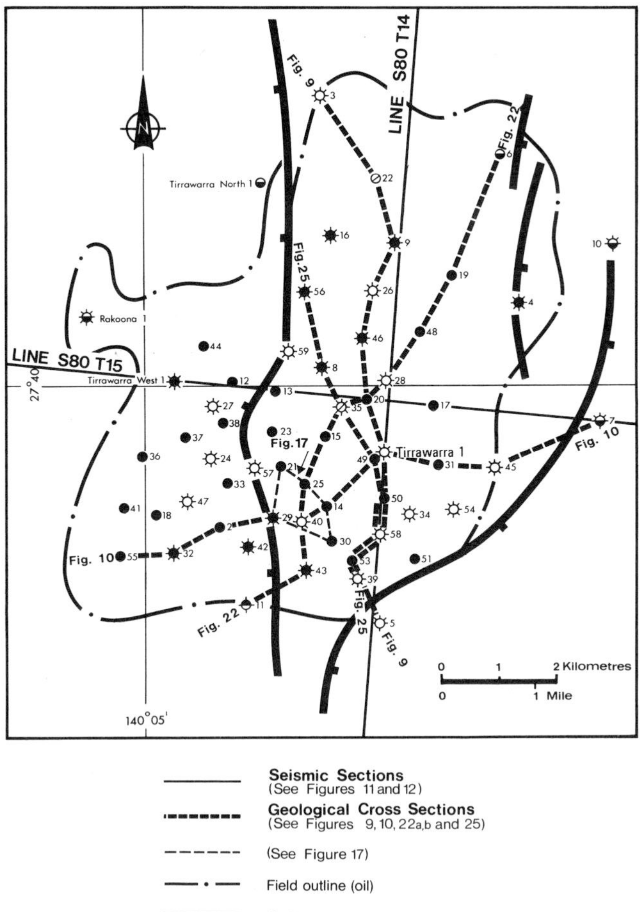

Figure 2. Tirrawarra field outline and well locations. The locations of selected seismic sections and geological cross sections appearing in this paper are also shown.

In 1900, natural gas was discovered in the eastern Great Artesian Basin in a water well at Roma, Queensland. During the next two decades, thousands of artesian wells were drilled in central Australia. Despite the presence of natural gas in many of these wells and traces of oil in Triassic shale cores from the Leigh Creek No. 2 coal bore drilled in 1908 in east-central South Australia (Ward, 1944), there were few early indications of significant oil in the region during the first quarter of the century. Some of these early wells were drilled to depths of up to 6500 ft (1980 m). The first oil recovery from the Eromanga basin was made at Longreach Bore No. 2 in 1927, following discovery of waxy crude in the town water bore (Sprigg, 1986). Despite the growing indications of hydrocarbons, however, the consensus of the geological community by the late 1930s was that the nonmarine source rocks known to exist in the region were incapable of significant oil generation, and this hindered the exploration effort during the pre-World War II years. Many geologists considered the continent of Australia to be too old and stable and its basins too shallow to favor significant hydrocarbon accumulations (Sprigg, 1986).

During World War II and the following years, regional geological mapping and geophysical surveys were conducted by several organizations, most notably Frome Broken Hill Company (Frobilco), Shell Development Queensland Pty Ltd, Zinc Corporation Pty Ltd, Vacuum Oil Company of New York, the Australian Bureau of Mineral Resources, and the South Australian Department of Mines. Between 1947 and 1951, seven stratigraphic wells were drilled by Frobilco. Freeman (1964) documented the major remote sensing exploration efforts of the 1950s and early 1960s. While these studies and stratigraphic wells delineated basin margins and clearly identified a number of regional structural highs, the crucial discovery remained elusive.

The first deep well, Innamincka No. 1 in the SANTOS-Delhi license areas, was drilled in the central portion of the Eromanga basin in 1959. Despite its failure to discover petroleum, it substantiated earlier geophysical interpretations that older sedimentary basins existed beneath the Great Artesian Basin and thus gave incentive for further comprehensive exploration.

After Innamincka No. 1, a further seven wells were drilled before natural gas was discovered by Gidgealpa No. 2, which spudded in December 1963. The main target in that well was Cambrian limestone that had given indications of hydrocarbons in Gidgealpa No. 1. The natural gas in Gidgealpa No. 2, however, was discovered in sandstone reservoirs in a Permian sequence also containing coal seams and shale beds.

Subsequent exploration established this sequence as the most important hydrocarbon prospect known in the license areas. The extent of the Permian and Triassic sediments that form the Cooper basin are now well established from exploration data (Figure 1).

After the discovery of the Moomba gas field in 1966, negotiations commenced for the supply of gas to Adelaide. A plant at Moomba to remove excess carbon dioxide and a pipeline to Adelaide were completed in 1969, and gas transmission began in 1969. Several further gas and gas-condensate discoveries were made in the Cooper basin; however, the most significant event at that time was the discovery of oil at Tirrawarra, which dispelled the earlier opinion that the Cooper basin was capable only of generating gas.

Discovery

The Tirrawarra No. 1 wildcat (Figure 2) was drilled in 1970 by Bridge Oil Limited under a farm-out agreement with Santos Limited, Delhi International Oil Corporation, and Vamgas NL, and discovered a total of 36 ft (11 m) of gas pay in several thin sandstones of Lower Permian Patchawarra Formation and 110 ft (34 m) of oil pay in the underlying Lower Permian Tirrawarra Sandstone (Figure 3). The well was drilled crestally to test the gas potential

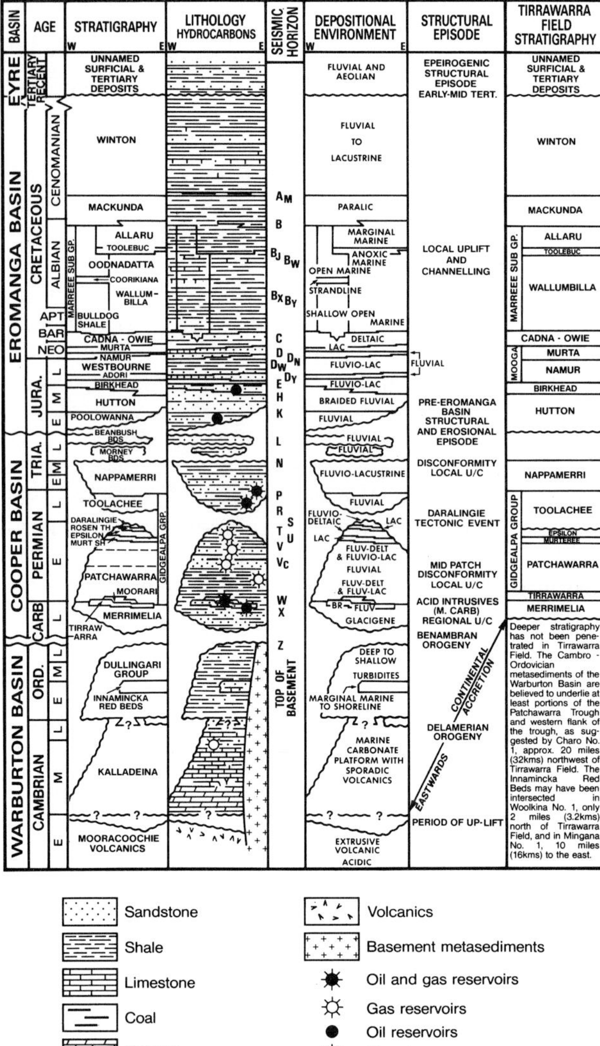

Figure 3. Generalized Cooper/Eromanga basin stratigraphy showing the distribution of recoverable oil and gas. The diagram also shows how this scheme applies to Tirrawarra field.

of a large seismically defined domal anticline. The reservoir rocks are fluvial sandstones and are located at depths ranging from 8700 ft (2650 m) to 9900 ft (3020 m). Hydrocarbons are intraformationally sourced from interstratified coal seams and associated organic-rich fine-grained clastic rocks. Open-hole testing of Tirrawarra No. 1 resulted in a maximum combined flow rate from the gas reservoirs of 10 MMcf (280 thousand Sm^3) per day through a 1-in. (2.5 cm) choke. Subsequent cased-hole testing of the oil pay zone yielded up to 650 bbl (100 Sm^3) of oil per day. Tirrawarra oil is a highly volatile, undersaturated oil with a gas-to-oil ratio (GOR) of 2700 ft^3 of C^{4-} per stock tank barrel of C^{5+} (approximately equivalent to 480 Sm^3 of C^{4-} per m^3 of C^{5+}).

Gas in the field has a high average liquid yield containing up to 45 bbl of condensate per MMcf of raw gas (250 m^3/MM m^3) and 32 bbl of liquified petroleum gas per MMcf of raw gas (180 m^3 per MM m^3). Tirrawarra No. 1 produced the first oil flow from the Cooper basin, and the field remains the principal known oil accumulation in the basin.

Post-Discovery

In the period from discovery to the end of 1972, eight appraisal wells were drilled on the flank of the Permian Tirrawarra structure to determine the areal extent of the field, and two crestal development wells were drilled (Tirrawarra No. 8 and No. 9) with dual completions in the oil and gas zones. Four of the appraisal wells were unsuccessful. Owing to the field's remoteness (500 mi or 800 km north of the nearest major city, Adelaide) and the low oil price at that time, development was delayed until 1981, when the Cooper Basin Unit Partners initiated the AUS$1200 million Cooper basin liquids project (Pecanek and Paton, 1984).

A reservoir engineering study undertaken upon completion of the initial appraisal phase (Osborne, 1974) estimated the in-place oil to be 70 MMSTB (11 MM m^3) with a maximum recovery factor of 28% under primary recovery. Simulation modeling conducted during Osborne's study indicated that recovery factors could be significantly enhanced by secondary production using high-pressure miscible gas injection. It was estimated that oil recovery in the better zones of the reservoir could be increased to as much as 70% (Pecanek and Paton, 1984).

An additional appraisal well (Tirrawarra No. 12) was drilled in 1980 as part of the liquids project feasibility study and resulted in an upgrading of the in-place oil estimate to 123 MMSTB (20 MM m^3) with associated gas reserves of 119 bcf (3.4 billion m^3) (Pecanek and Paton, 1984). Tirrawarra No. 14, drilled in July 1981, discovered gas in the Upper Permian Toolachee Formation.

The most intensive phase of development drilling began in 1981. Since then, a further 50 wells (Tirrawarra No. 13 to No. 62) have been completed on the main Tirrawarra structure. Three wildcat wells (Rakoona No. 1, Tirrawarra West No. 1, and Tirrawarra North No. 1) were drilled on small structures on the western flank of the field during this period.

The locations of all wells are shown in Figure 2. Both gas and oil development wells were located to optimize future potential enhanced oil recovery. In 1984, a pilot gas injection program was initiated to assess the feasibility of the miscible gas injection enhanced oil recovery (EOR) project and to provide pressure support for oil wells in the central part of the field. These wells had shown rapidly declining primary productivity since being brought on line (Brown and Barley, 1986). The ethane injection program in Tirrawarra field was expanded beginning in 1985. The current full-scale EOR project is the first of its kind in Australia to be operated on the basis of miscible gas flooding.

Producing wells in the field are completed through either 7-in. (17.8 cm) or 5.5-in. (14 cm) production casing. Of 54 completed wells in the field, there are 16 single gas completions, 22 single oil completions, and 11 dual oil and gas completions.

One of the oil wells, Tirrawarra No. 35, is currently used to monitor reservoir pressure and gas break-through as part of the injection project. In addition to the producing wells, there are five single completion gas injection wells (Tirrawarra Nos. 15, 18, 29, 38, and 50). The Tirrawarra Sandstone oil reservoir is routinely hydraulically fracture-stimulated on completion, resulting in up to fivefold increases in production (Pecanek and Paton, 1984, their table 2). The main gas reservoir sands are fracture-stimulated where they are of sufficient thickness to be economically viable (usually in excess of 20 ft, or 6 m) and where the lithology is suitable for vertical constraint of the fracture (G. Roberts, 1987, personal communication).

In July 1987, after the drilling of Tirrawarra No. 59, nonassociated gas-in-place (proved plus probable gas-in-place) was volumetrically calculated to be 401 bcf (11.3 billion Sm^3). Original oil-in-place was estimated to be 112 MMSTBO (17.8 million Sm^3 of oil) with solution gas-in-place of 303 bcf (8.6 billion m^3) (Anthony, 1987).

Cumulative production from Tirrawarra field to mid-1988 totaled 8.6 MMSTB (1.4 million Sm^3) of oil (C^{5+}) from the oil field and 94 bcf (2.7 billion Sm^3) of raw gas from the gas field.

DISCOVERY METHOD

Permian Cooper basin strata outcrop nowhere, being completely covered by younger strata. In the earliest phase of exploration, from 1957 to 1962, many wells were sited on large anticlines in overlying Mesozoic rocks of the Eromanga basin that are exposed at the surface and in which structures can be delineated by field mapping (Battersby, 1976). Since the discovery of the Cooper basin, geophysical techniques have been employed exclusively for exploration and development. Gravity-magnetic reconnaissance data were used to define major trends within the basin in the early days; however, seismic data have led to all the major discoveries in the basin and have served as the basis of detailed structural mapping and field development.

Commercially exploitable hydrocarbons were discovered in the Cooper basin in 1963. The discovery well, Gidgealpa No. 2, was the ninth wildcat drilled in the basin and was located on a regional seismically defined anticlinal trend now called the Gidgealpa-Merrimelia-Innamincka (GMI) ridge. The principal target of the well was Cambrian dolomite. After discovery of gas in Permian sediments, exploration efforts were directed toward identification of other Permian structural culminations above the large linear basement highs that subdivide the Cooper basin (Figure 1).

The main Permian reservoirs (Toolachee and Patchawarra formations and Tirrawarra Sandstone) are absent or thin over many of these regional highs (Martin, 1967) and success was therefore limited. In 1966, Moomba No. 1 discovered gas in a large relatively low-relief structure in the regional trough (now called the Nappamerri trough) to the east of the Gidgealpa high.

Once the prospectivity of deeper areas of the Cooper basin had been demonstrated, the exploration effort was refocused to address structural traps in similar geological settings. The discovery of several significant gas fields followed. These included Daralingie field in 1967 and Toolachee field in 1969. The structure and stratigraphic sequence of the Patcha-warra trough, in which Tirrawarra field is located, are similar to the Nappamerri trough but lie to the west of the Gidgealpa high. Tirrawarra field was discovered as wildcat drilling spread from the center of initial discovery.

Tirrawarra field has no surface expression and is located beneath the floodplain of the intermittently flowing Cooper Creek. The geomorphologic relief is dominated by large linearly extensive sand ridges oriented approximately north-south and separated by clay pans; however, much of the field is covered by swamplands, and drilling operations are occasionally interrupted by flooding. None of these features exhibits any discernible relation to the underlying Permian–Triassic structure and there are no tangible indicators, such as local oil seepage, of the field's existence.

STRUCTURE AND
BASIN HISTORY

There is presently no tectonic model providing an entirely satisfactory explanation of the origin of the Cooper basin. It is considered by some workers to be an intracratonic downwarp (Battersby, 1976), whereas others have suggested or implied a cratonic rift origin (Youngs, 1975; Mount, 1981; Gray and Roberts, 1984). The abundance of normal faults with strikes subparallel to the axis of the trough and the southwest-to-northeast regional paleocurrent axis (Thornton, 1978) refute simple epeirogenic subsidence.

On the other hand, it has been demonstrated by Stanmore and Johnstone (1988) that, in the south-western part of the basin, sediments locally onlap onto a gently basinward-dipping unconformity. Clearly, the depositional margin of the basin is preserved in these areas and the regional down-to-the-basin faults that typify the margins of rift basins elsewhere in the world are absent in this part of the Cooper basin. Gray and Roberts (1984) formulated a model for the development of normal faulting in the central part of the Cooper basin that comprises an initial protracted phase of epeirogenic subsidence during which time most of the stratigraphic fill of the basin accumulated, followed by a phase of extensional fault-controlled subsidence. While this model has merit locally, it cannot be ubiquitously applied because compressed stratigraphic sequences elsewhere show that normal faults were active at the inception of the basin. Thornton (1978) recognized six major structural zones in the southern part of the Cooper basin, which is dominated by a series of northeast-trending basement troughs and ridges (Figure 1).

The basement highs are bounded in places by steeply dipping normal faults that were active during the Early Permian (Kantsler et al., 1986). The axes of these structures plunge toward the northeast, with the Permian depocenter located in the south-central part of the basin. Structures in the northern part of the basin trend more easterly and, together with evidence of right-lateral en echelon offset of structures further to the south, suggest that a phase of wrench faulting followed initial extensional stress (Kantsler et al., 1986). By Triassic times the basin depocenter had migrated further north (Thornton, 1978).

Deposition in the Cooper basin began in the Late Carboniferous period, unconformably on a basement consisting of early Paleozoic epiclastic and carbonate sediments and volcanics with Carboniferous intrusives. Late Carboniferous to late Early Permian sedimentation resulted in the deposition of a fluvio-lacustrine megacycle that is up to 5000 ft (1525 m) thick. This cycle is terminated by a regional unconformity of early Late Permian age representing nondeposition in the south and uplift and erosion in the north of the basin (Kantsler et al., 1986). Sedimentation resumed in the latter part of the Late Permian with the deposition of a further maximum of 600 ft (180 m) of moderate-sinuosity mixed-load fluvial sandstone, shale, and coal (Toolachee Formation). This cycle of deposition continued into Early and Middle Triassic times with the accumulation of up to 2500 ft (760 m) of suspended-load fluvial and lacustrine sediments (Nappamerri Formation) (Battersby, 1976). Late Triassic uplift and erosion ended the sedimentary history of the Cooper basin and resulted in the loss of up to 1600 ft (488 m) of section (Kantsler et al., 1986).

The Patchawarra trough is the westernmost structural province of the southern Cooper basin (Figure 1). This part of the basin is characterized by a thick stratigraphic section from which only the Roseneath Shale and Daralingie beds are absent (Figure 3). Regional seals comprising the Murteree Shale and the Nappamerri Formation, as well as the major reservoir rocks of the Tirrawarra Sandstone and Patchawarra and Toolachee formations, are well represented in the Patchawarra trough. Its internal structure consists of a number of north-trending low-relief normal fault-bounded basement highs. The

faults were active during the early phases of deposition and the resultant horst blocks have acted as the loci of subsequent sediment draping. Tirrawarra is the largest structure of this type in the Patchawarra trough.

The only fault of consequence to appraisal and production is the north-striking normal fault that has bisected the Tirrawarra Sandstone and the lower part of the Patchawarra Formation (see Figures 4 and 10). This fault has a westerly downthrow of between 75 and 100 ft (23–30 m), thus placing the Tirrawarra Sandstone on either side out of direct abutment. Subdivision of the field by the fault has had a significant effect on the design and planning of the enhanced oil recovery project.

STRATIGRAPHY

The stratigraphy recognized throughout this paper is illustrated in Figure 3 and follows the model described by Thornton (1978). Hydrocarbons are produced from Permian and Triassic sandstones throughout the Cooper basin sequence. The main gas reservoirs are in the Patchawarra and Toolachee formations, while oil is produced principally from the Tirrawarra Sandstone. Minor amounts of oil also occur in thin sandstones in the basal Patchawarra Formation, as oil legs associated with liquids-rich gas in some mid-Patchawarra sandstone reservoirs and in thin sandstones in the Toolachee and Nappamerri formations. All commercial oil and gas in Tirrawarra field is trapped in Permian strata.

The Lower Permian Patchawarra Formation hosts 98% of the in-place nonassociated gas in Tirrawarra field. The remaining known gas is reservoired in the Upper Permian Toolachee Formation. Oil is found only in the Lower Permian Tirrawarra Sandstone.

TRAP

The Tirrawarra oil and gas field is a domed anticline positioned above a fault-bounded basement horst. The depth structures of the reservoir horizons are shown in Figures 4, 5, and 6.

Present structuring was initiated during deposition of the Tirrawarra Sandstone (Figure 7) and faulting was most active during the early phase of Patchawarra sedimentation (Figures 8A and 8B). Subsequently deposited beds, draped over this high, formed the structure as a result of differential compaction along the flanks (Figures 9 and 10). The structure of the field is shown by seismic lines in Figures 11 and 12.

Oil in the field is structurally trapped in a single reservoir, the Tirrawarra Sandstone, and lies stratigraphically beneath the gas-bearing rocks. The distribution of oil in the eastern part of the field is

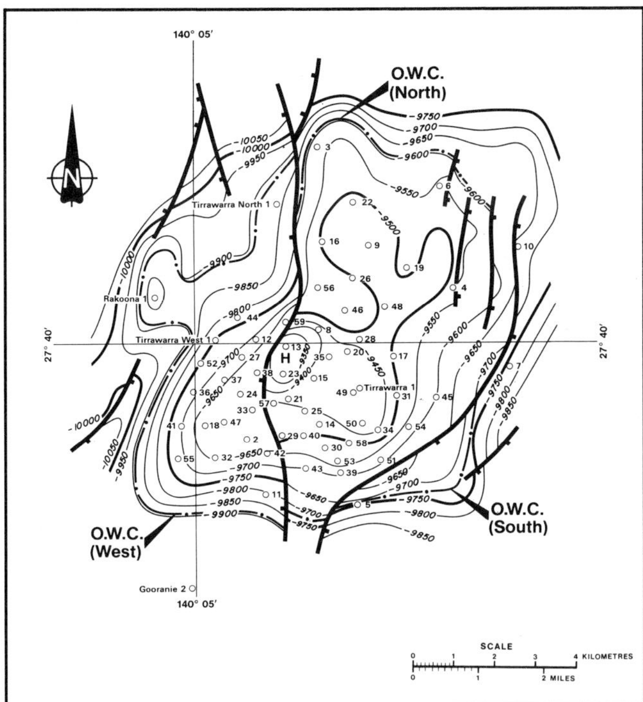

Figure 4. Depth structure contour map on top of Tirrawarra Sandstone oil reservoir. Locations of western field and eastern field reservoir limits are indicated by the dashed lines. Contour values are in feet below sea level. Contour interval, 50 ft.

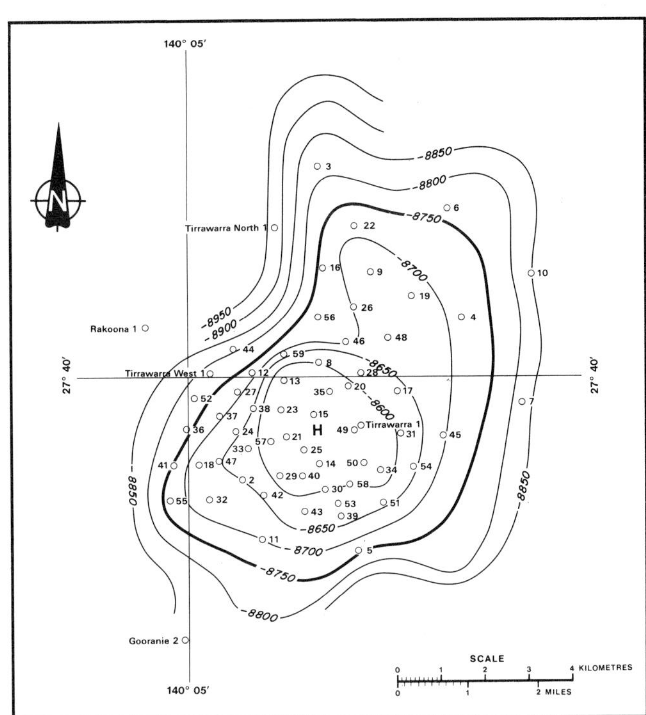

Figure 5. Depth structure contour map on top of Patchawarra Formation. Contour values are in feet below sea level. Contour interval, 50 ft.

restricted by a permeability barrier. The major north–south-striking fault in the central part of the field is impermeable, as demonstrated by the different reservoir limits and pressures. The northern part of the oil field is thought to have high capillary pressure in tight reservoir rocks that has raised the oil-water contact, but the relationship between reservoir contacts in the north and south of the field is not yet clearly understood. The Tirrawarra Sandstone has nearly 650 ft (195 m) of vertical structural closure (Figure 4) and, based on the well-defined oil-water contact in the western part of the field, is approximately 90% vertically filled with oil.

Multiple gas pay sands in Tirrawarra field are relatively thin (rarely thicker than 30 ft, or 9 m). Fourteen distinct reservoirs are recognized in the Patchawarra Formation, with 70% of the in-place gas hosted by four of these. An additional four gas sands are mapped in the Toolachee Formation.

The principal trap for gas accumulations is the anticlinal structure, but all Patchawarra Formation reservoir sandstones are depositionally restricted to some extent in their areal distribution and therefore incorporate a stratigraphic trapping component. The distribution of these sandstones is not necessarily correlatable with influences of the underlying structure.

Vertical closure at the top of the Patchawarra Formation is reduced to 270 ft (80 m) (Figure 5) with

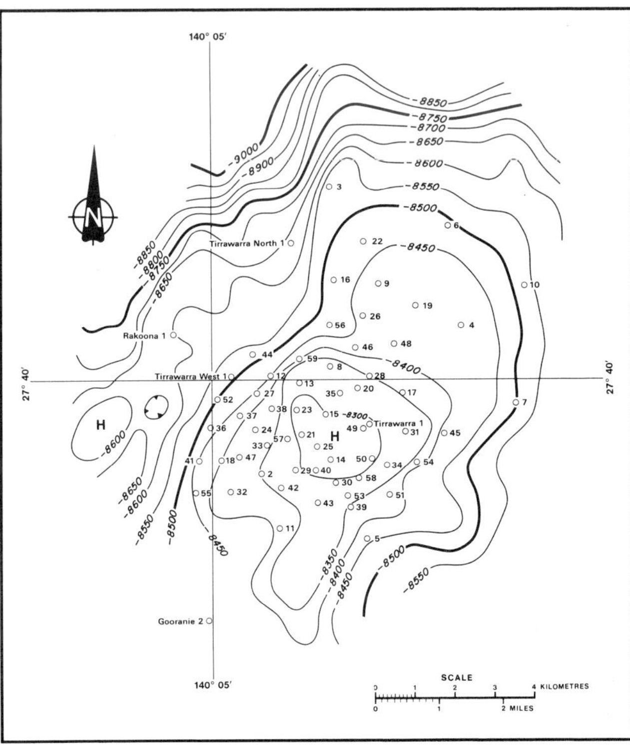

Figure 6. Depth structure contour map on top of Toolachee Formation. Contour values are in feet below sea level. Contour interval, 50 ft.

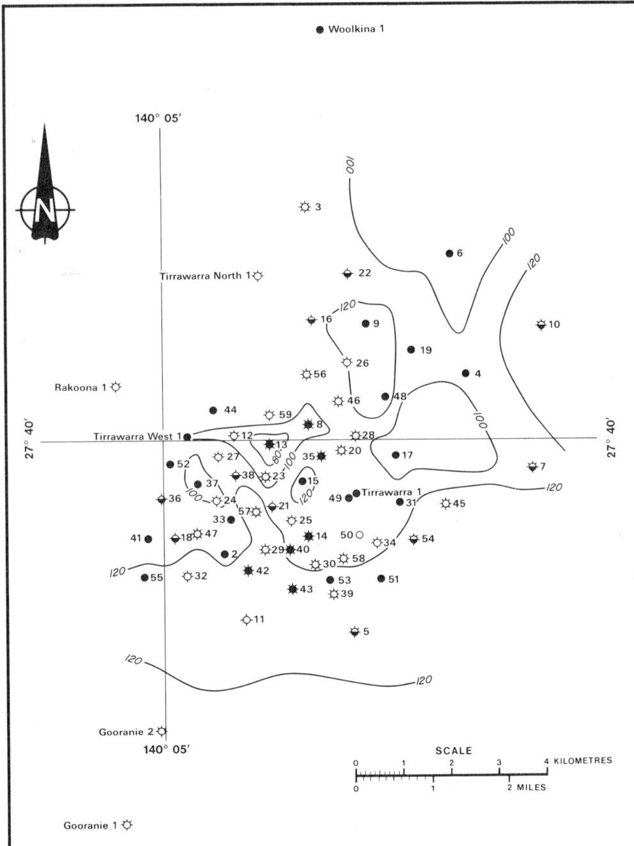

Figure 7. Isopach map of gross Tirrawarra Sandstone. Although the formation generally thins over the crest of the Tirrawarra structure, the relationship is not one of direct coincidence (see Figure 4). This map shows that structural growth had commenced by the end of deposition of the Tirrawarra Sandstone. Contour values are in feet. Contour interval, 20 ft.

a fill factor of up to 85% where individual sandstones are laterally continuous across the field. Structural relief at the top of the Toolachee Formation is 140 ft (43 m) (Figure 6). The reservoirs in this formation contain only 25% of their vertical hydrocarbon capacity.

Seals for all reservoirs in the field are provided by intercalated shales that also act as the source rocks. Oil in the Tirrawarra Sandstone is sealed by organic-rich shales at the base of the overlying Patchawarra Formation. Gas sands in the Patchawarra and Toolachee formations are sealed by intraformational shales that in places are only a few feet thick. A thick cap (greater than 800 ft, or 240 m) of predominantly fine grained Nappamerri Formation is believed to have prevented vertical communication between Permian source rocks and overlying potential reservoir rocks of Jurassic age. Elsewhere in the basin, where the Nappamerri Formation is considerably thinner, Jurassic strata contain several small oil fields.

RESERVOIR

Tirrawarra field contains 18 gas sands and one oil sand. The oil is trapped stratigraphically below the gas. Top of reservoir ranges from 8280 ft subsea (2520 m) for the shallowest gas reservoir to 9300 ft subsea (2840 m) for the top of the oil reservoir. The lowest known hydrocarbon occurrence is the western oil field oil-water contact at 9893 ft (3015 m) below sea level. The average gross thickness of the hydrocarbon-producing interval is 1130 ft (345 m).

Within this interval, the Tirrawarra Sandstone has an average gross thickness of 115 ft (35 m) with up to 85% of this thickness being net pay in any one well. In some wells on the northern and eastern flanks of the anticline the Tirrawarra Sandstone exhibits very low permeability and, although hydrocarbon-saturated, cannot be economically produced at the current oil price.

Fourteen gas sands comprise the reservoirs of the Patchawarra Formation. These are irregularly distributed over a stratigraphic interval of 900 ft (270 m). Productive sands within this interval have a total gross average thickness of 43 ft (13 m). The well intersection thicknesses of individual reservoir sands range between 1 ft (the minimum mapped) and 65 ft (0.3–20 m), averaging 11 ft (3.4 m) overall.

The four reservoir sands in the Toolachee Formation are distributed over an average column of 185 ft (55 m), with individual sands ranging up to 31 ft (9.4 m) in thickness. The average reservoir is 9 ft (2.7 m) thick. The stratigraphy of a representative well is shown in Figure 13.

Tirrawarra Sandstone

Lithology, Depositional Environment, and Development

The petrography and depositional environment of the Tirrawarra Sandstone, both regionally and in the Tirrawarra field vicinity, have been the subjects of studies by several investigators (Kapel, 1972; Gostin, 1973; Thornton, 1978; Williams, 1982; Williams and Wild, 1984; Martin, 1984). The unit was first recognized in Tirrawarra No. 1 and has been intersected by all subsequent wells in the field.

It has a maximum gross thickness of 166 ft (51 m) on the southern flank of the Tirrawarra dome in Tirrawarra No. 11 and thins crestally to 65 ft (20 m) in Tirrawarra No. 13. The formation isopach (Figure 7) does not closely correspond to the present field structure but differential subsidence was clearly affecting the area during deposition of the sandstone.

In most places the sandstone is primarily a quartzose arenite but locally can have a large lithic component (Williams, 1982). Subordinate amounts of conglomerate, mudstone, and coal constitute the remainder of the formation. The Tirrawarra Sandstone has been interpreted by most authors to be the deposits of a low-sinuosity bedload-dominated (braided) fluvial system (Kapel, 1972; Gostin, 1973;

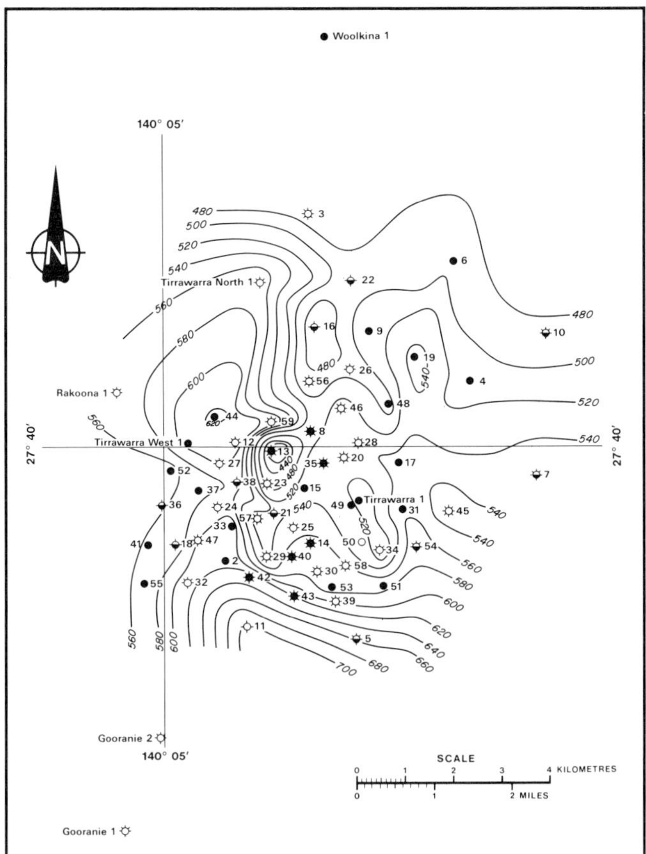

Figure 8A. Isopach map of lower Patchawarra Formation with subdivision at Patchawarra coal structural horizon (see Figures 13, 22A,B). Significant thinning of the lower part of the formation over the present structural crest suggests that the main phase of structural growth occurred during this period. No corrections have been made for differential compaction of shale. Contour values are in feet. Contour interval, 20 ft.

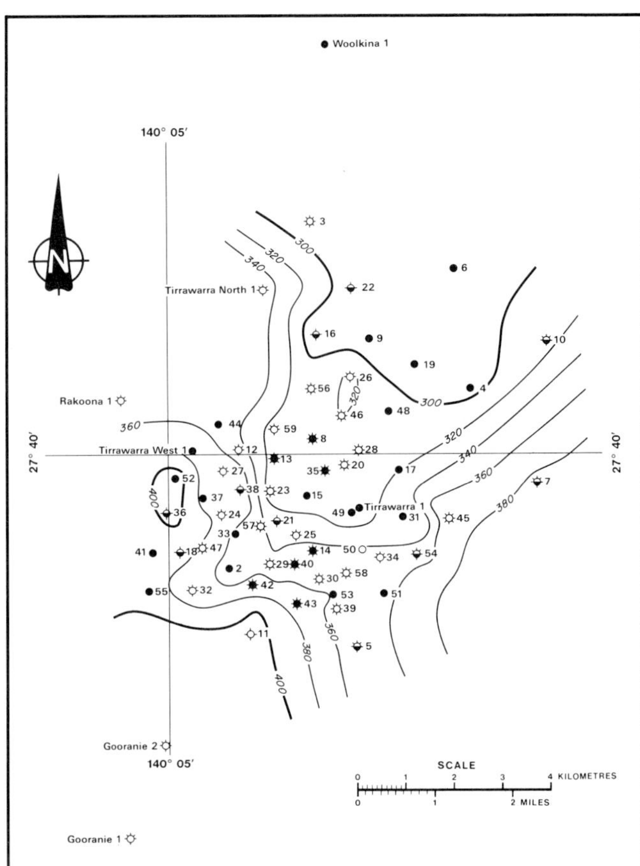

Figure 8B. Isopach map of upper Patchawarra Formation with subdivision at Patchawarra coal structural horizon (Figures 22A,B). Contour values are in feet. Contour interval, 20 ft.

Williams, 1982; Williams and Wild, 1984). Gostin (1973) recognized a vertical succession of three facies associations from cores taken in Tirrawarra field. His observations were confirmed and expanded by Williams (1982) to include a fourth regional association, but only the lower three of these are present in Tirrawarra field. The associations are numbered from 1 to 3 in ascending order and can usually be recognized on geophysical logs (Figure 14). Association 1 is not present everywhere in the field but is up to 28 ft (8.5 m) thick where it does occur. In cores, rocks of this association consist of interstratified conglomerate and coarse-grained sandstone with minor massive mudstone and coal. Bedforms include a range of cross-bedding types with reactivation surfaces locally prominent. Soft sediment deformation structures formed as a result of fluid expulsion are also present (Williams, 1982). Williams and Wild (1984) have interpreted the beds in this association to be the deposits of high-energy streams characterized by ephemeral discharge and rapid abandonment.

Association 2 is the most distinct and common subdivision and is present everywhere in the field. Williams (1982) reported a maximum thickness of 86 ft (26 m) in Tirrawarra No. 2. It typically comprises the stacked deposits of several types of fluvial bars that are represented mainly by flat and planar cross-bedded sandstones (Figure 15). Association 3 attains a maximum thickness of 49 ft (15 m) in Tirrawarra No. 3 but is not developed everywhere. It is typified by stacked upward-fining sequences that were interpreted by Williams (1982) to be compound bars in which both bar top and overbank deposits have been preserved. In Tirrawarra field this association represents the conformable transition from bedload-dominated to mixed load-dominated deposition, which is characteristic of the overlying Patchawarra Formation.

Permeability

The Tirrawarra Sandstone exhibits predominantly low permeabilities with very thin discontinuous high-permeability streaks. Permeability in the Tirrawarra Sandstone at surface conditions ranges from less than 1 md to 2 darcys and averages 1 md. Correlation

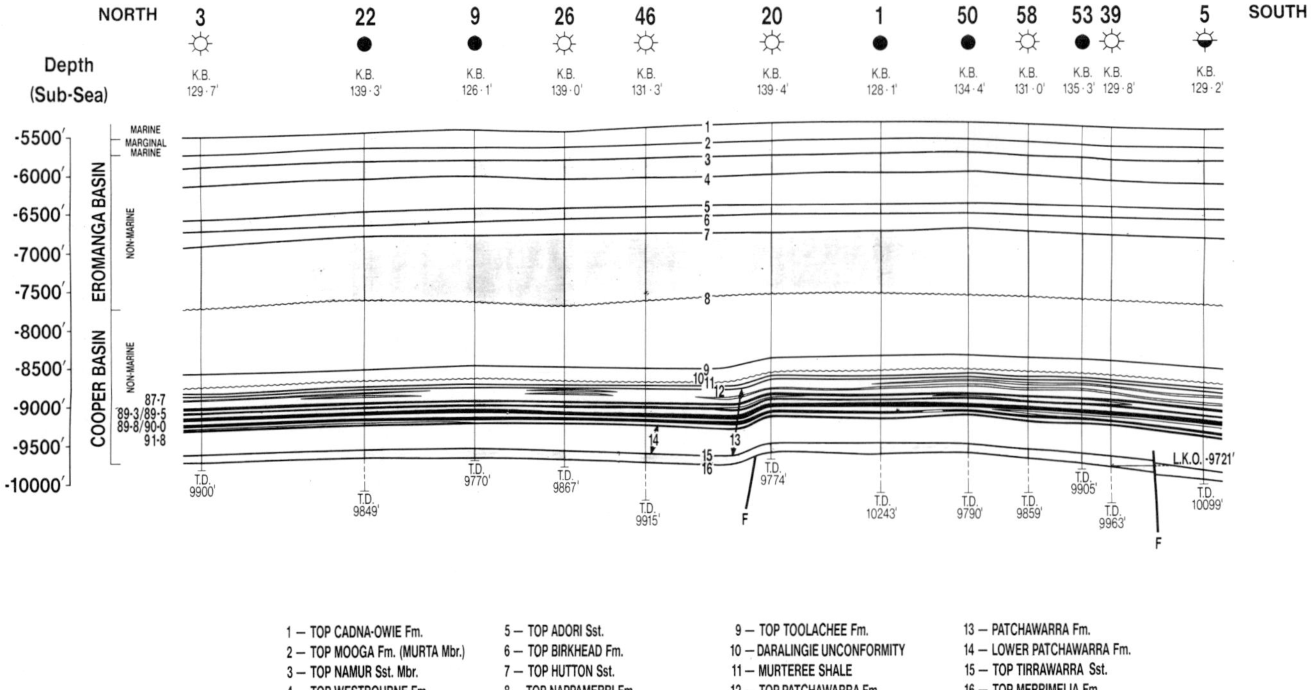

Figure 9. North-south structural cross section for Tirrawarra field showing the Cooper and nonmarine Eromanga basin sequence. Refer to Figure 2 for location. F, fault; L.K.O., lowest known oil; L.K.G., lowest known gas; 87-7 . . . 91-8 are individual sandstone names derived from their depth of occurrence in the discovery well, e.g., 87-7, a sandstone encountered between 8770 and 8780 in T#1.

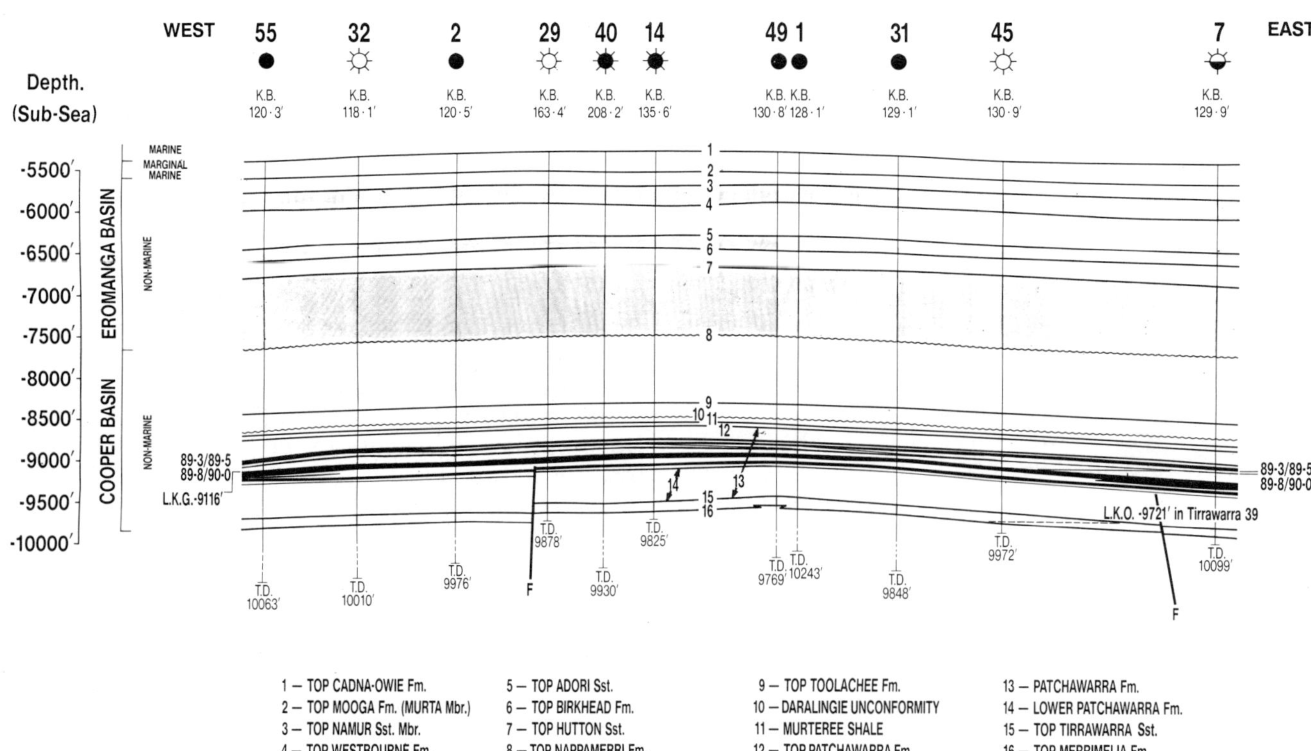

Figure 10. West-east structural cross section for Tirrawarra field showing the Cooper and nonmarine Eromanga basin sequence. Refer to Figure 2 for location. Note the nonjuxtaposing faulted offset of the Tirrawarra Sandstone in the vicinity of Tirrawarra No. 29. This fault divides the oil reservoir into western and eastern fields. See Figure 9 for explanation.

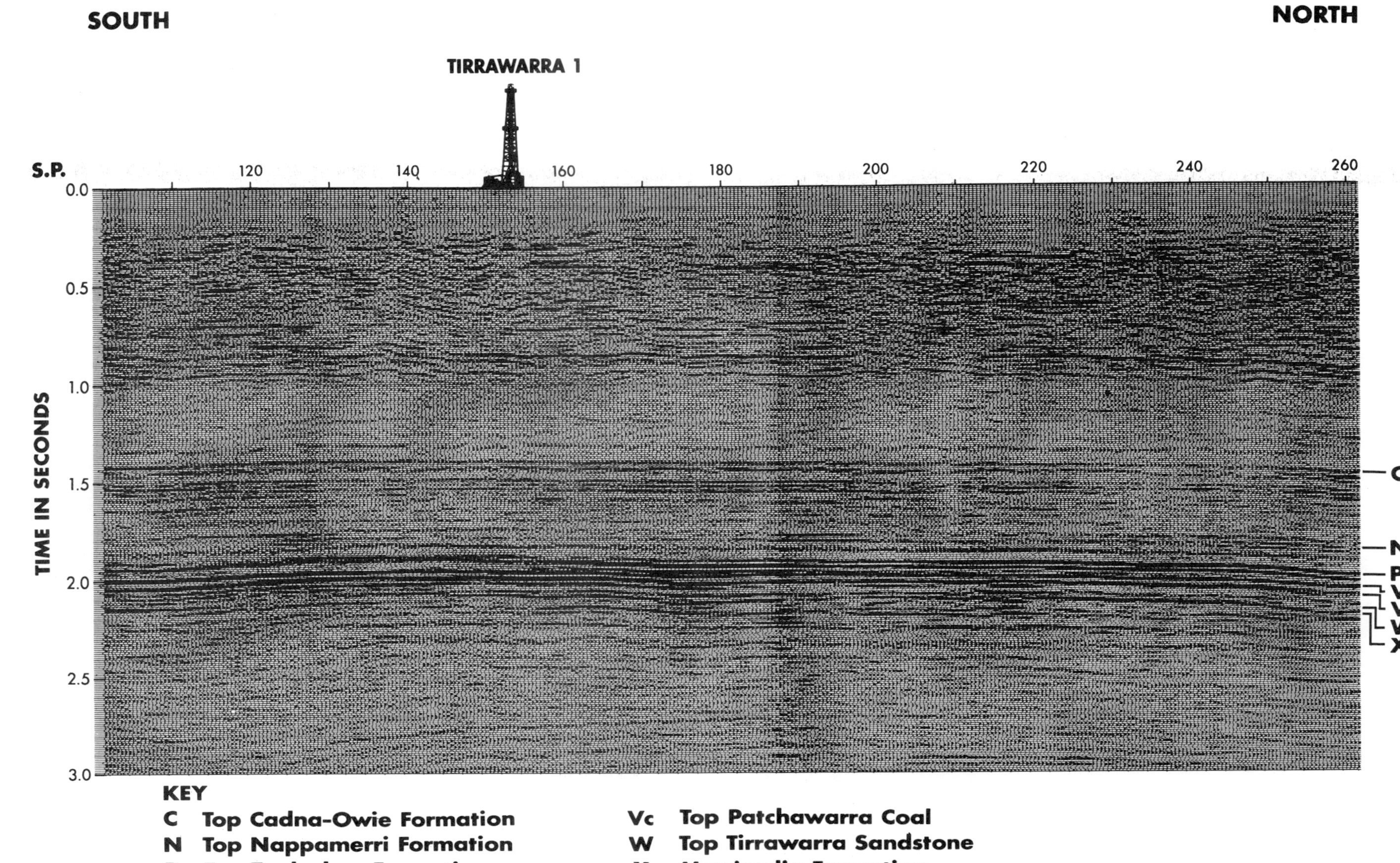

Figure 11. Seismic line (S80-T14X) from south to north across Tirrawarra field. Refer to Figure 2 for location.

261

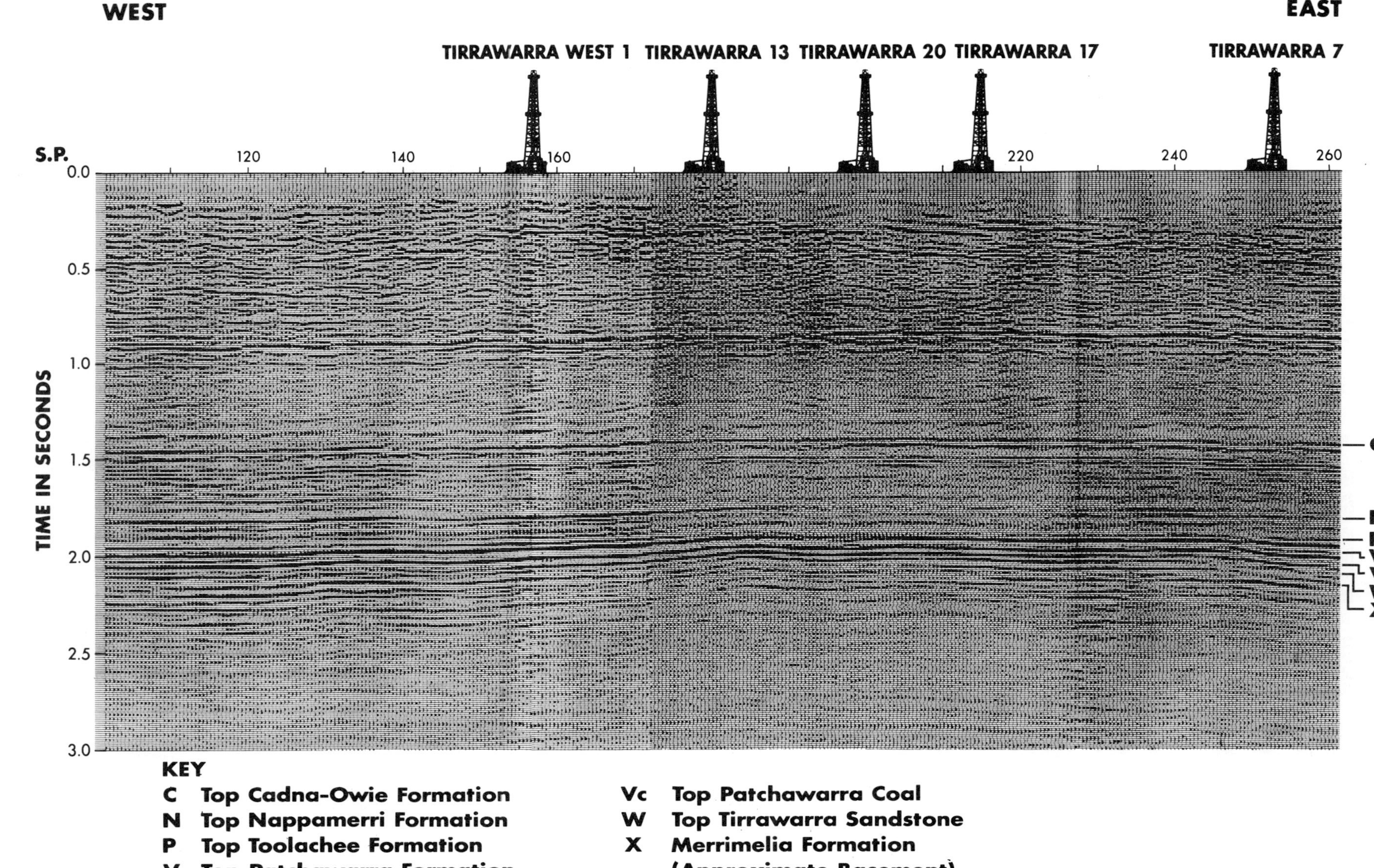

Figure 12. Seismic line (S80-T15) from west to east across Tirrawarra field. Seismic expression of the main Tirrawarra fault is shown on this line. Refer to Figure 2 for location.

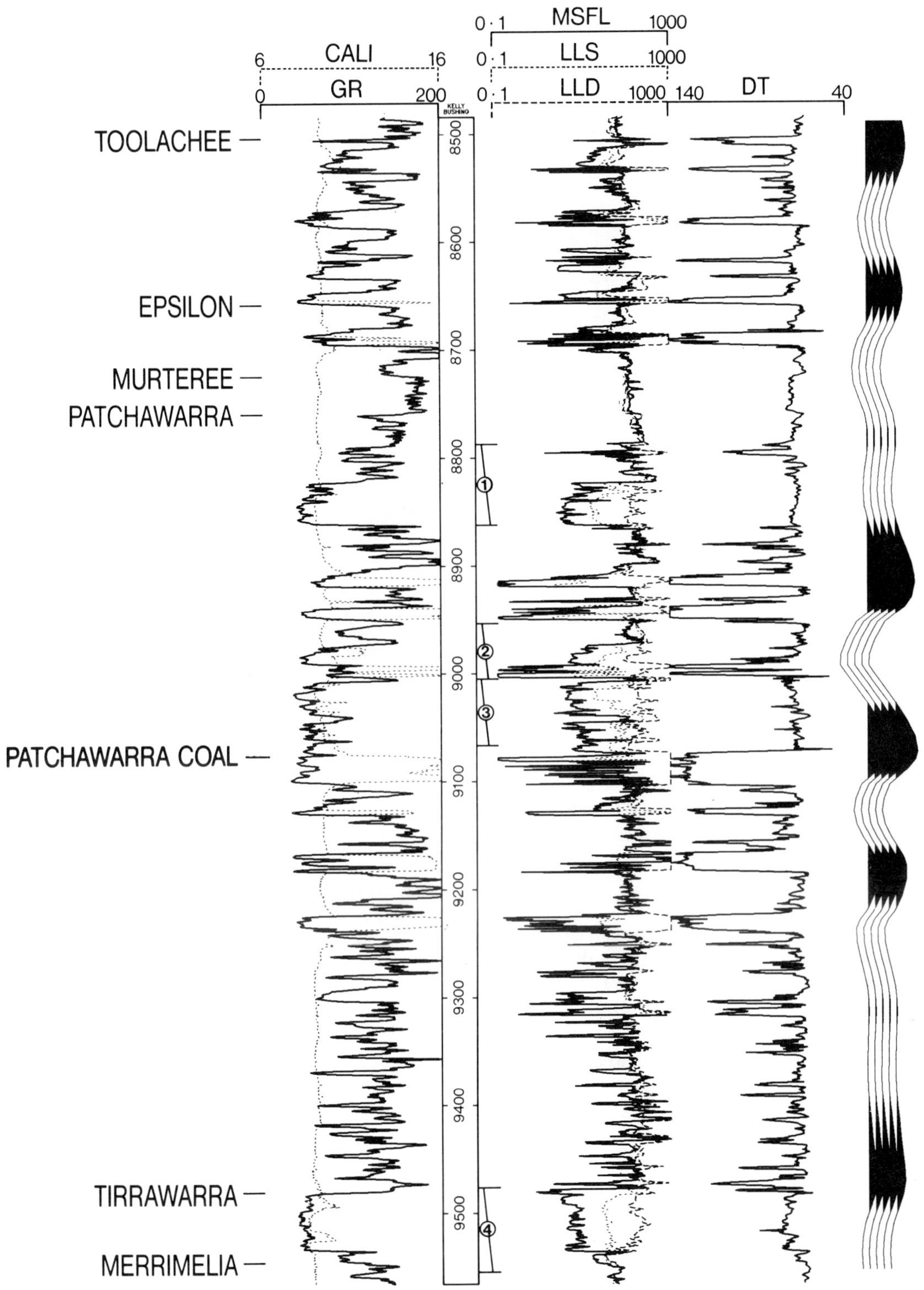

Figure 13. Representative Schlumberger geophysical log suite over the Permian section in Tirrawarra field. Data from gas producer Tirrawarra No. 13 (K.B., 183.5 ft; T.D., 9756 ft). Stratigraphic subdivision is indicated on the left. Track 1 contains caliper (CALI) and natural gamma ray (GR) logs with scales in inches and API units, respectively. Track 2 displays the normal electrical log suite including microspherically focused laterolog (MSFL), and the shallow and deep traces of the dual laterolog (LLS and LLD, respectively). All scales show resistivity measured in ohm-meters and are logarithmic. Track 3 shows the sonic log (DT) scaled in microseconds per foot. The synthetic seismic log on the right has been generated from the sonic log. Intervals tested in open hole are shown adjacent to the depth column. DST 1 recorded the maximum gas flow rate of 4.4 MMcf (125 thousand m^3) per day.

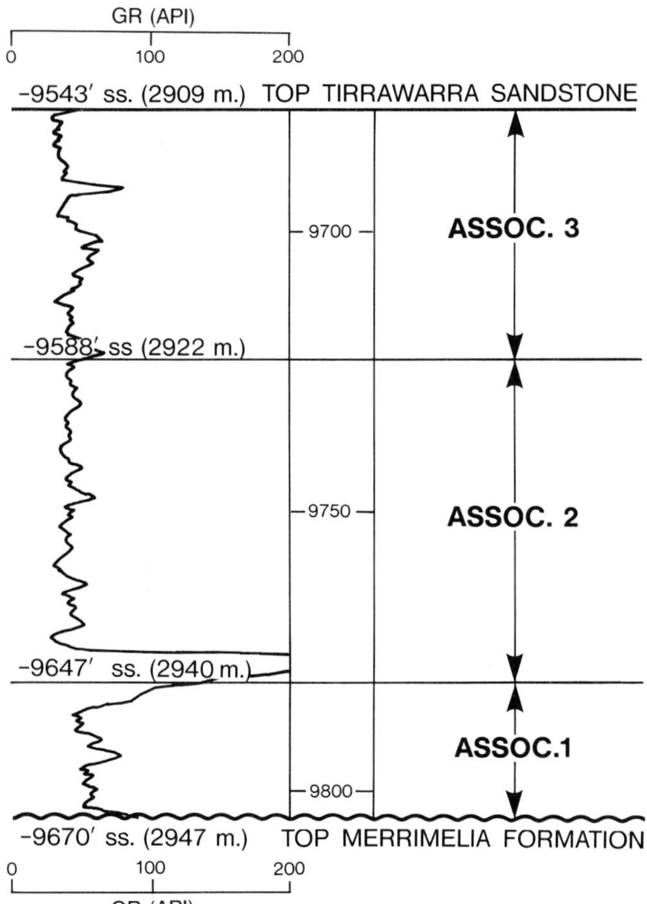

Figure 14. Representative gamma ray log over the Tirrawarra Sandstone showing the facies association subdivisions of Williams (1982). Data from Tirrawarra No. 53. See text for explanation.

Figure 15. Slabbed core of coarse- to very coarse grained, tabular cross-stratified quartzo-lithic sandstone overlain by medium- to coarse-grained horizontally stratified sandstone. Both are facies from a bedload-dominated fluvial system. Note: (1) severe reduction in original intergranular porosity caused by the precipitation of authigenic clay (most probably dickite); (2) preferential development of stylolites along bedding planes. The latter features result in reduced vertical permeability compared to horizontal permeability (Tirrawarra No. 21, 9658 ft, Tirrawarra Sandstone oil reservoir, Association 2 of Williams, 1982).

is good between the distribution of high (in excess of 20 md) and low (less than 2 md) permeability and the facies associations described above (Figures 16 and 17). Association 1 is characterized by low-permeability sands from which there is little or no contribution to production. Thin zones of high permeability are restricted to the lower and middle portions of Association 2, but in most cases it is difficult to correlate these between wells and to determine the proportion of the total reservoir comprising high permeability. These zones are everywhere less than 10 ft (3 m) thick and are commonly less than 4 ft (1.2 m) thick, with individual high-permeability streaks typically less than 1 ft (0.3 m) thick. Association 3 is also assumed to have good permeability.

Relative permeability characteristics of the Tirrawarra Sandstone are based on laboratory experiments on 26 core samples from seven early wells. Relative permeability to both gas and water was determined experimentally. From Figure 18A, it is apparent that the oil becomes immobile at water saturations of about 70%, leaving an irreducible oil saturation of 30% for the medium- to low-permeability

sands. For this group of sands the oil becomes immobile at a gas saturation of about 40% (Figure 18B), indicating that, under a primary production scenario, when the reservoir pressure falls below the bubble point and the combined oil and water saturations decrease to 60%, the oil will be immovable and only gas will flow in the reservoir.

Fracture Stimulation

The Tirrawarra field oil reservoir is contained within a 100 ft (30 m) thick low-permeability sandstone that has proved to be an ideal reservoir for hydraulic fracture stimulation treatment. All oil wells have been fracture-stimulated for significant improvements in oil production rates. A typical treatment involves pumping some 170,000 to 200,000 lb (80,000–90,000 kg) of frac sand and 65,000 to 80,000

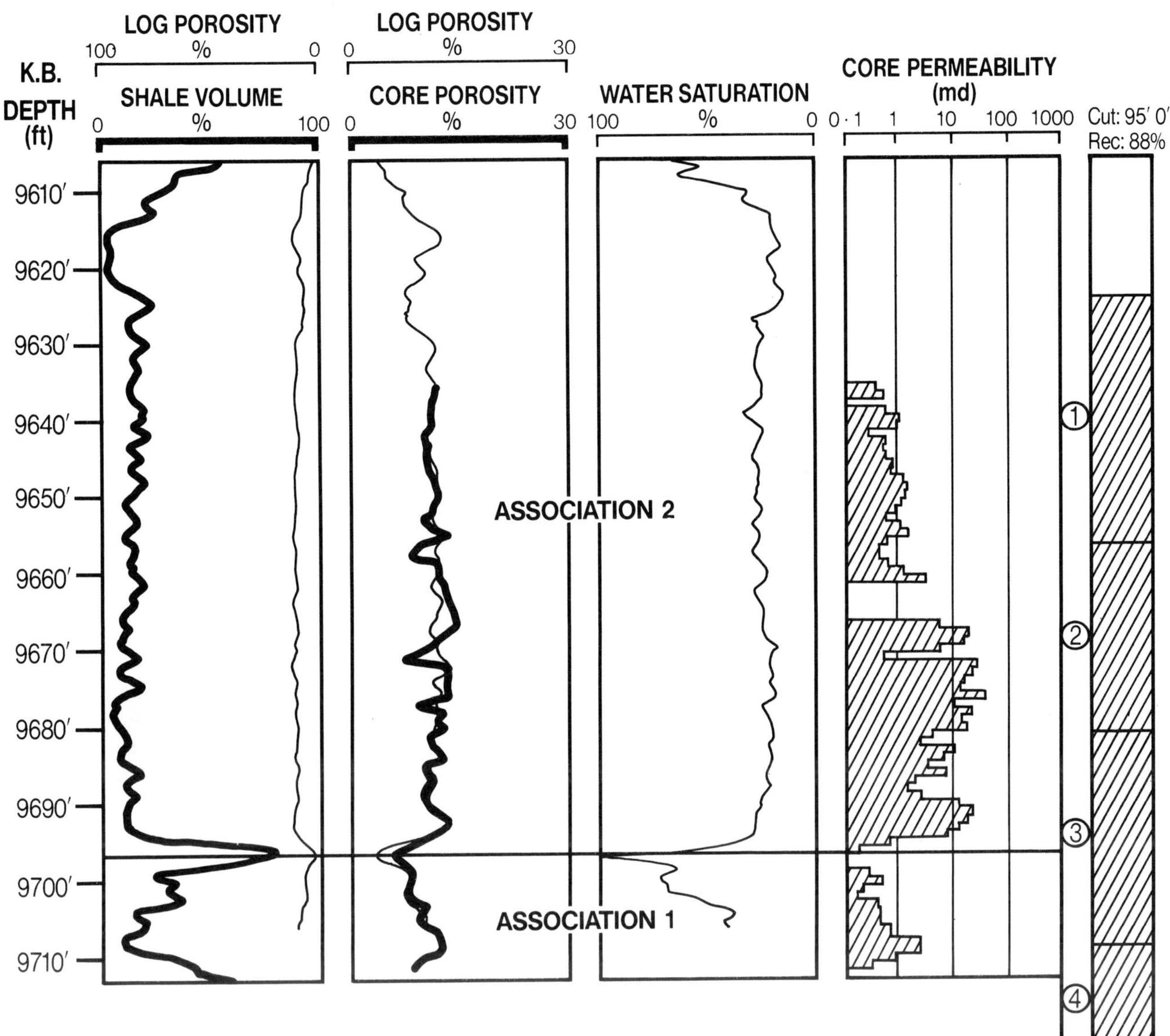

Figure 16. Diagram showing the relationship between clay content (shale volume), porosity and permeability, and water saturation in the Tirrawarra Sandstone. Data from Tirrawarra No. 21. Associations discussed in text.

U.S. gal (250–300 kL) of frac fluid into a well. Pressures at the reservoir sand face of approximately 8800 psia (60,700 kPa) are required to fracture the sandstone. Vertical fractures extending some 300 to 400 ft (90–120 m) from either side of the well bore are created, providing a high-permeability flow channel to the well bore. The productivity index (PI) generally increases fourfold on average after fracture stimulation treatment of a well; in other words, the production rate quadruples for an equal pressure drawdown.

Hydrocarbon Characteristics

The Tirrawarra reservoir oil is undersaturated at the initial reservoir pressure and temperature of 4300 psia (29,700 kPa) and 285°F (140°C) but is extremely volatile, with a formation volume factor of 2.8 reservoir bbl of C^{5+} per stock tank bbl of C^{5+} (2.8 Sm^3/m^3) and a solution gas-to-oil ratio of 2700 ft³ of C^{4-} per bbl of C^{5+} (480 m³ of C^{4-}/m^3 of C^{5+}). Figure 19 shows a gas chromatograph analysis of a Tirrawarra Sandstone liquid sample recovered from the flare line during an open-hole drill-stem test at Tirrawarra No. 18. The oil expands by 11% from the initial reservoir pressure to the bubble point pressure of 2900 psia (20,000 kPa). Once below the bubble point pressure, the oil vaporizes rapidly. At an average reservoir pressure of 2600 psia (17,900 kPa), 40% of the hydrocarbon pore space would be occupied by gas.

Production History and Development Considerations

Production in Tirrawarra oil field began in February 1983 at a rate of 4500 BOPD (720 m³) from eight wells. Production peaked at 6000 bbl (950 Sm³)

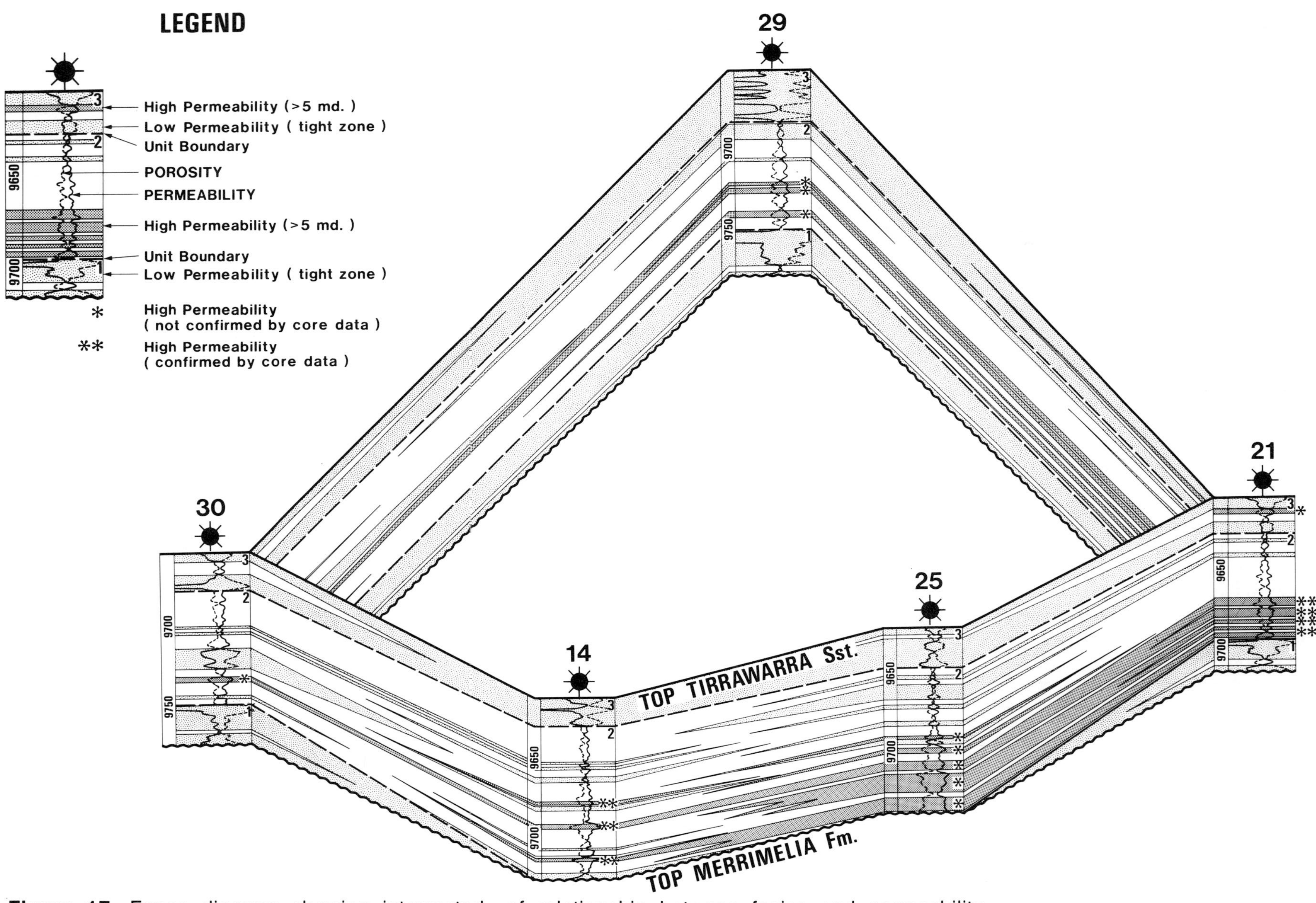

Figure 17. Fence diagram showing interpreted of relationship between facies and permeability distribution of permeability in the Tirrawarra Sandstone distribution. (After Harrison, 1984, unpublished.) in the central part of the field. Refer to text for explanation Location shown by Figure 2.

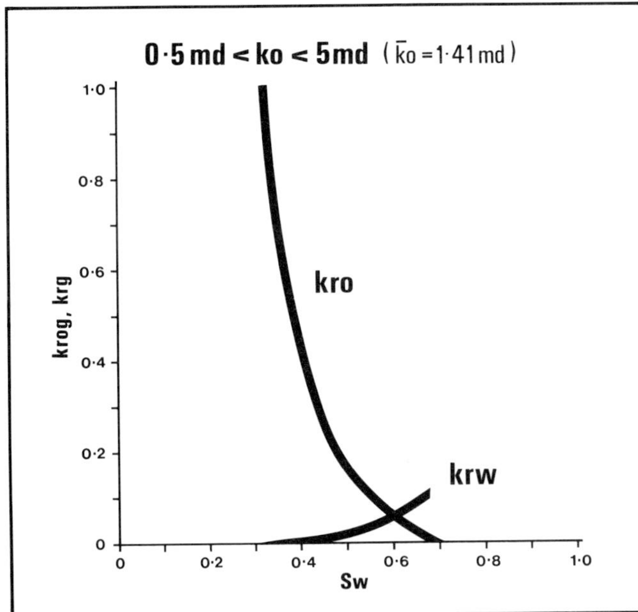

Figure 18A. Graph showing average oil/water- and water-relative permeability for the Tirrawarra Sandstone oil reservoir.

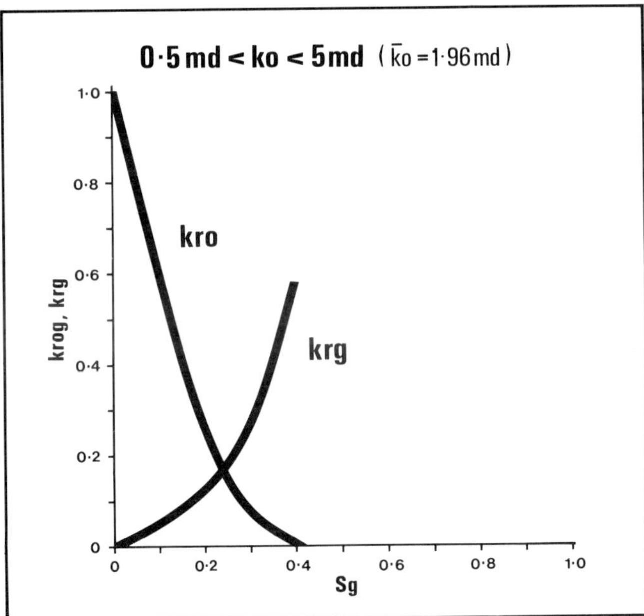

Figure 18B. Graph showing average oil/gas- and gas-relative permeability for the Tirrawarra Sandstone oil reservoir.

of oil/day from 13 wells in late 1983, and from that time until July 1985, production was maintained at approximately 4500 bbl (720 Sm³) of oil/day, primarily by infill drilling of a further 16 oil producers.

In February 1984, Tirrawarra No. 15 was converted from oil production to Tirrawarra satellite separator gas injection well, thus forming the inverted seven-spot pattern with surrounding wells with approx-

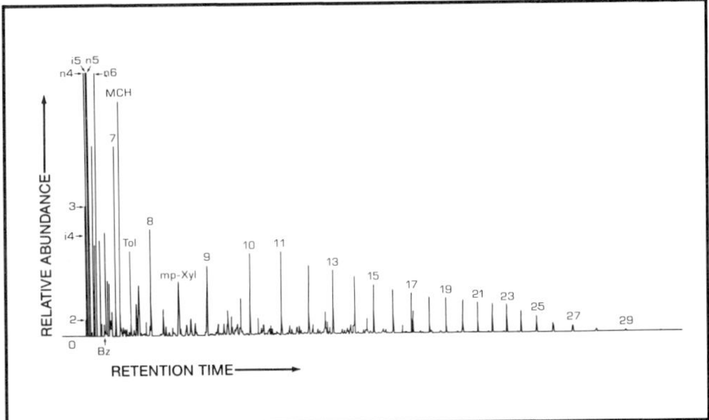

Figure 19. Gas chromatograph of liquid sample recovered from the flare line during open-hole drill-stem testing at Tirrawarra No. 18.

imately a 300 ac (0.5 mi², or 1.2 km²) spacing. The purpose of this pilot injection scheme was, first, to assess the feasibility of the miscible gas injection enhanced oil recovery (EOR) project and, second, to provide immediate pressure support to the pattern.

Where the miscible displacement process takes place in the reservoir, displacement efficiency is expected to be over 90%. The heterogeneous nature of the reservoir, however, may cause early injection gas breakthrough from high-permeability streaks that would have a detrimental effect on total oil recovery. Laboratory experiments have shown that favorable miscibility characteristics exist between the reservoir oil and the available injection gases (Tirrawarra satellite separator gas and ethane).

Based on the encouraging results of the pilot project, the injection program was later expanded with the conversion of Tirrawarra No. 18 from oil production to gas injection well status in October 1985. This was soon followed by recompletion of a Patchawarra Formation gas well (Tirrawarra No. 29) and a Tirrawarra Sandstone oil well (Tirrawarra No. 38) to injection wells.

The injection gas used initially was Tirrawarra satellite separator gas produced from Tirrawarra field and neighboring gas fields. Field separators at the Tirrawarra satellite facility were used to strip the raw gas of some of its liquids, leaving sufficient LPG's and condensate in the injection gas stream so that it was miscible with the oil at reservoir conditions.

In mid-1987, construction of a pipeline from the main plant at Moomba to Tirrawarra field was completed allowing injection of ethane as the miscible gas. Studies had been conducted to ensure that the injected ethane would not have deleterious effects on oil recovery, such as asphaltene precipitation in the reservoir. Both ethane and the Tirrawarra satellite separator gas are miscible with Tirrawarra oil at reservoir pressures above 3200 psia (22,000 kPa).

Since commencement of gas injection, field production rates have been maintained even though no new oil production wells have been drilled since

267

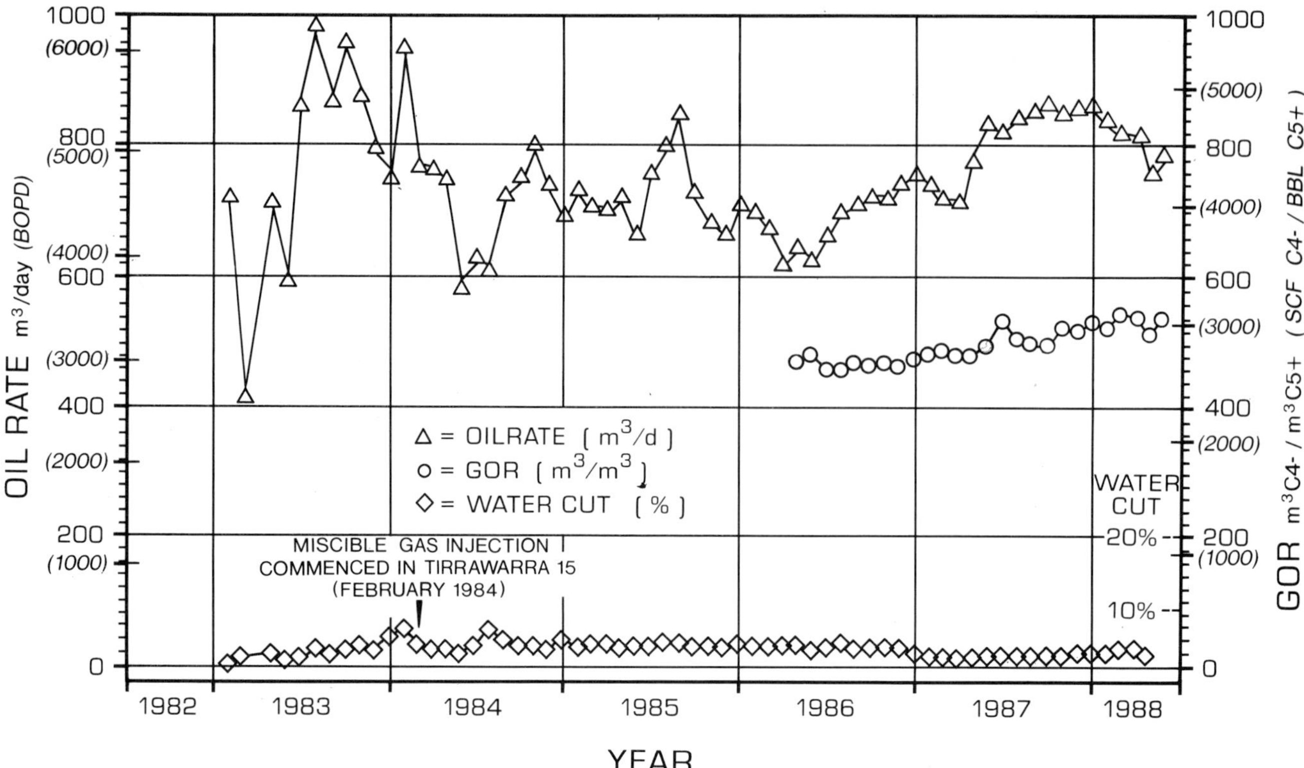

Figure 20. Graph showing the production history of the Tirrawarra Sandstone oil reservoir as production rate plotted against time. Gas injection commenced in February 1984.

mid-1985. Field production at mid-1988 averaged 5000 bbl (800 Sm³) of oil/day from 31 wells (Figure 20).

In terms of recovery efficiency of the 112 MMSTB (18 million Sm³) of in-place oil, only 22 MMSTB (3.5 MMSTB) would be recoverable under primary reservoir depletion. It is estimated that a further 20 million bbl (3.2 million Sm³) of oil could potentially be recovered through a full-scale EOR project.

The EOR scheme in 1988 involved 31 production wells and 5 injection wells. Ultimate development may comprise 39 production wells and 20 injection wells, depending on the continued technical and economic performance of the project.

Patchawarra Formation

Lithology, Depositional Environment, and Development

The petrology of reservoir rocks in the Patchawarra Formation was the subject of a study by Martin (1983). He found the detrital framework to consist predominantly of fine-grained moderately sorted quartz, together with minor amounts of rock fragments, chert, and mica. In some samples authigenic clay, principally illite and dickite, comprised almost 11% of the mode. Martin reported a significant reduction in primary depositional porosity that he attributed to a combination of compaction, authigenic clay precipitation, quartz overgrowth, and carbonate cementation (Figure 21).

Figure 21. (At right) Photomicrographs taken under: (A) plane polarized light, and (B) crossed polarizers, showing ankerite cement (A) infilling intergranular and secondary pore spaces in moderately sorted quartzo-lithic sandstone. Note corroded margins (arrowed) of many of the quartz grains (Q) due to partial replacement by ankerite, which also inhibits quartz overgrowth. A rock fragment (RF) that has been extensively replaced by ankerite is also marked. Porosity (from core analysis) 17.9%, permeability 241 md. (Tirrawarra No. 16, 9127 ft, 89-8 reservoir sand, Patchawarra Formation); (C) SEM photograph showing detail of a pore (P) partly filled by ankerite cement (A) and bordered by quartz grains (Q) with overgrowth faces. Illite (I) and dickite (D) are also distributed around the pore. Sample of moderately sorted fine-grained sandstone from mixed-load channel facies. Porosity (from core analysis) 13.2%, permeability 5.6 md (Tirrawarra No. 28, 9038 ft, 90-0 reservoir sand, Patchawarra Formation); (D) SEM photograph of same sample as in A and B showing detail of authigenic dickite (D) filling a pore. The platy, pseudohexagonal crystal form typical of the kaolinite group clays is clearly visible. The dickite crystals are very loosely packed and are readily dislodged during flow resulting in blockage of pore throats. The pore is bordered by quartz grains (Q) with small overgrowth crystal faces. (All photographs and captions are from an unpublished report for SANTOS Limited by Martin, 1983.)

A

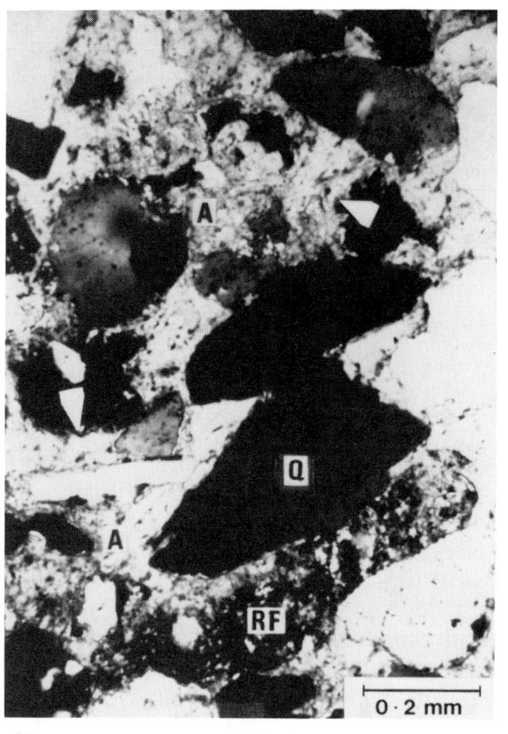

B

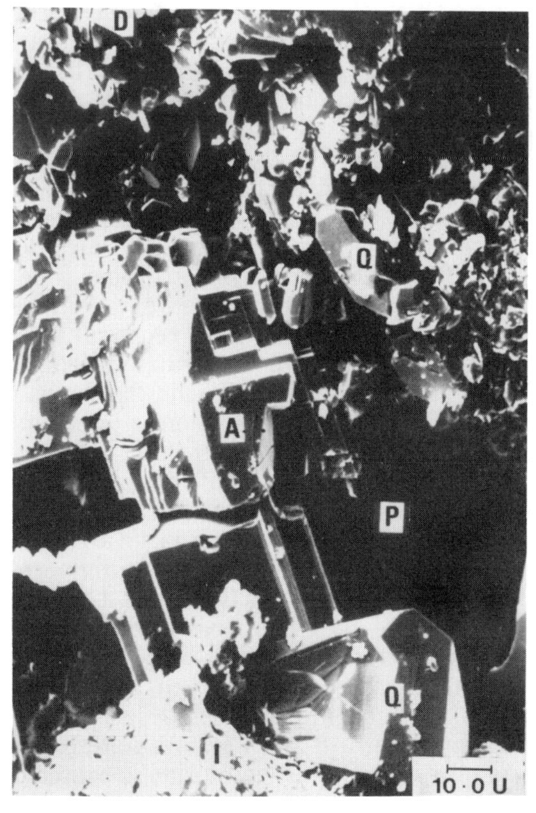

C

D

Abundant authigenic clay has produced a high microporosity but correspondingly low permeability. Minor secondary porosity has been created by the dissolution of some labile grains and of the carbonate cement; ambient permeabilities of over 4 darcys have been measured in some cores.

The Patchawarra Formation is interpreted to be the deposits of a mixed-load, moderate- to high-sinuosity fluvial or upper delta plain system (Battersby, 1976; Thornton, 1978). Kapel (1972) recognized three subdivisions within the formation, and these can be readily identified in Tirrawarra field (see Figures 22A and 22B). The lower unit is up to 400 ft (122 m) thick and consists of carbonaceous shale and argillaceous sandstone interbedded with coal stringers and thin sandstone. The seven reservoir sands located within this division are generally thin (less than 10 ft, or 3 m, thick) and laterally discontinuous. Thirteen percent of the formation's in-place gas is contained in these reservoirs. In contrast, thick, laterally continuous sandstone and coal beds dominate the middle part of the formation (Figures 22A and 22B), which is up to 300 ft (90 m) thick in places. Subordinate interstratified mudstones source and seal individual reservoirs. There are five reservoir sandstones mapped in this division and they contain 75% of the formation's 392 bcf (11 billion Sm^3) of in-place *proved* plus *probable* gas. The stratigraphy is dominated by upward-fining cycles comprising a basal sandstone unit overlain by thinly interbedded shale, fine-grained sandstone, and coal. Many of the cycles are vertically capped by coal seams ranging up to 50 ft (15 m) thick. The reservoir sandstones are typified by "boxcar" or "bell" (terminology of Selley, 1985, p. 79) gamma ray log signatures and are interpreted as fluvial point bar deposits.

The uppermost unit of the Patchawarra Formation hosts two sandstone reservoirs and 12% of the in-place gas. The dominant lithology is shale with mainly thin sandstone and coal beds. The upper unit represents a period of transition from alluvial plain sedimentation to lacustrine deposition (Murteree Shale).

Permeability

Better quality sands have porosities that average 11% and ambient permeabilities of about 80 md (Pecanek and Paton, 1984). The distribution of porosity and permeability through a representative part of the principal Patchawarra Formation reservoirs is shown in Figure 23.

Hydrocarbon Characteristics

The Patchawarra Formation reservoir gas is slightly retrograde, existing in a gas phase at the initial reservoir pressure and temperature of 3990 psia (27,510 kPa) and 270°F (130°C). The fluid has a dew point pressure of between 2922 and 3449 psig (20,150–23,780 kPa) at the reservoir temperature based on results of fluid studies of Tirrawarra No. 28 and No. 45, respectively. While producing gas at reservoir pressures above the dew point pressure, the liquids yield is 32 bbl of LPG per MMcf of raw gas (180 Sm^3 of LPG per MM Sm^3 of raw gas) and 45 bbl condensate/MMcf of raw gas (250 m^3 of condensate/MM Sm^3 of raw gas). At the expected maximum raw gas recovery from the field, some 13% points of condensate and 3% points of LPG are expected to condense in the reservoir and will not be recovered.

Production and Development

Production from Tirrawarra gas field started in August 1983, with seven wells delivering at an initial rate of 35 MMCFG/day (1 MM Sm^3 of gas/day) (Figure 24). Since then a further 24 wells have been completed.

Production from the field reached a peak rate of 80 MMCFG (2.3 MM Sm^3 of gas) per day during 1986. By mid-1988, cumulative production had reached 94 BCFG (2.6 billion Sm^3 of gas). Further drilling, recompletion of existing wells, fracture stimulation of new and existing wells, and installation of additional compression is required to maximize recovery from the field.

Toolachee Formation

Following uplift and erosion at the end of the Early Permian, the Late Permian Toolachee Formation represented a return to continental flood plain deposition (Battersby, 1976). In Tirrawarra field, the formation comprises as many as six vertically stacked upward-fining sequences, each up to 40 ft (12 m) thick (Figure 25). Sandstone is the principal lithotype in the formation. Reservoirs are laterally continuous and easily correlated between wells, although lateral facies transitions from relatively clean point bar sands to thinner levee and crevasse sands do occur. Sandstone correlation is facilitated by a notably consistent coal stratigraphy. As reported by Pecanek and Paton (1984), gas is restricted to a 700 ac area (1.1 mi^2, or 2.8 km^2) around the crest of the Toolachee Formation structure. The Toolachee Formation is a major reservoir elsewhere in the Cooper basin, but thick regional seals in the Patchawarra trough separate it from the main petroleum source rocks in the Patchawarra Formation, thus reducing its economic importance in that area.

Martin (1980) and Martin and Hamilton (1981) have studied the reservoir quality and diagenesis of the Toolachee Formation over a large area of the Cooper basin. They reported widely varying porosities (6 to 24%) and permeabilities (less than 1 md to more than 500 md). The average porosity of reservoir rocks in Tirrawarra field is 10.7%. Petrologically, the sandstones are comparable to those in the Patchawarra Formation, being composed mainly of quartz framework grains with subordinate rock fragments and muscovite. Authigenic clay is present.

The formation has been brought onto production and is being commingled with the Patchawarra

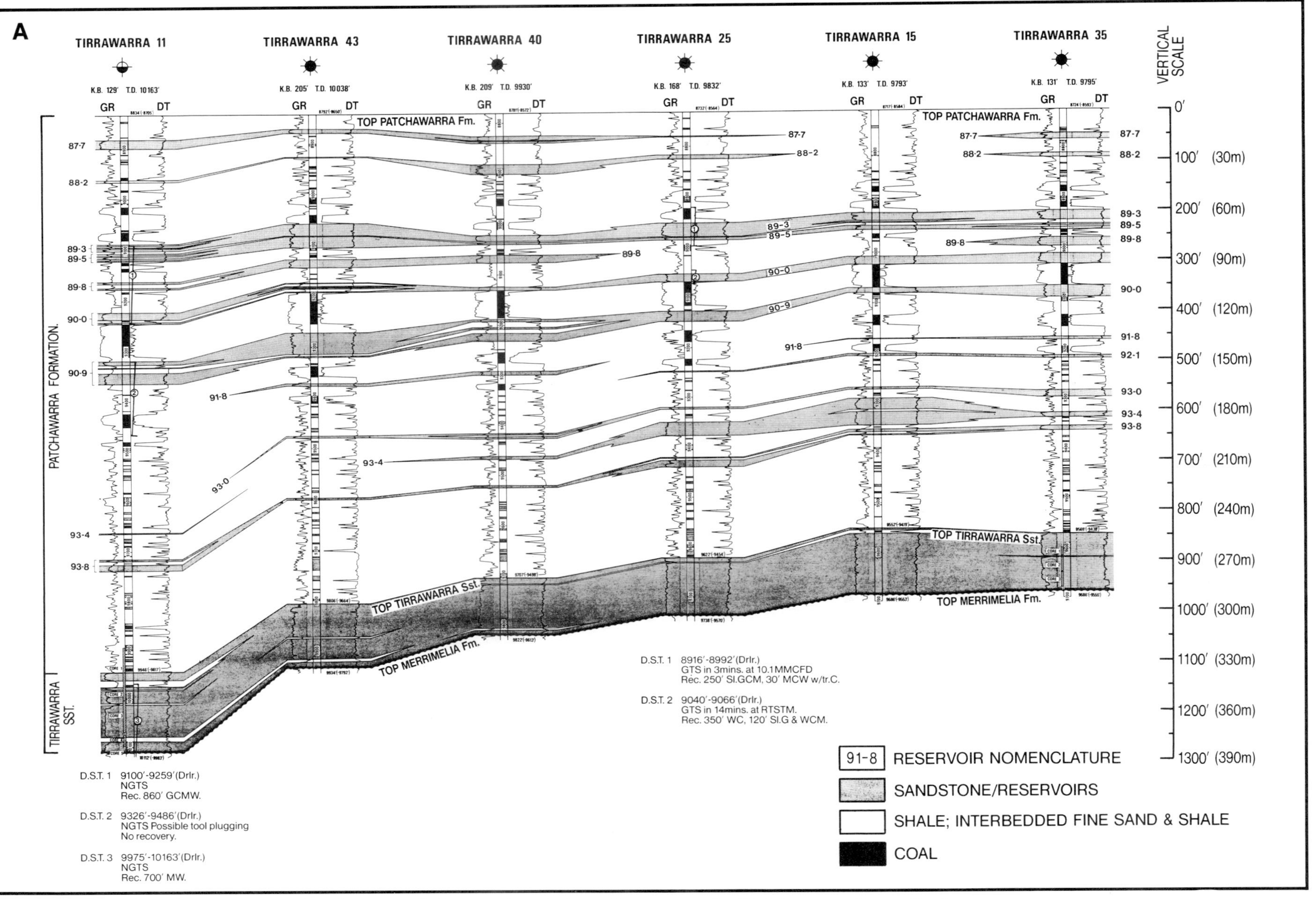

Figure 22. Stratigraphic cross section from south to north showing correlation of reservoirs in the Patchawarra Formation and the Tirrawarra Sandstone. Correlations based on a combination of gamma ray and sonic logs. Gas reservoirs (Patchawarra Formation) are designated according to their depth in the discovery well. Refer to Figure 2 for location. Diagram after Harrison (1985).

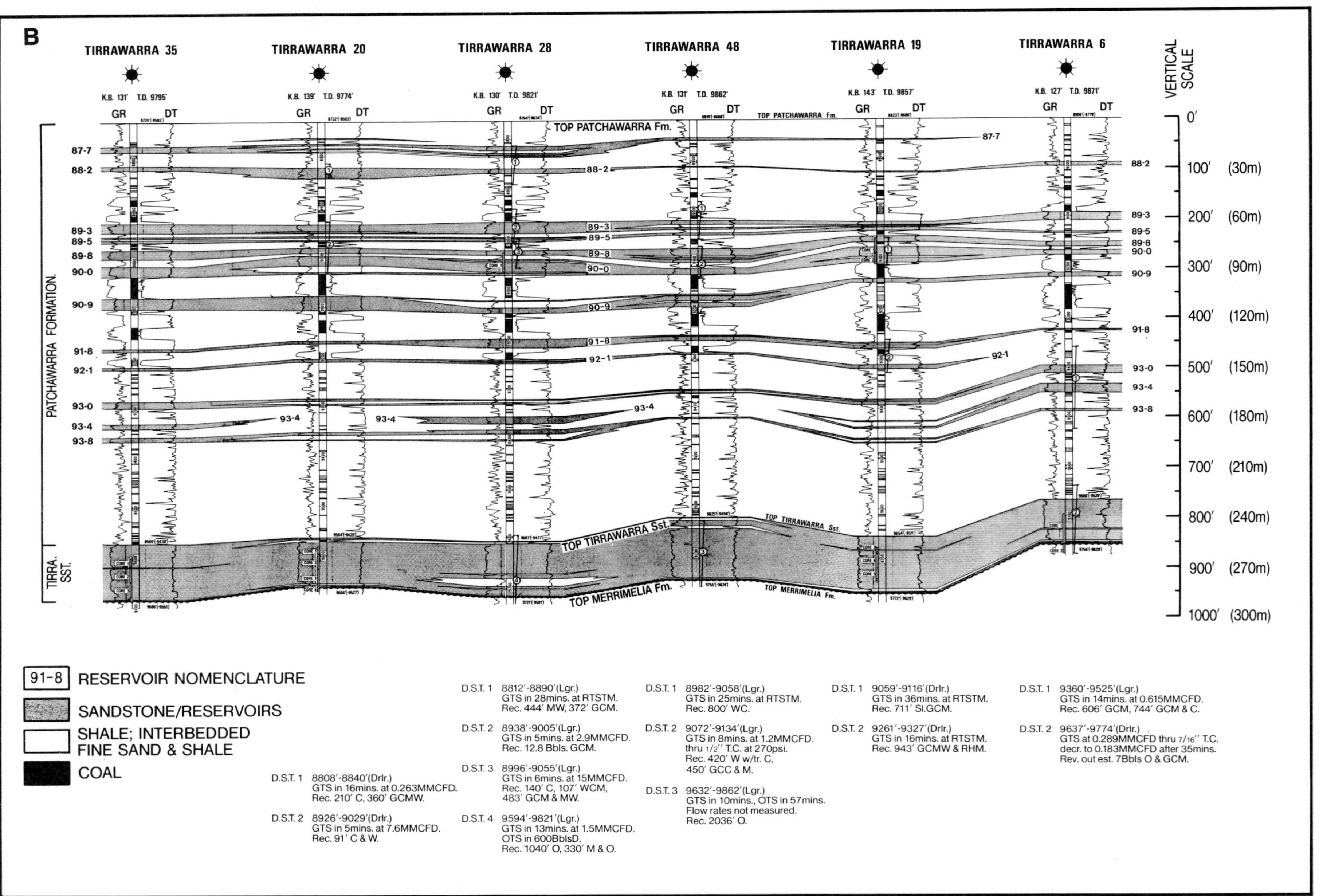

Figure 22. (Continued)

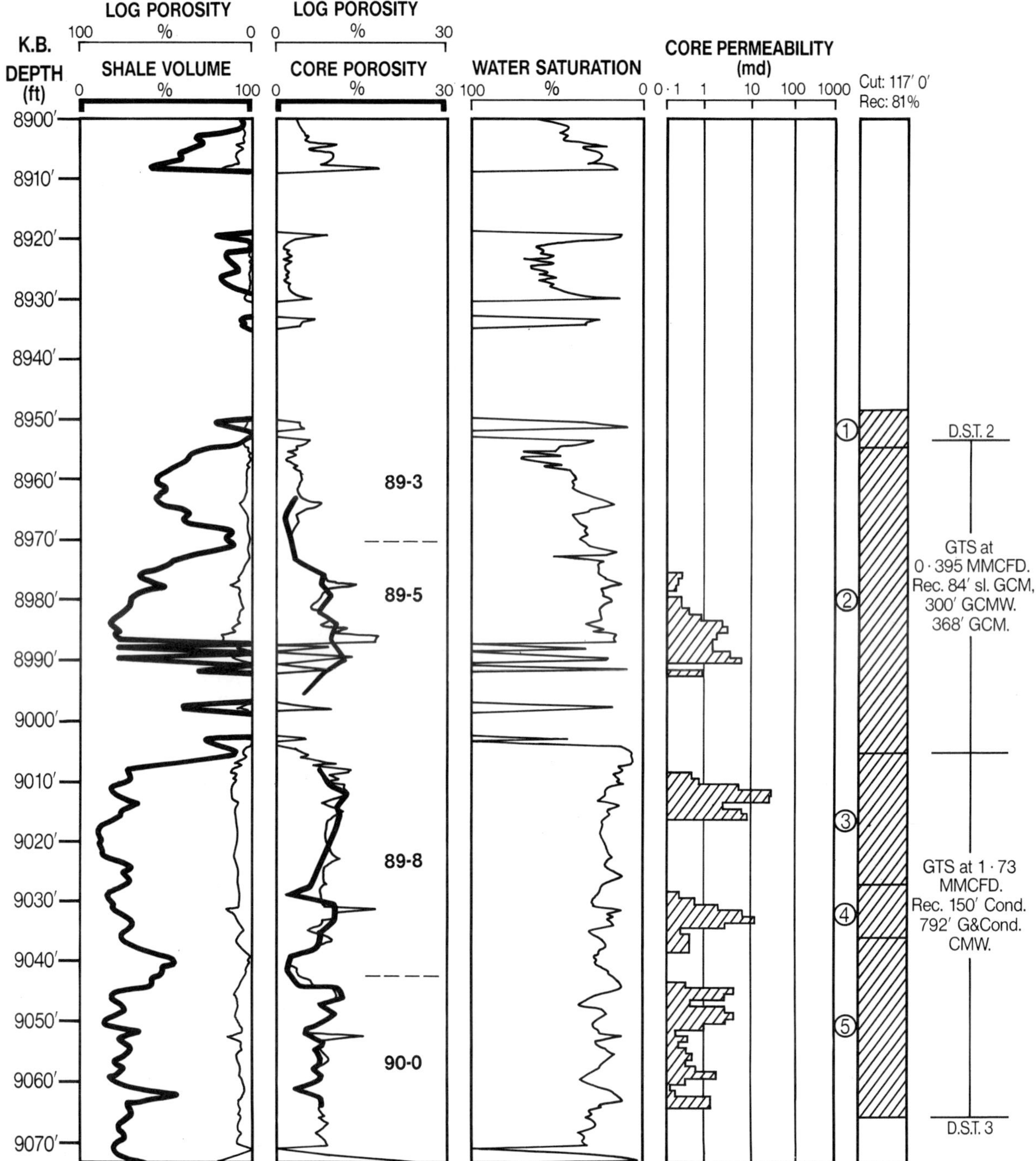

Figure 23. Diagram showing the relationship between clay content (shale volume), porosity and permeability, and water saturation in the principal gas reservoirs of the Patchawarra Formation. Data from Tirrawarra No. 13.

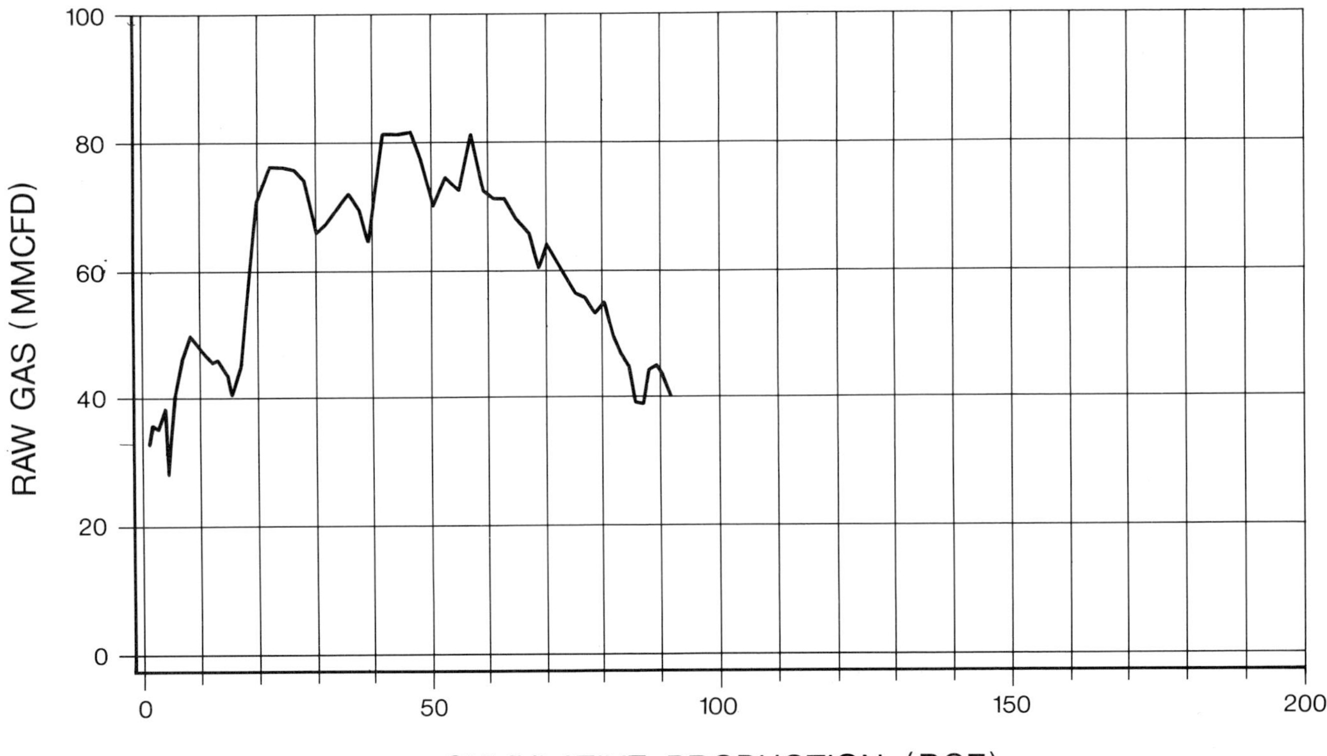

Figure 24. Graph showing cumulative production (in bcf) from Patchawarra Formation gas reservoirs plotted against rate (in MMCFG/day).

Formation gas. An open-hole gas flow rate of 6 MMCFG (170 thousand Sm³) per day was recorded in the discovery well, Tirrawarra No. 14.

SOURCE ROCKS, MATURITY, AND TIMING OF HYDROCARBON GENERATION

A summary of source quality for the Cooper basin is given in Figure 26.

Tirrawarra Sandstone

The most likely sources of oil reservoired in the Tirrawarra Sandstone are the shales and coals of the overlying Patchawarra Formation. The source of the oil has not been satisfactorily characterized because the oil is too mature to yield diagnostic biomarker data. Migration distances could be in excess of 6 mi (10 km) in the Tirrawarra Sandstone because of high transmissivity relating to its braided fluvial origin.

Vitrinite reflectance (R_v) in the basal Patchawarra Formation (seismic marker "W" in Figure 27) in Tirrawarra field is about 1.20%, increasing to 1.30% in the downdip potential source area. The source in the field is 9750 ft (2970 m) deep and reached oil maturity between 80 and 90 Ma (Figure 28), assuming the onset of significant oil accumulation at 0.80% R_v. In the downdip areas (Figures 29A and 29B), the source is about 10,250 ft (3125 m) deep and generation of significant liquids began about 90 to 95 Ma.

The Tirrawarra oil is a highly volatile, mature paraffinic oil with an average API gravity of 55°. The maturity of the oil as determined from aromatics composition is equivalent to 1.05% R_v with the adjacent source equivalent to 1.20% R_v. Although the oil is undersaturated with respect to gas, the GOR of the oil is high and the oil exhibits no evidence of water washing or biodegradation.

Patchawarra Formation

The source of gas reservoired in sands of the Lower Permian Patchawarra Formation is presumed to be the abundant organic matter in shales and coals of the Patchawarra Formation. Migration distances in the meandering fluvial reservoir sands are assumed to be small, on the order of 6 mi (10 km), because of the presence of interbedded fluvial shales and associated coals. The overlying lacustrine shale unit (Murteree Shale), while providing a semiregional seal

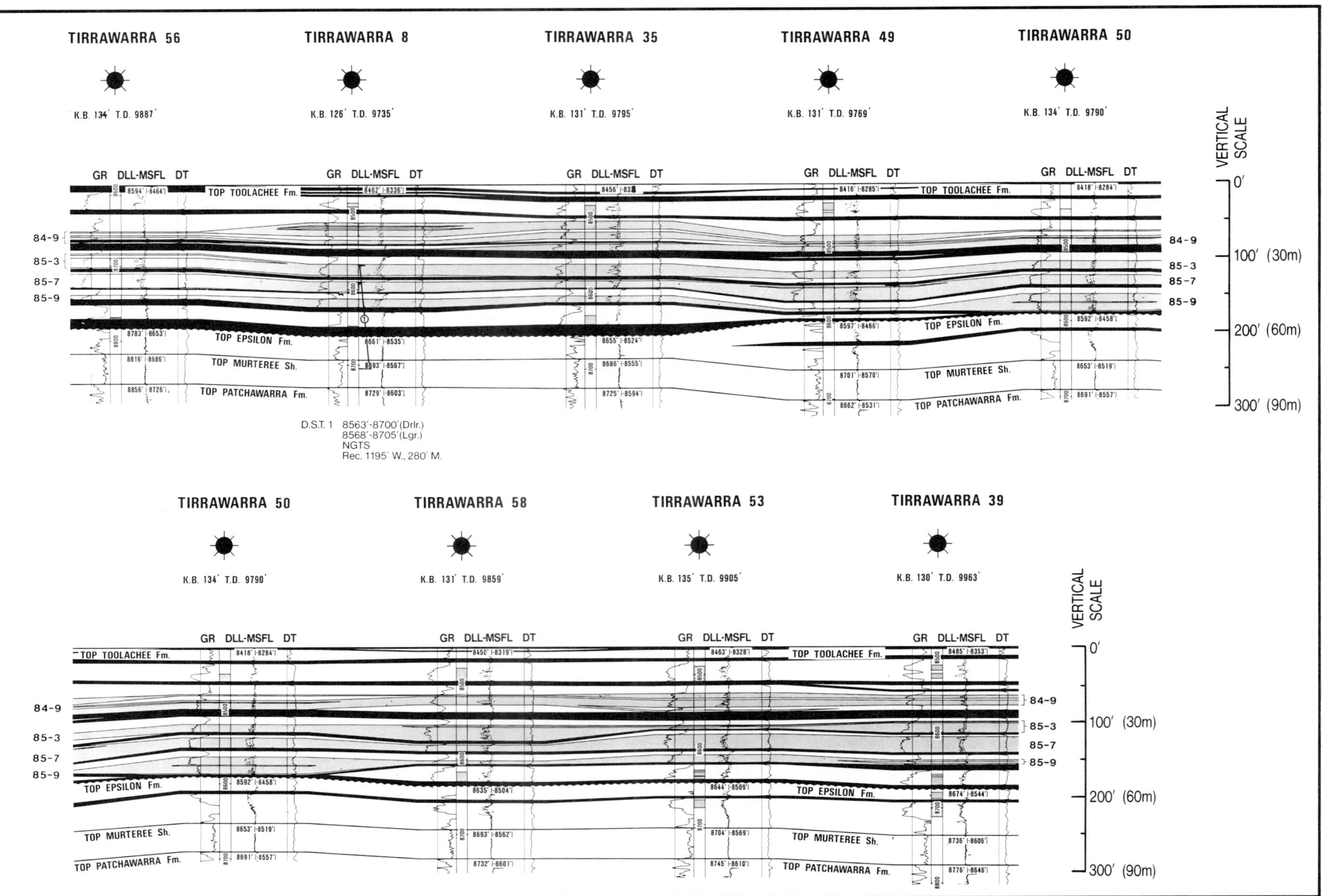

Figure 25. Stratigraphic cross section, north to south, showing correlation of reservoirs in the Toolachee Formation. Correlations based on a combination of gamma ray and sonic logs. Refer to Figure 2 for location.

275

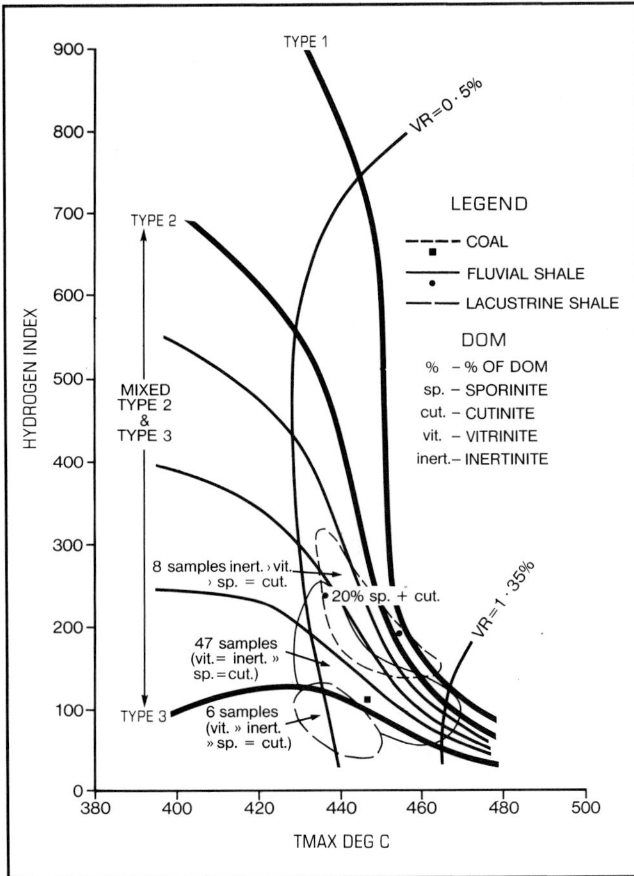

Figure 26. Source rock quality for Cooper basin core samples. The envelopes show the composition of most samples for Permian coal, fluvial shale, and lacustrine shale. Data lying outside the envelopes are shown as points. Type III organic matter is gas-prone higher plant material, type II organic matter is oil- and gas-prone higher plant material, and type I organic matter is oil-prone.

for the Patchawarra Formation, is thought to have a low source potential. Total organic carbon (TOC) in the Murteree Shale is 2 to 3% on average and the organic matter consists of vitrinite and inertinite (type III) with only minor sporinite. The lacustrine Murteree Shale has only limited gas potential based on Rock-Eval Pyrolysis of six rock samples (Figure 26).

Coals account for more than 30% of the sedimentary sequence in much of the Patchawarra trough and are typically inertinite-rich (60% average), with little exinite (5%) consisting of sporinite with minor cutinite (Cook and Struckmeyer, 1986). The composition of the Cooper basin coals measured on rock samples using Rock-Eval Pyrolysis is shown in Figure 26. The high hydrogen index of the coals is a result of analytical limitations owing to the high organic carbon content; the contribution of coal-derived hydrocarbons and the liquids potential of the coals is debatable (Cook and Struckmeyer, 1986). The

aggregate Patchawarra Formation coal thickness in the vicinity of the field is approximately 150 ft (45 m), increasing to 200 ft (60 m) in the downdip potential source area. TOC usually makes up 4% to 6% (by weight) of the Permian shales. Dispersed organic matter (DOM) in the fluvial shales consists of higher plant-derived coaly material; vitrinite, on average, comprises 50% of the DOM, inertinite 45% and exinite 5%. Vitrinite and inertinite, which are type III organic matter, and the exinite, which is type II organic matter, are also composed principally of sporinite and cutinite. The composition of the DOM measured on rock samples is shown in Figure 26. The fluvial shales are principally gas prone but with a significant though limited liquids potential in terms of their hydrogen index. Potential source lithologies that are transitional between shale and coal, shaly coals, and coaly sands are common. The aggregate Patchawarra Formation shale thickness in the vicinity of Tirrawarra field is 500 to 600 ft (150–180 m). Organic matter in the overlying lacustrine Murteree Shale is sparse, consisting mainly of oxidized humic material (inertinite).

Patchawarra Formation maturity in Tirrawarra field ranges from 1.00% to 1.20% R_v (source unit between the "V" and "W" seismic markers in Figure 27), increasing to between 1.10% and 1.30% in the downdip potential drainage area. Maturity gradients in the vicinity of the field are gentle, and maturity increases in the updip direction toward the GMI ridge to the east because of a substantially hotter geothermal regime in the Nappamerri trough. The gas in the Patchawarra Formation is relatively wet (22% C_{2-4}/C_{1-4}).

The source at the field is 8500 ft (2590 m) deep and reached gas maturity (R_v 1.00%) between 30 and 80 Ma (Figure 28). In the potential down-dip source areas, the source is 9250 ft (2820 m) deep and gas maturity was reached between 80 and 90 Ma (Figures 28, 29A, and 29B).

Toolachee Formation

The source of gas reservoired in sandstones of the Upper Permian Toolachee Formation is presumed to be the abundant organic matter in shales and coals of the formation. The contribution of coal-derived hydrocarbons to the reservoired gas is conjectural. Migration distances in the meandering fluvial reservoir sandstones are assumed to be comparable to those of the Patchawarra Formation.

The Toolachee Formation coals and shales exhibit characteristics similar to those of the Patchawarra Formation described above (Figure 26). The aggregate Toolachee Formation coal thickness in the vicinity of the field is about 40 ft (12 m), increasing to 50 ft (15 m) in the downdip potential source area. The aggregate Toolachee Formation shale thickness in the vicinity of Tirrawarra field is 90 ft (27 m).

The Toolachee Formation maturity in Tirrawarra field is about 0.90% R_v (source unit beneath the "P"

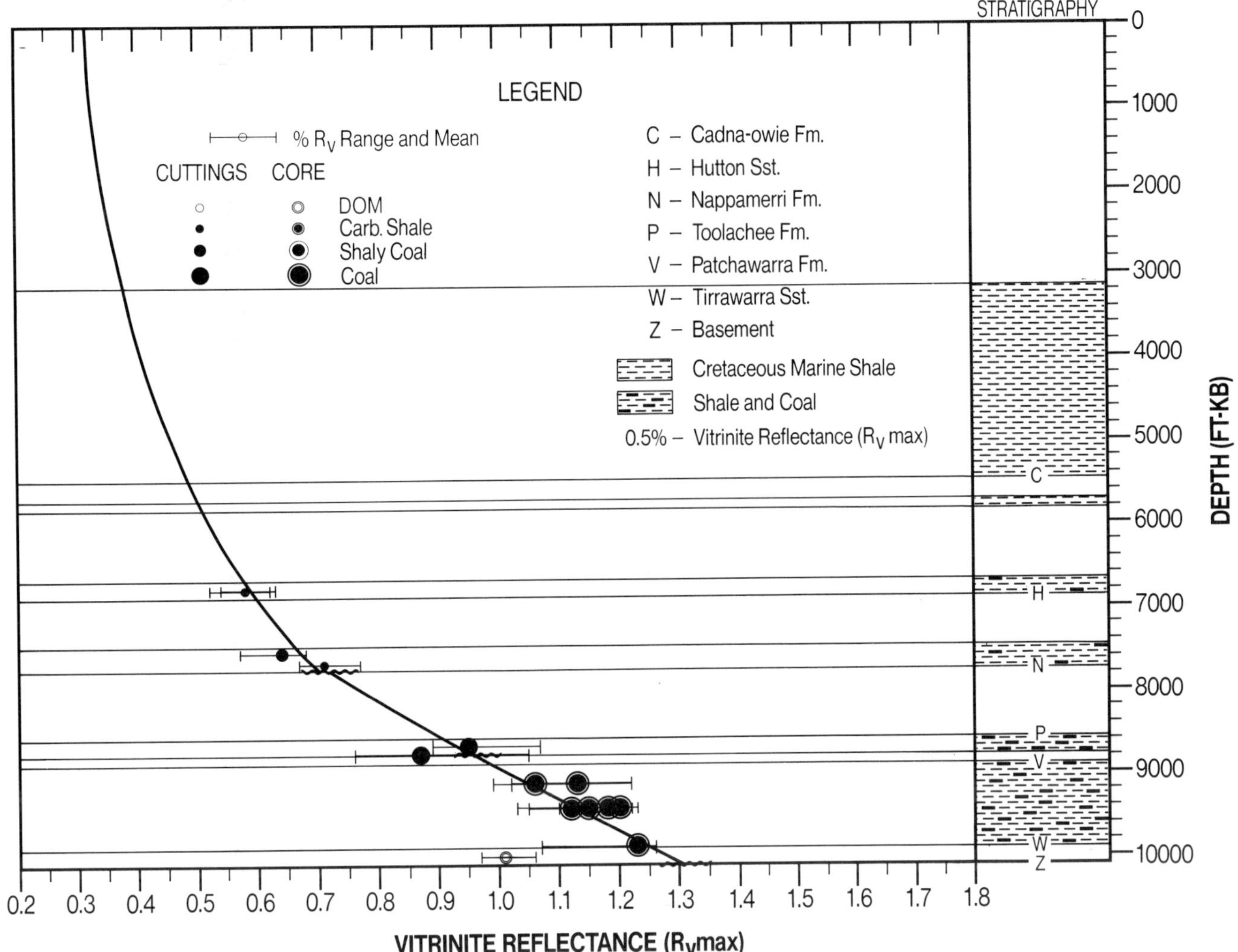

Figure 27. Vitrinite reflectance profile for Tirrawarra field, Tirrawarra 5. The legend differentiates samples type and organic matter type and the stratigraphic column shows potential shaly or coaly source units. The shallow part of the curve has been extrapolated on the basis of regional trends.

seismic marker in Figure 27) increasing to 0.95% in the downdip potential source area, and 1.10% in the deeper Patchawarra trough, approximately 12 mi (20 km) to the north. The regional association of source maturity and hydrocarbon accumulation in the Cooper basin marks the onset of significant gas accumulation at about 1.00% R_v. This suggests that the source in the vicinity of the field is marginally mature, which is consistent with the limited gas reserves and partial fill of the reservoirs in the Toolachee Formation.

The Toolachee Formation source is currently buried at 8500 to 9250 ft (2590–2820 m) but was probably uplifted about 500 ft (150 m) during the Tertiary period. Modern geothermal gradients in the vicinity of Tirrawarra field and the Patchawarra trough in general are on the order of 2.0°F/100 ft (3.6°C/100 m), which also accounts for the low maturity gradient near the field. An estimate of the timing of generation in this region can be made from the Lopatin diagrams in Figures 28 and 29. The temperature history used in the models was simply proportional to the present measured or estimated temperature gradients in or near the field. The source in the vicinity of the field reached gas maturity only recently, about 15 Ma (Figure 28), although in the downdip troughs to the west and east the onset of gas maturity was between 80 and 90 Ma (Figures 29A and 29B); however, the small reserves, partial fill of the reservoirs, and wetness of the gas (27% C_{2-4}/C_{1-4}) suggest that the reservoir is not in communication with a deeper, more mature source.

EXPLORATION AND DEVELOPMENT CONCEPTS

Siting of appraisal and development drilling in Tirrawarra field depends on the primary objective

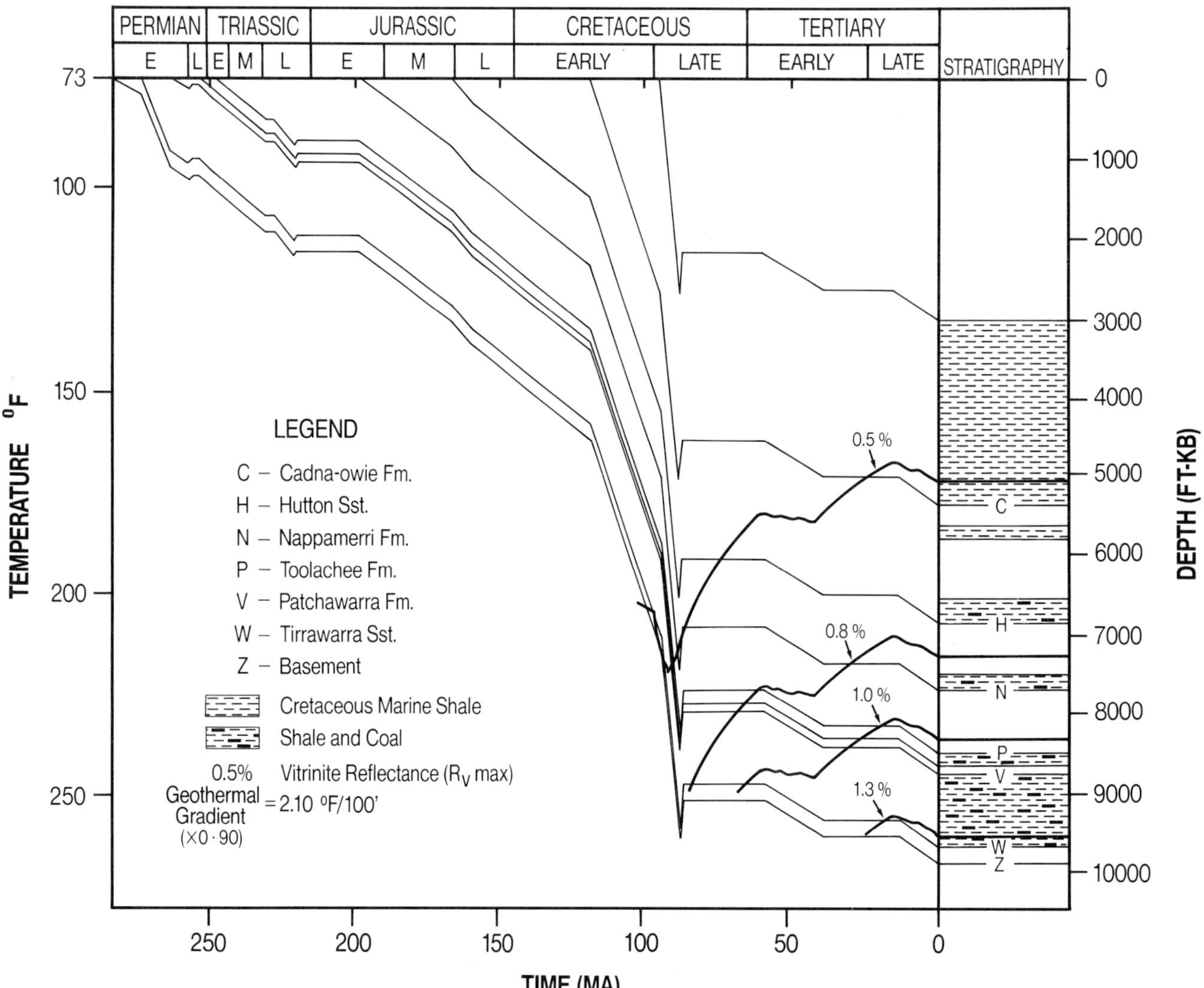

Figure 28. Burial history and timing of generation within Tirrawarra field, Tirrawarra 14. The temperature history used for the model was 0.9 of the present temperature gradient, so that the calculated and measured maturities agree.

of each well. In most cases it has been possible, with only minor compromise, to combine the gas and oil requirements for appraisal, drainage, and future enhanced oil recovery.

For siting designated gas wells, it has occasionally proved difficult to confidently predict the location of porous sandstone within the Patchawarra Formation. Even the thicker reservoirs are beyond the resolution of conventional seismic resolution and interpretation, and pay prediction in the past has been based on single sandstone isopach mapping. Many wells have unexpectedly intersected isolated gas-saturated lenticular lobes. Such instances demand redesign of theoretical drainage patterns; e.g., the impact of detailed isopach mapping based on genetic increments in the Patchawarra Formation (Kennedy, 1988) is being assessed. In contrast, designated oil development wells seldom fail to intersect predicted thicknesses of porous sandstone.

The study conducted prior to implementation of the second stage of the enhanced oil recovery scheme required close cooperation between the reservoir engineers responsible for simulation and the geological staff whose input of permeability and porosity distribution models was critical to the success of the project. Results of the study suggest that this has been effective. The highly volatile nature of the oil and the heterogeneity of the reservoir rock have presented the Cooper Basin Unit Partners with unique circumstances to be overcome in maximizing oil recovery.

ACKNOWLEDGMENTS

The Cooper Basin Unit Partners permitted publication of this paper. The authors would like to

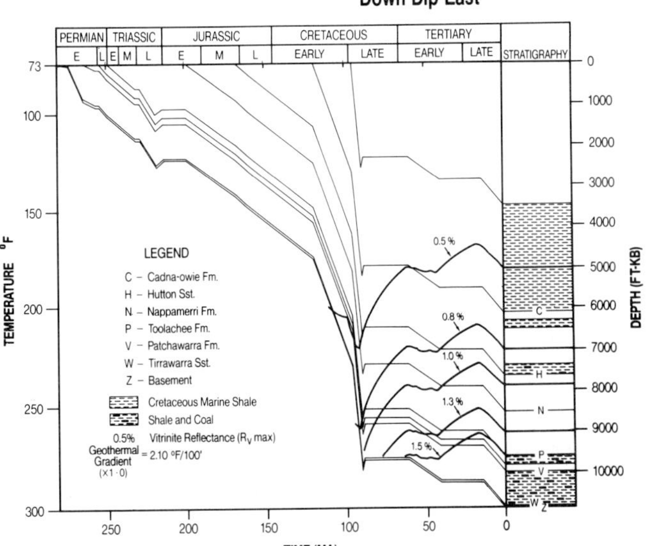

Figure 29A. Burial history and timing of generation in the downdip area of the Patchawarra trough 6 mi (10 km) east of Tirrawarra field. The temperature history for this model is the present temperature gradient in the area, so that the calculated and measured maturities agree.

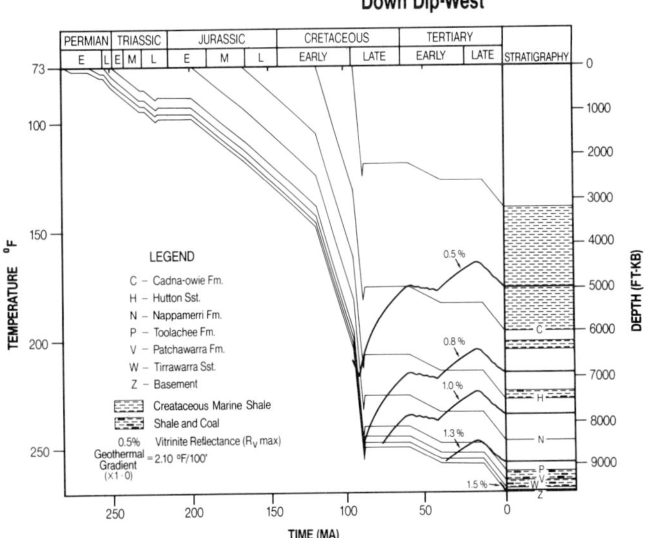

Figure 29B. Burial history and timing of generation in the downdip area of the Patchawarra trough 6 mi (10 km) west of Tirrawarra field. The temperature history for this model is the present temperature gradient in the area, so that the calculated and measured maturities agree.

thank Ian Johnstone, Liang Chen, and Fayaz Jamal for their contributions of engineering data for the summary tables comprising the appendices. The manuscript was read and improved by John Armstrong, Norrie Hamilton, Steven Kelemen, and John Rodda. Terry Denneny, Heather Reglar, and Tatjana Holtz skillfully drafted the diagrams. Merryl Press and Vi Davies typed the text. The authors would further like to express their indebtedness and appreciation to the many SANTOS Limited and Unit Partner geologists, geophysicists, and reservoir engineers who have made important contributions to the understanding of Tirrawarra field over the years.

REFERENCES CITED

Anthony, D., 1987, Proved and probable hydrocarbons-in-place for the Tirrawarra/Rakoona Field, Toolachee and Patchawarra Formations and Tirrawarra Sandstone: Unpublished report for the Cooper Basin Unit 1987 Review and Adjustment.

Battersby, D. G., 1976, Cooper Basin gas and oil fields, *in* R. B. Leslie, H. J. Evans, and C. L. Knight, eds., Economic geology of Australia and Papua New Guinea, 3: Petroleum, Australian Institute of Mining and Metallurgy Monograph 7, p. 321–368.

Brown, D. J., and M. R. Barley, 1986, Tirrawarra and Moorari enhanced oil recovery projects—"getting more oil out from down under": APEA Journal, v. 26(1), p. 389–396.

Cook, A. C., and H. Struckmeyer, 1986, The role of coals as source rock for oil, *in* R. C. Glenie, ed., Technical papers presented at the PESA Symposium, Melbourne, 1985.

Freeman, R. N., 1964, Oil exploration in the western Great Artesian Basin: Australasian Institute of Mining and Metallurgy, annual conference, 14–24 August 1963, Exploration Paper #6 (presented at Adelaide, South Australia).

Gostin, V. A., 1973, Lithological study of the Tirrawarra Sandstone: Unpublished report for Delhi Petroleum Pty Ltd.

Gray, R. J., and D. C. Roberts, 1984, A seismic model of faults in the Cooper Basin: APEA Journal, v. 24(1), p. 421–428.

Harrison, A., 1984, Geological report for the Tirrawarra Field reservoir simulation study: Unpublished report for Santos Limited.

Harrison, A., 1985, Proved and probable hydrocarbons-in-place for the Tirrawarra Field (Patchawarra Formation): Unpublished report for the Cooper Basin Unit 1985 Review and Adjustment.

Kantsler, A. J., A. C. Cook, and M. Zwigulis, 1986, Organic maturation in the Eromanga Basin, *in* D. I. Gravestock, P. S. Moore, and G. M. Pitt, eds., Contributions to the geology and hydrocarbon potential of the Eromanga Basin: Geological Society of Australia Special Publication 12, p. 304–322.

Kapel, A. J., 1972, The geology of the Patchawarra area, Cooper Basin: APEA Journal, v. 12, p. 53–57.

Kennedy, S., 1988, A study of the Patchawarra Formation, Tirrawarra Field, southern Cooper Basin, South Australia: Unpublished MSc. thesis, University of Adelaide.

Martin, C. A., 1967, Moomba—a South Australian gas field: APEA Journal, v. 7, p. 124–129.

Martin, K. R., 1980, A preliminary account of diagenesis and reservoir quality in the Toolachee Formation, Cooper Basin: Unpublished report to Santos Limited, 23 p.

Martin, K. R., 1983, Petrology of the Patchawarra Formation in Tirrawarra #16, #19 and #28, Cooper Basin, South Australia: Unpublished report for Santos Limited, 26 p.

Martin, K. R., 1984, Petrology, diagenesis and reservoir quality of the Tirrawarra Sandstone in the Tirrawarra, Fly Lake and Moorari fields: Unpublished report for SANTOS Limited, 57 p.

Martin, K. R., and N. J. Hamilton, 1981, Diagenesis and reservoir quality, Toolachee Formation, Cooper Basin: APEA Journal v. 21, p. 143–154.

Mount, T. J., 1981, Dullingari North #1—an oil discovery in the Murta Member of the Eromanga Basin: APEA Journal, v. 21(1), p. 71–77.

Osbourne, H. C., 1974, A reservoir engineering study of gas injection pressure maintenance in the Tirrawarra oil zone reservoirs: Unpublished report to Santos Limited.

Pecanek, H. T., and I. M. Paton, 1984, The development of the

Tirrawarra oil and gas field: APEA Journal, v. 24(1), p. 278-288.

Selley, R. C., 1985, Elements of petroleum geology: New York, W. H. Freeman and Company, 449 p.

Sprigg, R. C., 1986, The Eromanga Basin in the search for commercial hydrocarbons, *in* D. I. Gravestock, P. S. Moore, and G. M. Pitt, eds., Contributions to the geology and hydrocarbon potential of the Eromanga Basin: Geological Society of Australia Special Publication 12, p. 9-24.

Stanmore, P. J., and E. M. Johnstone, 1988, The search for stratigraphic traps in the southern Patchawarra Trough, South Australia: APEA Journal, v. 28(1), p. 156-166.

Thornton, R. C. N., 1978, Regional lithofacies and palaeogeography of the Gidgealpa Group: APEA Journal, v. 18, p. 52-63.

Ward, L. K., 1944, The search for oil in Australia: Bulletin of the South Australian Geological Survey, v. 22, p. 4-40.

Williams, B. P. J., and E. K. Wild, 1984, The Tirrawarra Sandstone and Merrimelia Formation of the southern Cooper Basin, South Australia—sedimentation and evolution of a glacio-fluvial system: APEA Journal, v. 24(1), p. 377-392.

Youngs, B. C., 1975, The hydrology of the Gidgealpa Formation of the western and central Cooper Basin: Geological Survey of South Australia Report of Investigations 43, 35 p.

SUGGESTED READING

Pecanek, H. T., and I. M. Paton, 1984, The development of the Tirrawarra oil and gas field: APEA Journal, v. 24(1), p. 278-288. Concise description of reservoirs and hydrocarbons prior to implementation of the enhanced oil recovery project.

Sprigg, R. C., 1986, The Eromanga Basin in the search for commercial hydrocarbons, *in* D. I. Gravestock, P. S. Moore, and G. M. Pitt, eds., Contributions to the geology and hydrocarbon potential of the Eromanga Basin: Geological Society of Australia Special Publication 12, p. 9-24. Riveting account of the count-down to oil discovery in central Australia.

Thornton, R. C. N., 1979, Regional stratigraphic analysis of the Gidgealpa Group, southern Cooper Basin: Bulletin of the South Australian Geological Survey 49, 140 p. Most comprehensive sedimentological analysis of the Cooper basin available in published form.

Williams, B. P. J., and E. K. Wild, 1984, The Tirrawarra Sandstone and Merrimelia Formation of the southern Cooper Basin, South Australia—sedimentation and evolution of a glacio-fluvial system: APEA Journal, v. 24(1), p. 377-392. Detailed sedimentological study of the principal oil reservoir in the Cooper basin.

Field name .... *Tirrawarra field*

Ultimate recoverable reserves .... *434 bcf, 42 MMSTB*

Field location:

    Country .... *Australia*

    State .... *South Australia*

    Basin/Province .... *Cooper basin*

Field discovery:

    Year first pay discovered .... *Lower Permian Patchawarra Formation   1970*

    Year second pay discovered .... *Lower Permian Tirrawarra Sandstone   1970*

    Third pay .... *Upper Permian Toolachee Formation   1981*

Discovery well name and general location:

    First and second pays .... *Tirrawarra-1, approx. 30 mi (50 km) north-northwest of Moomba, S. A.*

    Third pay .... *Tirrawarra-14, 0.75 mi (1.25 km) southwest of Tirrawarra-1*

Discovery well operator .... *Bridge Oil Pty Ltd*

    Second pay .... *Bridge Oil Pty Ltd*

    Third pay .... *Santos Ltd*

IP in barrels per day and/or cubic feet or cubic meters per day:

    First pay .... *3.1 MMCFD (DST)*

    Second pay .... *459 BOPD (production test)*

    Third pay .... *6.0 MMCFD (DST)*

All other zones with shows of oil and gas in the field:

Age	Formation	Type of Show
Early Cretaceous	*Mooga (Murta Member)*	*Oil fluorescence*
Jurassic	*Birkhead*	*Oil fluorescence*
	Hutton Sandstone	*Oil fluorescence*
Late Triassic	*Nappamerri*	*Gas*

Geologic concept leading to discovery and method or methods used to delineate prospect

Discovery well drilled crestally on seismically defined anticline to test four-way dip closure that had proven productive elsewhere in the basin. Structure has no surface expression.

Structure:

    Province/basin type .... *Bally, 1212; Klemme, III A/II B*

    Tectonic history

The Cooper basin is a continental rift basin that was initiated in the Late Carboniferous and had ceased receiving sediment by the latest Triassic. The sediments are entirely of clastic nonmarine origin. One major regional unconformity interrupts the stratigraphic sequence, spanning from Early to Late Permian. In the southern part of the basin the major structural elements consist of three troughs separated by two regional highs. These features are elongated approximately northeast/southwest. The Tirrawarra field is situated within the westernmost trough, called the Patchawarra trough.

Regional structure

The Patchawarra trough is regionally a broad, asymmetric syncline, the axis of which strikes approximately northeast and is located nearer the eastern margin. The trough is the westernmost extant structural element of the Cooper basin. Within the Patchawarra trough, detailed structure in the vicinity of the Tirrawarra field consists of parallel horst and graben blocks that were initiated at the commencement of Cooper basin sedimentation and were active early in basin history and intermittently active thereafter. Major fields, including Tirrawarra, developed by onlap and drape over the horst blocks.

Local structure

Domal drape anticline measuring approximately 6 mi (10 km) by 4 mi (7 km) located in the central western part of the Patchawarra trough adjacent to the Gidgealpa-Merrimelia-Innamincka ridge.

Trap:

Trap type(s)

Tirrawarra field of South Australia has three anticlinal traps, two of which have multiple pays

Basin stratigraphy (major stratigraphic intervals from surface to deepest penetration in field):

Chronostratigraphy	Formation	Depth to Top in ft (m)
Eromanga basin:		
Neocomian	Cadna-Owie (top of upper reservoir section)	–5223 (–1593)
Lower–Middle Jurassic	Hutton Sandstone	–6615 (–2017)
Cooper basin:		
Lower–Middle Triassic	Nappamerri	–7450 (–2272)
Upper Permian	Toolachee	–8284 (–2527)
Lower Permian	Patchawarra	–8550 (–2608)
	Tirrawarra	–9297 (–2836)
Cambrian–Ordovician	Basement (Kalladeina Formation)	–9783 (–2984)

Reservoir characteristics:

Number of reservoirs *16*

Formations .... *Total of 3: Toolachee Formation (4 reservoirs); Patchawarra Formation (11 reservoirs); Tirrawarra Sandstone*

Ages .... *Tirrawarra and Patchawarra, Early Permian; Toolachee, Late Permian*

Depths to tops of reservoirs .... *Tirrawarra, 9297 ft (2836 m) subsea; Patchawarra, 8550 ft (2608 m) subsea; Toolachee, 8284 ft (2527 m) subsea*

Gross thickness (top to bottom of producing interval) .... *Tirrawarra, 115 ft (35.0 m); Patchawarra, 896 ft (273.3 m); Toolachee, 184 ft (56.1 m)*

Net thickness—total thickness of producing zones

    Average .... *Patchawarra, 53 ft (16.2 m); Tirrawarra, 43 ft (13.1 m); Toolachee, 21 ft (6.4 m)*

    Maximum .... *Patchawarra, 137 ft (41.8 m); Tirrawarra, 97 ft (29.6 m); Toolachee, 30 ft (9.2 m)*

Lithology

All Permian reservoirs composed mainly of fine-grained moderately sorted quartz sandstone with subordinate amounts of rock fragments, chert, and mica; authigenic clays, principally dickite and illite, contribute up to 11% of the mode of some samples

Porosity type .... *Primary depositional porosity severely reduced by grain overgrowth and authigenic clay precipitation*

Average porosity .... *Tirrawarra, 10.6% (core), 11.3% (logs); Patchawarra, 10.4% (core), 10.7 (logs); Toolachee, no data (core), 10.7 (logs)*

Average permeability .... *Tirrawarra, 1–2 md; Patchawarra, 1–2 md; Toolachee, no data*

Seals:

Upper

    Formation, fault, or other feature .... *Intraformational*

    Lithology .... *Shale/coal*

Lateral

    Formation, fault, or other feature .... *Major reservoirs are anticlinal, some stratigraphic pinch-out of minor sands*

    Lithology .... *Shale/coal*

Source:

Formation and age

Toolachee Fm. (Toolachee reservoirs), Late Permian; Patchawarra Fm. (Patchawarra reservoirs), Early Permian; lower Patchawarra Formation (Tirrawarra reservoir), Early Permian

Lithology .... *Toolachee, sand > shale = coal; Patchawarra, shale > sand > coal; lower Patchawarra, shale > sand > coal*

Average total organic carbon (TOC) .... *Toolachee, Patchawarra, lower Patchawarra, 4.0%*

Maximum TOC .... *Toolachee, Patchawarra, lower Patchawarra, 6.0%*

Kerogen type (I, II, or III) .... *Toolachee, Patchawarra, lower Patchawarra, III>>II*

Vitrinite reflectance (maturation) .... R_o = *0.78–0.85 (Toolachee)*; 1.00–1.16 (Patchawarra); 1.23–1.28 (lower Patchawarra)*

Time of hydrocarbon expulsion .... *Toolachee gas 1.0%, 0–80 Ma; Patchawarra gas 1.0%, 31–80 Ma; lower Patchawarra oil 0.8%, 82–91 Ma***

Present depth to top of source .... *Toolachee, 8500–9250 ft*; Patchawarra, 8750–9500 ft; lower Patchawarra, 9750–10,250 ft*

Thickness .... *Toolachee, 50 ft shale, 90 ft coal; Patchawarra, 450 ft shale, 125 ft coal; lower Patchawarra, 200 ft shale, 70 ft coal*

Potential yield .... *NA*

**Well site, 10 km downdip (hypothetical).*
***0.8% R_v used for onset of Permian oil generation.*

Appendix 2. Production Data

Field name .... *Tirrawarra field*

Field size:

    Proved acres .... *12,500 ac (oil); 10,000 ac (gas)*

    Number of wells all years *62*

    Current number of wells
    18 single gas completions; 21 single oil completions; 10 dual oil/gas completions; 4 gas injection wells; 1 observation/gas dual completion; 3 suspended; 5 abandoned

    Well spacing .... *Approx. 300 ac (oil); approx. 360 ac (gas)*

    Ultimate recoverable .... *41.5 MMSTB, 240 bcf*

    Cumulative production .... *6.17 MMSTB (Feb. 1987); 68.5 bcf (Dec. 1986)*

    Annual production .... *1.76 MMSTB; 24.4 bcf*

    Present decline rate .... *Oil negative (gas injection); gas, 30%*

        Initial decline rate .... *Oil, approx. 50% per annum; gas, 25% per annum*

        Overall decline rate .... *Oil nil; gas, 27%*

    Annual water production .... *Oil, approx. 49,000 bbl; gas, approx. 146,000 bbl*

    In place, total reserves .... *117.9 MMSTB oil; 298 bcf nonassociated gas*

    In place, per acre-foot .... *230 bbl; 927 MMcf*

    Primary recovery .... *21.6 MMSTB oil; 240 bcf gas*

    Secondary recovery .... *NA*

    Enhanced recovery .... *6.0 MMSTB oil*

    Cumulative water production .... *193,000 bbl (oil reservoir); 420,000 bbl (gas reservoir)*

Drilling and casing practices:

    Amount of surface casing set .... *Approx. 600 ft*

    Casing program
    13⅜-in. surface casing, 9⅝-in. intermediate casing to approx. 5500 ft, 7-in. production casing to TD

    Drilling mud .... *Bentonite lignosulphonate (Gel ligno system)*

    Bit program .... *17½-in. × 1 to approx. 600 ft; 12¼-in. × 2 to approx. 5500 ft; 8½ × 5 to TD*

    High pressure zones .... *Nil*

Completion practices:

Interval(s) perforated ..... *Single completions: entire thickness of Tirrawarra sst (oil), all pay sands (gas); dual completions: both of above*

Well treatment .... *Hydraulic fracture stimulation of oil sands/major gas sands*

Formation evaluation:

Logging suites
CAL-GR-BHC Sonic; MSFL-DLL-GR-CAL LDL-CNL-PEF-GR (subject to hole conditions); resistivity run from intermediate casing level to TD; neutron density run only over Permian section; Sonic-GR run from surface casing to TD

Testing practices .... *Open hole DST, cased hole DST, RFT, production testing, intermittent pressure surveys*

Mud logging techniques
Monitor and log lithology, gas detection and chromatography, penetration rate, pit level and pump stroke rate from surface casing shoe to TD; logged cuttings collected and logged at 30 ft (9 m) intervals from surface casing to 150 ft (46 m) above top Cadna-Owie Formation, and thereafter at 10 ft (3 m) intervals to TD

Oil characteristics:

Type .... *Oil and gas liquids paraffinic*

API gravity .... *Oil, 50–52°; condensate, 55°*

Base .... *NA*

Initial GOR .... *2714 scf C^{4-} per bbl c^{5+}*

Sulfur, wt% *0.03–0.05*

Viscosity, SUS .... *0.1 cp (at 4300 psi and 302°F)*

Pour point .... *–3 to –2°C*

Gas-oil distillate .... *NA*

Field characteristics:

Average elevation .... *130 ft (40 m)*

Initial pressure .... *Oil, 4260 psi (29,370 kPa) at 9500 ft (2900 m); gas, 3990 psi (33,360 kPa) at 9500 ft (2900 m)*

Present pressure .... *Oil, 3500 psi (24,130 kPa); gas, 2800 psi (19,306 kPa)*

Pressure gradient .... *Oil, 0.44 psi/ft (9.94 kPa/m); gas, 0.15 psi/ft (2.71 kPa/m)*

Temperature .... *Oil, 285°F (140.5°C); gas, 270°F (132.2°C)*

Geothermal gradient .... *4°C/100 m (2.2°F/100 ft) TD to surface*

Drive .... *Oil: expansion/EOR gas injection to maintain pressure; gas: expansion*

Oil column thickness .... *Oil, 43 ft (13 m) (avg); gas, 53 ft (16 m) (avg); gas in multiple sands*

Oil-water contact
Oil: –9893 ft (–3017 m) west; –9606 ft (–2930 m) east-north; –9721 ft (–2965 m) lowest known oil east-south
Gas: multiple sands with lowest known gas from –8841 to –9607 ft (–2696 to –2930 m)

Connate water .... *Oil zones, 32.2%; gas zones, 27.4% (both averages from log analysis)*

Water salinity, TDS .... *Oil, 9300 ppm NaCl equiv.; gas, 13,000 ppm NaCl equiv.*

Resistivity of water .... *Oil, 0.63 ohm-m at 75°F; gas, 0.46 ohm-m/m² at 75°F*

Bulk volume water (%) .... *Oil, 3.4%; gas, 2.8%*

Transportation method and market for oil and gas:
Oil: pipeline to satellite, thence to Moomba and Port Bonython for sale on international market.
Gas: individual or bunched flow line to satellite station, from thence to main plant at Moomba and then pipeline to Sydney or Adelaide market for domestic use and electricity generation.

Sacha Field—Ecuador
Oriente Basin

ROBERT W. CANFIELD
Texaco Petroleum Company
Quito, Ecuador

FIELD CLASSIFICATION

BASIN: Oriente
BASIN TYPE: Foredeep
RESERVOIR ROCK TYPE: Sandstone
RESERVOIR ENVIRONMENT OF DEPOSITION: Fluvial (Braided Stream)
TRAP DESCRIPTION: Elongate, asymmetric anticline with multiple pays

RESERVOIR AGE: Cretaceous
PETROLEUM TYPE: Oil
TRAP TYPE: Anticline

LOCATION

Sacha field, located in the north-central Oriente region of Ecuador about 112 mi (180 km) east of the capital city, Quito, is the second largest field discovered in the Oriente basin (Figure 1). Other significant fields lying in the same region include Shushufindi-Aguarico, Lago Agrio, Auca, and Libertador.

Areal extent of Sacha field at the top of the main producing horizon is 74 mi² (192 km²). There are three important producing horizons and one of lesser importance. Sacha is in the giant class, having an estimated ultimate recovery of 743 million bbl, which is somewhat higher than the 650 million bbl (rank 358) estimated by Carmalt and St. John (1986).

HISTORY

Pre-Discovery

The first important geological studies in the Oriente region of Ecuador were made by Theron Wasson and Joseph H. Sinclair (1922, 1927) in the 1920s for the Leonard Exploration Company. This reconnaissance work was generally restricted to stratigraphic investigations along streams in the western outcrop belt between the town of Macas and the Sumaco volcano. Heavy oil seeps and tar sands were mapped during these studies.

The Shell Company of Ecuador Ltd., a subsidiary of the Royal Dutch Shell group, was very active in the region from 1938 to 1950. Shell was joined by Esso Standard Oil Company (Ecuador) S.A. in 1948. Based on geological and geophysical studies (reflection and refraction seismic methods and gravity), six unsuccessful exploratory wells were drilled from 1944 to 1950. Five of these wells were drilled south and southwest of Sacha field and the sixth about 95 mi (153 km) to the east. All had shows of oil. An excellent comprehensive paper of the Oriente regional geology based on Shell's efforts was compiled and published by H. J. Tschopp (1953).

Texaco and Gulf Oil in a joint venture discovered oil in commercial quantities at Orito in the Putumayo region of southern Colombia in 1963 and immediately filed for a large concession in Ecuador in the same geological setting. After being awarded this block, the companies farmed in adjoining acreage, the so-called Coca Concession, from another operator. Surface mapping was started by Texaco, the operator for the consortium, along the mountain front in late 1964, and reflection seismic operations were begun in mid-1965. The consortium's first well, Lago Agrio-1 located about 11 mi (18 km) south of the international border with Colombia, was completed in April 1967 as the first commercial discovery in the Oriente region of Ecuador.

Discovery

The Sacha anticline, located in the northwest corner of the former Coca Concession, was mapped in 1967 and 1968 with analog and digital seismic recordings in a seismic grid continuing south from the Lago Agrio field. Sacha-1, the discovery well, was drilled on the crest of the structure at long. 76°52'46.4"W and lat. 00°19'47.6"S using a helicopter supported rig (Figure 2). The well was completed on 25 February 1969, flowing 1328 bbl/day of 29.9° API oil through a ¼-in. choke from 53 ft (16 m) of selected perforations between 9816 ft (2992 m) and 9885 ft (3013 m) in the Hollin Formation of later Early Cretaceous age (probably late Aptian to Albian). The Late Cretaceous Napo "T" and "U" sands and the basal Tena (Maastrichtian) were also oil-bearing by

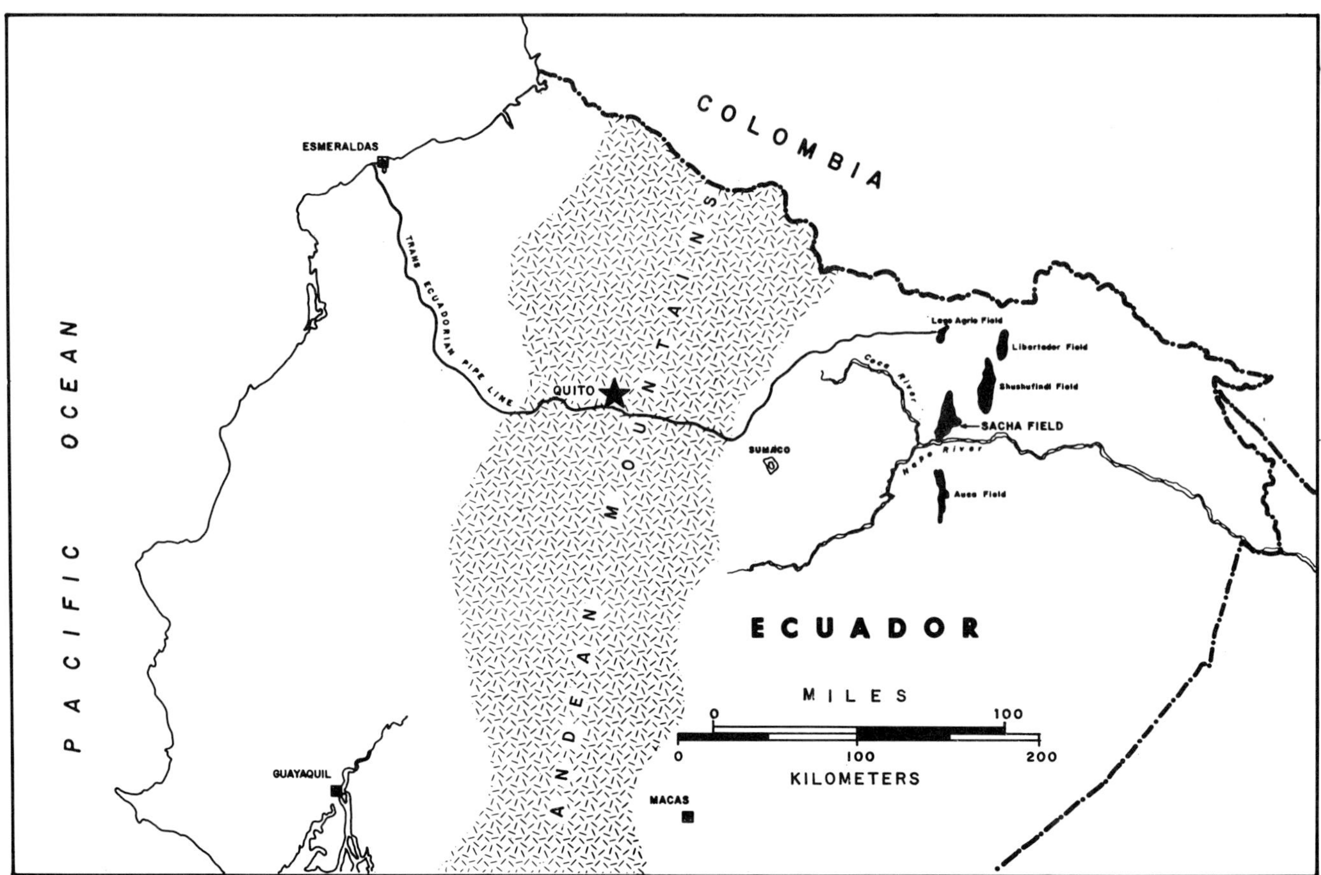

Figure 1. Index map of a portion of Ecuador showing location of major oil fields in northern Oriente basin.

log analysis but were not tested at the time of the discovery. Subsequently during a workover, both the Napo "T" and "U" sands were successfully completed in the discovery well.

The main Hollin sandstone at Sacha is generally fine to medium graincd but occasionally coarse grained, quartzose, and both porous and permeable. In Sacha-1 it extends from 9815 ft (2991 m) to 10,058 ft (3066 m) and has a well-defined oil-water interface at 9910 ft (3020 m).

Post-Discovery

During 1969 and 1970, three on-structure wildcats were drilled using the helicopter-supported rig to determine if the discovery was large enough to warrant developing in the remote area. Sacha-2 was drilled 4.8 mi (7.8 km) south of the discovery well, Sacha-3, 4 mi (6.5 km) northeast of Sacha-1 and Sacha-4, 4.7 mi (7.5 km) north of Sacha-3 (Figure 2). Because all of these wells were successful, the giant size of Sacha was confirmed, and a road from Lago Agrio was justified to develop the field. The Napo "T" and "U" sands were first tested at Sacha-3 where they yielded 605 bbl/day and 480 bbl/day, respectively, through ¼-in. chokes. Though the basal

Tena sand was not tested until 1971 at Sacha-16, log calculations indicated that it was oil-bearing in many of the earlier wells.

Development was started in March 1971 on 250 ac (100 ha) spacing. Drilling was concentrated along the anticlinal axis and in the central portion of the field. Three rigs were used until a road to the Shushufindi field was completed in early 1972 when one of them was diverted for that field's development. By mid-1972, when producing facilities and a pipeline were available, 48 Sacha field wells (including the exploratory wells) had been completed and the remaining two rigs were working at Shushufindi. From early 1973 until mid-1974, two drilling rigs were again active in Sacha generally completing the 250 ac development phase. Most of the earlier wells were drilled with the drilling rig to total depth and cased; then the drilling rig was moved to a new location and a smaller, less expensive workover rig was used for the completion.

The earlier wells had 9⅝-in. surface casing set to a depth between 240 ft (73 m) and 2400 ft (731 m), depending upon their location in the field, and 7-in. production casing cemented to total depth. In wells drilled after well 57, 10¾-in. surface pipe has been cemented to about 2000 ft (610 m). In the earlier wells, the long casing string had cement behind pipe to just

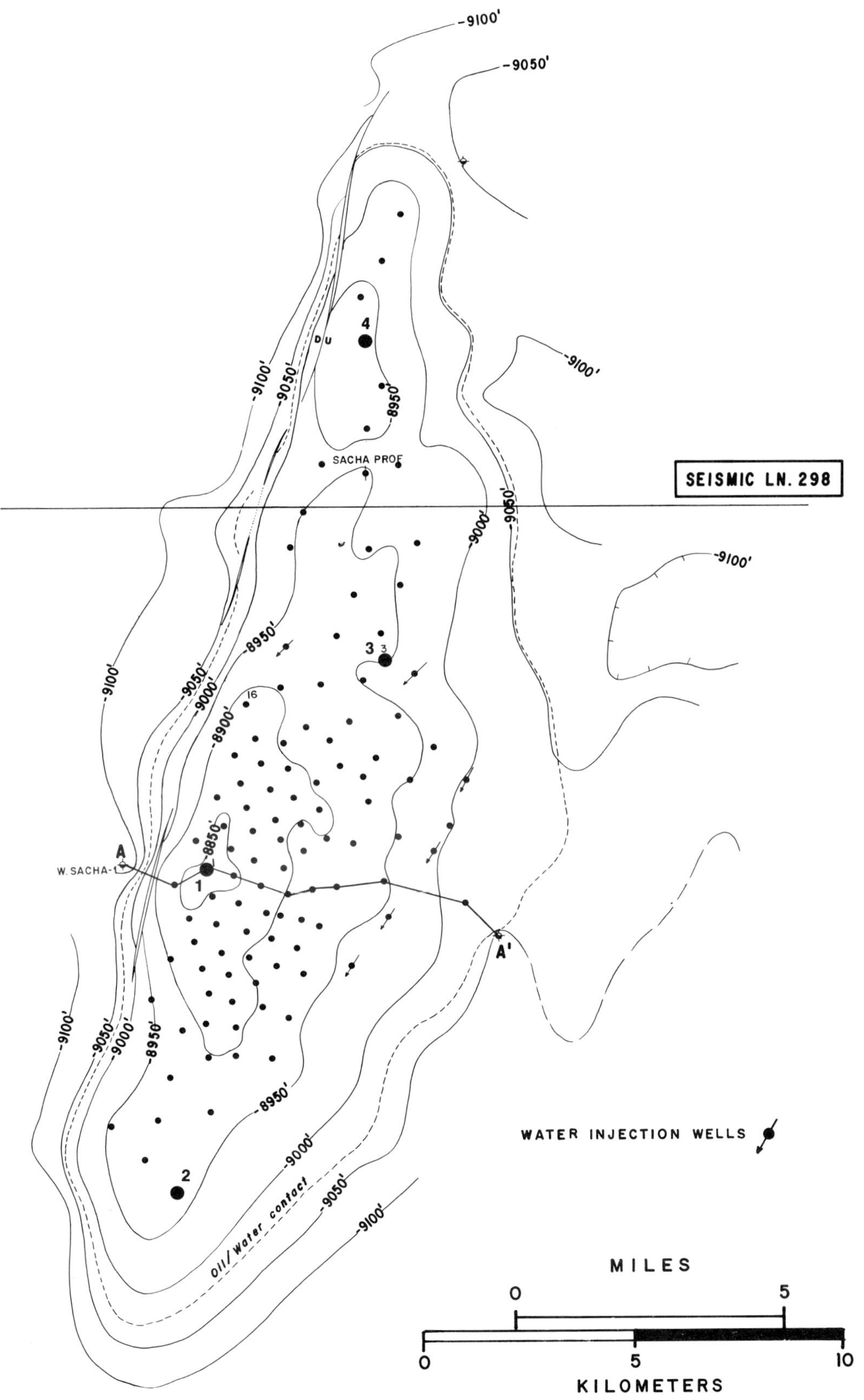

Figure 2. Structural contours on top of Hollin Formation, Sacha field. A–A' is the location of the cross section shown by Figure 8. Seismic profile line 298 is shown by Figure 4A and 4B.

high enough to cover the basal Tena sand, the uppermost producing horizon, but now the entire casing string is cemented to the surface. Casing corrosion has occurred in some of the older wells and the additional cement is expected to avoid this problem. Production tubing is 2⅞-in. or 3½-in. The earlier wells were usually drilled very close to or through the Hollin oil-water contact, but in wells since 1979, care has been taken to make a limited penetration of the main Hollin sandstones. The Hollin, Napo "T," and Napo "U" sandstones are gun perforated, isolated by packers, and produced through sliding sleeves. Initially all zones were opened but since 1975 the Hollin and Napo sands are produced separately. When all zones were opened, spinner surveys indicated that only the main Hollin sandstone was contributing to production, apparently because of high differential pressure resulting from its very strong water drive.

A few in-fill wells were drilled in 1976 but most of this work has occurred since 1979. The in-fill drilling is being done on 87 ac (35 ha) spacing in the central portion of the field. In 1979, two systems of artificial lift were employed: Hydraulic pumps are being used in the central part of the field and submersible pumps in the northern and southern extremes.

In November 1986, a waterflood project was started for the two Napo sandstones in the central portion of the field. One water injection well was drilled and five previous producers were converted to injection wells. The locations of the water injection wells are indicated on the structure map (Figure 2).

Through 1987 there have been 110 Sacha Cretaceous wells drilled. Of these, two have been dry holes and one was drilled as a water injection well. There has also been one dry exploratory well drilled immediately west of the field. The in-place crude oil reserves are estimated to be 2711 million bbl of which approximately 743 million bbl are recoverable. Cumulative production through 1987 has been 338 million bbl of which 245 million bbl (or about 72.5%) has been from the Hollin sandstones. The field is presently producing about 24 million bbl of oil and 5.5 million bbl of water annually. Most of the water is from the Hollin Formation. Essentially there has been no decline in the producing rate since production was initiated as the result of in-fill drilling and, more recently, water flooding.

In 1984 the Ecuadorian government oil company, CEPE, drilled a deep test in the Sacha field for its own account. Sacha Profundo-1 had a total depth of 16,143 ft (4920 m) in Paleozoic sediments. There were no hydrocarbon shows reported from the pre-Hollin sediments and several tests of selected intervals in these rocks were dry. Correlation of pre-Hollin section with similar sediments in a deep Shushufindi well immediately following drilling suggested the presence of the Jurassic Chapiza and Santiago formations and the Paleozoic Macuma and Pumbuiza formations (Trujillo et al., 1985). Later studies, however, have indicated that the interval originally assigned to the Jurassic Santiago Formation is Upper Devonian and apparently only the Jurassic Chapiza and Paleozoic Pumbuiza and possibly Macuma were encountered in the pre-Cretaceous section of the well.

DISCOVERY METHOD

The field was delineated by reflection seismic surveys. No surface expression of the structure was noted from aerial photographs and surface mapping was not attempted in the low-lying jungle-covered area. Good potential reservoir Cretaceous sandstones had been mapped along the uplifted mountain front to the west and had been seen in wells to the north in Colombia. Potential source beds posed no problem since the generation of hydrocarbons had been proved by the oil fields in southern Colombia, in the unsuccessful wells drilled in Ecuador by Shell during the late 1940s, and by the presence of tar sands and asphalt deposits in outcrop.

The seismic grid was carried south from the Colombian border and structural trends were located that were later shot using a smaller grid. Initially two seismic horizons were mapped, the base of the Cretaceous and an upper horizon, the top of the Orteguaza Formation. In later surveys, mapping of the base of the Cretaceous horizon was dropped and an event near the producing horizons, top of Napo "U" sand zone, was mapped. It was noted early in the seismic exploration that thinning was present between the two horizons over prospective structures and an isochron map was also constructed. Most of the thinning is apparently due to early structural growth, which is believed to be necessary for oil accumulations in the basin.

Seismic exploration is still being actively carried on in the Oriente basin. Though structures as large as Sacha or Shushufindi are not likely to be found, smaller anticlines are being mapped that have the same basic characteristics as the large ones and smaller oil accumulations are being discovered.

STRUCTURE

Seismic records from pre-Cretaceous rocks are generally not of good quality in the Oriente basin. Angular unconformities are known to exist in the older sediments and it is apparent from seismic that major faults are present that do not extend into the relatively undisturbed Cretaceous sediments. The older rocks were essentially peneplaned prior to the deposition of the Cretaceous. Minor tectonic pulses occurred in Cretaceous time and a major uplift took place during the Senonian to the west and southwest of Sacha. During this uplift, the Cretaceous seas disappeared from the Oriente basin and the mostly continental Tena (Maastrichtian to Paleocene) deposition ensued. As the Andean Mountains to the west were uplifted, minor movements in the basin also occurred due to compression. Using the compilation of a list of hydrocarbon provinces of the world by St. John et al. (1984) the Oriente (Putumayo) basin is classified as 221 under the modified scheme of Bally and Snelson (1980), i. e., perisutural basin on a rigid lithosphere associated with formation of compressional megasuture; foredeep underlying platform sediments, or moat on continental crust adjacent to A-subduction margin; ramp with buried grabens but with little or no block-faulting.

The classification using the scheme of Klemme (1971) is II A (i. e., continental multicycle basin; craton margin-composite).

A regional structure map contoured on the base of the Cretaceous Hollin Formation (Figure 3) shows the Oriente basin to be asymmetric and south-plunging. A major fault zone separates the basin from the Andes. Between this faulting and the basinal axis are foothill structures that are dominated by the Napo and Cutucu uplifts. The northern part of the basin does not have a simple synclinal axis. The presence of low relief anticlines located on a subtle cross-basin arch has distorted the synclinal area and formed a strip about 30 mi (48 km) wide that separates the gently dipping east flank of the basin from the more steeply dipping west flank (Canfield et al., 1982). Most of the oil fields of the Oriente are located in this broad synclinal area. The synclinal axis is much better defined as it plunges southward toward Peru.

Figure 2 shows the Sacha structure at the top of the Hollin Formation. The map is based on seismic and well control. The productive area of the Hollin reservoir covers 47,350 ac (19,160 ha). The faulting on the west side of the anticline is apparently not continuous; it is recognized on some seismic lines but on others closure appears to be controlled only by abnormally "steep" dip. The fault is apparently not present in the seismic line shown by Figure 4, and only "steep" west dip has been interpreted. On the lines where faulting has been mapped, the fault

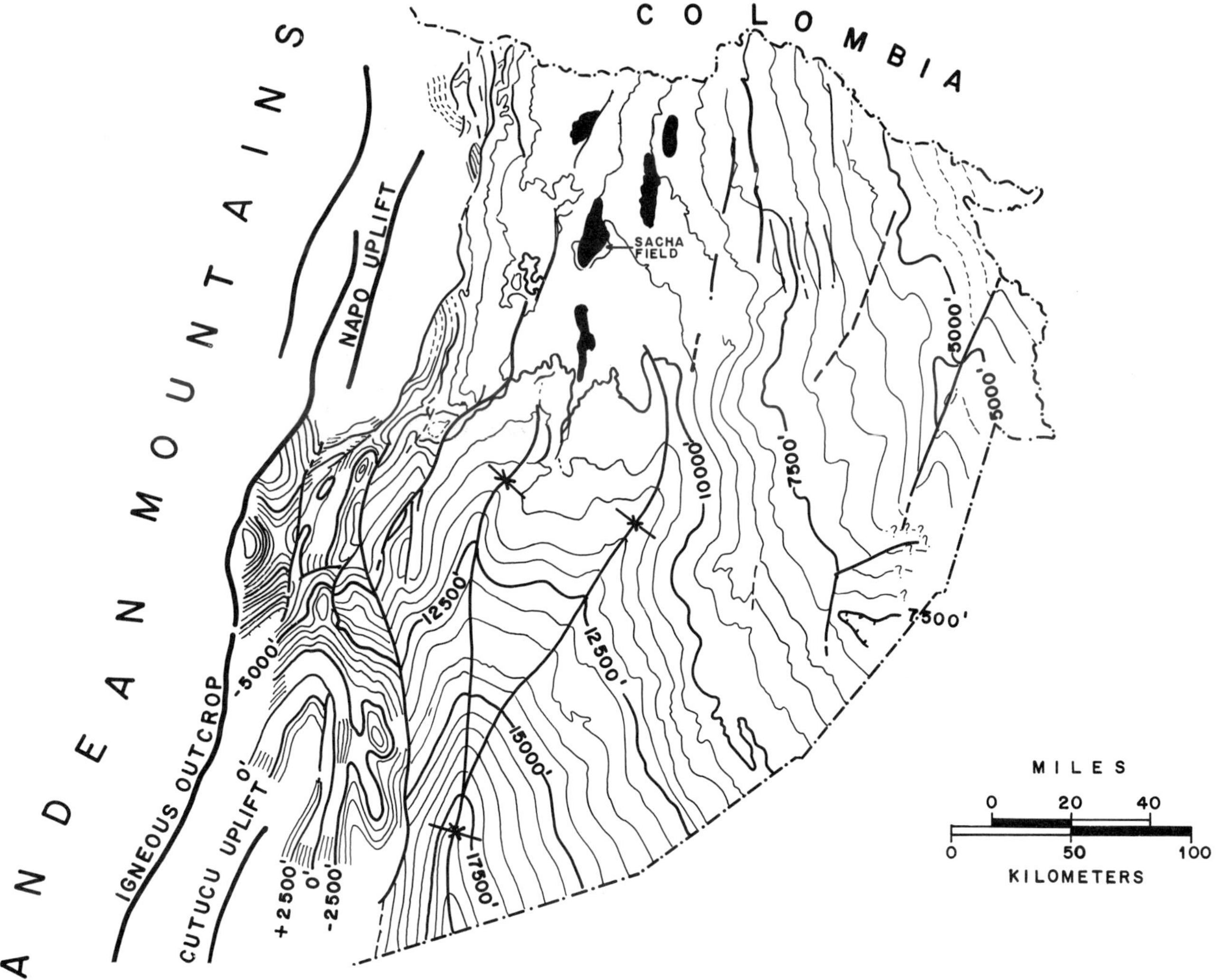

Figure 3. Map showing structural contours on base of Cretaceous Hollin Formation, Ecuadorian Oriente. C.I. 500 ft (150 m). (Modified from Canfield et al., 1982.)

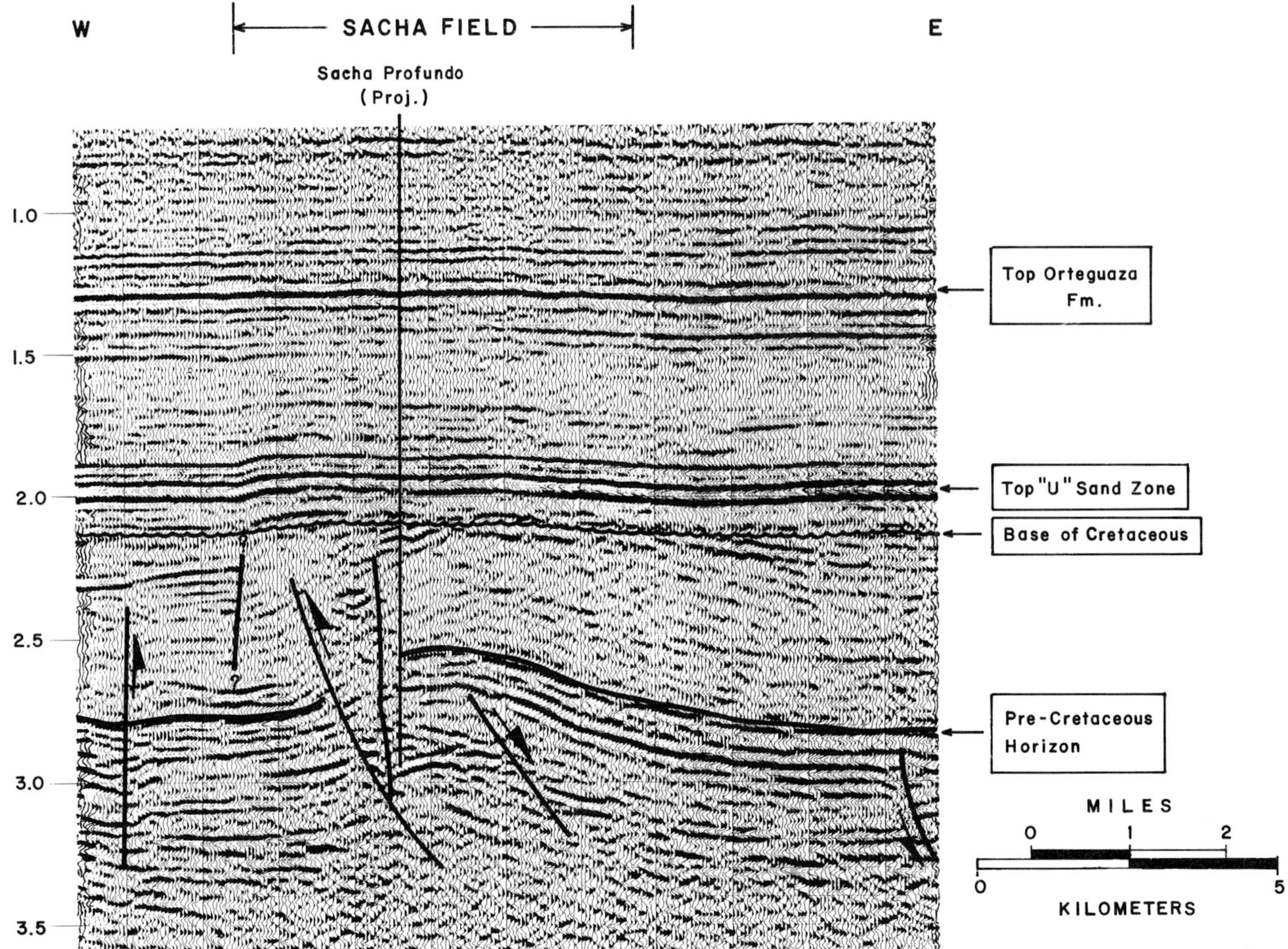

Figure 4A. East-west migrated digital seismic line 298 across Sacha structure showing (1) top of Orteguaza Formation, (2) top of Napo "U" sand zone, (3) base of Cretaceous, and (4) a pre-Cretaceous horizon. Note the lack of structural relief and faulting in the section above the base of Cretaceous unconformity as compared to the pre-Cretaceous section. The location of the line is shown by Figures 2, 5, and 6. (Modified from Alvarado et al., 1985, and Canfield et al., 1982.)

appears to be nearly vertical and to die out in the upper Napo shales. Fault movement apparently occurred during Albian to possibly Senonian time, perhaps contemporaneously with the initial forming of the Sacha Cretaceous structure. The seismic section illustrates the small amount of east-west vertical relief across the structure at the top of the productive "U" sand zone. Even less relief is present at the upper mapped horizon, the Orteguaza Formation. The pre-Cretaceous structure, however, shows a large amount of vertical relief as well as considerably more faulting than is present above the base of Cretaceous unconformity. It is noted that the deeper pool exploratory well, Sacha Profundo, was drilled about half a mile (800 m) north of the seismic line.

An isochore map of the Napo Formation (Figure 5) based on well data is essentially the same as an isopach of the formation because the dip of the strata is very low. Erosion at the top of the Napo accounts for some of the irregularities, but the map still shows that the structure was forming during Napo deposition. A second isochore map (Figure 6) of the combined Tena, Tiyuyacu, and Orteguaza formations shows that the structural relief at the end of Orteguaza deposition was nearly the same as it is today. The oil had probably been generated and trapped by that (Oligocene) time.

STRATIGRAPHY

The oldest sedimentary rocks found to date in the Oriente basin are shales of Lower Devonian to Upper Silurian age (Figure 7). These Paleozoic sediments were found at total depth of a deep test in the Shushufindi field and in the Sacha Profundo well. Shales and slates somewhat younger in age (Upper Devonian to Lower Mississippian) have been

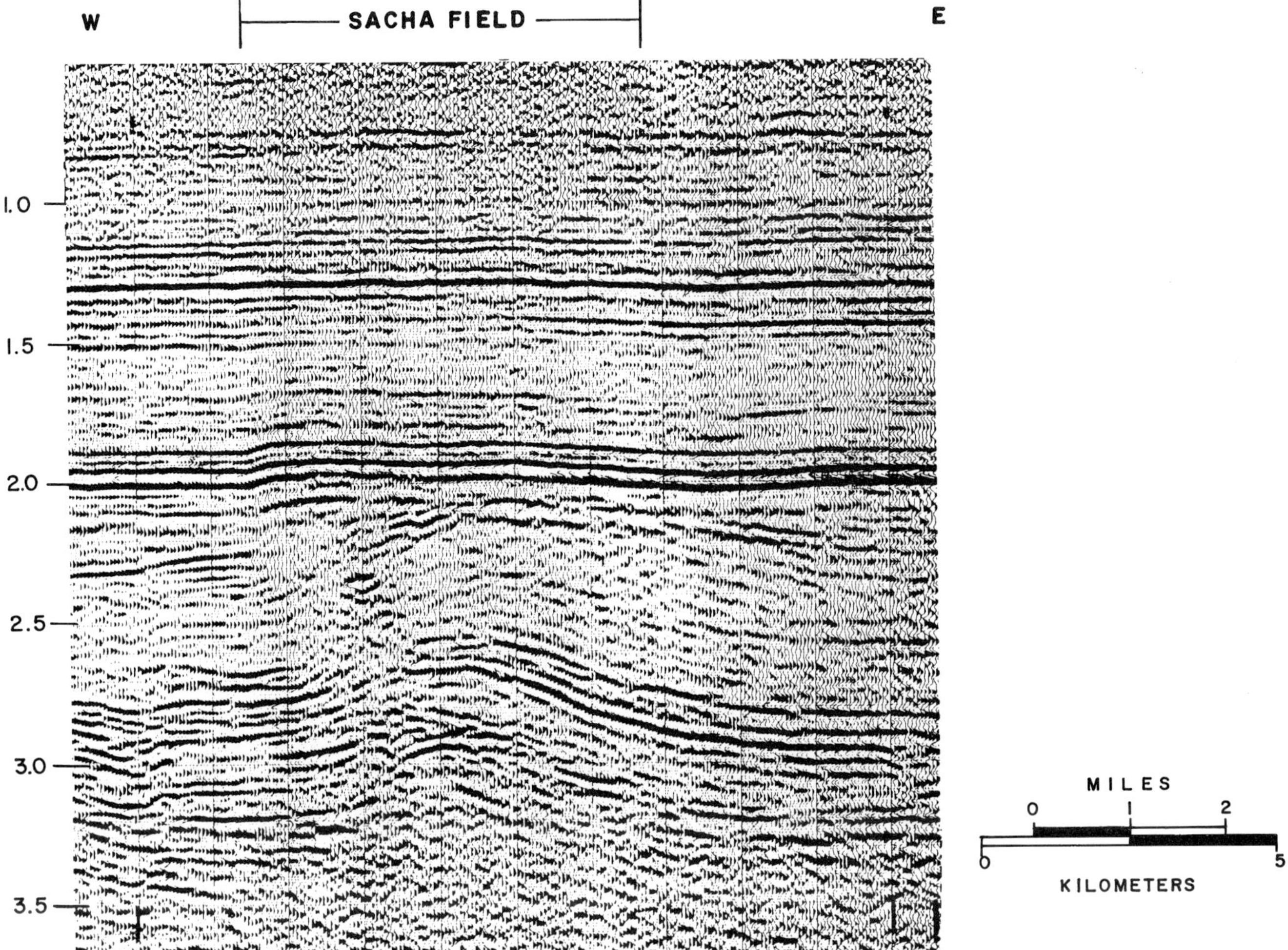

Figure 4B. Uninterpreted east-west migrated seismic line 298 across Sacha structure.

described in outcrop. Unconformably overlying these sediments, referred to as the Pumbuiza Formation, is the Pennsylvanian–Permian Macuma Formation consisting mostly of fossiliferous marine limestones with some grayish-green sandstones, gray siltstones, and thin calcareous shales. Geochemical studies of these older rocks indicate that, though at one time in their history they may have been suitable source beds, they are no longer.

The Early Jurassic Santiago Formation consists of marine limestones and shales with minor sandstones and siltstones in the upper portion. Though its presence is suspected, the Santiago Formation has not been identified in wells, perhaps because of the lack of sufficient penetration into the pre-Hollin sediments or because of its generally poor fossil content. Outcrop samples suggest that the Santiago, which has TAI values of about 2.5, is a good petroleum source.

Unconformably overlying the Santiago is a continental red-bed series, the Jurassic Chapiza Formation. The upper part of the formation usually has a volcanic sequence named the Misahualli member, which is apparently very early Cretaceous in age.

Unconformably overlying the older rocks is the transgressive Cretaceous Hollin Formation. This blanket sandstone thins northeastward toward the Guayana Shield and becomes indistinguishable from other Cretaceous sandstones. Most of the formation was apparently deposited in a fluvial braided stream environment, but near the end of Hollin deposition the basin subsided and marine conditions existed, as shown by an increase in shaliness, the presence of glauconite, and a few marine fossils.

The Napo Formation conformably overlies the Hollin and is a series of marine shales, limestones, and sandstones. Two of the regressive sandstones in the lower third of the formation, together with the Hollin sandstone, are the most important reservoirs in the Oriente. These three units account for about 95% of the Oriente production. Eastward toward the source area these sands merge and are indistinguishable from the Hollin. Younger regres-

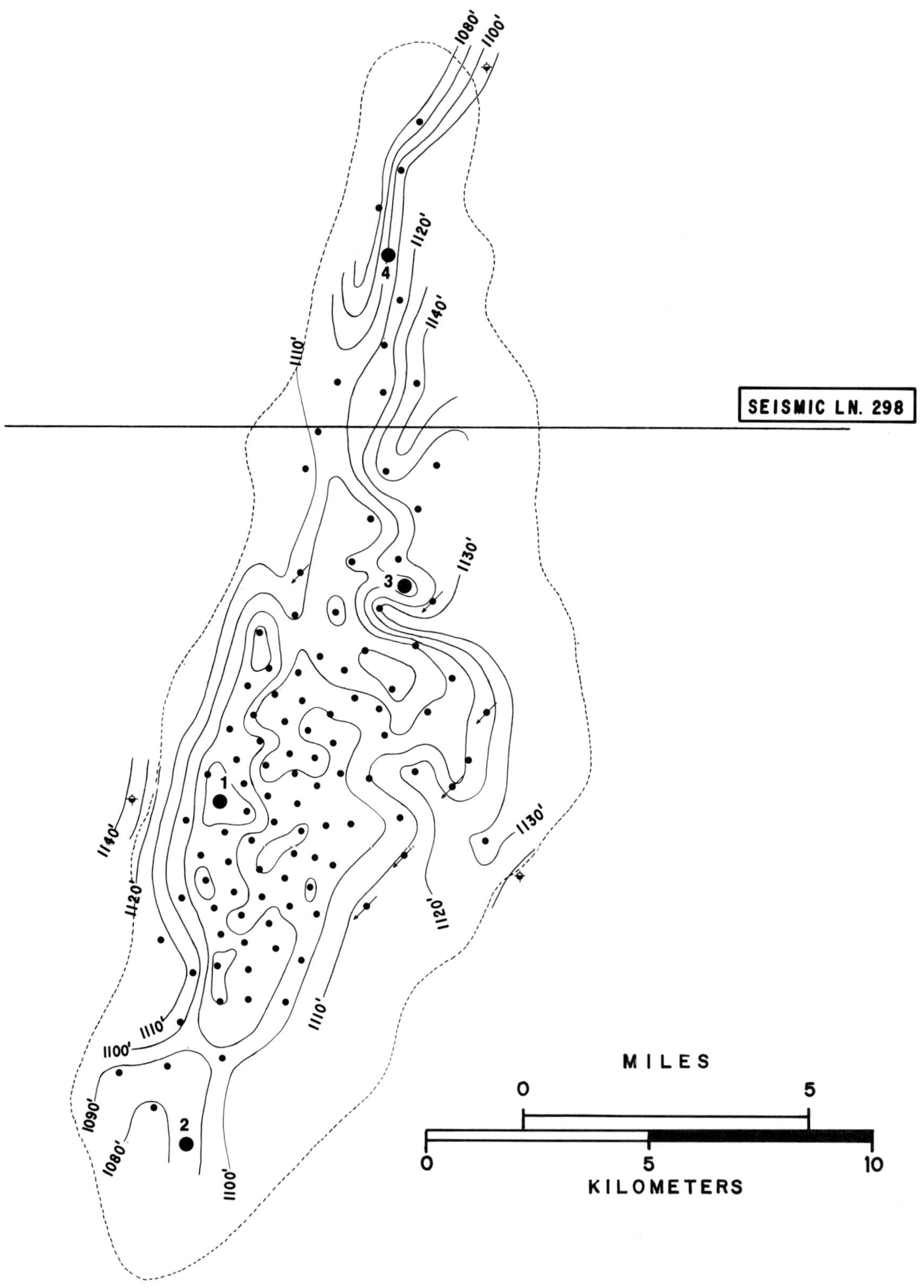

Figure 5. Isochore map of Napo Formation, Sacha field. The unconformity at the top of the formation has caused some irregularities, but structural closure is apparent. The seismic profile is shown by Figure 4. (Modified from Canfield et al., 1982.)

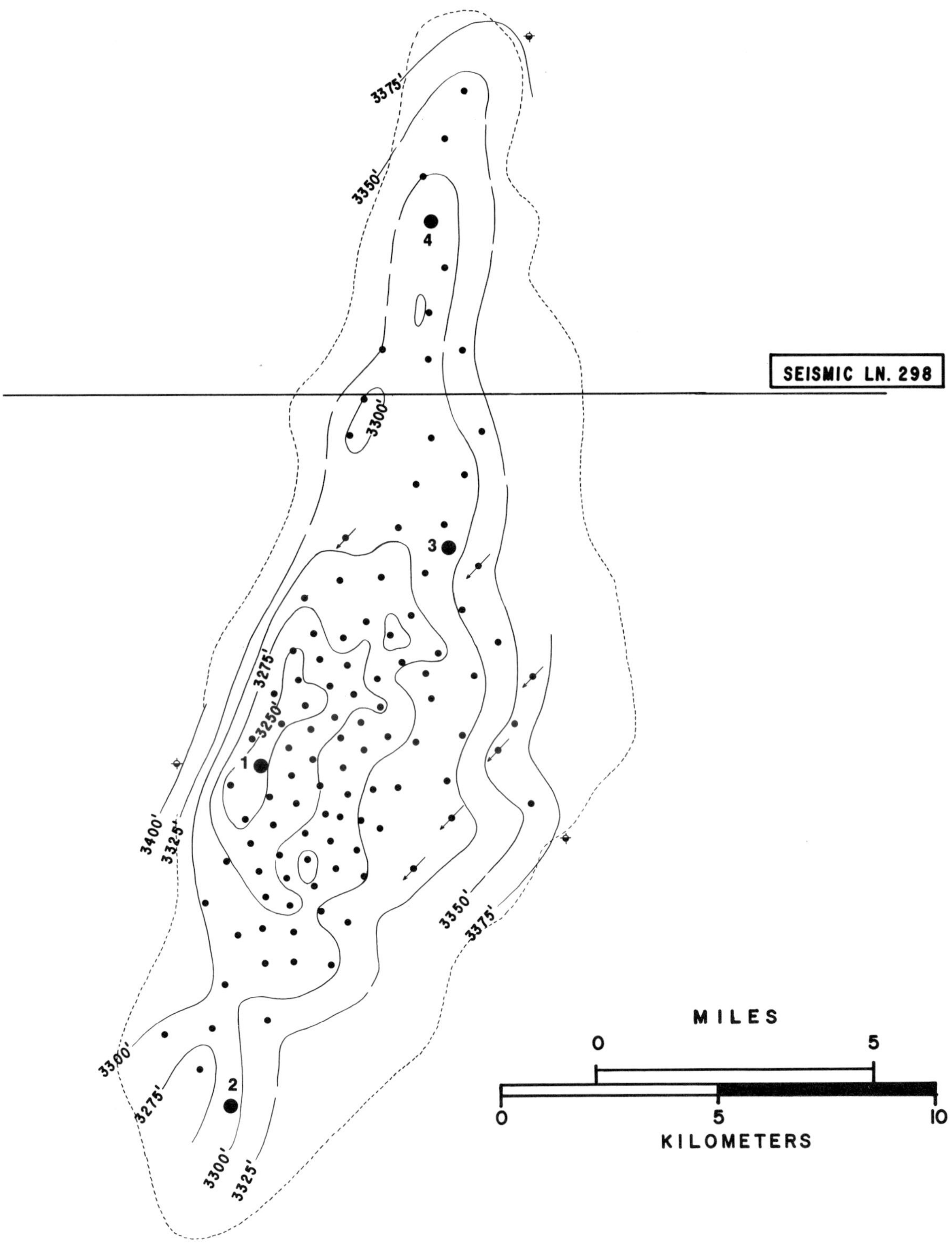

Figure 6. Isochore map of Orteguaza, Tiyuyacu, and Tena formations, Sacha field. The seismic profile is shown by Figure 4. (Modified from Canfield et al., 1982.)

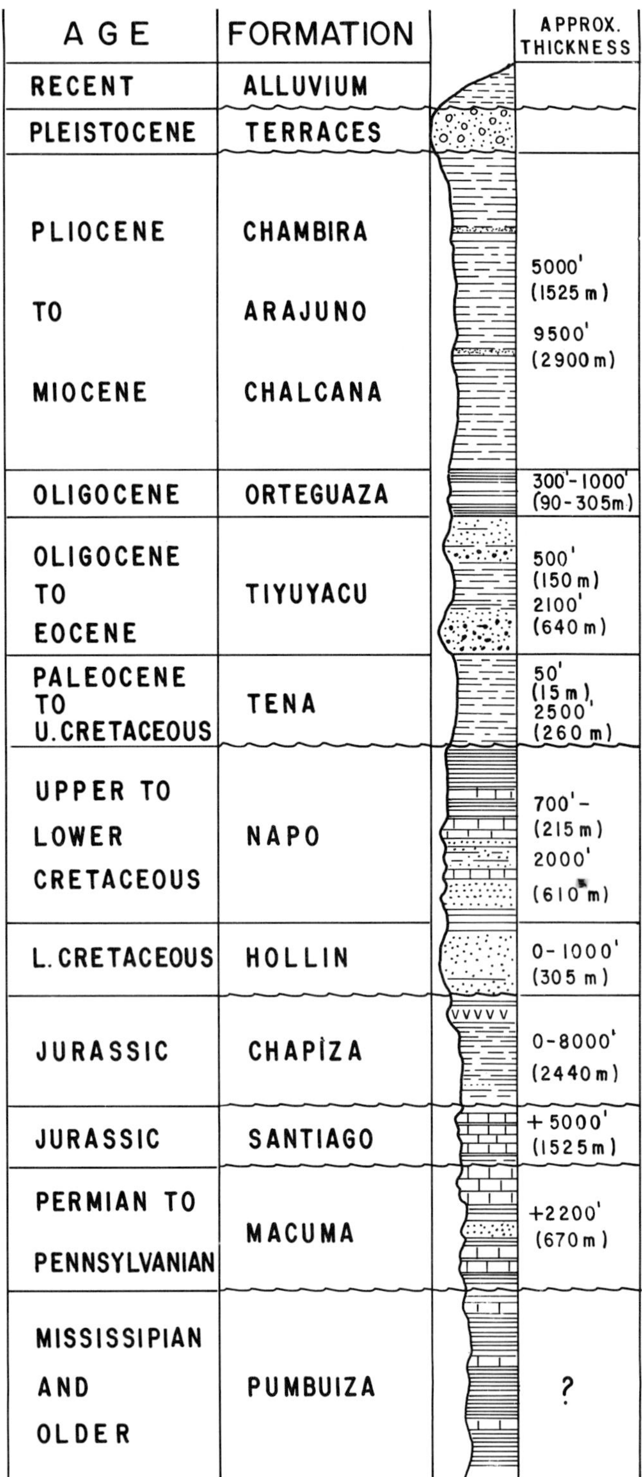

AGE	FORMATION		APPROX. THICKNESS
RECENT	ALLUVIUM		
PLEISTOCENE	TERRACES		
PLIOCENE	CHAMBIRA		5000' (1525 m)
TO	ARAJUNO		9500' (2900 m)
MIOCENE	CHALCANA		
OLIGOCENE	ORTEGUAZA		300'-1000' (90-305m)
OLIGOCENE TO EOCENE	TIYUYACU		500' (150 m) 2100' (640 m)
PALEOCENE TO U.CRETACEOUS	TENA		50' (15 m) 2500' (260 m)
UPPER TO LOWER CRETACEOUS	NAPO		700'- (215 m) 2000' (610 m)
L.CRETACEOUS	HOLLIN		0-1000' (305 m)
JURASSIC	CHAPIZA		0-8000' (2440 m)
JURASSIC	SANTIAGO		+5000' (1525 m)
PERMIAN TO PENNSYLVANIAN	MACUMA		+2200' (670 m)
MISSISSIPIAN AND OLDER	PUMBUIZA		?

Figure 7. Generalized stratigraphic column, Oriente basin, Ecuador.

sive Napo sandstones are productive in wells east of Shushufindi and may become more important objectives as exploration in the basin extends east and southeast from the known fields. The organically rich black marine shales and limestones are probably the main source of petroleum in the Oriente, though the Lower Jurassic Santiago Formation may also contribute. TAI values for the Napo source beds range from 2.0 to 3.5.

Unconformably overlying the Napo is a red-bed series named the Tena Formation. Though most of the unit consists of varicolored claystones and siltstones, a few argillaceous sandstones are usually present. A thin basal sandstone is commonly present that frequently has good porosities and permeabilities and is the uppermost producing sandstone in the Oriente of Ecuador. The basal part of the Tena is uppermost Cretaceous (Maastrichtian), but most of the unit is probably Paleocene.

The upper boundary of the Tena is marked by the abrupt appearance of the Tiyuyacu conglomerates. Though a disconformable contact is suspected, careful studies in the outcrop belt have failed to prove one. The Tiyuyacu Formation can be divided into a lower conglomerate section; a middle series of red claystones, siltstones, shales, and minor sandstones; and an upper conglomerate section. The Tiyuyacu equivalent in Colombia produces oil at the Orito field but no oil is produced from the formation in Ecuador. A very minor amount of oil was tested from the Tiyuyacu in one well in the northern Oriente. The Tiyuyacu is considered Eocene, but the uppermost portion may extend into the Oligocene.

The Tiyuyacu apparently grades into the Orteguaza Formation, which consists of gray to blue-gray shales with fine-grained sandstones in the lower part of the unit. The sandstones are locally glauconitic. Overlying the Oligocene Orteguaza are several thousand feet of Miocene to Holocene, fine to coarse continental clastics.

TRAP

Sacha field is an anticlinal trap. There is faulting on the west flank of the structure, but as stated above, the faulting is probably not continuous and is not responsible, nor needed, for oil entrapment. There may be a stratigraphic influence on the oil accumulations in the Napo "T," "U," and basal Tena sandstones, but this has not been proved by drilling. The accumulation in the main reservoir, Hollin sandstone, is structurally controlled. There is a well-defined oil-water contact having a slight northeasterly tilt as determined from well logs that penetrate the interface. Locally, the contact is affected by differences in sand permeabilities or thin shale beds.

An east-west structural section (Figure 8) across the field illustrates thickness variations and local discontinuity of sand in the Napo "T," "U," and upper Hollin. The original oil-water contact in the main Hollin reservoir is also shown. Connate water differences, i. e., saline or fresh, are also indicated on the section.

The Cretaceous folding of the Sacha anticline occurred over an older north–south-oriented anticline. Part of the folding is probably due to drape,

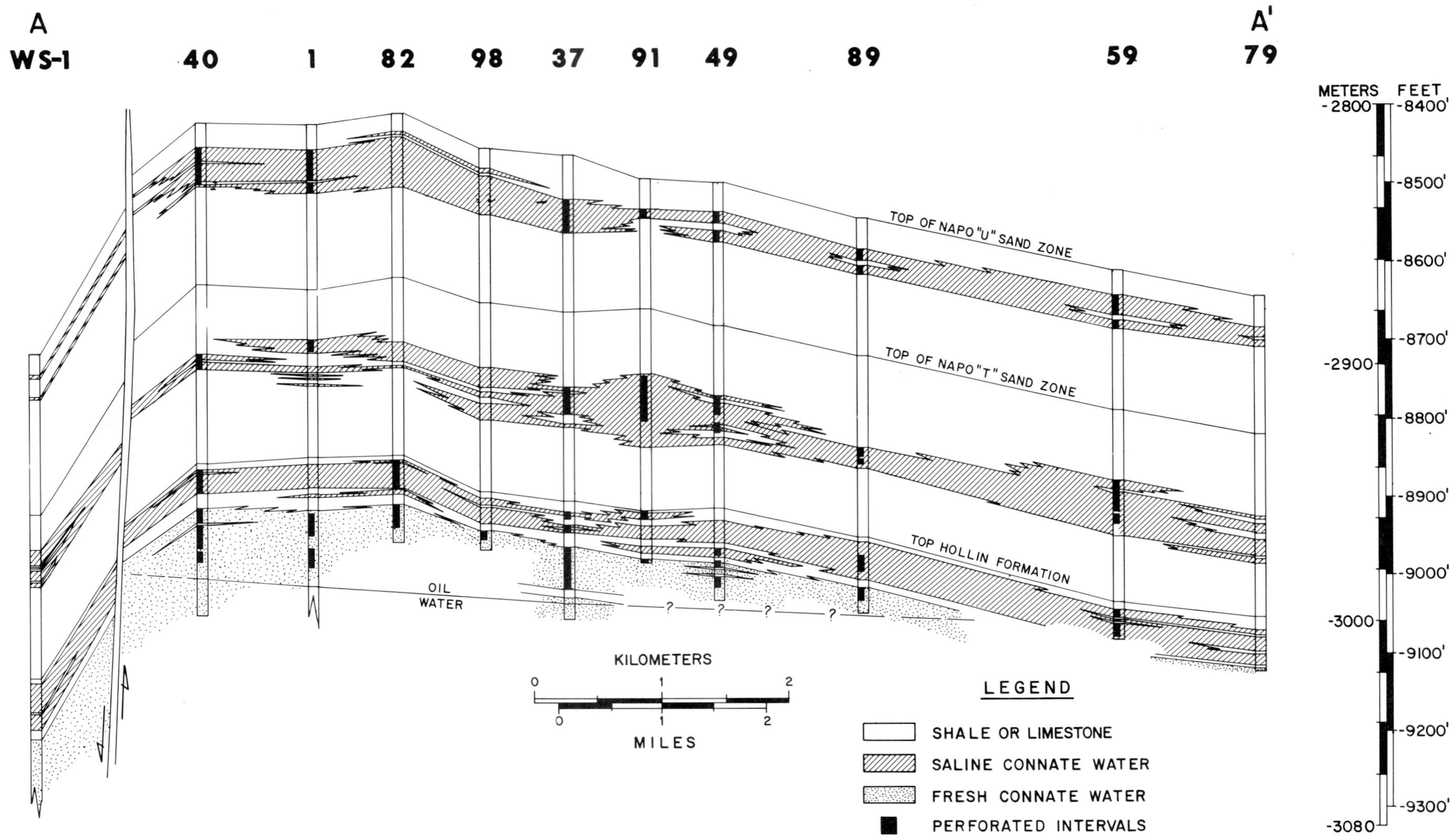

Figure 8. East-west structural and stratigraphic section of lower Napo and upper Hollin formations, Sacha field. Easternmost well was abandoned without testing as Hollin and Napo "T" sands were "tight" and Napo "U" sand was thin and oil was not movable according to log calculations. The location of the cross section is shown by A-A' in Figure 2.

but compressional forces were also apparently active. Trapping of the oil was essentially completed by Orteguaza (Oligocene) time.

RESERVOIRS

As stated earlier, there are four producing sands in the Sacha field. These reservoirs are shown on the typical electric log (Figure 9) and will be discussed individually.

Hollin Reservoir

Of the 743 million bbl of recoverable oil reserves in Sacha, nearly 454 million or 61% are assigned to the Hollin Formation. The estimated recovery factor is 30%. About 90% of Hollin reserves are in the main freshwater-bearing sandstone. This thick transgressive sandstone was apparently deposited in a fluvial braided stream environment. Near the end of Hollin time, however, the basin subsided and marine conditions existed during deposition of the upper sandstones.

The main Hollin sandstone at Sacha is composed mostly of fine- to medium-grained quartzose sandstones. X-ray analyses have indicated small amounts of interstitial silt (±4%) and clay (±3%). Most of the clay is composed of authigenic kaolinite and quartz with minor amounts of montmorillonite and illite. Porosities average about 16.5%, and both horizontal and vertical permeabilities vary widely but average about 500 md. A representative relative permeability curve is shown in Figure 10. A very strong freshwater drive exists in the reservoir. The original reservoir pressure was 4426 psi, and at the end of 1987, after 14½ years of production, it was still about 4200 psi. The gravity of the oil is approximately 29.5° API with negligible sulfur content. The GOR is less than 10.

The marine upper Hollin sandstones also produce oil but much less than the underlying main reservoir. About 50% of the upper Hollin consists of resistive shales, "tight" hard siltstones, and/or arenaceous limestones. The sandstones are generally glauconitic to very glauconitic, lowering the resistivities and possibly leading to inaccurate log calculations (Canfield et al., 1982). Lateral facies changes occur rapidly, as can be seen on the accompanying cross section (Figure 8). The upper sands contain saline connate water, similar to that of the overlying Napo sandstones, as opposed to the freshwater main Hollin sandstones. Porosities of perforated sandstones average about 12% and permeabilities range from very low to about 350 md.

Hollin production covers 47,350 ac (19,160 ha) and amounted to 244.9 million bbl or about 72.5% of the total Sacha production through 1987. At the end of 1987, there were six flowing Hollin wells, 40 producing using hydraulic pumps, and nine using electric submersible pumps.

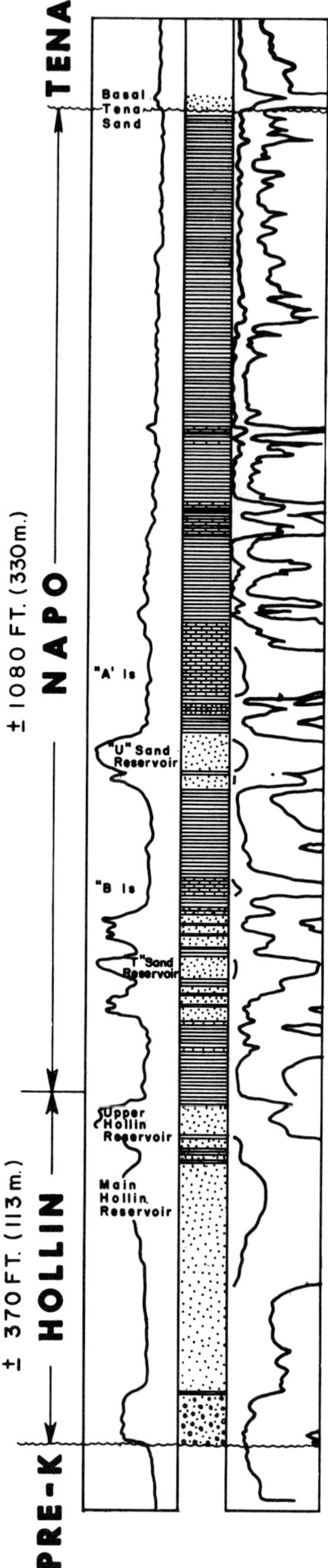

Figure 9. Electric log of Hollin and Napo formations with reservoirs identified, Sacha field. (Left side, SP curve. Right side, 16-in. normal. Scales: 0–10, 0–50, and 0–500 ohms · m²/m.)

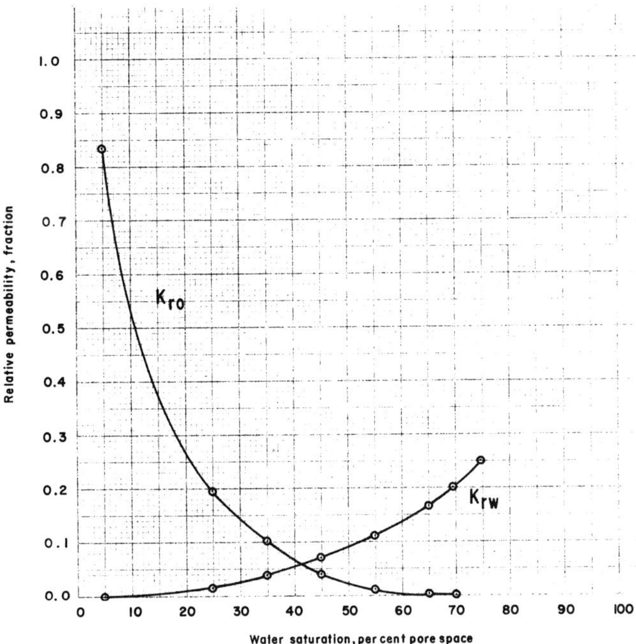

Figure 10. Representative relative permeability curves, Hollin sand, Sacha field.

Figure 11. SEM photo of open-packed, "clean," well-sorted Napo "T" sand cemented by authigenic quartz and containing locally abundant pore plugging kaolinite (K). Note the open porosity (P). Porosity of sample is 17.7%.

Napo "T" Sand Reservoir

Not all of the sands in the "T" sand zone are shown on Figure 8. Most of the upper glauconitic to very glauconitic marine sandstones have been purposely omitted from the figure because they show limited continuity and are only very rarely perforated and tested. Though they are oil-bearing, their porosities and permeabilities are quite low and no recoverable reserves have been assigned to them.

The lower or main "T" sand is usually gray to tan but locally may be greenish owing to the presence of glauconite. The sandstones are usually fine to medium grained but locally coarse grained, subangular to subrounded and quartzose. Sorting is usually good. The sandstones are mostly continental and conventional cores of them frequently have plant remains and locally abundant amber. X-ray diffraction analyses indicate clay content varies up to about 2.5%. About 85% of the clay is composed of kaolinite with the remaining being mostly illite. Figure 11 is a SEM photograph showing a typical Sacha "T" sand having good open packing and porosity with locally abundant pore-plugging authigenic kaolinite. Though kaolinite is present, there have been no problems encountered that can be attributable to fine kaolinite movement in the process of injecting treated surface water for reservoir flooding.

Porosities of the producing sandstones generally range from 12% to about 23% and average about 15%. Permeabilities also vary over a wide range (from 1.6 to over 1500 md) but average about 300 md. A representative relative permeability curve is shown in Figure 12. Oil gravity is about 31° API, again with negligible sulfur. The GOR is about 325. "T" sand production covers approximately 19,825 ac (8025 ha) and recoverable oil reserves are estimated to be 93.7 million bbl or about 12% of the Sacha field total. The recovery factor for "T" sand production approaches 34%. Through 1987, the reservoir had produced 38.3 million bbl or approximately 11% of total production. At the end of 1987 four wells were producing solely from the "T" sand reservoir—one by an electrical submersible pump and three by hydraulic pumps.

Napo "U" Sand Reservoir

Though the "U" and "T" sand zones are very similar, the sands of the "U" zone, although fewer in number, are generally more continuous. The "U" zone is a more important reservoir. Nearly all of the production comes from the lowermost and thickest main "U" sand reservoir. The upper, glauconitic sandstones are only rarely perforated because they have lower porosities and permeabilities, frequently due to the presence of silt, argillaceous material, and/or calcareous cement.

The main "U" sand is light gray to tan, very fine to medium grained and occasionally coarse grained, subrounded to subangular, and quartzose. Minor shale stringers locally having very small amounts of pyrite have been described in cores. Sorting varies from poor to fairly good. The main sandstone is considered to be continental and frequently contains plant remains. X-ray diffraction analyses indicate the principal clay mineral to be kaolinite, with the second most important usually being montmorillonite but sometimes very fine quartz or illite. Figure 13 is a SEM photograph showing a typical Sacha "U" sand having authigenic kaolinite, particularly noticeable in the upper right-hand portion of the photograph.

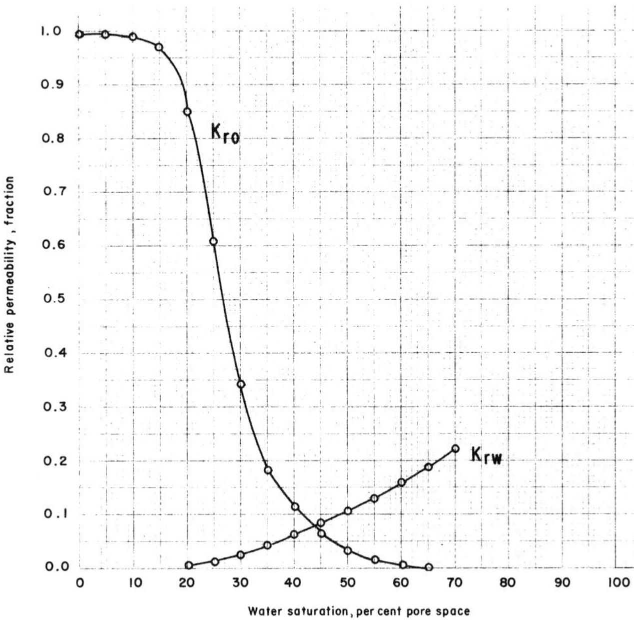

Figure 12. Representative relative permeability curves, Napo "T" sand, Sacha field.

Figure 13. SEM photo of open-packed, silica-cemented Napo "U" sand with abundant authigenic kaolinite (k). Open porosity (P) is again noteworthy. The sorting of the sample is much poorer than that of the Napo "T" sand shown in Figure 11. Porosity of sample is 18%. Though kaolinite is present in both sands, there has been no problem encountered in injecting treated surface water for reservoir flooding.

Flooding the main "U" sand with treated surface water has encountered no problems caused by the clay content.

The porosity of the main "U" sandstone varies from well to well and vertically in individual well bores. Conventional core analyses generally yield porosities in the range from 12.5 to 26%, averaging about 18%. Permeabilities also vary considerably but average about 520 md. A representative relative permeability curve is shown in Figure 14. Oil gravity averages 27° API and the GOR is about 220. "U" sand production covers some 35,250 ac (14,265 ha), and no oil-water contact has been observed on electric logs nor proved by tests. Recoverable oil reserves amount to 183.3 million bbl or nearly 25% the Sacha's total. The recovery factor is only about 21% for "U" sand production because of the lower gravity of the crude, especially in flank wells. Through 1987, the "U" sand had produced 50.8 million bbl of oil or some 15% of total production. At the end of 1987 there were 11 wells producing solely from the "U" sand and 13 wells producing from the combined "U" and "T" sands. One of these wells was producing using an electrical submersible pump whereas 23 were using hydraulic pumps.

Basal Tena Sand Reservoir

The basal Tena sand is irregularly distributed disconformably over upper Napo Formation shales and, where present, varies in thickness from a few feet up to about 25 ft (7.5 m). The sand is fine to medium grained, angular to subangular, quartzose, and commonly argillaceous. It is usually gray to tan but sometimes reddish owing to a locally high clay content. Where the sand is well developed, its porosities can be as high as 23%, but normally the porosity is about 18% as calculated from electric logs. No conventional cores have been taken in the unit at Sacha. Average permeability is 220 md.

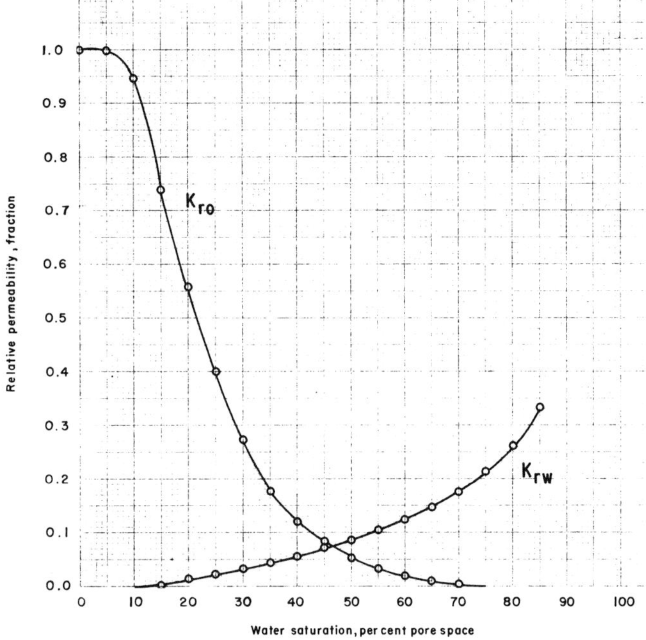

Figure 14. Representative relative permeability curves, Napo "U" sand, Sacha field.

298

Proved basal Tena sand recoverable reserves of 12.7 million bbl oil come from four separate sand bodies covering 7960 ac (3220 ha) in the central portion of the field. Additional sandstone lenses are present in the northern part of the field that calculate to be oil-bearing but have not been tested and no reserves have been assigned them. At the end of 1987 only six Sacha wells were producing from the unit, five by hydraulic pumps and one by using an electrical submersible pump. Through 1987, 2.5 million bbl of about 26° API oil had been recovered from the basal Tena sand at Sacha.

SOURCE

The source for the oil found at Sacha and other fields in the northern Oriente appears to be the Napo shales and limestones. The formation correlates with the La Luna Formation of northern Colombia and western Venezuela, long known to be an excellent source for hydrocarbons. The Napo Formation, however, is generally immature in the producing area of the northern Oriente and only reaches the onset of oil generation much deeper in the basin toward Peru. Feininger (1975) proposed that the Cretaceous source beds were to the west of the present-day basin and have since been metamorphosed and are no longer recognizable because of the later Andean mountain building. There is also a possibility that the source beds lie to the north in Colombia where a thick Cretaceous section exists. Additional source studies are required. In any case migration of the oil must have taken place by Oligocene time and apparently over a considerable (possibly 200 to 300 km) distance. Since Oligocene time remigration probably occurred with renewed tectonic activity.

There is also the possibility that the Jurassic Santiago Formation contributed to the Oriente oil. The formation has not been defined in the subsurface but outcrop samples have TAI values of about 2.5, which is very similar to those of the Napo Formation.

The Napo shales and limestones have an average TOC of about 2.5% near the producing fields but locally well over 6.0% TOC has been recorded (Rivadeneira, 1986). Generally the higher TOC values are found to the west. The kerogen type is probably II.

EXPLORATION AND DEVELOPMENT CONCEPTS

The principal producing fields in the Ecuadorian Oriente are low relief anticlines showing early structural growth. Most of these fields lie in the broad axial region on a subtle cross-basin arch (Figure 3). Apparently a short distance south of the southern-most producing field, Cononaco, the effects of the cross-basin arch are diminished as the synclinal axis becomes much better defined where the basin plunges southward toward Peru (Canfield et al., 1985).

It is apparent from the seismic section (Figure 4) that there was structural deformation prior to Cretaceous deposition at Sacha. Most of the oil fields in the basin are also located on early positive structures. Seismic exploration is now concentrated on locating anomalies overlying pre-Cretaceous structural trends and showing growth during the Cretaceous.

Since the lower producing reservoir at Sacha, the Hollin, has good vertical (and horizontal) permeability, the completion program had to be changed shortly after production was initiated because of the influx of formation water. The present program of penetrating the reservoir only a few feet and staying well above the oil-water interface has slowed down water production considerably. Also, other production techniques have been employed to help restrict water production.

ACKNOWLEDGMENTS

Texaco Petroleum Company, a subsidiary of Texaco International Ltd., and Corporación Estatal Petrolera Ecuatoriana allowed the publication of this paper. Sacha Oil Field of the Ecuadorian Oriente by Canfield et al. (1982) and unpublished studies by Texaco Petroleum Company, the operator of the Sacha field, were referred to for the preparation of the paper. The exhibits were done by Texaco Petroleum's drafting section.

REFERENCES

Alvarado, A., J. Cisneros, and M. Santos, 1985, Modelos Geológicos del Pre-Cretácico del Oriente Ecuatoriano: Presented in "Exploración y desarrollo de hidrocarburos en horizontes Pre-Cretácico" at LV reunión a nivel de expertos de "Arpel" held in Quito, Ecuador, in March 1985.

Canfield, R. W., G. Bonilla, and R. K. Robbins, 1982, Sacha oil field of Ecuadorian Oriente: American Association of Petroleum Geologists Bulletin, v. 66, p. 1076–1090.

Canfield, R. W., R. N. Diaz, and J. Montenegro, 1985, El campo Cononaco del Oriente Ecuatoriano: Paper presented at II Simposio Bolivariano "Exploración petrolera en las cuencas subandinas" at Bogotá, Colombia, August 1985.

Carmalt, S. W., and B. St. John, 1986, Giant oil and gas fields, in M. T. Halbouty, ed., Future petroleum provinces of the world: American Association of Petroleum Geologists Memoir 40, p. 11–54.

Feininger, T., 1975, Origin of petroleum in the Oriente of Ecuador: American Association of Petroleum Geologists Bulletin, v. 59, p. 1166–1175.

Rivadeneira, M. R., 1986, Evaluación Geoquimica de Rocas Madres de la Cuenca Amazónica Ecuatoriana: Paper presented at II Congreso Andino de Petróleo at Bogotá, Colombia, October 1986.

Rosania, G., H. F. San Martin, and R. W. Canfield, 1983, Geologia de los campos gigantes en el Oriente Ecuatoriano Petróleo Internacional, v. 41, n. 3, p. 34–41.

Sinclair, J. H., and T. Wasson, 1922, Exploration in eastern Ecuador: The Geographical Review, v. 13, p. 190–210.

St. John, B., A. W. Bally, and H. D. Klemme, 1984, Sedimentary provinces of the world—hydrocarbon productive and nonproductive: American Association of Petroleum Geologists Map Series, p. 1–35.

Trujillo, B., E. Mejia, and J. Huacho, 1985, Evaluación geólogica, hidro-carburifera del pre-Cretacico: Presented in "Exploración y desarrollo de hidrocarburos en horizontes Pre-cretácicos" at LV reunión a nivel de expertos de "Arpel" held in Quito, Ecuador, in March 1985.

Tschopp, H. J., 1953, Oil exploration in the Oriente of Ecuador: American Association of Petroleum Geologists Bulletin, v. 37, p. 2303–2347.

Wasson, T., and J. H. Sinclair, 1927, Geological explorations east of the Andes in Ecuador: American Association of Petroleum Geologists Bulletin, v. 11, p. 1253–1281.

SUGGESTED READING

Canfield, R. W., G. Rosania, and H. F. San Martin, 1982, Geologia de los campos gigantes del Oriente Ecuatoriano: Paper presented at Simposio "Exploración Petrolera en las cuencas subandinas de Venezuela, Colombia, Ecuador y Perú," held in August 1982 in Bogotá, Colombia. Describes other major oil fields in eastern Ecuador and is more complete than referenced publication of Rosania et al. (1983).

Tschopp, H. J., 1953, Oil exploration in the Oriente of Ecuador: American Association of Petroleum Geologists Bulletin, v. 37, p. 2303–2347. Good description of Oriente stratigraphy.

Appendix 1. Field Description

Field name .... *Sacha field*

Ultimate recoverable reserves .... *743.346 million bbl*

Field location:

    Country .... *Ecuador*

    State .... *Napo*

    Basin/Province .... *Oriente*

Field discovery:

    Year first pay discovered .... *Cretaceous Hollin Formation sandstone 1969*

    Year second pay discovered .... *Cretaceous Napin sandstone "T" 1969*

    Third pay .... *Cretaceous Napin sandstone "U" 1969*

    Fourth pay .... *Paleocene to Cretaceous Tena sandstone 1971*

Discovery well name and general location:

    First pay .... *Sacha-1, 11.9 mi (19 km) northeast of the confluence of the Coca and Napo rivers*

    Second pay .... *Sacha-3, 4 mi (6.4 km) northeast of Sacha-1*

    Third pay *Sacha-3*

    Fourth pay .... *Sacha-16, 2.5 mi (4 km) N15°E of Sacha-1*

Discovery well operator .... *Texaco*

    Second pay .... *Texaco*

    Third pay .... *Texaco*

    Fourth pay .... *Texaco*

IP in barrels per day and/or cubic feet or cubic meters per day:

    First pay .... *1348 BOPD, ¼-in. choke*

    Second pay .... *605 BOPD, ¼-in choke*

    Third pay .... *480 BOPD, ¼-in. choke*

    Fourth pay .... *250 BOPD, ¼-in. choke*

All other zones with shows of oil and gas in the field:

Age	Formation	Type of Show
Cretaceous	*Napo "A" Limestone*	*Fair to good oil show*

Geologic concept leading to discovery and method or methods used to delineate prospect, e.g., surface geology, subsurface geology, seeps, magnetic data, gravity data, seismic data, seismic refraction, nontechnical:

Oil had been discovered to the north in Colombia near the synclinal axis. The Ecuadorian acreage was obtained by concession and a seismic grid was shot using analog and, later, digital instruments to locate prospective structures in the broad synclinal area.

Structure:

    Province/basin type .... *Bally 221, Klemme IIA*

    Tectonic history

    Tertiary uplift of Andean Mountains formed south-plunging Oriente basin. The basin is asymmetric with west flank much steeper than gently dipping (+2°) east flank.

    Regional structure

    Field is located in broad synclinal region of the north-central Oriente basin on a subtle cross-basin arch.

    Local structure

    North–south-trending anticline having four-way closure. West flank is steeper than east flank and is partially faulted.

Trap:

Trap type(s)

Field is a low relief anticlinal trap having multiple pays (three excellent pays and one minor). The field is partially faulted on west flank but faulting does not affect oil accumulation.

Basin stratigraphy (major stratigraphic intervals from surface to deepest penetration in field):

Chronostratigraphy	Formation	Depth to Top in ft (m)
Pliocene–Miocene	*Undivided*	*At or near surface*
Oligocene	*Orteguaza*	*5400 (1645)*
Oligocene–Eocene	*Tiyuyacu*	*5925 (1805)*
Paleocene to Late Cretaceous	*Tena*	*7750 (2360)*
Late to Early Cretaceous	*Napo*	*8675 (2645)*
Early Cretaceous	*Hollin*	*9750 (2970)*
Jurassic	*Chapiza*	*10,125 (3085)*
Devonian	*Pumbuiza*	*13,230 (4030)*

Location of well in field

Reservoir characteristics:

Number of reservoirs 4

Formations .... *Tena (basal), Napo, Hollin*

Ages .... *Tena (Maastrichtian), Napo (Cenomanian), Hollin (Albian)*

Depths to tops of reservoirs .... *Basal Tena, 8650 ft (2635 m); Napo "U" sand, 9325 ft (2840 m); Napo "T" sand, 9550 ft (2910 m)*

Gross thickness
(top to bottom of producing interval) .... *Basal Tena, 25 ft (8 m); "U" sand, 100 ft (30 m); "T" sand, 120 ft (37 m); Hollin, 125 ft (38 m)*

Net thickness—total thickness of producing zones

Average .... *Basal Tena, 9 ft (3 m); "U" sand, 25 ft (8 m); "T" sand, 19 ft (6 m); Hollin, 39 ft (12 m)*

Maximum .... *Basal Tena, 18 ft (5 m); "U" sand, 77 ft (23 m); "T" sand, 105 ft (32 m); Hollin, 117 ft (36 m)*

Average
Maximum

Lithology

Basal Tena: fine- to medium-grained, commonly argillaceous, quartzose sandstone
Napo "U" sand: very fine to medium-grained, poorly to fairly well sorted quartzose sandstone
Napo "T" sand: fine- to medium-grained, well-sorted quartzose sandstone
Hollin: fine- to medium- and locally coarse grained, fairly well sorted, friable quartzose sandstone

Porosity type .... *Intergranular porosity, locally fractured*

Average porosity .... *Basal Tena, 18%; "U" sand, 18%; "T" sand, 15%; Hollin, 16.5%*

Average permeability .... *Basal Tena, 220 md; "U" sand, 520 md; "T" sand, 300 md; Hollin, 500 md*

Seals:

Upper ..... *Tena claystone overlies basal Tena sand; Napo shales overlie other three producing horizons*

Formation, fault, or other feature
Lithology

Lateral .... *Anticlinal fold*

Formation, fault, or other feature
Lithology

Source:

Formation and age .... *Cretaceous Napo (?)**

**Note: In addition to Cretaceous Napo Formation, there is a possible source from Jurassic Santiago Formation. The Napo shales and limestones become thicker to south and west in basinal deep (±18,000 ft; 5485 m) and are thought to be source. Actual "fingerprinting" of source has not been done.*

Lithology .... *Shale and limestone*
Average total organic carbon (TOC) .... *2.5 wt%*
Maximum TOC .... *>6.6 wt%*
Kerogen type (I, II, or III) .... *probably II*
Vitrinite reflectance (maturation) .... $R_o = 0.873–0.976$
Time of hydrocarbon expulsion
Present depth to top of source
Thickness
Potential yield

Appendix 2. Production Data

Field name .... *Sacha field*

Field size:

    Proved acres .... *Basal Tena, 7960 ac (3220 ha); "U" sand, 35,250 ac (14,265 ha); "T" sand, 19,825 ac (8025 ha); Hollin, 47,350 ac (19,160 ha)*

    Number of wells all years *110*

    Current number of wells .... *89 at end of 1987 in production*

    Well spacing .... *70 and 87 ac on top of structure*

    Ultimate recoverable .... *743.346 million bbl*

    Cumulative production (through 1987) .... *337.930 million bbl*

    Annual production (1986) .... *22.475 million bbl*

    Present decline rate (1979–1987) *4%*

        Initial decline rate (1972–1978) *29%*

        Overall decline rate *12%*

    Annual water production (1986) *8,148,451*

    In place, total reserves *2,711.496*

    In place, per acre-foot .... *3445 bbl*

    Primary recovery .... *574.914 million bbl*

    Secondary recovery .... *168.432 million bbl*

    Enhanced recovery

    Cumulative water production .... *197.324 million bbl*

Drilling and casing practices:

    Amount of surface casing set .... *2300 ft (700 m)*

    Casing program .... *10¾-in. to 2300 ft; 7-in. to TD with DV tool at ±7500 ft (2285 m)*

    Drilling mud .... *Water-based polymer, L.S., L.W. 9.8 #/gal, vis. 42 sec.*

    Bit program .... *13¾-in. DSJ (1), 9⅞-in. SD (3), SDGH (2), DSJ (4), F-3 (2), 4 13J (1), 4 JS (1), F-2 (1) total of 149⅞-in*

    High pressure zones .... *None*

Completion practices:

    Interval(s) perforated .... *Hollin, 30 ft (9 m); "U" sand, 40 ft (12 m); "T" sand, 20 ft (6 m)*

    Well treatment .... *None*

Formation evaluation:

    Logging suites .... *1:200 ISF-GR, MSFL-GR, merged MSFL-ISF-GR, FDC-GR; 1:500 ISF-GR (linear)*

    Testing practices .... *Zones are tested individually open to flow or evaluated with a jet pump; final completion is dependent upon test results and government authorization*

    Mud logging techniques .... *None on development wells*

Oil characteristics: *(Mixture of crude oil from all four reservoirs except where indicated)*

Type .... *Mixture naphthenic-paraffinic*
API gravity .... *29.9 C (60°F)*
Base .... *KUP 12–13*
Initial GOR .... *8 SCF/STO at 60°F**
Sulfur, wt% *0.99*
Viscosity, SUS .... *2.97 cp at saturation pressure**
**Hollin only.*
Pour point *3.5*
Gas-oil distillate .... *18% by volume*

Initial GOR .... *328 SCF/STO at 60°F**
Sulfur, wt% *0.99*
Viscosity, SUS .... *0.80 cp at saturation pressure**
**Napo "T" sand only.*
Pour point *3.5*
Gas-oil distillate .... *18% by volume*

Initial GOR .... *246 SCF/STO at 60°F**
Sulfur, wt% *0.99*
Viscosity, SUS .... *1.53 cp at saturation pressure**
**Napo "U" sand only.*
Pour point *3.5*
Gas-oil distillate .... *18% by volume*

Field characteristics: *(Hollin data)*

Average elevation .... *–8975 ft*
Initial pressure .... *4426 psi*
Present pressure .... *4203 psi*
Pressure gradient .... *0.37 psi/ft*
Temperature .... *195°F at 9891 ft*
Geothermal gradient .... *0.012°F*
Drive .... *Water drive*
Oil column thickness .... *38 ft*
Oil-water contact .... *–9055 ft*
Connate water *27.1%*
Water salinity, TDS .... *1100 ppm Cl⁻ (main Hollin)*
Resistivity of water .... *1.1 ohm-m at 195°F*
Bulk volume water (%) *4.4%*

Transportation method and market for oil and gas:
Trans-Ecuadorian Pipeline System to Ecuadorian coast, tankers to markets of South Korea, Taiwan, Chile, and United States.

Field characteristics: *(Napo "T" sand data)*

Average elevation .... *–8765 ft*
Initial pressure .... *4131 psi*
Present pressure (12/31/1987) .... *2112 psi*
Pressure gradient .... *0.37 psi/ft*
Temperature .... *191°F at 9781 ft*
Geothermal gradient .... *0.012°F*
Drive .... *Fluid expansion and partial water encroachment*
Oil column thickness .... *19 ft*
Oil-water contact .... *–8820 ft*
Connate water *20%*

    Water salinity, TDS .... *17,400 ppm Cl⁻ (average)*
    Resistivity of water .... *0.136 ohm-m at 191°F*
    Bulk volume water (%) *3%*

Transportation method and market for oil and gas:
See Hollin.

Field characteristics: *(Napo "U" sand data)*
    Average elevation .... *–8530 ft*
    Initial pressure .... *4028 psi*
    Present pressure (12/31/1987) .... *2205 psi*
    Pressure gradient .... *0.37 psi/ft*
    Temperature .... *188°F at 9546 ft*
    Geothermal gradient .... *0.012°F*
    Drive .... *Fluid expansion and partial water encroachment*
    Oil column thickness .... *25 ft*
    Oil-water contact .... *None*
    Connate water *12.8%*
    Water salinity, TDS .... *22,200 ppm Cl⁻ (average)*
    Resistivity of water .... *0.11 ohm-m at 188°F*
    Bulk volume water (%) *2.4%*

Transportation method and market for oil and gas:
See Hollin.